ELECTRONICS
A COURSE FOR ENGINEERS

R J MADDOCK

Principal Lecturer
Department of System and Communication Engineering
Southampton Institute of Higher Education

D M CALCUTT

Senior Lecturer
Department of Electrical and Electronic Engineering
Portsmouth Polytechnic

Longman
Scientific &
Technical
Copublished in the United States with
John Wiley & Sons, Inc., New York

Longman Scientific and Technical,
Longman Group UK Limited,
Longman House, Burnt Mill, Harlow,
Essex CM20 2JE, England
and Associated Companies throughout the world

Copublished in the United States with
John Wiley & Sons, Inc., 605 Third Avenue, New York, NY 10158

First published 1988

British Library Cataloguing in Publication Data

Maddock, R.J. (Robert John)
  Electronics—a course for engineers.
  1. Electronic circuits
  I. Title   II. Calcutt, D.
  621.3815'3   TK7867

  ISBN 0-582-99476-4

Library of Congress Cataloging in Publication Data

Maddock, R.J.
  Electronic—a course for engineers.
  Includes index.
  1. Electronic circuits.   I. Calcutt, D.M.,
1935– .   II. Title.
TK7867.M319   1987   621.3815'3   87-2738
ISBN 0-470-20811-2 (Wiley, USA only)

Set in 10/11 Monophoto Times Roman

Produced by Longman Group (FE) Limited
Printed in Hong Kong

Book Plan

Electronics, and education in electronics, have been in flux for 25 years. Education has tended to lag behind current technology; many would say rightly so as engineers are concerned with existing equipment and the 'latest development' may rapidly fade into obscurity. However, the fundamentals of electronics education can now be identified and there is a need for an up-to-date and comprehensive text outlining the broad range of electronics at an analytical level. This book aims to satisfy that need.

Students following first year undergraduate courses and courses leading to higher technician and technician engineer qualifications in electrical and electronic engineering, such as the BTEC Higher National Certificate and Higher National Diploma, will find this book a vital companion to other sources of educational material such as lectures and laboratory work. The book will also serve as a good introduction for trained engineers who wish to bridge the gap between the old technology and present-day integrated circuit technology.

Many of the students who will be attracted to this book will have had an introduction to electronics including a descriptive treatment of the operation and characteristics of devices and basic circuits. They may, however, have little concept of why electronics is important in the modern world and why the analysis of electronic circuits is an important aid to understanding their behaviour and uses. This book will help students realise the importance of electronics, its limitations and its potential.

The first two chapters are introductory, describing basic concepts and passive components and devices. Chapter 3 develops the techniques of analysis used in the book including graphical methods, analysis by equivalent circuit techniques, analysis using phasor diagrams and complex algebra, frequency response analysis, transient analysis and digital analysis. The remaining chapters cover specific classes of electronic circuits: digital circuits, analogue circuits, waveform generation and conversion, analogue and digital conversion, power supplies and controlled rectifier systems.

Every effort has been made to make the text and the mathematics in the text easily comprehensible. There are numerous worked examples in the text and a large number of problems at the end of the chapters. Significant new terms and concepts throughout the text are emphasised by **bold type** and care has been taken to ensure that these are adequately defined and described. Chapters start with a list of learning objectives, which the student will find to be a helpful key to the text. At the end of chapters important points are listed out and students will find these to be an invaluable revision aid.

Contents

Abbreviations

ALU	Arithmetic and logic unit	LED	Light-emitting diode
ADC	Analogue-to-digital converter	LASCR	Light-activated SCR
		LSB	Least significant bit
BCD	Binary coded decimal	LSI	Large scale integration
BJT	Bipolar junction transistor	LNI	Low noise immunity
BIFET	Bipolar FET		
		MOSFET	Metal oxide semiconductor FET
CB	Common base	MSI	Medium scale integration
CLD	Delayed clock output	MSB	Most significant bit
CML	Current mode logic		
CMOS	Complementary metal oxide silicon	NIC	Negative impedance converter
CMRR	Common mode rejection ratio	NMOS	Negative metal oxide semiconductor
CRO	Cathode ray oscilloscope		
CS	Common source	PA	Power amplifier
		PMOS	Positive metal oxide semiconductor
DAC	Digital-to-analogue converter	PROM	Programmable ROM
DIL	Dual-in-line	PUT	Programmable unijunction transistor
DRL	Diode resistor logic	PIV	Peak inverse voltage
DTL	Diode transistor logic		
		RAM	Random access memory
EAPROM	Electrically alterable PROM	ROM	Read only memory
EPROM	Erasable PROM	RTL	Resistor-transistor logic
ECL	Emitter coupled logic		
		SBS	Silicon bilateral switch
FET	Field effect transistor	SNFB	Simple negative feedback
		SCR	Silicon controlled rectifier
GCS	Gate controlled switch	SR	Set-reset
GIC	Generalised impedance converter	SUS	Silicon unilateral switch
GTO	Gate turn-off switch		
		TTL	Transistor-transistor logic
HNI	High noise immunity		
		VCO	Voltage controlled oscillator
IC	Integrated circuit		
		UJT	Unijunction transistor
JFET	Junction FET	UV	Ultra violet
LCD	Liquid crystal display		

1 Introduction

Why electronics?

The word 'electronics' is used in general conversation with little idea of the meaning of the term or of its importance. Electronics is that branch of electrical engineering in which the circuits include electronic devices such as transistors and integrated circuits. Electronic devices, circuits and systems are only important because of the functions they provide and, even then, the importance of a particular item depends on the relative cost, size, efficiency and reliability. It is possible that a new non-electronic technology might be discovered and make the electronic stereo hi-fi system as dead and archaic as is the Edwardian phonograph today.

However, in the technology of today, electronics is essential in many areas such as communications, automatic control, automation, computing and instrumentation. The function of the electronic parts in these systems is to receive input information, to process this information and then to produce an output.

For example, in a computer, the input information is provided by pressing the buttons on a keyboard, the processing may involve arithmetic or comparison with previous information in a memory and the output will be a printout or a display on a video display unit. In another example, both input and output of a communications system may be sound such as words or music. The processing in this case allows the information contained in the words and music to be transmitted from one side of the world (or of a room) to the other, where it is then reproduced in its original form.

Could such processes be achieved without electronics? To a limited degree the answer must be 'yes'. The first computers were mechanical analytical engines and, more recently, electromechanical calculating machines (comptometers) have been widely used in the business world. Non-electronic communications have been achieved by systems such as semaphore but these were very limited in the quantity of information that could be transmitted. In other areas such as instrumentation and control, there are many non-electronic systems widely used today. Both power steering and servo assisted brakes in automobiles are usually non-electronic, and mechanical clocks and watches still have a share of the market. However, many of the advances and benefits of modern technology have only been possible as a result of electronic information processing.

Information and signals

Electronic systems are required to process or react to **information**. The information may have many forms, including physical quantities such as temperature, velocity or mass, simple on/off information resulting from a switch being operated, or the highly concentrated and detailed information in speech, music and pictures. All these different forms of information have one common factor, i.e. both amplitude and frequency may vary with time. This means that they can be illustrated by means of graphs, and in many cases they can be defined as functions of time with mathematical expressions.

Electronic circuits can only react to information in the form of time-varying voltages and currents. We can conveniently refer to these forms of information as **signals**. There must therefore be some form of interface or converter between the real world and the electronic world. The interface may simply be a **transducer**. Transducers are devices that convert one form of energy to another. For example, a microphone converts sound energy containing information into electrical energy in the form of a signal which contains the same information.

An electronic system will have input signals from transducers and ouput signals which are reconverted by other transducers to produce energy in various required forms. Between these two processes, within the system, there will be other signal forms.

Consider the form of a range of typical signals which may be represented graphically as waveforms. The first group shown in Fig. 1.1 are **steady-state** signals such as a single note on an electronic organ. This means that the waveforms have been unchanged for such a long time that any transient effects due to an earlier change can no longer be detected. The second group in Fig. 1.2 are transient signals resulting from a sudden change, such as the switching on of a signal. Other forms are shown in Figs 1.3 and 1.4.

The signal shown in Fig. 1.1(a) is a simple direct positive voltage that might have been obtained from a battery. The polarity is positive and the amplitude is 2 V. The symbol $v(t)$ means that the instantaneous value of the voltage v is expressed as a function of time; in this case $v(t)$ is a constant and does not vary with time. The second example, in Fig. 1.1(b), is for the current $i(t)$ which does vary with time according to the relationship

$$i_1(t) = 5 \sin 5000t \text{ mA}$$

Thus the current varies between plus and minus 5 mA with a sinusoidal wave having a frequency of 5000 radians per second or approximately 796 Hz. If there is more than one sinusoidal signal within a system, it may be necessary to define the time or phase relationship between them. If, for example, the signal described in Fig. 1.1(b) was the input to a system and the output was

$$i_2(t) = 200 \sin (5000t - \pi/6) \text{ mA}$$

then we can see that when $t = 0$, $i_2(t)$ is 200 sin $(-\pi/6)$ mA. This means that it is lagging behind $i_1(t)$ by $\pi/6$ radians or 30°. The time lag equivalent is one-twelfth of a cycle (i.e.30°/360°) or approximately 0.1 ms.

Fig. 1.1 Waveforms for some typical steady-state signals.

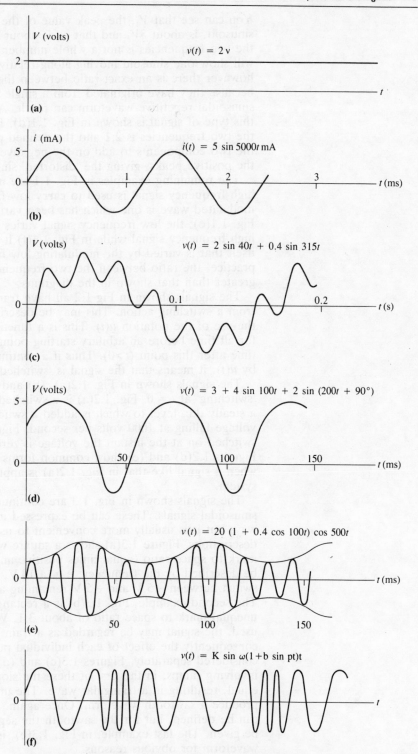

The signal shown in Fig. 1.1(c) is the sum of two sine waves and will have the form

$$v(t) = V_1 \sin \omega_1 t + V_2 \sin \omega_2 t$$

You can see that V_1, the peak value of the lower frequency sinusoid, is about $5V_2$ and that ω_2 is about $8\omega_1$. If the ratio between the two frequencies is not a whole number, an oscilloscope display will show one sinusoid moving along relative to the other. If however there is an exact ratio between the two frequencies (usually because they have originated from a single frequency), a non-sinusoidal repetitive waveform can result. A common example of this type of signal is shown in Fig. 1.1(d). In this case, the ratio of the two frequencies is 2:1 and the defined phase relationship causes the two components to add on the negative peaks and subtract on the positive peaks, giving the 'distorted' sinusoid shown.

The remaining examples in Fig. 1.1 are **modulated** waves where a high frequency signal is used to **carry** low frequency information (a modulated wave is one which has been varied by another signal). In Fig. 1.1(e), the low frequency signal varies the amplitude of the high frequency signal while in Fig. 1.1(f) it is the high frequency itself that is varied by the modulating low frequency signal. In practice, the ratio between the two frequencies is usually much greater than that shown in the diagrams.

The signals shown in Fig 1.2 all have transient properties resulting from a switching action. This may be described mathematically by the use of the notation $u(t)$. This is a function of time that is *zero* for all time before an arbitary starting point ($t = 0$) and *one* for all time after this point ($t>0$). Thus if a continuous signal is multiplied by $u(t)$, it means that the signal is 'switched on' at $t = 0$.

The signals shown in Fig. 1.2(a), (b) and (c) are all the result of 'switching' at $t = 0$. Fig. 1.2(a) is a switched d.c. level, Fig. 1.2(b) is a steady d.c. level to which is added a switched negative ramp voltage falling at 1000 volts per second. Figure 1.2(c) is a sinusoid, switched on at the instant the voltage is zero and going negative. Figures 1.2(d) and (e) show common forms of transient that result when a signal like that in Fig. 1.2(a) is applied to some types of system.

The signals shown in Fig. 1.3 are continuous repetitive non-sinusoidal signals. These can be expressed mathematically in various ways but it is usually more convenient to use simple verbal descriptions. Figure 1.3(a) shows a square wave signal having equal mark to space ratio (equal times positive and negative). It will also be described by the frequency, e.g. a 10 V, 2 kHz square wave will switch between +5 V and −5 V remaining at each level for 0.25 ms. The second example, Fig. 1.3(b) is a rectangular wave with an unequal mark to space ratio of about 3:1. When large ratios are used, the signal may be regarded as a train of pulses and consequently the effect of each individual pulse can often be considered separately. Figures 1.3(c) and (d) are both signals involving ramps; in the first of these the slopes of the ramps are equal, resulting in a triangular wave. The unequal slopes in (d) produce a sawtooth waveform. Once again, frequency and amplitude can be defined, but for the sawtooth the separate slopes must also be given. The last example, in Fig. 1.3(e), is known as a staircase waveform for obvious reasons.

It can be shown mathematically that all the signal waveforms in Fig. 1.3 consist of sums of many sine waves having related frequencies (compare Fig. 1.3(b) with Fig. 1.1(d)).

Fig. 1.2 Waveforms for
some typical transient
signals.

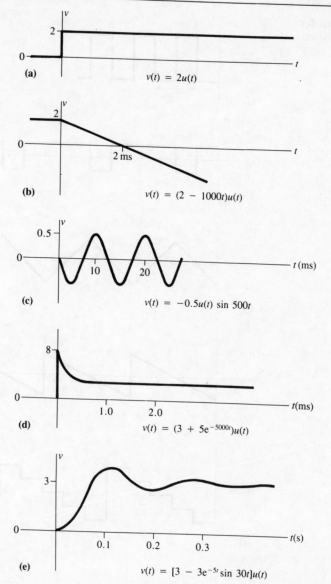

(a) $v(t) = 2u(t)$

(b) $v(t) = (2 - 1000t)u(t)$

(c) $v(t) = -0.5u(t) \sin 500t$

(d) $v(t) = (3 + 5e^{-5000t})u(t)$

(e) $v(t) = [3 - 3e^{-5t} \sin 30t]u(t)$

The three groups of signals described above are all **analogue**
signals. This means that the information carried in these signals
depends on the amplitude of the signal and on the way in which this
amplitude varies with time. An alternative and very important form
of signal is known as **digital**. The information contained in a digital
signal can only have a *discrete number of levels*. This may be
compared with measuring a quantity of liquid with an ungraduated
cup. We could say that we have more than 10 cups or less than 11
cups, but we have no way of getting an accurate measure in
between. For greater accuracy we could use a much smaller cup and
obtain perhaps 104 cups for the same quantity of liquid. Digital
quantities can be expressed as decimal numbers, as in this example,
but for electronic purposes digital signals are nearly always based on

Fig. 1.3 Waveforms for some typical repetitive non-sinusoidal signals.

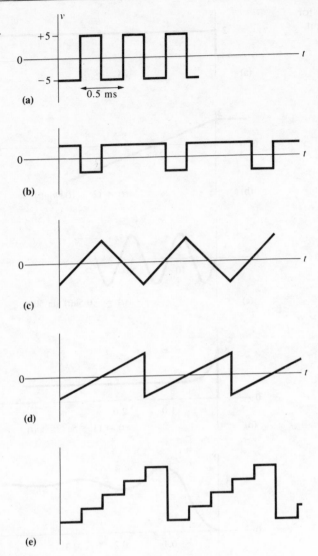

Fig. 1.4 Digital signals: (a) parallel; (b) serial.

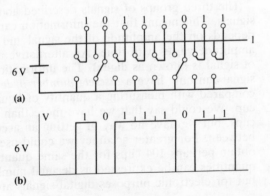

binary numbers using only 0s and 1s. The two numbers, 0 and 1, can be represented by two voltage or current levels. For example, the information given by the number 187 converts to binary information 10111011. The equivalent signal contains eight 'bits' of information, each of which must be at one of two voltage (or current) levels. This can be provided in **parallel** with eight wires or switches to a battery or other d.c. voltage as shown in Fig. 1.4(a). Alternatively, a **serial** signal carries the bits one after the other as a waveform such as that shown in Fig. 1.4(b). Another similar form of signal is a control signal, which may consist of a single voltage pulse on one line, initiating some action or it may be the continuous train of pulses known as the clock which synchronises actions within an electronic system.

Examples of signal processing

The function of electronic circuits and systems is to process signals. Many different processes are possible and useful but, before considering a wide range of processes, it will be helpful to examine some simple systems and to consider what type of processes may be required.

A radio communication system

Requirement   To transmit speech and music from a concert hall to a place 100 miles away while other similar transmissions are occurring in the same area.

Problems   Alternating signals (a.c.) can result in radiated electromagnetic waves (radio signals) but signals at audio frequencies do not radiate efficiently. Also, as there would be similar transmissions in the same area, there would be interference as all the radiated signals would be received more or less equally.

Solution   Use higher frequencies, which can be radiated readily, in order to carry the signal information on these higher frequencies. Use different high frequency carriers for each separate transmission to avoid interference.

Processes required   The following processes are the minimum needed for a simple radio communication system and can be represented in the block system diagram shown in Fig. 1.5.

(a) Transducers to convert sound information to signals.
(b) High frequency signal generator (oscillator).
(c) Modulator, to vary the amplitude (or frequency) of the high frequency signal to carry the audio frequency signal to the transmitter (see Fig. 1.1(e) and (f)).

Fig. 1.5 A radio communication system.

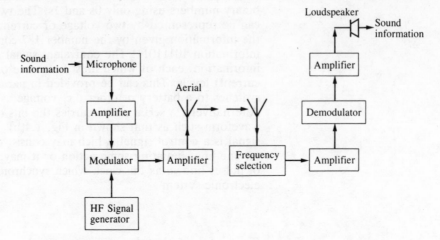

(d) Amplifier, to provide large signal with sufficient energy to radiate over 100 miles.

(e) Aerials for efficient transmission and reception of signals.

(f) Frequency selection at the receiving end to pass required high frequency (with audio information) and to stop or reject other high frequencies from other transmitters.

(g) Amplifier to increase the signal amplitude to a level suitable for further processing.

(h) Demodulator to extract the audio signal from the carrier signal.

(i) Amplifier to increase the power level of the audio signal.

(j) Loudspeaker or output transducer to convert electrical signal power into audible sound having the same information as in (a) above.

An automatic washing machine

Requirement To fill a washing chamber with water and detergent, to raise and maintain the temperature of the water to a preselected level, to agitate the water for a preselected time, to remove the hot soiled water, to refill with cold water, agitate and replace several times, to rotate the chamber at high speed for a preselected time and, finally, to prevent opening of the chamber unless it is stationary and with no water in it.

Solution To operate the machine in a series of logically controlled states, e.g. water ON, heater OFF, motor OFF; water OFF, heater ON, motor ON (agitating); water OFF, heater OFF, motor OFF, pump ON; etc. The sequence of states is to be predetermined by the chosen program. The duration of each state will be determined by sensor signals for temperature or water level and by a preselected timer.

Processes required A major part of the system outlined consists of electrical and electromechanical components such as motors and heaters. These will be controlled by electromechanical or electronic relays which in turn will be operated by signals from an electronic controller. Other signals will be fed to the controller from temperature and level sensors (transducers) and from manually selected inputs. The electronic controller must have a **memory** to know which state it is in, which state it must change to (as a result of program signals) and when it is to change (as a result of sensor signals or a timer). As the controller changes state, it must generate and direct control signals to the relays operating motors and heaters.

A full block diagram for this system would be very complex, but the major features are shown in Fig. 1.6.

Fig. 1.6 An automatic washing machine system.

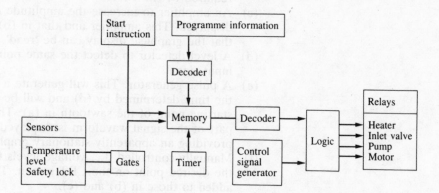

- An electronic memory
- An electronic timer
- Electronic gates whose outputs change on signals from sensors
- A control signal generator
- Logic to direct control signals to required devices
- Encoder to convert manual input signals to the required digital signals for the memory
- Decoder to take digital signals from the memory and control the logic.

A cathode ray oscilloscope

Requirement   To provide an automatic graphical display of time-varying voltage signals so that their form may be inspected or measured.

Solution   A cathode ray tube (CRT) is a device having a screen, one point of which fluoresces (emits light) when it is struck by an electron beam from the rear. The point of impact of the electron beam can be moved either horizontally or vertically by suitable signals. Since the position of the point depends on the actual voltages applied, a rapidly changing voltage results in a rapidly moving point of light. The result is a line of light, the shape of which is dependent on the voltage signals. Thus the required system will consist of a cathode ray tube with any associated circuits necessary for its operation, together with circuits to process the signals that are to be measured or inspected.

Processes required   A block diagram illustrating the arrangement of these components to form a simple CRO is shown in Fig. 1.7.

(a) A variable frequency sawtooth voltage generator. As this voltage changes at a constant rate from a negative value to a positive value, it can be used to move the point of light or spot at a constant velocity from one side of the screen to the other. This provides the **time base** for the graphical display and at the end of each sweep the voltage returns rapidly to its starting value, causing the spot to **fly back** to its starting point at one side of the screen.

(b) An amplifier to increase the time base voltage to the level required by the CRT.

(c) An amplifier to increase the amplitude of the signal to be displayed. This amplifier and that in (b) can be **calibrated** so that the graphical display can be 'read' in V/cm and ms/cm.

(d) A level detector to detect the same point on each cycle of the input signal.

(e) A pulse generator. This will generate a **synchronising** pulse at the time determined by (d) and will be used to synchronise the starting point of the sawtooth in (a). This ensures that the same part of the signal waveform is displayed on each sweep, providing an apparently stationary graphical display.

(f) Manually controlled d.c. voltage levels to **shift** the display to the desired point on the CRT screen. These voltages will be **added** to those in (b) and (c).

(g) Manually controlled d.c. voltage levels to adjust the brilliance and size of the light spot.

(h) Fixed d.c. voltage levels or power supplies for the correct operation of the CRT and the various electronic circuits.

Fig. 1.7 A cathode ray oscilloscope system.

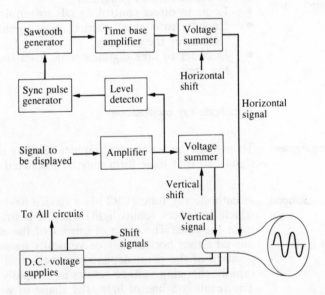

Signal processes

Three widely differing electronic systems have been examined, each of which will, in general terms, be familiar. Within these descriptions, a number of different **signal processes** have been

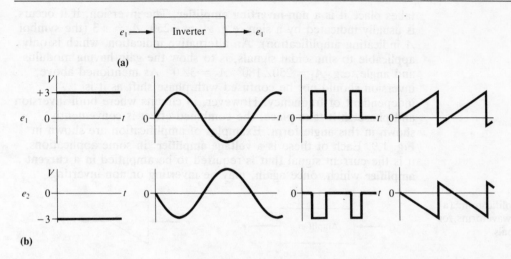

Fig. 1.8 Inversion: (a)
symbol; (b) waveforms.

introduced. Each of these processes, and many others, will be
encountered in the world of electronic systems. In this section we
shall examine, with more detail, a wide range of processes
commonly employed in electronic systems. The processes to be
considered can be loosely classified into analogue processes, digital
processes and a group concerned with waveform conversion which
comes between the other two.

Analogue processes

Inversion

This is the most fundamental process, involving simply a sign
change. If the input to an inverter is $+3$ V, the output will be
-3 V; if the input is $10 \sin 500t$, the ouput will be $-10 \sin 500t$.
(The output should not be confused with $10 \sin (500t + \pi)$. This
looks the same on a CRO, but π represents a phase shift which
usually varies with frequency while inversion affects all frequencies
equally.) Examples of inversion are shown in Fig 1.8. In every case,
the only change to the waveform is that due to the sign change.

Amplification

This essential process involves an increase in the amplitude or size
of a signal without any change to the waveform. If, for example, a
voltage signal of $2 \sin 200t + 0.5 \sin (300t + \pi/6)$ mV is applied to an
amplifier with a voltage **gain** of 50, then the output signal will be
$100 \sin 200t + 25 \sin (300t + \pi/6)$ mV. Notice that only the
amplitude elements 2 and 0.5 are modified to 100 and 25
respectively (i.e. multiplied by the gain); the remaining elements
involving the waveform, sine, angular frequency ω, and phase shift
$\pi/6$, are not changed.

In general, both voltage and current signals are involved in the
amplification process so the **signal power** at the output will be
greater than that at the input. (A transformer is not an amplifier
since, disregarding the losses, the power levels are the same at input
and ouput.) Various different amplifiers are described below.

Amplification is frequently accompanied by inversion and the
resulting circuit is known as an **inverting amplifier**. If no inversion

takes place it is a **non-inverting amplifier**. The inversion, if it occurs, is usually indicated by a sign, e.g. $A_1 = -250$, $A_2 = +3$ (the symbol A indicating amplification). An alternative indication, which is only applicable to sinusoidal signals, is to show the gain having modulus and angle, e.g. $A_1 = 250\angle 180°$, $A_2 = 3\angle 0°$. As mentioned above, inversion should not be confused with phase shift as it is independent of frequency. However, in circuits where both inversion and phase shift take place, the combined effect is conveniently shown in this angle form. Examples of amplification are shown in Fig. 1.9. Each of these is a **voltage amplifier**: in some applications, it is the current signal that is required to be amplified in a **current amplifier** which, once again, may be inverting or non-inverting.

Fig. 1.9 Amplification: (a) symbol; (b) waveforms for different signals.

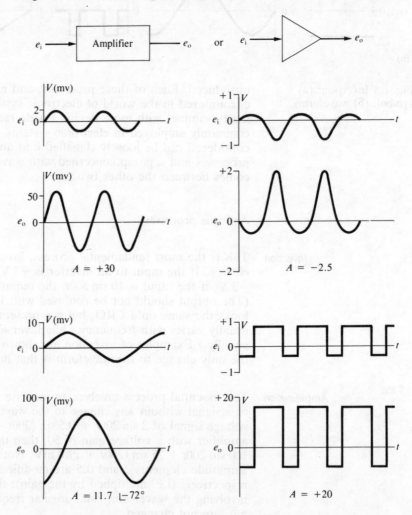

In many systems, a necessary final process is to supply signal power to a particular load such as a loudspeaker or other output transducer. In such amplifiers, known as **power amplifiers** (P.A.), the levels of both current and voltage signals may be important; their ratio will depend on the impedance of the load in question. Power amplifiers should, more correctly, be referred to as **large signal amplifiers**.

Signal amplification can be difficult if a following circuit has a very low impedance. In such cases, **a buffer amplifier** may be used. This type of amplifier has a low voltage gain (about 1), but it may be used to **match** (interface) a high impedance signal source to a low impedance load. This may be thought of as conversion from a voltage signal to a current signal, i.e. a *V* **to** *I* **converter**. A similar system may be required for the reverse process of current to voltage conversion.

In some systems, the **addition** of two or more signals is required, sometimes with a different gain applied to each input. This can be expressed in the form

$$v_O = A_1 e_1 + A_2 e_2 + \text{etc.}$$

Two examples of this process are shown in the waveforms in Fig. 1.10.

Fig. 1.10 Input and output signals for a summing amplifier.

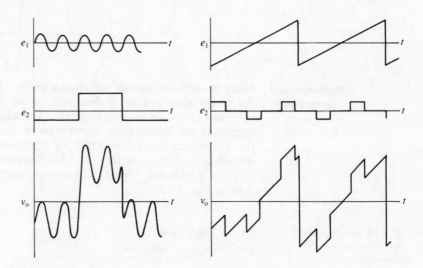

Differential or **difference amplifiers** have two input signals, but it is the difference between them that is amplified. The output of a differential amplifier is given by

$$v_O = A_d(e_1 - e_2) + A_c \frac{(e_1 + e_2)}{2}$$

A_d is the **differential gain** and may be very large. A_c is the **common mode gain** and should ideally be zero. If the two signals e_1 and e_2 have a large equal d.c. component and one has additionally a small a.c. component, then the common d.c. component will not be amplified (zero d.c. output) while the small a.c. signal will be.

All amplifiers will be limited in some way as to the range of signal frequencies which they can amplify with a constant gain. **Direct coupled amplifiers** have constant gain from zero frequency (d.c.) to a high frequency above which the gain falls steadily towards zero. Capacitor-coupled amplifiers (a.c. coupled) have a similar high frequency performance but also have a low frequency limit below which the gain falls to zero at zero frequency. The range of signal frequency for which the gain is approximately constant is known as the amplifier **bandwidth**. **Frequency selective amplifiers** are designed to amplify only a certain range of signal frequencies. For example, in many communication applications, the selection of a certain channel of information is achieved by using an amplifier which amplifies only those signals having frequencies in the desired range. Another requirement can be for the gain of the amplifier to be proportional to frequency in some way. These processes are best illustrated by graphs showing gain variation with signal frequency. Some examples are shown in Fig. 1.11.

Fig. 1.11 Frequency selective amplifiers: gain frequency response: (a) low pass; (b) band pass; (c) $A \propto 1/f$.

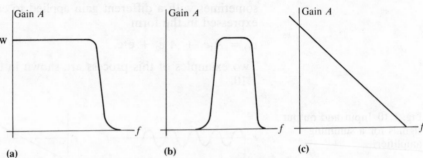

(a)   (b)   (c)

Modulation and demodulation

Some modulated signals are shown in Fig. 1.1 and an application of the modulation process is discussed in the description of the radio communication system on page 00. A **modulator** is a circuit which combines the information signal with the carrier signal to produce the modulated wave (amplitude or frequency modulated). A **demodulator** (detector for AM, discriminator for FM) recovers the original modulating signal. These two processes are illustrated for AM in Fig. 1.12.

Fig. 1.12 Amplitude modulation and demodulation.

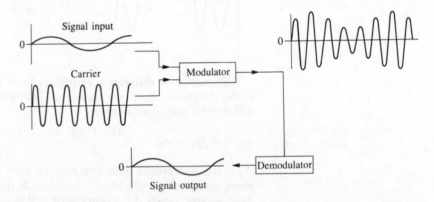

Frequency changing

Another similar process often required in communication systems is that of frequency changing. The input signal to a frequency changer will be a modulated carrier. The function of the unit is to change the carrier frequency without changing the low frequency information signal. This is illustrated in Fig. 1.13.

Oscillators

In modulators and frequency changers, the information signal is derived from a microphone or other transducer outside the system. The carrier, however, must be **generated** within the system. Such a generator is known as an oscillator and its function is simply to

Fig. 1.13 Input and output signals for a frequency changer.

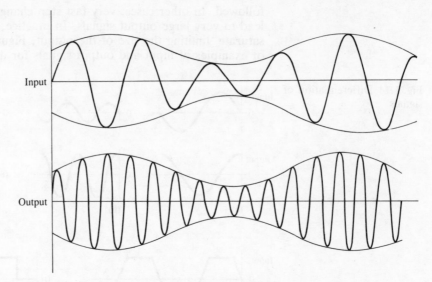

Input

Output

produce a constant frequency, constant amplitude sinusoidal signal. In some applications the frequency may be adjustable to select a particular station or channel of information.

Converters These are usually concerned with the supply of energy in a particular form. The most widely used is an a.c.–d.c. converter, which provides d.c. power from an a.c. source, usually the public mains electricity supply. This is frequently associated with a **regulator** which ensures that the d.c. supply is at a constant d.c. voltage (alternatively current) level regardless of changes in the a.c. supply voltage or changes in the system to which the d.c. is supplied. Other types of converter include d.c. to a.c., d.c. to d.c. with a change of voltage, and a.c. to a.c. with a change of frequency. The d.c. to a.c. converter is known as an **inverter** and, although it could be likened to an oscillator, the function is different in that the inverter changes the form of the power.

Waveform modification processes In the last section, the processes examined were mainly applied to sinusoidal waveforms although, in most cases, the processes can be equally applied to non-sinusoidal waveforms. The next group of processes are mainly concerned with the generation and modification of non-sinusoidal waveforms. Common forms are square waves, rectangular waves, triangular waves, sawtooth waves and staircase waves. Examples may be seen in Figs 1.8 and 1.10. Such waveforms can be generated either directly or by generation of a sine or square wave followed by a modification to give the required waveform.

Differentiation The signal output from a differentiator is proportional to the rate of change or slope of the signal input. Mathematically, this is expressed

$$v_O = K(dv_i/dt)$$

where K is a multiplying constant.

With some waveforms, this relationship can be very closely

followed. In other cases, very fast step changes in the input should
lead to very large output signals. In practice, the circuits will
saturate, limiting the size of the output. Figure 1.14 shows a number
of examples of input and output signals for differentiators.

Fig. 1.14 Differentiation of
signals.

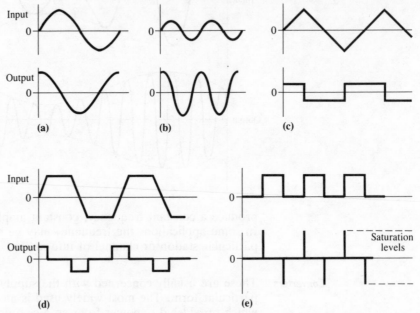

Figures 1.14(a) and (b) show differentiation of sinusoidal signals;
in both cases a cosine results, with the output leading the input by
90°. In addition, the ratio of output to input amplitudes increases
with the signal frequency. This is because the rate of change of
voltage increases with the frequency. Mathematically

$$\frac{\mathrm{d}}{\mathrm{d}t}V \sin \omega t = \omega V \cos \omega t$$

In Figs 1.14(c) and (d), differentiation of the sloping parts of the
input signal result in positive or negative d.c. levels. In (c) the
triangular wave is converted into a square wave; in (d) the input is
a truncated ramp and, for those parts of the wave when the input is
constant, the output is zero while the sloping parts produce positive
and negative output pulses. The square wave input in (e) is similar
to (d) with the slopes of the ramps tending to infinity. The output
pulses are now very short with the amplitude limited by the
operational range of the differentiator.

Integration   This is mathematically the inverse of differentiation and is written

$$V_O = \int v_i \mathrm{d}t$$

It also represents a summation of the input over a period of time.
This may be appreciated by reference to the examples shown in Fig.
1.15.

In Fig. 1.15(a) the area, voltage × time, of each short pulse is
added to the output which then remains constant until the next

Fig. 1.15 Integration of signals.

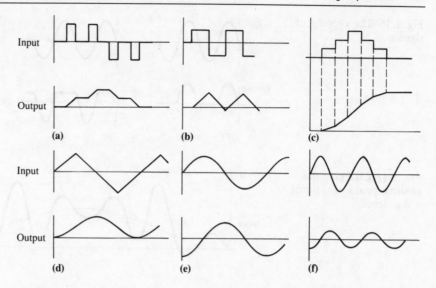

pulse. Negative pulses then subtract from the total in the same way. During the pulses the output changes, with a ramp having slope proportional to the input pulse amplitude. This may be observed in Fig. 1.15(b) when the input is a simple square wave (having alternate negative and positive half-cycles). During the positive half-cycle of input, the output ramp is positive and in the same way the negative half-cycle results in a negative ramp. The combined effect is to convert a square wave into a triangular wave (compare this with the differentiation in Fig. 1.14(c)). Mathematically, the integral of a constant is given by $\int K \mathrm{d}t = Kt$, a ramp with slope K proportional to the level of the constant input. In Fig. 1.15(c) each increase in input level increases the slope of the output. It is a simple step to the smooth change in slope in (d) resulting from a ramp input. The integral of a ramp is given by $\int Kt \mathrm{d}t = Kt^2/2$, which is the expression for a parabola. The output in Fig. 1.15(d) is a series of positive and negative parabolas which looks rather like a sinusoid. The last two examples (e) and (f) are both sinusoids for which the integral is given by $\int V \sin \omega t = -V/\omega \cos \omega t$. The $-\cos$ is shown by the 90° phase lag and the $1/\omega$ appears as a reduction in gain as signal frequency is increased.

In most practical differentiators and integrators the process is accompanied by inversion or sign change. Thus the practical output waveforms would have the opposite polarity to that shown in Figs 1.14 and 1.15.

Clipping   This is a very simple but useful process by which a part of a waveform is clipped off or removed. Some examples are shown in Fig. 1.16. In Fig. 1.16(a), all negative parts of the input signal are removed (this is also known as rectification). In (b), positive parts above a certain reference level are removed; in (c), the peaks of both positive and negative spikes are removed or clipped.

Clamping   In this process, positive or negative peaks of a waveform are clamped to a reference level, as shown in Fig. 1.17.

Fig. 1.16 The clipping of signals.

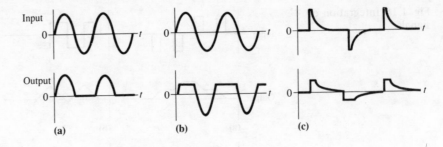

(a)   (b)   (c)

Fig. 1.17 Clamping the positive peaks of a signal to a d.c. level.

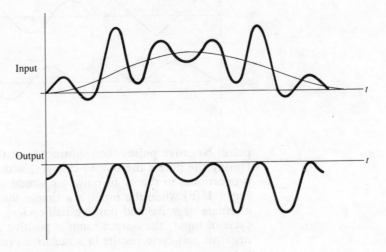

Delay   Delay in a signal can be either phase delay, which applies only to sinusoidal signals (Fig. 1.18) or time delay, which can be applied to any signal (Fig. 1.19).

Fig. 1.18 Phase delay for signals of different frequency.

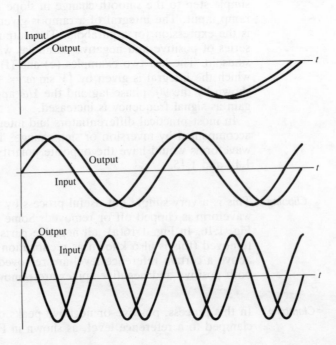

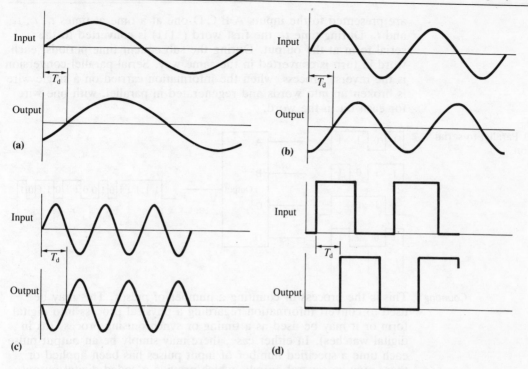

Fig. 1.19 Time delay for different signals.

Figure 1.18 shows three sinusoidal signals having different frequencies which have been phase delayed by the same delay circuit. The low frequency signal is delayed by only a few degrees: the second, medium frequency, signal is delayed by about 45° or one-eighth of a cycle: the highest frequency signal is delayed by nearly 90° or a quarter of a cycle. Thus, phase delay may be related to the signal frequency or period. Time delay, illustrated in Fig. 1.19, is applied equally to signals having any frequency or waveform. A particular time delay t_d may represent only a few degrees for a low frequency sinusoid, while for higher frequencies, the same t_d may be equivalent to half a cycle or several cycles. Figure 1.19(d) shows the effect of the same delay on a rectangular waveform.

Digital processes

Digital signals usually have only two values, 0 and 1. The digital processes to be described are only concerned with these two values. If a digital signal is to contain useful information there will be many *bits*, each of which will be either 0 or 1. Information which is represented by a group of bits known as a **word** is said to be **coded**.

A digital signal can be **parallel**, when all the bits of a word are available at the same time, or **serial**, when the bits become available one after the other (see Fig. 1.4). In most digital processing circuits, there will be a number of inputs and a number of outputs. A selection of typical digital processes is described in this section.

Parallel-serial conversion

It is frequently necessary to convert digital information from the parallel form to the serial form or vice versa. The first of these processes is shown diagrammatically in Fig. 1.20. The four-bit words are presented to the inputs A B C D one at a time at times t_1, t_2, t_3 and t_4. During time t_1, the first word (1111) is converted to the serial form at the output. During the subsequent time periods, each word in turn is converted in the same way. Serial-parallel conversion is the reverse process, when the information carried on a single wire is broken up into words and regenerated in parallel, with one wire for each bit in the word.

Fig. 1.20 Parallel to serial conversion.

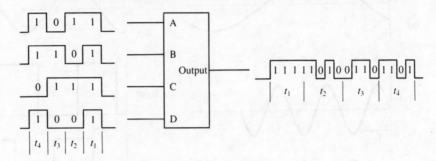

Counting This is the process of counting a number of pulses. This may be used to convert information regarding a physical process into digital form or it may be used as a timing or synchronising process (as in digital watches). In either case, there may simply be an output pulse each time a specified number of input pulses has been applied or there may be several outputs which provide a coded digital output representing the number of pulses that has been applied. Both forms of counting are illustrated in Fig. 1.21.

Fig. 1.21 A counting circuit.

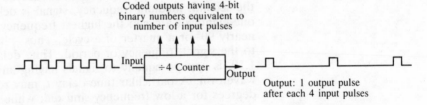

Storage The storage of digital information is an essential part of many processes. A single word can be stored in a **register** for further processing after a short time interval, or hundreds of thousands of words can be stored in a **random access memory** for seconds, months or years. Each word is stored at a **location** with an address in the memory and is instantly available when the code for the required address is supplied.

Logical detection Logical relationships are concerned with relationships like AND, OR and NOT applied to several variables. In electronic processing, the variables are digital signals which may only have the values 1 and 0 (represented by two voltage or current levels). A logical detector will recognise a particular combination of 0s and 1s applied to a set of inputs and as a result produce an output. This could be 0 for unrecognised codes and 1 for the others, or vice versa. An example of this process is shown in Fig. 1.22. This detector will produce a 1 at ouput F, if and only if

P AND Q are both 1, OR P is 0 AND (either Q is 1 OR R is 0).

Fig. 1.22 Inputs and output
for a logic detector.

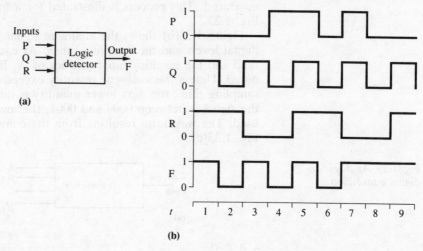

(a)

(b)

This statement may also be written as a **Boolean** expression

$$F = P.Q + \overline{P}(Q + \overline{R})$$

Now examine the serially applied digital signals during the time
periods t_1 to t_9. Notice that Q and R alternate between 1 and 0,
with Q changing for each period while R only changes on alternate
periods. P, however, appears to have a random variation 000110100.
Next examine the three signals at a particular time period with
reference to the defined process. During period t_1, P = 0, so the
first condition, P AND Q is not satisfied. The alternative condition
however is satisfied with P = 0 provided, either Q = 1 OR R = 0.
During period t_1, Q = 1, so the condition is satisfied (in spite of R
being 1 as well) and the output F is therefore 1. During period t_2,
P = 0 still, but Q = 0 and R = 1; the condition is not satisfied so
the result is F = 0. If you examine the conditions for the remaining
time periods, you can confirm the digital waveform at F.

This type of digital process may involve many different input
signals, and as a result the range of combination patterns is
enormous.

Analogue to digital
conversion

Both analogue and digital signals have their uses in electronic
processing and conversion between the two forms is frequently
required. To convert an analogue voltage signal to a digital signal,
the instantaneous level of the voltage is measured and the result is
converted to a binary number. The number of bits used for this
number depends on the accuracy required. For example, an eight-bit
binary number can have 256 different values, representing 256
different voltage levels in the analogue signal. Thus if the maximum
analogue voltage is going to be 10 V, the minimum level difference
will be 39 mV (10/256). All convertible levels will be multiples of
39 mV. For example, 2.7 V is approximately 69 × 39 mV and the
binary equivalent for 69 is 1000101.

The analogue signal to be converted is not usually a d.c. signal: it
will have one of the waveforms described earlier in the chapter. If a
time-varying analogue is to be converted, it must be sampled at
regular intervals and the level of each sample converted as it is

measured. This process is illustrated for a four-bit conversion in Fig. 1.23.

Figure 1.23(b) shows the analogue signal to be converted, the digital levels and the sampling times. At each time, the level can be read and the resulting binary values of A, B, C and D can be noted. Unless the analogue quantity exceeds a binary quantity at the sampling time, the next lower quantity is taken. For example, at t_3 the signal is between 0000 and 0001; the lower value is therefore used. The waveform resulting from the conversion is shown in Fig. 1.23(c).

Fig. 1.23 Analogue-to-digital conversion.

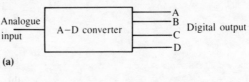

(a)

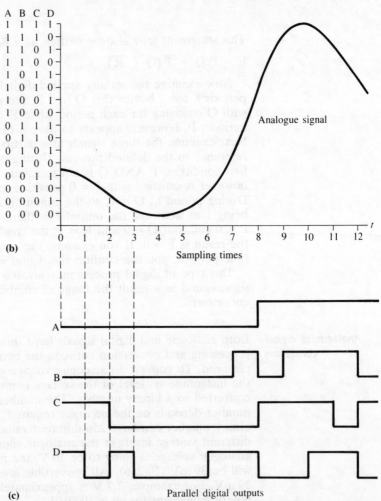

(b)

(c)   Parallel digital outputs

The alternative conversion, from digital to analogue, will have parallel inputs, one for each bit in the word used for a sample. A new digital word will be available for conversion after a time interval equivalent to the sampling time. This word is then converted to an equivalent analogue voltage level which is maintained until the next word is converted. The result for the same signal as that shown in Fig. 1.23 is shown in Fig. 1.24.

The analogue output waveform shown in Fig. 1.24(b) is stepped. This is exaggerated in this example as only four-bit words have been used for the digital signal. In practice, eight or more bits would be used and the resulting steps would be only one-sixteenth of the size or less. The resulting 'ripple' on the waveform can then be removed by filtering with a frequency selective circuit.

Fig. 1.24 Digital-to-analogue conversion.

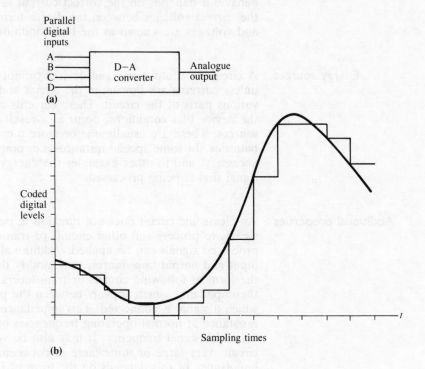

Electronic circuits

The preceding section was concerned with processes that are commonly encountered in electronic systems. A unit which provides one of these processes may be referred to as an **electronic circuit**. Some more complex processes may involve several electronic circuits and the resulting unit can be referred to as a system or a sub-system. In this section, we shall consider some general features that are common to all electronic circuits.

Components and devices

Nearly all electronic circuits consist of interconnected passive **components** and electronic **devices**. The passive components are resistors, capacitors and inductors; these may be actual components included in the circuit fabrication or they may be residual effects such as capacitance between conductors, inductance of wires or the resistance of a poor insulator. Passive components can be described by simple mathematical relationships, which are usually independent of signals associated with them. Electronic devices (e.g. diodes, transistors, thermionic valves) can have many forms, of which the most widely used are constructed from the semiconductor element silicon. Other devices use other semiconductors, insulators, magnetic materials and electrons passing between conductors mounted in a vacuum or in an inert gas. One common factor in their behaviour is that they cannot be described by simple mathematical relationships. Another factor in many devices is that the useful or required behaviour depends on the correct current levels in the device and on the correct voltages between the device terminals. These currents and voltages are known as the **bias** conditions.

Energy sources

A circuit of components and devices cannot perform any function unless currents are flowing in the circuit and voltages exist across various parts of the circuit. These currents and voltages, including the device bias conditions, occur as a result of external **energy sources**. These are usually one or more d.c. power supplies such as batteries. In some special instances a.c. power supplies may be necessary, and in other examples the energy can be supplied as the signal that is being processed.

Additional properties

An electronic circuit does not function in isolation: it requires signals to process and other circuits or transducers to which the processed signals can be applied. Additional circuit properties, the **input and output impedances**, may modify the input signal or limit the form of following circuits or transducers. The input impedance is the impedance which appears between the pair of input terminals to which a signal is connected. This impedance may be a simple resistance at normal operating frequencies or it may be complex and vary with signal frequency. It may also be very small, nearly a short circuit, very large or somewhere in between. The relative importance of this depends on the form of the signal to be processed (voltage or current) and upon the internal impedance of the transducer or other circuit providing the signal. If a voltage signal is being used, an input impedance that is much larger than the source internal resistance will give the largest signal to process. The largest current signal, however, will be obtained with a very small input impedance. In some cases, **maximum power transfer** from source to circuit is desirable; in such cases, the input resistance should be the same as the source resistance. A similar situation arises when the signal is supplied by a **transmission line**, which has a property known as its **characteristic impedance**. It will then be essential that the input impedance has the same value as this characteristic impedance, as otherwise the **mismatch** can cause signals to be reflected back along the transmission line, which may interfere with the correct operation of other parts of the system.

The output impedance of an electronic circuit is the effective internal impedance that is in series with the processed output signal. The relationship between this internal impedance and a following load is the same as the relationship between the source impedance and input impedance described above. The whole combination is illustrated is Fig. 1.25. These relationships are summarised below.

- If Z_{in} is large compared with Z_s, the voltage signal will be maximum.
- If Z_{in} is small compared with Z_s, the current signal will be maximum.
- If $Z_{in} = Z_s$, the power supplied to the circuit will be maximum.
- If Z_L is large compared with Z_{out}, the voltage signal output will be maximum.
- If Z_L is small compared with Z_{out}, the current signal output will be maximum.
- If $Z_L = Z_{out}$, the signal power supplied to Z_L will be maximum.

Fig. 1.25 The input and output impedances of an electronic circuit.

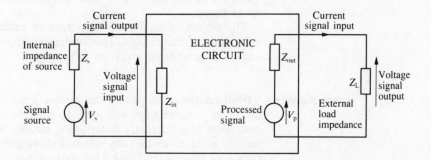

Signal limitations

The correct performance of an electronic circuit also depends upon the amplitude, frequency and waveform of the input signal. For example, an amplifier (see page 00) might perform correctly, i.e. amplify or increase the signal amplitude by a factor of 200 for sinusoidal signals between 0.1 mV and 20 mV in amplitude. The resulting output signals would be respectively 20 mV and 4 V and the output waveform would be sinusoidal. An input signal of 40 mV, however, might produce an output of 6 V (amplified by 150) and the waveform may be changed or distorted. With further increase in signal amplitude, the output saturates at perhaps 8 V and appears to be square wave. At the other extreme, when the input is very small, the output shows **electrical noise**. This is a random mixture of all frequencies (white noise) and has an average amplitude in the example of perhaps 1 mV at the output. An input signal of less than 5 μV should produce an output of less than 1 mV: since such an output cannot be separated from the noise, the amplifier does not perform correctly for very small signals. The signal frequency and waveform are also important and only those in the correct range will be properly processed.

Environmental limitations

The environment of an electronic circuit must include the electrical environment provided by the energy sources or power supplies mentioned above. With some circuits, the voltage or current levels of these power supplies is very critical. Other circuits are more tolerant and perform correctly with a wide range of power supply conditions. Other environmental factors include temperature, humidity, vibration and radiation. Most circuits will operate correctly within a comparatively wide range for these factors, but if this range is exceeded the resulting breakdown may well be catastrophic.

Analysis of electronic circuits

The term **analysis** can have various meanings: to break down into constituent parts; or a critical examination of system components; or the resolving of problems by reducing them to equations. when speaking of the analysis of electronic circuits, we may mean any or all of these.

The object of analysis is to obtain an understanding of the behaviour of the circuit. This understanding or analysis can be at one of a number of levels as follows.

Verbal analysis

Verbal analysis is a purely descriptive approach which in some instances is more than adequate to provide an understanding of the operation of a correctly functioning circuit. Such a description will usually need to to be fully illustrated with circuit diagrams and timing or waveform graphs for various parts of the circuit including input and output signals.

Graphical analysis

Graphical analysis can often usefully accompany verbal analysis. It allows for the examination of some limitations, the choice of values for some critical components and perhaps the determination of signal levels or timing for some processes. Although graphical analysis is limited in application, it is a most useful aid in the understanding of the behaviour of many types of electronic circuit.

Partial mathematical analysis

Mathematical analysis may be applied to a critical feature of an electonic circuit, assuming correct operation, design and ideal components. This might include such properites as the delay time introduced by the circuit, the frequency of oscillation of the circuit or the maximum amplification for stable operation. This form of analysis will often accompany verbal analysis so that the selection of suitable values of critical components can be made.

Ideal mathematical analysis

An ideal analysis of a circuit can be made assuming that the components and devices are ideal. This will determine all d.c. bias and signal levels throughout the circuit. Simplifications and approximations can often be made so that only the essential features are studied. The results of such analysis allow comparison between theoretical and measured performance. This can also be of great assistance in fault finding as, if a particular bias or signal level is not as predicted, a faulty device or component can easily be identified. The first stages of a design exercise will frequently involve analysis at this level. This will determine whether the desired objective is possible and allow the selection of components and devices for fabrication of a prototype circuit.

Further mathematical analysis

Analysis of limiting conditions resulting from the use of non-ideal devices and components can be made. This will be particularly concerned with signal frequency and transient limitations. It may also indicate design improvements and preferred fabrication techniques.

Statistical analysis

Statistical analysis builds on the previous levels of analysis to allow for device and component tolerances, thus permitting investigation of 'worst case' situations, reliability and similar problems. Analysis at this level is essential if a circuit or system is to be properly *engineered* for production.

A technician engineer will require an understanding of electronic circuits and systems at most of these levels for most of the circuits that he encounters. He should certainly be aware of the importance of good design and engineering practice.

How the book is structured

The remainder of this book is intended to provide the appropriate level of understanding of electronic circuits to a technician engineer. The first essential will be to review the various 'descriptions' of the passive components and active devices that are the building blocks of all electronic circuits. These descriptions must match the types of analysis required i.e. verbal, graphical or mathematical, and, where applicable, limitations of signal and environment will be discussed. Internal operation of the devices and components however will not be examined in any detail.

The next requirement is a study of the technique of analysis. In many areas these techniques will be familiar, but their revision will be aimed specifically at the analysis of electronic circuits where the particular analysis leads to a better understanding of circuit operation.

These two areas, device and component description and techniques of analysis, will then be brought together to examine the whole range of important electronic circuits. Throughout, the ideas are illustrated by numerical examples. The final stage of combining the electronic circuits into systems is outside the scope of this book, but with the background obtained, you should be able to proceed to the study and understanding of such systems.

2 Description of Components and Devices

Electronic circuits consist of **networks** of passive components and electronic devices. These circuits are supplied with energy from external sources and are used to process the time-varying voltages and currents that we refer to as signals. Before investigating the operation of such circuits, it is essential that the properties of the components and devices are *described* in the various terms necessary for analysis at any required depth. This chapter is concerned with such descriptions and not with the physical background to the device and component behaviour. In the case of passive components in isolation, a mathematical description is not only possible but is usually all that is required. With active devices, non-linearity makes graphical descriptions essential for many analytical procedures. Since devices and components are to be incorporated in the same circuits, graphical descriptions for the components will also be helpful.

Notation

Voltage and current signals are applied to circuit inputs and, as a result, processed output signals will appear at another part of the circuit. These signals will have to be *measured* or *calculated*, not in isolation but with reference to each other and to the circuit. For this purpose, the **sense of measurement** or **reference direction** must be indicated on the circuit diagram. Various notations for the sense may be found but the most widely used is the **arrow notation**.

Consider the circuit shown in Fig. 2.1. The voltage arrows e and v_O show the sense of measurement of these two voltages: e is the voltage *measured at point A with respect to point B* and similarly v_O is the voltage at C measured with respect to D. Since v_O will probably be related to e, any time or phase relationship between them must take into account the senses of measurement shown.

Fig. 2.1 Arrow notation for currents and voltages.

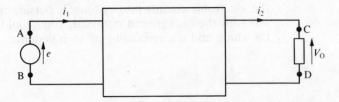

Similarly, the input and output current senses of measurement are indicated by the i_1 and i_2 arrows. A final important concept concerns the current/voltage relationship in a single component: this is shown in Fig. 2.1 with the i_2, v_O relationship. The convention is that if v is produced by i *in a passive component*, and the voltage and current arrows are *opposite in direction* (as shown) then the relationship will be positive. If the arrow directions are the same, then the realtionship will be negative.

Components

The basic passive components in electronic circuits are resistors, capacitors and inductors. These are lumped constant components having ideally the properties only of resistance, capacitance and inductance respectively. The term **lumped constant** implies that the particular property can be assumed to exist at a point and that the observable behaviour is independent of the physical dimensions. Two other components must be considered: these are mutual inductors or transformers, and transmission lines. Transmission lines or delay lines are examples of **distributed constant** components in which the physical dimensions do contribute to the behaviour.

Component descriptions   The basic component properties are described mathematically in terms of the instantaneous values of voltage and current. Further descriptions can be given if the voltage and current signal waveforms are limited (for example to steady-state sinusoids). Such descriptions are best understood graphically, either as functions of time or of frequency or as direct *v-i* relationships.

Definitions   Figure 2.2 shows the three components, each with the sense of instantaneous voltage v and current i indicated by the arrow notation. Table 2.1 gives the fundamental relationships in each case, either for v in terms of an applied current i or vice versa.

Table 2.1 Fundamental relationships for resistance (R), capacitance (C), and inductance (L)

		R	C	L
Applied i	$v =$	iR	$\dfrac{1}{C}\displaystyle\int \, dt$	$L\dfrac{di}{dt}$
Applied v	$i =$	$\dfrac{v}{R}$	$C\dfrac{dv}{dt}$	$\dfrac{1}{L}\displaystyle\int v \, dt$

Fig. 2.2 Components with notation for fundamental *v/ i* relationships.

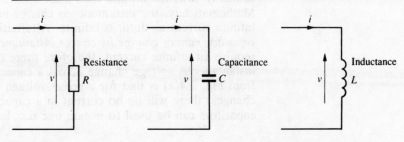

All the relationships are *positive* if the voltage and current senses or directions are as shown in Fig. 2.2.

These relationships are for the properties of resistance, capacitance and inductance and these relationships can be usefully stated in another way.

Resistance   If, for a physical component, the instantaneous voltage is directly proportional to the instantaneous current, then the component has the property of resistance and the constant ratio v/i is the resistance, normally measured in ohms. There is no differential relationship (involving rates of change of v or i); thus, if a signal current of any waveform is passed through a resistance R, there will be an identical voltage waveform across R. The amplitude of this waveform will be the amplitude of the current waveform multiplied by R. These relationships are shown graphically in Fig. 2.3.

Fig. 2.3 Current and voltage waveforms for a resistance.

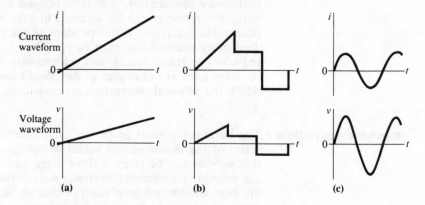

(a)   (b)   (c)

Capacitance   If, for a physical component, the instantaneous current is proportional to the rate of change of instantaneous applied voltage, then the component has the property of capacitance. The constant ratio $(dv/dt)i$ is the capacitance C measured in farads and v and i are in volts and amps respectively. The properties of capacitance are usually associated with two conducting surfaces or regions separated by an insulating region with an electric field between the surfaces. Alternatively, if a current of any waveform is applied to the component and the voltage across it is proportional to the integral (with respect to time) of that current then the constant ratio is $1/C$.

These differential and integral relationships mean that the current and voltage waveforms cannot, in general, be the same. Some examples involving both applied voltages and applied currents are shown in Fig. 2.4. The first, in Fig. 2.4(a), illustrates some important principles: the applied voltage represents a direct voltage suddenly switched on and then after a time switched off again. Mathematically, the instantaneous change in v would require an infinite current as dv/dt is infinite. Alternatively, we can say that a *capacitor cannot change its charge instantaneously* as this would require an infinite current, therefore there cannot be an instantaneous voltage change across a capacitor. A further point from Fig. 2.4(a) is that for a direct voltage (after any transient changes) there will be no current in a capacitor, and for this reason capacitors can be used to isolate one d.c. level from another in

Fig. 2.4 Voltage and
current waveforms for a
capacitance.

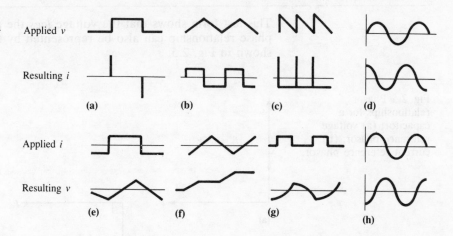

electronic circuits. The waveforms in Figs 2.4(b) and (c)
demonstrate the differential relationship $i = C\mathrm{d}v/\mathrm{d}t$. In each case,
the current amplitude is proportional to the rate of change of
voltage and the sign of the current ($+$ or $-$) is given by the
direction of the voltage slope. In the important case of a sinusoidal
voltage (Fig. 2.4(d)) the instantaneous current at any time is
proportional to the rate of change of voltage. For example, when
the voltage is at a maximum, its rate of change and the resulting
current are zero. Similarly, when the voltage is zero its rate of
change is a maximum as is the resulting current. The remaining
diagrams show the integral relationship ($1/C\int i\mathrm{d}t$). In Fig. 2.4(e) the
integral of a constant is a ramp, positive or negative according to
the polarity of the constant. In Fig. 2.4(f), however, each pulse or
short period of d.c. has the same polarity. During each pulse the
capacitor charges a little further, with the series of ramps shown.
The integral of a ramp is a parabola ($\int at\mathrm{d}t = at^2/2$); this is shown in
Fig. 2.4(g), which is a series of positive and negative parabolas and
not a sine wave. The sinusoidal current and voltage waveforms in
Fig. 2.4(h) show the reverse relationship to that in Fig. 2.4(d).

Sinusoidal signals in
capacitance

If the instantaneous voltage applied to a capacitor is given by

$$v = \hat{V} \sin \omega t$$

Then $i = C\dfrac{\mathrm{d}v}{\mathrm{d}t} = C\hat{V}\omega \cos \omega t$

This current is also sinusoidal (see Fig.2.4) and we may write

$$i = \hat{I} \sin (\omega t + 90°) = \omega C \hat{V} \sin (\omega t + 90°)$$

This gives the ratio between the voltage amplitudes as ωC (which
depends on the signal frequency) and shows that the current leads
the voltage by 90°, or a quarter of a cycle. These two relationships
are combined in the complex notation

$$I = \mathrm{j}\omega C V \tag{2.1}$$

$$V = \frac{I}{\mathrm{j}\omega C} = -\frac{\mathrm{j}}{\omega C} I \tag{2.2}$$

This last form shows that the voltage *lags* the current by 90°. This phase relationship can also be represented by the phasor diagrams shown in Fig. 2.5.

Fig. 2.5 Phasor relationships for a capacitor; (a) voltage reference phasor; (b) current reference phasor.

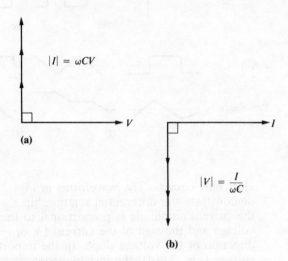

For a phasor diagram to be meaningful, a reference phasor must be defined. For simple series circuits, this reference will be the current *I* and for simple parallel circuits, a voltage reference must be used. For a single component, either variable can be used and the alternatives are shown in Fig. 2.5. In Fig. 2.5(a), the voltage *V* is the reference and the current is shown leading this reference. The scales of the two phasors must be different as they represent the different dimensions, voltage and current. The length of the current phasor will be directly proportional to the frequency of the sinusoidal signal. In Fig. 2.5(b), the current is the reference and the resulting voltage phasor lags by 90°. The length of this phasor is inversely proportional to frequency: it tends to zero at very high frequency (short circuit) and tends to infinity at zero frequency (open circuit). The ratio *V/I* is known in general as the impedance of the component or circuit but in this case, as the 90° relationship is included, it is referred to as the circuit reactance. The expression for capacitive reactance X_c is given by

$$X_c = \frac{1}{\omega C} \qquad [2.3]$$

The relationships between capacitor current and reactance with frequency are shown in Fig. 2.6. In Fig. 2.6(a), we can see that for a constant amplitude voltage the current amplitude is proportional to frequency, starting at zero for d.c. when $\omega = 0$. In Fig. 2.6(b) the ratio of *V/I*, the reactive impedance, can be seen to be infinite at d.c. and falls to zero as the frequency tends to infinity. These properties of capacitance are widely used in electronic circuits; a capacitor can be used to separate two d.c. levels (zero d.c. current) while passing high frequency signals with only a very low impedance.

Fig. 2.6 Frequency response
for a capacitor.

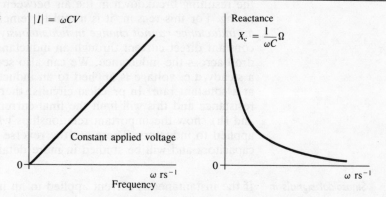

Inductance   If for a physical component the instantaneous voltage is proportional
to the rate of change of instantaneous current, then the component
has the property of inductance. The constant ratio $v/(\mathrm{d}i/\mathrm{d}t)$ is the
inductance L measured in henries where i and v are in volts and
amps respectively. The property of inductance is usually associated
with a coiled current-carrying conductor with a magnetic field
linking the individual turns. The magnetic field will be intensified if
the coil is wound on a core of ferrous material and the result is a
larger value of inductance. At the other extreme, a short straight
current-carrying conductor will still have an associated magnetic field
and will thus possess a small value of inductance.

Ｔhe alternative view of inductance is that if a voltage of any
waveform is applied to an inductor, then the current will be
proportional to the integral of that voltage (with respect to time)
and the constant ratio between the waveforms is $1/L$. The
differential and integral relationships for inductance mean that the
current and voltage waveforms cannot be the same.

Some examples of applied voltages and currents are shown in Fig.
2.7. These waveforms should be compared with those for a capacitor
in Fig. 2.4. as this demonstrates the **duality** of the two properties. In
Fig. 2.7, one important concept is illustrated in Fig. 2.7(e). Here,
the applied current is switched on and off again, instantaneous
changes. The differential of such changes is infinite and should
therefore produce infinite voltage 'spikes'. This does not happen as

Fig. 2.7 Current and
voltage waveforms for
inductance.

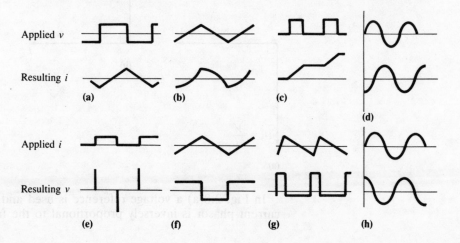

the resulting breakdown in the air between the switch points causes arcing. For this reason, it is useful to remember that *the current in an inductance cannot change instantaneously*. Furthermore, for a constant direct current through an inductance, there will be no volt drop across the inductance. We can also see from Fig.2.7(a) that if a steady d.c. voltage is applied to an inductor, the current will rise at a constant rate. In practical circuits, there will always be some resistance and this will limit the final current value. Figures 2.7(d) and (h) show the important relationships when sinusoidal signals are applied to inductors. These are the reverse of those applicable to capacitors and will be studied in more detail in the next section.

Sinusoidal signals in inductors

If the instantaneous current applied to an inductance is given by

$$i = \hat{I} \sin \omega t$$

Then $v = L\dfrac{di}{dt} = L\hat{I}\omega \cos \omega t$

This voltage is also sinusoidal (see Fig. 2.6) and we can write

$$v = \hat{V} \sin (\omega t + 90°) = \omega L \hat{I} \sin (\omega t + 90°)$$

This gives the ratio between the voltage and current amplitudes as ωL (which depends upon the signal frequency) and shows that the voltage leads the current by 90°, or a quarter of a cycle. These two relationships are combined using the complex notation as

$$V = j\omega L I \qquad\qquad [2.4]$$

or

$$I = \frac{V}{j\omega L} = -\frac{jV}{\omega L} \qquad\qquad [2.5]$$

The last form shows that the current lags the voltage by 90°. The phasor diagrams for inductance are similar to those for capacitance with the angles reversed. These are shown in Fig. 2.8

Fig. 2.8 Phasor relationships for an inductor: (a) voltage reference; (b) current reference.

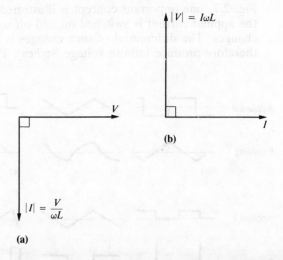

In Fig. 2.8(a) a voltage reference is used and the length of the current phasor is inversely proportional to the frequency of the applied voltage. In Fig. 2.8(b) the current reference produces a leading voltage phasor whose length is proportional to the signal frequency. Thus, at low frequency the inductor tends to a short circuit and at high frequency it tends to an open circuit. The ratio V/I is known in general as the impedance of the component or circuit, but in this case, as the 90° relationship is included, it is known as the reactance. Hence inductive reactance

$$X_L = \omega L \tag{2.6}$$

The signal frequency relationships for an inductor are shown in Fig. 2.9. In Fig. 2.9(a) we can see that for a constant amplitude applied current, the voltage amplitude is proportional to frequency, starting at zero for d.c. when ω is zero. Also, from Fig. 2.9(b), the ratio V/I, the reactive impedance, rises from zero and increases directly with the frequency.

Fig. 2.9 Frequency response for an inductor.

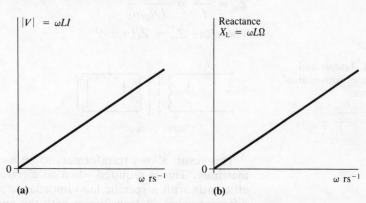

(a)   (b)

Mutual inductance   This is a special form of inductance concerning two conductors linked by a magnetic field. The property of mutual inductance can be recognised when a changing current in one component not only produces a voltage across that component, but also across another unconnected component. This is illustrated in Fig. 2.10. The induced voltage v_2 is the result of the magnetic field due to i_1 linking with the second coil. The relationship is similar to that for **self-inductance** and this can be extended to the various signal waveforms shown in Fig. 2.7, where the applied current would be i_1 and the mutually induced voltage v_2 would have the waveforms shown. For sinusoidal signals. v_2 is given by

$$v_2 = \pm j\omega M I_1 \tag{2.7}$$

Fig. 2.10 The v/i relationship for a mutual inductor M.

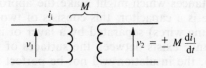

The alternative signs $\pm$ arise as the relative directions of winding of the two coils must be taken into account. If these are chosen correctly, either the positive or the negative sign can be provided. The value of M is related directly to the individual inductors and is given by

$$M = k\sqrt{L_1 L_2} \tag{2.8}$$

where k is the coupling factor and has a maximum value of unity. A figure near this maximum is obtained when the two coils are wound on the same iron core. The resulting component is known as a **transformer**, for which certain additional properties can be stated.

The transformer in Fig. 2.11 has two windings of n_1 and n_2 turns respectively. These are known as the **primary and secondary windings**, where the source of power or signal is applied to the primary and a load is connected to the secondary. Since the efficiency approaches 100%, the VI product is the same for each winding but the volt drop across each winding is proportional to the number of turns on that winding. Thus $V_1/V_2 = n_1/n_2$ and $I_1/I_2 = n_2/n_1$, the inverse ratio, but $V_2/I_2 = Z_L$ the load impedance, and Z_{in} the input impedance is given by

$$Z_{in} = \frac{V_1}{I_1} = \frac{V_2 n_1/n_2}{I_2 n_2/n_1}$$

Therefore $Z_{in} = Z_L(n_1/n_2)^2$ [2.9]

Fig. 2.11 Transferred impedances with mutual inductance.

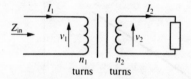

This result allows transformers to be used for **impedance matching**. This is required when an electronic circuit operates most efficiently with a specific load impedance and the actual load has a different value. A transformer with the appropriate turns is then interposed between the circuit and the load. With high frequency signals, iron-cored transformers have undesirable properties and less efficient air-cored transformers are used.

Non-ideal components

All the components described so far are of the lumped constant form. This assumes that their properties can be considered as existing at a point. Practical components can usually be approximated to this form but we do not need to consider circumstances which might make the approximation invalid. A good example is a capacitor: this consists of two conducting plates (with connecting wires) separated by a layer of insulating material. The capacitance exists between the surfaces of the two plates. In practice, the insulation will not be perfect and there will be a very small leakage current between the plates. This may be represented by a very high resistance in parallel with the capacitance. With a.c. signals, the energy losses which this resistance represents can vary with frequency. Thus this effective loss resistance may be frequency dependent. The plates and connecting wires of the capacitor will not be perfect conductors and will have a low resistance and a very low inductance. A possible resulting equivalent circuit or model is shown in Fig. 2.12.

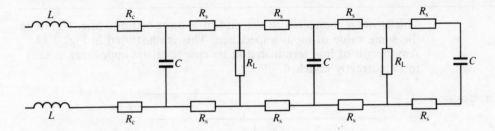

Fig. 2.12 A model for a non-ideal capacitor.

In practical situations, it is only in very special circumstances (high signal frequencies or high voltages) that these additional factors need to be taken into account. However some **distributed constant capacitors** are fabricated with appreciable series and shunt resistance for special high frequency applications.

Inductors, on the other hand, can rarely be considered ideal. The resistance of the coil windings can seldom be neglected: it determines the **selectivity** and **Q-factor** of frequency selective circuits. The Q-factor is a measure of the selectivity, the ability to select a narrow band of frequencies. This resistance is itself dependent on the signal frequency and rises with signal frequency for frequencies above about 100 kHz. With iron-cored inductors and transformers, additional losses represented by shunt resistance will be appreciable for all a.c. signals and again increase with frequency (shown by a reducing shunt resistance). There will also be capacitance between the coil windings. An equivalent circuit or model, together with a simplified version, are shown in Fig. 2.13. The values of the components in the simplified version will all be very frequency dependent.

Fig. 2.13 Models for non-ideal inductors.

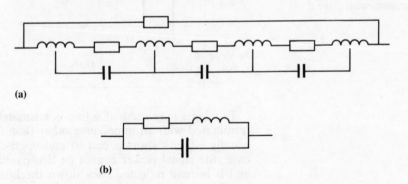

(a)

(b)

Transmission lines

A **transmission line** is the general name for distributed constant components whose physical dimensions are such that the time taken for signals to transfer from one end to the other is not negligible. Two essential properties of transmission line behaviour are described below.

The first important property concerns the characteristic impedance of a line, Z_O. This can be regarded as the input impedance of an infinitely long line. Alternatively, it is the value of impedance which, when connected to the end of a short or finite line, results in the same value of input impedance. This is illustrated in Fig. 2.14. Any length of line **terminated** in its characteristic impedance is said to be correctly **matched**.

Fig. 2.14 The characteristic impedance of a transmission line.

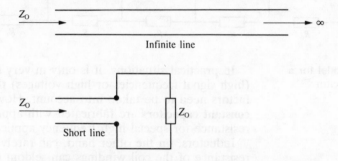

The second important property concerns a signal travelling along the line. The **propagation constant** has two elements; the first is concerned with the **time delay** (or phase delay for sinusoidal signals) between the signal entering the sending end of the line and arriving at the receiving end or termination. The second element is concerned with the **attenuation**; some signal power is lost as it travels along the line and the received signal will be smaller than the transmitted signal. These properties are illustrated with a signal pulse in Fig. 2.15.

Fig. 2.15 The delay of signal pulses in a transmission line.

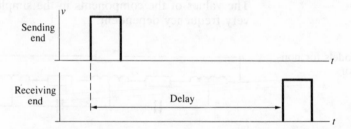

Problems may occur if a line is **mismatched**, which happens if it is terminated with an impedance other than Z_O. The worst cases are usually either a short-circuit or and open-circuit termination. In each case, the signal power cannot be dissipated in the terminating load and is instead reflected back down the line to reappear at the sending end. The form of the reflection depends both on the nature of the termination and upon the length of the line. In most circumstances, these effects of mismatch are undesirable but in some applications they can be used for matching purposes or in certain pulse generating circuits.

Devices

Electronic devices are specially fabricated structures which exhibit various useful non-linear properties. Most devices are **solid state** in that they are made from solid semiconducting materials such as silicon and germanium, but other special devices operate with beams of electrons in vacuum or low pressure inert gas. The essential description of device behaviour is necessary so that the combined behaviour of devices, components and power supplies can be predicted.

The devices that are described include the basic two- and three-terminal devices such as diodes, transistors and thyristors, and also the basic integrated circuit elements such as operational amplifiers, logic gates and flip-flops. In each case, a range of descriptions will be given. These will define the variables (current and voltage) associated with each device, show graphical relationships (characteristics), specify essential parameters and introduce equivalent circuits or models.

Diodes   An ideal **diode** is a switch that can be open circuit or short circuit according to the polarity of the voltage applied to it. The symbol and definition of voltage direction is shown in Fig. 2.16(a). As a first approximation, if V in the direction shown is positive, the diode will conduct with the current I limited by external conditions. Under these circumstances, the diode is said to be **forward biased**. If the polarity of the voltage is reversed, with the diode **reverse biased**, the current will be zero. The graphical representation of this behaviour is shown in Fig. 2.16(b). For many applications, this approximate model is sufficient to predict circuit behaviour, but in other cases a

Fig. 2.16 A diode: (a) symbol and notation; (b) ideal characteristic.

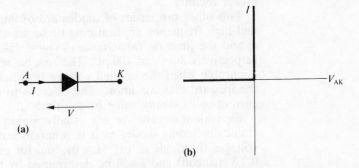

more accurate description will also be required. A graphical description or characteristic for a diode is given in Fig. 2.17. This shows that, for positive voltages greater than a small **threshold** value, the current increases rapidly. With negative voltages there is only a very small current (leakage current) until a high negative voltage is reached. The diode then **breaks down** and the reverse current increases rapidly. This diode characteristic can be approximately described (in the normal operating region) by the **diode equation**

$$I = I_0(e^{kV}-1) \hspace{4cm} [2.10]$$

where I_0 is the (constant) reverse leakage current and k is a constant which at normal operating temperatures is about $1/0.025$. The value of I_0 depends on the material from which the diode was manufactured, on the temperature and on the physical dimensions of the diode. Typical values of I_0 for silicon and germanium respectively are a few nanoamps and a few microamps. In equation

Fig. 2.17 The V/I
characteristic of a practical
semiconductor diode.

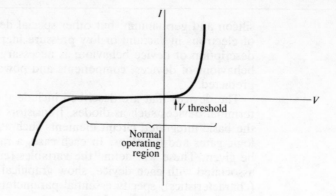

[2.10], when V is negative I is approximately $-I_0$, as the exponential
term is negligible. For positive V, the exponential term dominates
the expression and the current rises exponentially after a threshold
value of V has been exceeded. This threshold for forward
conduction may be taken as 0 V for germanium diodes and 0.5 V
for silicon devices. The normal conducting volt drop V_F depends
upon the current but for low power silicon devices a figure of 0.7 V
is usually taken, while for high power rectifiers a figure in excess of
1 V may be applicable with a forward current of 100 A. The reverse
breakdown voltage V_R also varies considerably with the specific
device type ranging from a few volts to more than 1000 V for a
power rectifier.

Two other properties of diodes are of importance for switching
and high frequency applications; these are the reverse recovery time
t_{rr} and the junction capacitance. Typical values for these are
respectively 2 μs and 100 pF. The reverse recovery time t_{rr} is
applicable when the applied voltage is switched from the reverse to
the forward bias condition. The effect is to delay the rise in forward
current to its steady value by the time t_{rr}.

Equivalent circuits are not usually required in the analysis of
circuits including diodes as it is generally sufficient that for negative
voltages the diode is 'off' ($I = 0$), and for positive voltages V will be
0.7 V (silicon) and I will be determined by the external circuitry.
The threshold between these two conditions is about 0.5 V.
However, changes in voltage in the conducting region can, for small
signals, be represented by a resistor as shown in Fig. 2.18. The
junction capacitance can also be included in this model as shown.

Fig. 2.18 A possible model
or equivalent circuit for a
diode.

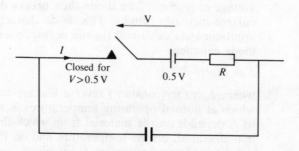

Typical diode types and their essential properties

Rectifying diodes   Used for a.c.–d.c. conversion. V_{RRM} the maximum reverse voltage before breakdown is 200 V to 1600 V. I_F the average forward current is 1 A to 30 A.

Signal rectifying diodes   For operation at very high frequencies. These will usually have a low forward resistance and low capacitance. Operating currents and voltages will be much less than for the last group.

Switching diodes   For high-speed switching between the on and off conditions. The important feature here is the reverse recovery time t_{rr} which will range from a few nanoseconds to a microsecond.

Variable capacitance diodes   Used in the reverse-biased state where the capacitance may be controlled by the applied voltage. A typical range would be from 2 pF to 12 pF with reverse voltage values between 28 V and 3 V respectively. The leakage current in this range is about 0.1 μA.

Reference or zener diodes   These are also used in the reverse-biased state. The breakdown characteristic shows a very sharp 'knee' after which the volt drop is approximately constant for a wide range of currents. A wide range of reference voltages are available, typically between 2 V and 75 V. The small voltage variation in the reference state is described by the slope resistance R_Z. This depends on the diode type and on the reference voltage V_Z. Typical values lie in the range 4 Ω for a nominal 6 V diode to 200 Ω for a 75 V reference. Figure 2.19 shows the symbol, characteristics and equivalent circuit for a reference diode.

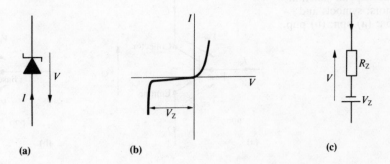

Fig. 2.19 The zener diode: (a) symbol and notation; (b) characteristic; (c) a possible equivalent circuit for zener operation.

(a)   (b)   (c)

In Fig. 2.19(a) the conventional directions for V and I are shown for comparison with the characteristics in Fig. 2.19(b). For reference operation, V and I will be negative (i.e. current flowing downwards in Fig. 2.19(a)). Once V exceeds V_Z, there will be little change in voltage and the current is limited by external circuitry. The maximum safe operating current is determined by the power rating of the device, typical values being 0.5 W to 5 W. The equivalent circuit in Fig. 2.19(c) is applicable beyond the knee voltage and the direction of V and I shown are the normal d.c. values. The actual voltage is then given by

$$V = V_Z + IR_Z$$

Transistors   These are the basic amplifying and switching devices in all electronic systems. They are used either as separate devices or as internal components in integrated circuits such as operational amplifiers or flip-flops. Integrated circuits may include only a few transistors or many thousands of these devices. There are two basic transistor types, the **bipolar junction transistor** (BJT) and the **field effect transistor** (FET). Their properties will be described separately.

Bipolar junction transistors   A BJT has three external connections known respectively as the emitter, the collector and the base. Internally the BJT consists of two back-to-back 'diodes' or semiconductor pn junctions (p and n stand for the different types of semiconductor materials). This construction can be made in two ways, npn or pnp: however, in practice, the majority of transistors used are of the npn configuration. In each case the middle layer is the **base**; one of the outer layers is designed to be the **emitter** and the other the **collector**. Transistors can be operated back to front but the properties are not as satisfactory for most applications. The symbols and the voltage and current variables are shown in Fig. 2.20. In each case the conventional d.c. currents are shown. The first relationship, which is true for all transistors under all conditions is

$$I_E = I_C + I_B$$

Fig. 2.20 Bipolar junction transistors; symbols and notation: (a) npn; (b) pnp.

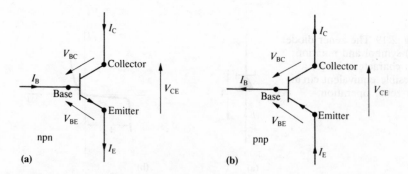

There are three normal operating states for BJTs and these depend upon the bias conditions applied to the two pn junctions. These three states are defined in Table 2.2.

The OFF and ON states are used mainly in switching circuits in digital systems while the amplifying state (normal transistor operation) is used in linear amplifying circuits.

More can be understood about these states by reference to the d.c. characteristics. Many sets of interrelated characteristics can be drawn but the most useful are those for the *input* variables I_B, V_{BE}

Table 2.2 Bias conditions for bipolar junction transistors

				npn		pnp	
	State	*Base-emitter junction*	*Base-collector junction*	V_{BE}	V_{BC}	V_{BE}	V_{BC}
1.	OFF	Reverse biased	Reverse biased	−ve	. −ve	+ve	+ve
2.	ON	Forward biased	Forward biased	+ve	+ve	−ve	−ve
3.	Amplifying state	Forward biased	Reverse biased	+ve	−ve	−ve	+ve

and the *output* variables I_C and V_{CE}. In each case, a third variable (the parameter of the characteristic) is kept constant for one curve. These parameters are, respectively, V_{CE} for the input characteristic and I_B for the output characteristics.

The two sets of characteristics shown in Fig. 2.21 are for a low power silicon npn transistor. On the input characteristic, for V_{BE} greater than a threshold value, the base current increases rapidly for only a small increase in voltage. This characteristic is very similar to the diode characteristic in form but the position is modified slightly with changes in the collector voltage V_{CE}. In the linear or amplifying region, these changes are usually neglible but if the collector voltage V_{CE} becomes less than the base voltage V_{BE}, the transistor is in the ON state with the increase in I_B shown. The OFF state on the input characteristic is when V_{BE} is less then the threshold value or negative. The base current is then known as the reverse leakage current I_{CEO} which is typically only a few nanoamps. Referring now to the output characteristics, consider first the case when I_B is set to I_{B1}. As V_{CE} is increased from zero, it is initially less than V_{BE}, so the transistor is in the ON state. The collector current I_C rises rapidly with increase in V_{CE} until V_{CE} exceeds V_{BE}. The collector–base junction becomes reverse biased and the transistor moves into the *linear* state. In this region, further increase in V_{CE} results in only a small change in I_C and the characteristic becomes nearly horizontal.

Fig. 2.21 The d.c. characteristics of an npn BJT: (a) input, I_B/V_{BE}; (b) output, I_C/V_{CE}.

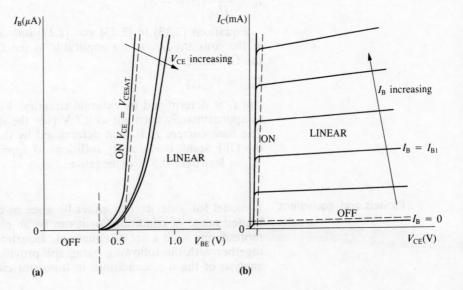

The particular value of I_C under these conditions depends on the particular value of the base current I_B. The relationship between the two currents in this state is given by

$$\frac{I_C}{I_B} = H_{FE} \qquad\qquad [2.11]$$

where H_{FE} is the d.c. current gain or amplifying factor.

Typical values of H_{FE} lie in the range of 10 to 1000. Each value of I_B results in a different 'horizontal' characteristic. In the ON state, however, all these characteristics follow the same line and the transistor is said to be **saturated**.

V_{CE} under these conditions is known as V_{CESAT}, with typical values of less than 0.1 V. For the OFF state on the output characteristic, I_B is reduced to zero. I_C will not be zero since the base leakage current is still amplified by H_{FE} to give I_{CBO} which for this transistor type would be less than a microamp. When I_B is zero the transistor is said to be **cut off**.

Various useful relationships can be specified for the d.c. conditions as follows

$$I_E = I_C + I_B \qquad\qquad [2.12]$$

$$I_C = H_{FE}I_B + I_{CBO} \simeq H_{FE}I_B \qquad\qquad [2.13]$$

Also, since $H_{FE} \gg 1$

$$I_E = I_B(1 + H_{FE}) \simeq I_C \qquad\qquad [2.14]$$

or

$$I_B = \frac{I_C}{H_{FE}} \qquad\qquad [2.15]$$

It is sometimes useful to define the input characteristic in terms of the diode equation [2.10]

$$I_C \simeq I_E = I_0(e^{kV_{BE}} - 1) \qquad\qquad [2.16]$$

and

$$I_B = \frac{I_0}{H_{FE}}(e^{kV_{BE}} - 1) \qquad\qquad [2.17]$$

Equations [2.13] to [2.15] and [2.17] *only apply in the linear state* as the constant H_{FE} is not applicable to the ON or OFF states. In the ON state

$$V_{CE} = V_{CESAT} \qquad\qquad [2.18]$$

and I_C is determined by external circuitry. V_{BE} may usually be taken as approximately constant at 0.7 V (see the diode description, p 00). The base current I_B is then determined by the external circuitry. In the OFF state, it is usually sufficient to approximate I_C to zero if V_{BE} is less than 0.5 V or negative.

Models and equivalent circuits

A model for a device can either be a set of equations covering a limited range of conditions or it can be an electrical network representing such a set of equations. Equations [2.12] to [2.17] together with the following paragraph provide a model for the analysis of the d.c. conditions in transistor circuits. This model includes the ON, the OFF and the linear states; it is also sufficient for many calculations on switching circuits involving transistors. Analysis for signals in the linear region is more involved. In some situations, analysis directly from the characteristics is useful (see Ch. 3), but with most circuits a **small-signal equivalent circuit** provides the only practical approach. Such a circuit, consisting of electrical components and *dependent generators*, behaves in the same way as the transistor for *small changes* in voltages and currents. This raises the question 'how small is small?'. In practice, a reasonable approximation to the transistor behaviour will be obtained provided the changes do not take the transistor into cut-off or saturation for any part of a signal cycle. Many equivalent circuits are possible but only a few have widespread practical application. These are the *h*-parameter equivalent and its simplified versions which are most useful at low and medium signal frequencies; the hybrid π equivalent and the *y*-parameters are useful for high frequency applications.

The h-parameter equivalent circuit

Figure 2.22 shows an npn BJT and the *h*-parameter equivalent. The notation for these parameters, h_{ie}, h_{re}, h_{fe} and h_{oe} can be described as follows: *h* stands for hybrid, as the set of parameters have a mixture of units, impedance, voltage gain, current gain and admittance respectively. The first suffix i, r, f and o stands for input, reverse, forward and output respectively. The second suffix e is for the common or reference terminal used in the measurement of the parameters. In this case, the emitter is used; sets of parameters using the base or collector as the reference terminal (h_b and h_c parameters) can be obtained but the h_e parameters are the most widely quoted and the most useful. The natural properties of the transistor to the four separate parameters can be usefully described as follows.

Fig. 2.22 An npn bipolar junction transistor: (a) symbol; (b) an *h*-parameter small-signal equivalent circuit.

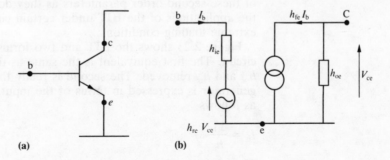

- h_{ie} is the input impedance, which is the slope of the input characteristic. In Fig.2.21(a), it can be seen that this slope and therefore h_{ie} varies with the level of current (typically varying from perhaps 100 kΩ with an I_B of 1 µA to a few ohms when I_B is 10 mA).
- h_{fe} is the current gain, which is the ratio of collector to base current shown in the output characteristic in Fig. 2.21(b). This is the most important parameter, and in fact is the *only* parameter generally quoted in manufacturers' published data. There is some variation of h_{fe} over the linear operating range

but the sample-to-sample variation is much greater for a particular transistor type. Typical specifications are 10 to 35, 100 to 500 or greater than 1000. There are differences between the small signal h_{fe} and the d.c. current gain h_{FE}, but in practice the two values can be taken to be the same.

- h_{oe} is the output admittance and represents the slope of the output characteristic in Fig. 2.21(b). This can also be seen to be current dependent (slope increases as I_C increases). However, in most practical situations, external component tolerances cause a larger variation of predicted performance than that resulting from ignoring h_{oe}. Circumstances for which h_{oe} must be included are limiting conditions such as maximum gain with large external resistance values or tuned *RLC* circuits. h_{oe} values range typically from 10 $\mu s(z = 100$ k$\Omega)$ with I_C of 1 mA to about 1 $\mu s(z = 1$ M$\Omega)$ for high current large signal transistors.
- h_{re} represents the reverse effect, by which the output voltage variation modifies the input characteristic in Fig. 2.21(a). This effect is very small and can in most cases be neglected. Typical values for h_{re} are between 10^{-4} and 10^{-3}.

Simplified equivalent circuits for the BJT

The *h*-parameter equivalent circuit described above provides an accurate description of the behaviour of the device if the parameters are known and if the external components have precise values. In practice, the available relevant information will usually be limited to the range of the small-signal current gain h_{fe} (which will be wide), component values having a tolerance of $\pm20\%$ or at best $\pm5\%$ and the d.c. supply voltages. The information required from analysis will also be limited to the expected order of voltage and current gain and the expected order of input and output impedances. The equivalent circuit in Fig. 2.22(b) may be simplified by ignoring h_{oe} and h_{re} as this will have less effect upon the results than the tolerances mentioned above. It is important, however, to be aware of these second-order parameters as they do provide limitations on the applications of the BJT under certain circumstances, e.g. with extreme loading conditions.

Figure 2.23 shows the BJT and two forms of simplified equivalent circuit. The first equivalent is the same as that in Fig. 2.22(b) with h_{oe} and h_{re} removed. The second is really the same but the current generator is expressed in terms of the input voltage V_{be} instead of I_b as follows

$$I_b = \frac{V_{be}}{h_{ie}}$$

Therefore

$$h_{fe}I_b = \frac{h_{fe}}{h_{ie}}V_{be} = g_m V_{be} \qquad [2.19]$$

The parameter g_m is not an *h*-parameter but a *y*-parameter. It does, however, have a particular advantage in that its value is approximately related to the quiescent or d.c. collector current of the transistor by

$$g_m = 40I_C$$

(This relationship is obtained by differentiating equation [2.16].)

Fig. 2.23 Simplified equivalent circuits for a BJT.

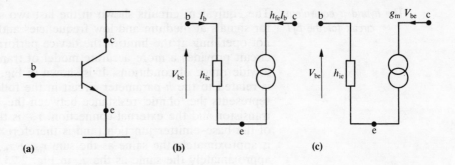

(a)   (b)   (c)

This means that if the d.c. collector current I_C is measured, estimated or calculated, g_m (the **mutual conductance**) is known and with typical h_{fe} values from the manufacturers' quoted data, h_{ie} can be found from Equation [2.19]

$$h_{ie} = \frac{h_{fe}}{g_m}$$

Example 2.1   *A transistor type has h_{fe} quoted as 150 to 400. Two transistors are used with d.c. collector currents of 0.2 mA and 5 mA respectively. In each case calculate the values of the simplified equivalent circuit components.*

Transistor Tr1: $g_m = 40 \times 0.2 = 8.0\ \text{mA/V}$

Minimum $h_{fe} = 150$; minimum $h_{ie} = \dfrac{150}{8.0} = 18.9\ \text{k}\Omega$

Maximum $h_{fe} = 400$; maximum $h_{ie} = \dfrac{400}{8.0} = 50\ \text{k}\Omega$

Transistor Tr2: $g_m = 40 \times 5 = 200\ \text{mA/V}$.

Minimum $h_{ie} = \dfrac{150}{200} = 750\ \Omega$

Maximum $h_{ie} = \dfrac{400}{200} = 2.0\ \text{k}\Omega$

The resulting two equivalent circuits with average h_{ie} values are shown in Fig. 2.24. The current generator can be shown as either $g_m V_{be}$ or $h_{fe}I_b$, whichever is the most convenient for the analysis in hand.

Fig. 2.24 Equivalent circuits for Example 2.1.

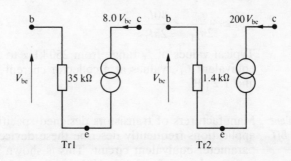

Tr1   Tr2

The hybrid π equivalent circuit for the BJT

The equivalent circuits shown in the last two sections are satisfactory for signals at medium and low frequencies and for circuits which are not operating at the limit of the device performance. The hybrid π circuit provides a more accurate model of transistor behaviour over a wide range of conditions. It is shown in Fig. 2.25(b). This circuit is related to the *h*-parameter circuit in the following way: $r_{bb'}$ represents the 'ohmic' resistance between the active part of the transistor and the external connection. $r_{b'e}$ is the effective resistance of the base–emitter junction (and is therefore dependent on I_C). h_{ie} is approximately the same as the sum $r_{bb'} + r_{b'e}$. g_m is $h_{fe}/r_{b'e}$ and is approximately the same as the g_m in Fig. 2.23. r_{ce} represents the output impedance effect in the same way as h_{oe}, and $r_{b'c}$ (which is very large) provides the h_{re} effect. Since an approximate description will be valid if $r_{b'c}$ and r_{ce} are removed and at low frequency, the two capacitors will be effectively open-circuit; the remaining parts of Fig. 2.25(b) can be seen to be the same as the circuit in Fig. 2.23(c).

Fig. 2.25 The hybrid π equivalent circuit for a BJT.

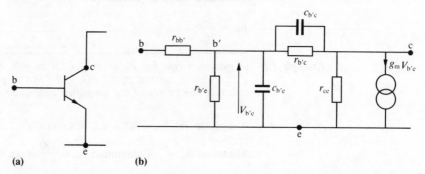

(a)   (b)

The two capacitors $C_{b'c}$ and $C_{b'e}$ represent the junction capacitance for the emitter–base junction and the collector–base junction respectively. $C_{b'c}$ is small (a few picofarads) but may have an important effect upon the input capacitance due to **Miller effect** (see p. 369). $C_{b'e}$ is dependent on the d.c. collector current and its effect upon the short-circuit gain h_{fe} is as follows.

h_{fe} falls at high frequencies due to the effect of $C_{b'e}$ and f_T (manufacturers' published data always include a parameter f_T, the **transition frequency**) is the frequency at which h_{fe} is reduced to 1, a convenient parameter as it is easy to measure. The high frequency reduction in h_{fe} is the result of the parallel combination of $C_{b'e}$ and $r_{b'e}$. It may be shown that for a given f_T and I_C

$$C_{b'e} \simeq \frac{g_m}{2\pi f_T} = \frac{40 I_C}{2\pi f_T}$$

Typical values of f_T range from 250 kHz to 5000 MHz and the equivalent $C_{b'e}$ values at a collector current of 1 mA are 0.025 μF and 1.2 pF.

The y-parameter equivalent circuit for the BJT

Manufacturers of transistors designed specifically for high frequency applications frequently describe these devices in terms of the *y*-parameter equivalent circuit. This is shown in Fig. 2.26. The notation ie, re, fe and oe is the same as that used for the *h*-parameters (see p. 45). However, in this case, each of the four

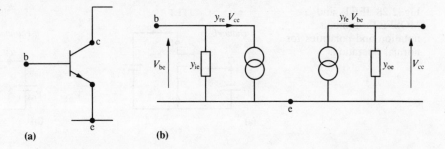

Fig. 2.26 The *y*-parameter equivalent circuit for a BJT.

(a)   (b)

parameters has the dimension of admittance. In the high frequency range where these parameters are used, all four parameters are complex. y_{ie} is usually expressed in terms of the parallel conductance and capacitance g_{ie}, c_{ie}; similarly y_{oe} is expressed in terms of g_{oe}, c_{oe}. The forward and reverse transfer parameters y_{fe} and y_{re} are usually given by modulus and angle. A typical set of *y*-parameters at a signal frequency of 35 MHz is as follows:

$$y_{ie} = g_{ie} + j\omega C_{ie} \qquad g_{ie} = 3 \times 10^{-3}\,\text{S} \qquad C_{ie} = 40\,\text{pF}$$
$$y_{oe} = g_{oe} + j\omega C_{oe} \qquad g_{oe} = 5 \times 10^{-5}\,\text{S} \qquad C_{oe} = 1.3\,\text{pF}$$
$$y_{fe} = y_{fe}\angle\phi_{fe} \qquad |y_{fe}| = 0.1\,\text{S} \qquad \phi_{fe} = 340°$$
$$y_{re} = y_{re}\angle\phi_{re} \qquad |y_{re}| = 5 \times 10^{-6}\,\text{S} \qquad \phi_{re} = 268°$$

y_{ie} and y_{oe} are expressed in this way as it is more convenient to add in the effect of external components, as shown in Fig. 2.27.

Fig. 2.27 A loaded *y*-parameter equivalent circuit.

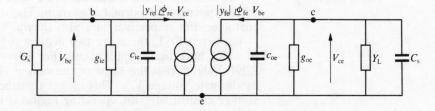

Field effect transistors

There are two main types of **field effect transistor** or FET; these are junction or **JFET**s and insulated gate or **MOSFET**s. In both cases, there are further subdivisions the most important of which are described in this section. All types of FET have a conducting path or **channel** between the **source** and the **drain**. They also have a **gate** which is used to control the current flow between source and drain. With the JFET, the gate and the channel form a reverse biased pn junction and the voltage on the gate widens or narrows the channel so controlling the current flow. With MOSFETs, the gate is insulated but the voltage applied to it modifies the conductivity of the channel adjacent to it, again controlling the source to drain current. This modification may either reduce the conductivity or increase it, depending upon the construction of the device. The two types are known respectively as **depletion** type and **enhancement** type MOSFETs. Finally, both JFETs and MOSFETs may be constructed with n-channel (drain positive with respect to source) or p-channel (drain negative with respect to source). Thus, altogether there are six different types, and some symbols are shown in Fig. 2.28. In each case, the necessary polarity of the bias voltages is indicated by

Fig. 2.28 JFETs and
MOSFETs; symbols,
notation and polarities for
normal operation.

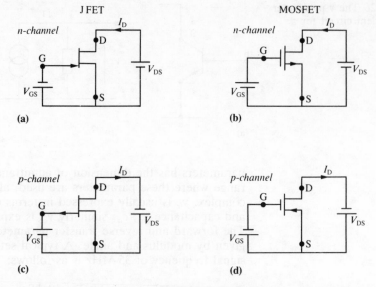

batteries. In Fig. 2.28(c) and (d) simplified symbols are shown for n-
and p-channel MOSFETs which do not indicate whether the device
is of the enhancement or the depletion type. Special symbols can be
used to identify the two types but in most situations use of the
simplified symbol is adequate.

The n-channel JFET

Figure 2.29 shows the **n-channel JFET** with d.c. supplies of the
correct polarity for normal operation. The output and transfer
characteristics respectively are also shown.

   With the JFET, there are two distinct regions of operation, both
of which have useful applications. For low values of V_{DS}, Fig.
2.29(b) shows that the *slope* of the output characteristic is controlled
by the gate voltage V_{GS}. This means that the channel resistance is
voltage controlled. This operating region is known as the **triode
region** (from the similarity with the triode valve characteristic). For
higher values of V_{DS}, the FET is said to be in the **pinch-off region**.
This region is characterised by little variation of drain current with
large changes in applied V_{DS}. The level of current, however, is
controlled by V_{GS} the gate voltage. This control is shown in the
transfer characteristic in Fig. 2.29(c). Note that from both
characteristics the drain current I_D is cut off completely if the
negative V_{GS} is sufficiently large. Note also that there is no
characteristic showing gate current: I_G is negligible as it is only the
reverse leakage current for the gate–channel junction.

JFET d.c. relationships

Two constants are normally defined for a JFET. These are I_{DSS} and
V_P. I_{DSS} is the saturated drain current level, above pinch-off, with
$V_{GS} = 0$. The pinch-off voltage V_P is the d.c. level of V_{GS} which
makes $I_D = 0$ with V_{DS} small. In practice, this negative voltage
corresponds to the positive level of V_{DS} when I_D is just equal to I_{DSS}
(see Fig. 2.29(b)). Neither of these parameters (V_P, I_{DSS}) is usually
specified within tight limits and for I_{DSS} it is often only a minimum
value that is quoted. However, approximate results in terms of V_P
and I_{DSS} can be given for various d.c. relationships.

Fig. 2.29 Symbol and
characteristics for an n-
channel JFET: (a) symbol
and notation; (b) output I_D/
V_{DS} characteristic; (c)
transfer I_D/V_{GS}
characteristic.

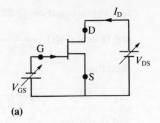

(a)

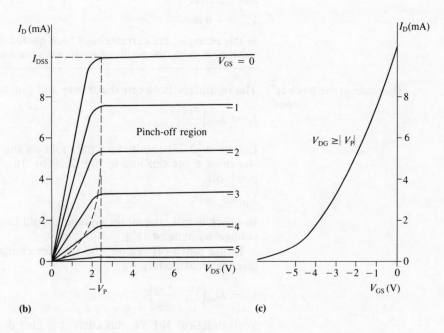

(b)　　　　　　　　　(c)

The triode region　In this region, the main features of interest are the value of the
controlled resistance r_{DS} and the drain current I_D. Approximate
formulae for these are given by

$$I_D = \frac{2I_{DSS}}{-V_P}\left(1 - \frac{v_{GS}}{V_P}\right)V_{DS} \qquad [2.20]$$

$$r_{DS} = \frac{v_{DS}}{i_D} = \frac{1}{\frac{2I_{DSS}}{-V_P}\left(1 - \frac{V_{GS}}{V_P}\right)} \; \Omega \qquad [2.21]$$

$$r_{DS} = \frac{V_P^2}{2I_{DSS}(V_{GS} - V_P)} \; \Omega \qquad [2.22]$$

Example 2.2　*A JFET has I_{DSS} and V_P specified as 2 mA and −2.2 V respectively. If it is
operating in the triode region with v_{DS} of 0.7 V, calculate (a) the drain
current I_D when V_{GS} is −1.2 V and (b) the value of V_{GS} if the required r_{DS} is
900 Ω.*

(a)　From equation [2.20]

$$I_D = \frac{2 \times 2}{2.2}(1 - \frac{1.2}{2.2})0.7 = 0.58 \text{ mA}$$

(b) From equation [2.22]

$$900\ \Omega = 0.9\ K\Omega = \frac{2.2^2}{2 \times 2(V_{GS} + 2.2)}$$

therefore

$$V_{GS} + 2.2 = \frac{2.2^2}{2 \times 2 \times 0.9} = 1.34$$

and therefore

$$V_{GS} = -0.86\ V$$

In this example, the currents have been quoted in milliamps and the voltages in volts; the resulting resistance values will therefore be in kilohms.

Operation in the pinch-off region

The boundary between the triode and pinch-off regions is given by

$$I_D = I_{DSS}\!\left(\frac{v_{DS}}{V_P}\right)^{\!2} \qquad [2.23]$$

Equation [2.23] results in a parabola on the output characteristic, shown as a broken line in Fig. 2.29(b). In general, to operate in pinch-off,

$$V_{DG} > -V_P \qquad [2.24]$$

In other words, the drain voltage should be greater than the gate voltage by at least $|V_P|$.

In the pinch-off region, the transfer characteristic (Fig. 2.29(c)) is given approximately by

$$I_D = I_{DSS}\!\left(1 - \frac{v_{GS}}{V_P}\right)^{\!2} \qquad [2.25]$$

With practical JFETs, this current is also dependent on the drain voltage V_{DS}. This is shown by the slope of the output characteristics in Fig. 2.29(b).

The p-channel JFET

The characteristics and relationships for the **p-channel JFET** differ from the n-channel described above only in the polarity of the voltages and the directions of current flow. Thus, if required, the same formulae may be used taking these factors into account.

The n-channel MOSFET

Since MOSFETs have an insulated gate, one feature of their behaviour is that they have an even smaller gate current than that found in the JFET. Figure 2.30 shows the symbol and characteristics for an n-channel, depletion type MOSFET.

These characteristics can be seen to be very similar to those for the n-channel JFET in Fig. 2.29. Two differences can be observed; V_{GS} can be made positive to increase I_D above the I_{DSS} value (in the enhancement mode) and the characteristics in the triode region show more curvature than those of the JFET. Equations [2.20] to [2.22] may be applied for approximate calculations of I_D and r_{DS} as before.

The alternative enhancement type n-channel MOSFET has the output and transfer characteristics shown in Fig. 2.31. In this case, a major difference is apparent. If V_{GS} is zero, there is no current flow. The device must be enhanced by a positive V_{GS} which causes a

Fig. 2.30 Circuit and characteristics for an n-channel depletion MOSFET: (a) circuit and polarities; (b) output I_D/V_{DS} characteristic; (c) transfer I_D/V_{GS} characteristic.

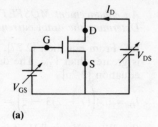

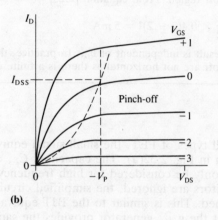

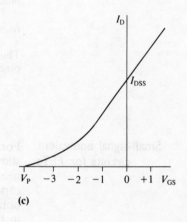

Fig. 2.31 Characteristics for an n-channel enhancement type MOSFET: (a) output; (b) transfer.

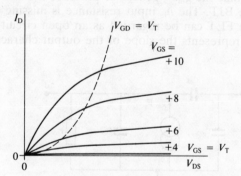

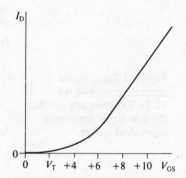

channel to be formed, thus allowing drain current to flow. The threshold value of gate voltage V_T (equivalent to V_P) is the positive value below which i_D is zero. Typical values of V_T range between 1 V and 3 V. I_D cannot be expressed in terms of I_{DSS} as this would be zero. In the triode region (below pinch-off) the drain current I_D is given by

$$I_D = \beta\left[(V_{GS} - V_T)V_{DS} - \frac{1}{2}V_{DS}^2\right] \qquad [2.26]$$

β is a constant applicable to the particular device and has a typical value of 0.5 mA/V². The triode region is defined by:

$$V_{GS} \geq V_T \text{ and } V_{DS} \leq (V_{GS} - V_T) \qquad [2.27]$$

In the pinch-off region

$$I_D = \frac{1}{2}\beta(V_{GS} - V_T)^2 \qquad [2.28]$$

Example 2.3   *An enhancement MOSFET has V_T and β of 2 V and 0.4 mA/V² respectively. Determine the drain current I_D if V_{GS} is 7 V and V_{DS} is (a) 3 V and (b) 8 V.*

(a) From equation [2.27], v_{GS} is greater than V_T and $V_{GS} - V_T = 5$ V which is greater than V_{DS}. The device is therefore in the triode region. From equation [2.26]

$$I_D = 0.4\left((7 - 2)3 - \frac{9}{2}\right) = 4.2 \text{ mA.}$$

(b) When V_{DS} is 8 V, it is greater than $(V_{GS} - V_T)$ and the device is in the pinch-off region. From equation [2.28]

$$I_D = \frac{1}{2} \times 0.4(7 - 2)^2 = 5 \text{ mA.}$$

This result is independent of v_{DS}. In practice, the output characteristics in pinch-off are not horizontal as there is a finite output impedance.

Small-signal equivalent circuits for FETs

For all types of FET, the small-signal equivalent circuit is that shown in Fig. 2.32(a). The capacitors C_{gs}, C_{gd} and C_{ds} are small and need only be considered for high frequency calculations. If these capacitors are ignored, the simplified circuit in Fig. 2.32(b) is obtained. This is similar to the BJT equivalent circuit (Fig. 2.23(c)) in that the $g_m v_{gs}$ generator provides the same function as $g_m V_{be}$ for the BJT. The h_{ie} input resistance is missing since the equivalent in the FET can be regarded as an open circuit. The output resistance r_O represents the slope of the output characteristics (Fig. 2.29(b)).

Fig. 2.32 Small-signal equivalent circuits for FETs: (a) complete circuit; (b) low signal frequency equivalent circuit.

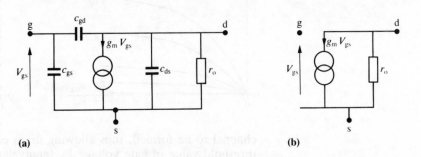

This shows the same features as h_{oe} in the BJT and can usually be neglected when compared with the tolerances of external components. The relationship between g_m and the d.c. conditions is not as simple as that for the BJT($g_m = 39I_C$), but for the JFET an approximate expression for the value of g_m is given by

$$g_m = \frac{2I_{DSS}}{|V_P|}\sqrt{\frac{I_D}{I_{DSS}}} \qquad [2.29]$$

This is the slope of the transfer characteristic (Fig. 2.28(c)) which has a maximum value when $I_D = I_{DSS}$. The maximum value of g_m will occur when V_{gs} is zero and is given by

$$g_{mo} = \frac{2I_{DSS}}{|V_P|} \qquad [2.30]$$

Manufacturers' data usually includes minimum values for $g_m(Y_{FS})$ together with values of V_P and I_{DSS}.

For depletion type MOSFETs, results [2.29] and [2.30] can still be applied, but with enhancement types the constant I_{DSS} is no longer applicable and the equivalent expression for g_m is given by

$$g_m = \beta(V_{GS} - V_T) \tag{2.31}$$

Typical values for the g_m of FETs are smaller than those for BJTs and range from 1 mA/V to 5 mA/V for small signal amplifiers with I_D of a few milliamps to 100 mA/V for power devices operating at an I_D of 2 A.

Op-amps

Op-amps are integrated circuits that are widely used in analogue electronic systems. The internal construction includes a number of BJTs and, in some types, FETs but these internal devices are inaccessible and in the context of this book we shall regard the complete op-amp as a device. The term op-amp is derived from **operational amplifier**, which is an amplifier designed to implement certain mathematical operations, particularly in analogue computing. The op-amp could certainly be used in such computing systems but it has many other important applications. It is probably best regarded as a universal analogue device. Most applications of op-amps involve external feedback components and the properties of the op-amp itself are designed to make this feedback effective.

Properties of an ideal op-amp

An op-amp is a direct coupled, differential amplifier. The term **direct coupled** means that there will be no variation in gain at low frequencies down to and including changes in d.c. signal level. The term **differential** amplifier means that the signal that is amplified is the difference between two separate input signals. The basic symbol for an op-amp is shown in Fig. 2.33 where there is a single output V_O and two inputs V_1 and V_2. The two inputs are respectively known as non-inverting (+) and inverting (−). It must be stressed here that the + and − signs do *not* indicate positive and negative signals.

Fig. 2.33 Symbol for an op-amp with differential inputs.

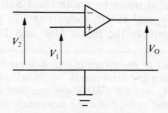

Thus the basic relationship for an op-amp is given by

$$V_O = A_d(V_1 - V_2) \tag{2.32}$$

where A_d is the differential gain which, for an ideal op-amp would be infinite for all signal frequencies. The impedance levels are also important and ideally are zero for the output resistance r_O and infinity for the input resistance r_i at either input. A final essential property for the ideal amplifier is zero d.c. offset which means that if $V_1 = V_2 = 0$ V, then V_O must also be 0 V. In practice this means that op-amps are usually operated from a dual power supply system, one positive and one negative.

Properties of practical op-amps

The properties of practical op-amps approach the ideal in most respects but some additional properties and limitations must also be considered. Typical values for gain and impedance levels are

A_d 10^5 or 100 dB
r_i 5 MΩ(10^6 MΩ for JFET input types)
r_O 60 Ω

These figures apply to low frequency or d.c. signals.

Additional properties

Common mode gain A_c

With an ideal op-amp, any signal or signal component which is common to both inputs will not be amplified since, from equation [2.32], $(V_1 - V_2)$ will eliminate or reject any common parts. For example, if an op-amp has A_d 10^4 and V_1 and V_2 are respectively 10 mV and 9.8 mV; from equation [2.32]:

$$V_O = 10^4 (10 - 9.8) \text{ mV} = 2 \text{ V} \qquad [2.33]$$

The common mode or average signal is $(10 + 9.8)/2$ or 9.9 mV. If the same amplifier has a common mode gain A_c of 10, the total output would be given by

$$V_O = 10^4 (10 - 9.8) \text{ mV} + 10\left(\frac{10 + 9.8}{2}\right) \text{ mV} = 2099 \text{ mV}.$$

This result shows a 5% error compared with result [2.33] for an amplifier with no common mode gain. In general

$$V_O = A_d (V_1 - V_2) + A_c\left(\frac{V_1 + V_2}{2}\right) \qquad [2.34]$$

Specifications for commercial devices do not quote values for A_c but instead, give values for A_d/A_c the **common mode rejection ratio (CMRR)**. It is also usual to quote the CMRR using the logarithmic unit, the dB. Typical values range from 70 dB to 100 dB which represent respectively the ratios 3160 and 10^5.

Example 2.4 *The signals applied to the inverting and non-inverting terminals of a differential amplifier are respectively −0.38 mV and −0.385 mV. The differential gain A_d and the CMRR are specified as 3×10^5 and 65 dB. Calculate the output voltage and the percentage error resulting from the common mode gain.*

The ratio equivalent to 65 dB is given by

$$\text{ratio} = \text{antilog}_{10}\left(\frac{65}{20}\right) = 1778$$

Thus, common mode gain

$$A_c = \frac{3 \times 10^5}{1778} = 168.7$$

The differential input

$$e_d = -0.385\,mV - (-0.38\,mV) = -5 \times 10^{-6}\,V$$

and common mode input

$$e_c = \frac{-0.385\,mV + (-0.38\,mV)}{2} = -0.3825\,mV$$

Therefore total output

$$\begin{aligned}V_O &= 3 \times 10^5 \times (-5 \times 10^{-6} \pm 168.7 \times 0.385 \times 10^{-3}\,V \\ &= -1.44 \text{ or } -1.56\,V.\end{aligned}$$

The alternative answers arise as A_c may be inverting or non-inverting. In either case the percentage error is $0.06/1.5 \times 100$ or 4.3%.

The effects of A_c and A_d can be represented by the equivalent circuit shown in Fig. 2.34.

Fig. 2.34 An op-amp equivalent circuit showing differential and common mode gains.

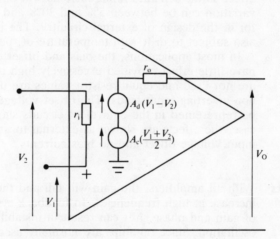

Bias and offset properties

The input voltage offset   In the ideal amplifier, if $V_1 = V_2$ then V_O will also be zero. In practice, under these conditions, there will be a finite V_O. This is due to small differences between internal transistors within the op-amp. It is convenient to think of this behaviour resulting from an equivalent input voltage offset V_{IO}, which when multiplied by the gain A_d results in the observed output voltage. Typical figures for V_{IO} range from 0.5 mV to 10 mV. If these figures are multiplied by typical values for A_d, the resulting V_O values are 50 to 1000 V! Such voltages are greatly in excess of the d.c. supply voltages and in practice the amplifier will saturate with V_O at a voltage somewhat less than these d.c. voltages. In most circuit applications involving op-amps, this apparent problem of voltage offset is eliminated by the use of external feedback circuitry (see Ch. 6). In other situations, op-amps having an **offset null provision** should be used. This involves the use of an external variable resistance which can be adjusted to reduce the output voltage offset to zero. There may still be a problem as V_{IO} is subject to **drift** or variation with temperature. A typical figure for this drift is 10 μV/C°.

Input bias currents   The two input terminals to an op-amp are connected directly to the bases of two internal BJTs (or the gates of two JFETs for FET input types of op-amp). This means that a path must be provided for the base currents to flow, as the transistors can only operate in the linear mode with base currents in the correct range. It also means that any external resistance must be balanced between the two inputs so that an *input voltage difference* (due to equal bias currents in unequal resistors) does not occur as this would cause an output voltage offset.

Typical bias currents are 20 nA to 100 nA for BJTs and 2 pA to 100 pA for FET inputs. These figures are also subject to an offset effect as the separate input currents will differ from each other. This variation can be between 5% and 20% and may have to be allowed for in the design of external circuitry. The input current offset is also subject to drift with temperature of, perhaps, 1 nA/C°.

In most applications, the bias and offset effects described above have little effect provided excessively high or low resistance values are not used and equal resistor values are used in the inverting and non-inverting input leads. The offset voltage and current effects can be represented in the equivalent circuits shown in Fig. 2.35. In each case, the effects are shown as external to an ideal op-amp having no input voltage offset and no bias currents.

Signal frequency effects   With all amplifiers, the gain will fall and the phase shift will increase at high frequencies. In feedback systems, the combination of gain and phase shift can result in instability or unwanted oscillation. Since op-amps are normally used with feedback, features for avoiding this problem are included in the design. Two techniques are used: external compensation or internal compensation. For external compensation, the manufacturer recommends values of external capacitance and resistance which should be connected to the appropriate terminals for specific feedback circuitry. The alternative internal compensation causes the gain to fall steadily for all signal frequencies above about 1 Hz. This ensures that, no matter how much feedback is applied, the gain will be too low to cause the unwanted oscillation. In both cases, the frequency response will depend upon the external feedback components.

Slew rate   The maximum output current of an op-amp limits the maximum rate of change of output voltage. Typical values for slew rate range between 1 and 10 V/μs. The maximum slope of a particular output signal depends upon the waveform, the frequency and the amplitude. This may be seen by considering ideal and practical output signals for a sinusoidal input to an op-amp having a slew rate of 1 V/μs when the signal frequency is 100 kHz. In Fig. 2.36, the straight line shows the slew rate of 1 V/μs. The smallest sine wave,

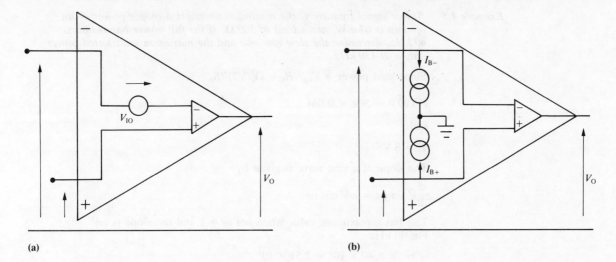

(a)   **(b)**

Fig. 2.35 Op-amp
equivalent circuits showing
offset and bias effects.

having a peak amplitude of 1 V, has a maximum slope or rate of
change of voltage of 0.63 V/μs. Since this is less than the slew rate,
the waveform is not modified. If the signal amplitude is now
increased by a factor of three, the ideal output would be the larger
sine wave shown. The maximum slope of this sinusoid is 1.9 V/μs,
which cannot be produced by the op-amp. Thus for those periods
when the ideal output *would* exceed the slew rate, the practical
output will follow the slew rate. The slew-distorted waveform is
shown between the other two. This is no longer sinusoidal and the
amplitude is also limited. An alternative way of defining slew rate
limitation is by quoting the **full power bandwidth**: this is illustrated
in example 2.5.

Fig. 2.36 Slew rate
limitation on gain and
waveforms.

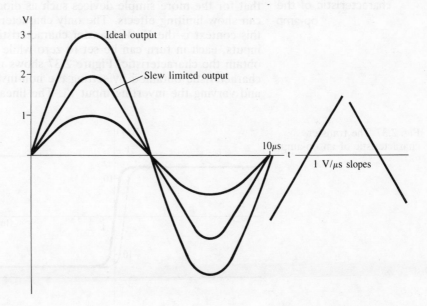

Example 2.5 *At low signal frequency, the maximum undistorted output power of an op-amp is 64 mW into a load of 500 Ω. If the full power bandwidth is 40 kHz, determine the slew rate and and the maximum undistorted power output at 150 kHz.*

Since load power $= V_{\text{rms}}^2/R_L = (\hat{V}/\sqrt{2})^2/R_L$

$(\hat{V}/\sqrt{2})^2 = 500 \times 0.064$

and

$\hat{V} = 8$ V

The slope of a sine wave is given by

$$\frac{\mathrm{d}}{\mathrm{d}t}\hat{V}\sin\omega t = \omega\hat{V}\cos\omega t$$

This has a maximum value when $\cos\omega t = 1$ and the slope is $\omega\hat{V}$.
For 40 kHz

$\omega = 2\pi \times 40 \times 10^3 = 2.51 \times 10^5$

Therefore

Slew rate $= 8 \times 2.51 \times 10^5 = 2 \times 10^6$ V/μs.

At 150 kHz

$\omega = 9.4 \times 10^5$

Therefore

$9.4 \times 10^5\hat{V} = 2 \times 10^6$ or $\hat{V} = 2.1$ V.

and

Output power $= \dfrac{(2.1/\sqrt{2})^2}{500} = 4.5$ mW

The transfer characteristic of the op-amp

Graphical representation of op-amp behaviour is not as useful as that for the more simple devices such as diodes and transistors but it can show limiting effects. The only characteristic that is useful in this context is the V_O/V_i transfer characteristic. Since there are two inputs, each in turn can be set to zero while the other is varied to obtain the characteristic. Figure 2.37 shows the transfer characteristic obtained by setting the non-inverting input V_1 to zero and varying the inverting input V_2. The linear operating region is

Fig. 2.37 The transfer characteristic of an op-amp.

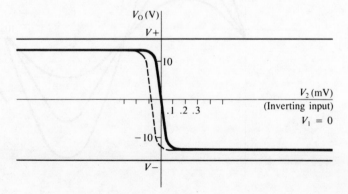

the straight line portion passing through the origin, and the slope of this line is the open-loop differential gain A_d. In this case, it can be seen that an input change of 0.1 mV results in an output change of 10 V so the A_d in this case is 10^5.

Again, referring to the transfer characteristic, the maximum positive and negative output levels are less than the d.c. supplies V_+ and V_- and, as the output approaches saturation, the gain becomes non-linear.

If any voltage offset was present, the linear part of the characteristic would no longer pass through the origin but to one side, as shown by the broken line.

A simplified small-signal equivalent circuit

For most applications, analysis of op-amp circuits can disregard common mode gain and offset effects. The resulting low frequency equivalent circuits are shown in Fig. 2.38. In Fig. 2.38(a), the input and output resistances r_i and r_o have been included, but with practical values and components these can be neglected in most circumstances (as open circuit and short circuit respectively), leaving the simplified circuit in Fig. 2.38(b).

Fig. 2.38 Equivalent circuits for op-amps.

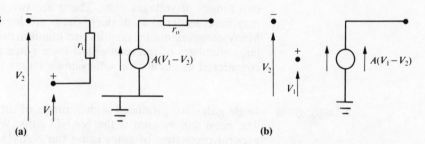

Range of op-amps

Although most types of op-amp are general purpose the manufactured range includes types that are upgraded for particular parameters. Examples include: high slew rate, low offsets, low bias current. Other types offer a wider working range in either signal frequency, signal voltage or output power.

In addition, there are many special purpose devices based on op-amps. These include such circuits as comparators and voltage regulators which are discussed in context later (see Ch. 6).

Digital devices

Digital systems operate with signals having only two values, 0 and 1. In a particular system or sub-system, these two values will be represented by two voltage (or current) levels. In practice, the precise levels will not be used but there will be an acceptable range of voltage which will represent 0 and another range representing 1. Two examples are shown in Fig. 2.39; the voltage range between the 0 and 1 ranges is said to be **indeterminate**. Voltages beyond the limits of 0 and 1 are not used as this may result in damage to the devices.

Fig. 2.39 Voltage levels and
digital logic levels.

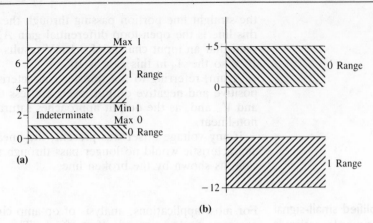

(a)

(b)

The range used to represent a 1 need not be the higher voltage
range. A system designer could use the voltage ranges in Fig.
2.39(a) and choose to make the 2.4 V to 7 V range represent 0 and
the 0 V to 1 V range to represent 1.

Digital devices are integrated circuits that operate or react to the
two ranges of voltages only. There are two basic types of circuit that
may be considered as devices: these are logic gates and flip-flops.
Many complex digital circuits are manufactured but they consist of
large numbers of interconnected logic gates and flip-flops. These are
considered as circuits or sub-systems rather than devices.

Logic gates
Logic gates are produced with a range of different logic functions.
The most widely used is the NAND gate. We will discuss the
general properties of gates using the NAND gate as an example and
then consider the range of different functions and gate technologies.

Logic NAND gates
In general, logic gates have several inputs (one to eight) and one
output. The function of a particular gate can be shown by a truth
table or defined by a Boolean statement. Figure 2.40 shows the
symbol for a three-input NAND gate together with these alternative
representations of its function. A **truth table** shows the value of the
function or output for each possible combination of inputs.

The statement in Fig. 2.40(b) can be written F = NOT((A) AND
(B) AND (C)), which simply means that F will be 0 only if A is 1,
B is 1 and C is 1, which is indicated on the bottom line of the truth
table in Fig. 2.40(c).

This truth table could also be drawn showing the input and output
voltage levels associated with a particular device. This is somewhat
clumsy and, in any case, the input and output levels will lie within a
range (see Fig. 2.39). It is more useful to examine the **voltage
transfer characteristic** showing the variation of output voltage at F
as a critical input is varied over the operating range. A **critical input**
is one for which a change from 0 to 1 (or vice versa) results in a
change in the output. For example, in the NAND gate in Fig.
2.40(a), if A and B are both at 1 (or the equivalent voltage) then
any change in C results in a change in F. Thus under these
circumstances, C is a critical input.

Fig. 2.40 A logic NAND gate: (a) symbol; (b) Boolean statement: (c) truth table.

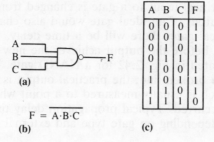

A
B
C
(a)

$$F = \overline{A \cdot B \cdot C}$$
(b)

A	B	C	F
0	0	0	1
0	0	1	1
0	1	0	1
0	1	1	1
1	0	0	1
1	0	1	1
1	1	0	1
1	1	1	0

(c)

The voltage transfer characteristic shown in Fig. 2.41 could be used to provide the logic levels indicated in Fig. 2.39 since inputs of less than 1 V result in an output within the 1 range of 3 V to 7 V and an input of greater than 3 V results in an output well within the 0 range. The indeterminate range is in fact smaller than that indicated in Fig. 2.39, being only from 1.2 V to 1.8 V.

Fig. 2.41 The voltage transfer characteristic for an inverting gate.

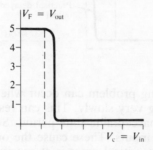

Output loading effects Logic gates are rarely used in isolation and the output of a particular gate may provide the input to a number of other gates. If too many other gates are connected in this way, the 0 and 1 voltage levels can be changed so much that the indeterminate condition may apply, causing a system malfunction. To avoid this, a specified maximum number is quoted for a particular gate type. This number is known as the **fan-out** and can only be applied to the interconnection of gates of the same type. Typical values of fan-out are 10 to 30.

An alternative approach to loading is to specify the maximum current sinking and current sourcing of a gate. This is more convenient than fan-out specifications when different types of circuit or device are being interfaced. The two conditions are relevant to the output 'low' and output 'high' conditions respectively (the high and low voltage conditions). When the output of the gate is low and it is connected to other devices or circuits, current will flow in through the gate output to earth. This is **current sinking** and, if a maximum value is exceeded, the low level (usually logic 0) cannot be maintained. When the gate output is high, current will flow from the gate, through the external circuit to earth. This is **current sourcing** and, if the maximum value is exceeded, the output high level (usually logic 1) will fall below the specified minimum.

Propagation delay   If a critical input to a gate is changed from 0 to 1 instantaneously, the output of an ideal gate would also change instantaneously. With practical gates there will be a time delay, known as the **propagation delay**, before the output achieves the new condition. This is illustrated in Fig. 2.42 for a NAND gate. The ideal output is shown with a broken line; the practical output is shown delayed by a time t_d which is usually measured to a point where 90% of the change has occurred. Typical propagation delay times range from 1 ns to 1 μs depending on gate type and external loading.

Fig. 2.42 Propagation delay for a logic gate.

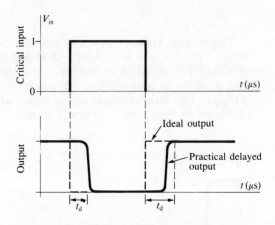

Schmitt input gates   A different timing problem can occur when the voltage of a critical input is changing very slowly. This can result in an indeterminate output for an unacceptable time period. Some gates are constructed to have **Schmitt inputs**. These cause the output to **switch** state very quickly when the critical input is at specific d.c. levels. This is illustrated in Fig. 2.43. The slowly changing input is shown in Fig. 2.43(a) and the resulting standard gate output in (b). Figure 2.43(c) shows the equivalent output for a Schmitt input gate. The *two* Schmitt switching levels are shown as broken lines on Fig. 2.43(a).

Fig. 2.43 Waveforms for slow changing gate signals: (a) input; (b) normal gate output; (c) output for gate with Schmitt inputs.

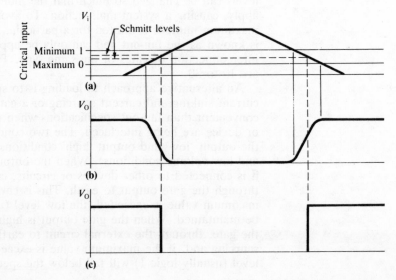

Two levels are necessary as, if a single level were used, slight voltage variation due to electrical noise could cause multiple switching of the output as the input passed through the single critical level several times.

Noise immunity of logic gates

The noise immunity of a logic gate is a measure of how much electrical noise can be added to a logic signal before an incorrect or indeterminate logic output will result from a following gate. This may be best understood by considering the two gates and their voltage transfer characteristics shown in Fig. 2.44.

The characteristic for gate A in Fig. 2.44 shows a shaded operating range. This allows for the specified maximum fan-out (resulting in the maximum 0 output level) and for specified variation in d.c. power supply level. The characteristic for gate B is the same but it has been drawn with the axes reversed so that the available output signals from gate A can be compared with the required input signals for gate B. We can now see that the threshold for gate B is at a higher voltage than the maximum 0 output from gate A; the **threshold** is the voltage at which the output starts to change. Thus if any noise was added to the output from A, this would have no effect upon the output of B provided the peak noise amplitude was less than the difference between the two broken lines shown. This difference is therefore known as the **low noise immunity (LNI)**. A similar difference occurs between the minimum 1 output level and the 0 threshold. This is the **high noise immunity (HNI)**. Typical figures for noise immunity are about 1 V.

Fig. 2.44 Noise immunity of logic gates: (a) circuit arrangement; (b) transfer characteristics.

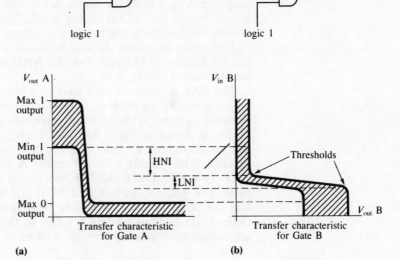

Logic families

A number of logic families have been developed using different device technologies. Only four of these need be considered: CMOS using complementary symmetry MOSFETs, standard TTL using BJTs in the saturated and cut-off states, Schottky TTL and ECL using BJTs in the unsaturated state.

CMOS (complementary metal oxide silicon) offers medium speed (30 ns), improved noise immunity and a wide range of functions. Power supply requirements are less restrictive than for TTL (3 V to 18 V) and the power consumption is *very* small.

Standard **TTL** (transistor-transistor logic) offers high speed (10 ns), good noise immunity (1 V) and a wide range of functions. The Schottky form offers even higher speed with propagation delay down to 5 ns. Power supply requirements (4.5 V to 7 V) and logic levels have a narrow range; power consumption is 'medium' compared with the other families.

ECL (emitter-coupled logic) offers very high speed (1 ns) but has poor noise immunity and a limited range of functions. Power supply and logic level requirements are very strict and the power consumption is relatively high.

Logic functions

The three basic logic functions are **NOT** (or invert), applied to a single variable, **OR** and **AND**, which may be applied to any number of logic signals. In the basic gate range, we also have **NOR** and **NAND** in which the OR and AND are combined with a NOT. In practice, these inverting functions are more useful as they can be combined in different ways to provide any other function that may be required. The families of logic circuits discussed above provide in their most basic form either NAND or NOR functions.

Figure 2.45 shows the symbols and truth tables for the five functions. The multiple input gates are shown limited to two inputs, but gates are avaliable with two, three, four or eight inputs as required.

The extension of the function to more inputs can be seen by inspection of the truth tables. For example, with the NOR gate, the output is only 1 if *all* the inputs are 0 and for the AND gate, the output is only 1 if *all* the inputs are 1. These descriptions can be equally applied with any number of inputs.

Two other functions are also available as gate circuits. These are the Exclusive-OR (**EXOR**) and the AND or INVERT (**AOI**). They are illustrated in Fig. 2.46.

The EXOR gate can only have two inputs and the output is 1 if *only one* of the inputs is 1. It is widely used in arithmetic and error checking systems. The AOI gate provides several functions in one device as shown in Fig. 2.46(b). A truth table has not been given as this would require 64 lines (2^6) but the function shown can be described: the output will be 1 unless A AND B are 1 s OR C AND D are 1 s OR E AND F are 1 s, in which case the output will be 0. The gate shown in Fig. 2.46(b) is 2 input 3 wide. This means that the AND sections each have two inputs and that there are three AND sections feeding the OR section. A variety of functions are available, 4 input 2 wide, 2, 2, 3 input 3 wide, etc.

The range of logic gates

As with other devices, gates are available with a range of special features; low power consumption, high fan-out, high speed, etc. In addition, alternative output facilities can be obtained; open-circuit collector or tri-state output.

The open-circuit collector is provided so that the outputs of a number of gates may be interconnected without loss of fan-out.

Fig. 2.45 Symbols and truth tables for basic logic gates.

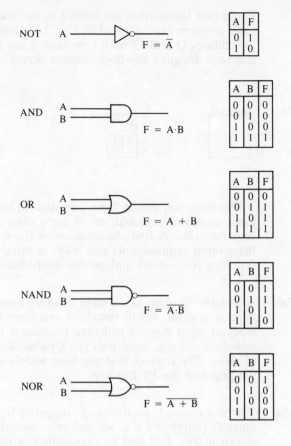

A	F
0	1
1	0

NOT $F = \overline{A}$

A	B	F
0	0	0
0	1	0
1	0	0
1	1	1

AND $F = A \cdot B$

A	B	F
0	0	0
0	1	1
1	0	1
1	1	1

OR $F = A + B$

A	B	F
0	0	1
0	1	1
1	0	1
1	1	0

NAND $F = \overline{A \cdot B}$

A	B	F
0	0	1
0	1	0
1	0	0
1	1	0

NOR $F = \overline{A + B}$

Fig. 2.46 Symbols and truth tables for other common gate forms: (a) Exclusive-OR gate; (b) AOI gate.

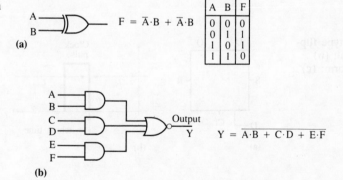

(a) $F = \overline{A} \cdot B + \overline{A} \cdot B$

A	B	F
0	0	0
0	1	1
1	0	1
1	1	0

(b) $Y = \overline{A \cdot B + C \cdot D + E \cdot F}$

With tri-state output devices, the output can be 1 or 0, following the usual logic, or it can be effectively open circuit (very high impedance). This allows a number of sub-systems to be connected in parallel with only one at a time being used.

Flip-flops   **Flip-flops** have, in general, two logic outputs, which for normal operation will have the opposite logic value. These two outputs are usually known as Q and $\overline{Q}$. Thus if Q is 1, $\overline{Q}$ is 0 or vice versa.

These two alternatives are known as the **states** of the flip-flop. This basic property is illustrated in Fig. 2.47. Note that the other possibilities Q and Q̄ both 0 or both 1 are not states, and, in correctly designed flip-flops, cannot occur.

Fig. 2.47 Flip-flop symbol and truth table.

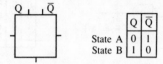

	Q	Q̄
State A	0	1
State B	1	0

Flip-flops will switch from one state to the other if appropriate input signals are applied, or, in some cases, after predetermined time intervals. A first classification of flip-flop types is based on these input requirements and leads to bistable flip-flops, monostable flip-flops (one-shots) and astable multivibrators.

Bistable flip-flops   A bistable flip-flop is a simple memory element in that, until an input is applied, it will remain in one state indefinitely. Various types of input may be provided to cause it to change state: these include d.c. logic, logic plus clock pulse, d.c. level, or a combination of these. The devices that are most widely used are the D-type flip-flop and the JK flip-flop.

The D-type flip-flop   This is a clocked, positive-edge-triggered flip-flop with a single logic input D (data) and d.c. set and reset facilities. A symbol for this is shown in Fig. 2.48 and an explanation of the facilities is given below.

Fig. 2.48 A D-type flip-flop: (a) symbol; (b) clocking waveform; (c) truth table.

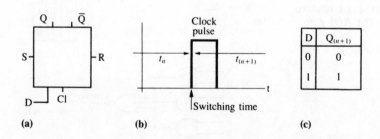

D	$Q_{(n+1)}$
0	0
1	1

(a)   (b)   (c)

In the normal mode of operation, switching can only occur on application of a clock pulse, and then only according to a logic input. This D-type flip-flop is positive-edge-triggered which means that, if it switches state, it will do so at the same time as the clock pulse is going from 0 to 1 (Fig. 2.48(b)). The logic for the D-type is shown in Fig. 2.48(c) and can be interpreted in the following way: t_n is the time period after the *n*th clock pulse and $t_{(n+1)}$ is the time period after the next clock pulse; if D has the logic value shown during t_n, then $Q_{(n+1)}$, the value of Q during $t_{(n+1)}$, will have the value indicated in the table. This can be stated more simply: Q will switch to, or remain at, the value of D prior to the clock pulse.

Fig. 2.49 Timing waveforms
for a D-type flip-flop.

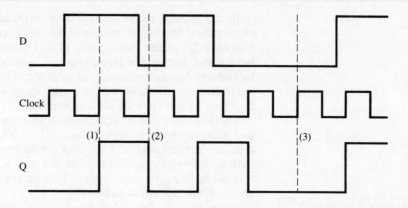

This can be seen from the timing waveforms in Fig. 2.49. Note
that Q only changes on the leading edge of the clock; then note the
value of D prior to this time and see that Q switches to (1, 2) or
remains at (3), the value of D. $\overline{Q}$ will, of course have the opposite
waveform to Q

The S and R inputs are used to **preset** a circuit and operate
without the clock, overriding all other signals. They are normally
held at logic 1 and either S *or* R is temporarily switched to 0 in
order either to set Q to 1 or reset Q to 0. If both S and R are
operated simultaneously the resulting state of the flip-flop will be
indeterminate.

In a simplified form, the D-type is known as a **latch**. In this case,
several flip-flops (usually four) are provided in a single integrated
circuit with a common clock input. Each latch has a single D input
and a Q output; a common Set may also be provided.

The master-slave JK flip-flop

This is a trailing-edge-triggered flip-flop with two logic inputs (J and
K) and d.c. set and reset facilities. A symbol for this is shown in
Fig. 2.50 and an explanation of the facilities is given below.

Fig. 2.50 A master-slave JK
flip-flop: (a) symbol; (b)
clocking waveform; (c)
truth table.

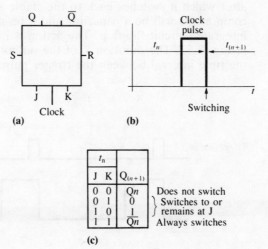

t_n		
J	K	$Q_{(n+1)}$
0	0	Qn	Does not switch
0	1	0	Switches to or
1	0	1	remains at J
1	1	$\overline{Q}n$	Always switches

(c)

In the normal mode of operation, switching can only occur on a clock pulse. In the JK master-slave flip-flop, switching occurs on the trailing edge of the clock pulse (1 to 0) according to the logic inputs before and during the clock pulse. These inputs are, however, **locked out** before switching takes place. This avoids an indeterminate situation which could occur if the inputs were changing at the time of the clock pulse.

The logic is more comprehensive than that for the D-type: if J and K have opposite logic values (01 or 10) then Q switches to or remains at the value of J (equivalent to the D input); if J and K are both 0, no switching occurs, and if J and K are both 1, switching will occur in every clock pulse. This information can usefully be given in the **transition table** (Fig. 2.51).

From the first line, Q is required to remain at 0. Since it is already at 0, the 'does not switch' input 00 would be satisfactory; alternatively, the 01 'switches to or remains at J' would also provide the required transition. Thus provided J is a 0, it does not matter whether K is a 0 or a 1. This 'don't care' condition is indicated by the *. Similar arguments can be applied to obtain the remaining transitions.

The Set and Reset inputs operate in the same way as those on the D-type flip-flop, overriding all other inputs to set and reset Q to 1 or 0 respectively.

In simpler forms, several flip-flops with a common clock are produced on a single integrated circuit. Each flip-flop will have J and K inputs and a Q output. Common S or R inputs can also be provided. Some more comprehensive JK flip-flops provide additional logic in the J and K inputs. For example, three inputs may be provided for each of J and K (J_1 J_2 J_3, K_1 K_2 K_3). The logic is then $J = J_1.J_2.J_3$ and $K = K_1.K_2.K_3$. In some types , one of the additional inputs may be inverted to make $J = J_1.J_2.\bar{J}_3$ etc. Positive-edge triggered flip-flops can also be obtained with JK logic for special high speed applications.

Required change in Q	J	K
0 to 0	0	*
0 to 1	1	*
1 to 0	*	1
1 to 1	*	0

Fig. 2.51 The transition table for a JK flip-flop.

Monostable flip-flops (one-shots)

A monostable flip-flop has a single stable state: when a trigger pulse is applied, the flip-flop switches to a second **quasistable state** where it remains for a time determined by external timing components, after which it switches back to the stable state. The timing components will be a capacitor and a resistor external to the integrated circuit flip-flop. The action is illustrated by the waveforms in Fig. 2.52. The duration T of the output pulse is independent of the time interval between the trigger pulses and of their duration.

Fig. 2.52 Waveforms for a monostable flip-flop (one-shot).

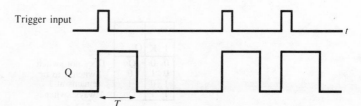

Trigger input

Q

The astable flip-flop   An astable is a flip-flop that has two quasistable states with time intervals determined by external components. An astable multivibrator is therefore a generator of a rectangular waveform. In some cases, a single timing circuit will be used and the resulting output will be symmetrical, having equal times for 0 and 1. Other examples can have independent timing for the two periods T_1 and T_2 as shown in Fig. 2.53, but there may be practical limitations on the ratio T_1/T_2 (see p. 00).

Fig. 2.53 Waveforms for an astable multivibrator.

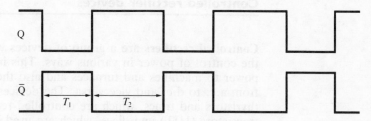

Timers   The functions of either monostable or astable flip-flops are available in integrated circuits known as **timers**. External components have to be provided to determine the required times which can be selected from a few microseconds to several minutes. Other external connections are used to select free-running (astable) or one-shot (monostable) and an additional input may be provided to allow voltage control of the time periods.

The Schmitt trigger   The **Schmitt trigger** is a bistable flip-flop which switches at two d.c. levels on a trigger input. This is illustrated in Fig. 2.54 which shows the two d.c. levels (V_1 and V_2) together with an applied changing signal input.

Fig. 2.54 Waveforms for a Schmitt trigger bistable flip-flop.

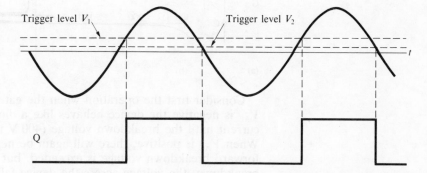

Note that the output switches first at the higher level V_1 and switches back at the lower level V_2. The difference between these two input levels is known as the **circuit hysteresis** and, although this may be small, it is essential to the circuit properties. If a single switching level were used and the input signal was held at this level, the output Q would oscillate between the 0 and 1 states in an uncontrolled manner.

Schmitt triggers are available as integrated circuits but the function is also provided at the inputs of some logic gates and other digital integrated circuits which are used with analogue input signals to interface between analogue and digital systems.

Controlled rectifier devices

Controlled rectifiers are a group of devices which are mainly used in the control of power in various ways. This includes the supply of power to machines and furnaces and also the conversion of power from a.c. to d.c. and vice versa. The devices to be described are thyristors and triacs, which are controlled rectifiers, and unijunction transistors (UJT) and diacs, which are used in the trigger circuits for the rectifiers.

The thyristor

The **thyristor** is a three-terminal device, the terminals being known as the anode, cathode and gate. The anode and cathode provide the normal conducting path in the same manner as a simple diode rectifier. The gate controls the conditions under which forward conduction takes place. Figure 2.55 shows the symbol for the thyristor together with the VI characteristic.

Fig. 2.55 The thyristor: (a) symbol; (b) d.c. characteristics.

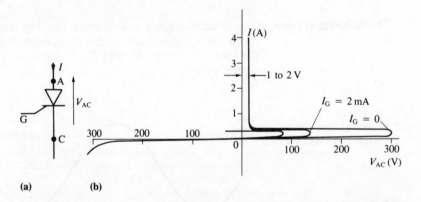

Consider first the operation when the gate current I_G is zero. If V_{AC} is negative the device behaves like a diode, passing negligible current until the breakdown voltage (400 V to 1200 V) is exceeded. When V_{AC} is positive, there will again be negligible current until the forward breakdown voltage is exceeded. but in this case, after breakdown, the voltage across the device falls to a very low level (1 V to 2 V) and the current is then limited only by external resistance (see the circuit in Fig. 2.56). Supposing the thyristor shown to have a forward breakdown of 300 V, if V is increased to 300 V the thyristor will switch on and V_{AC} will fall to the conducting level of 2 V. The circuit current I will then be $(300 - 2)/20$ or approximately 15 A.

Fig. 2.56 A basic thyristor circuit.

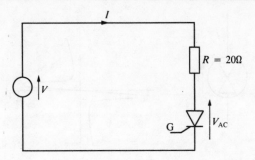

To turn the device off again, it is necessary to reduce the current to below a minimum threshold level of a few milliamps. Thus, in this example, taking I_{th} as 10 mA, V must be reduced to a value given by $(V - 2)/20 = 0.01$, or $V = 2.2$ V.

The effect of the gate current I_G is simply to control the turn-on voltage. In Fig. 2.55(b), a gate current of 2 mA is shown as reducing the turn-on voltage to 150 V. The higher the gate current, the lower will be the turn-on voltage and if I_G is above a certain value, the thyristor will turn on for any positive V_{AC}. A control characteristic can be drawn showing forward breakdown voltage against gate current, but such a characteristic is very temperature sensitive and variable from device to device. It is normal practice therefore to pulse the gate current at the time when the thyristor is required to turn on. The turn-off condition is independent of gate current, as is the reverse bias behaviour.

The triac

The **triac** is similar to the thyristor except that it can be turned on with either polarity voltage applied and with either direction of gate current in both cases. The symbol and characteristics are shown in Fig. 2.57. Since the triac can conduct in either direction it is not normally used as a rectifier but as a device to control a.c. power.

Fig. 2.57 The triac: (a) symbol; (b) d.c. characteristics.

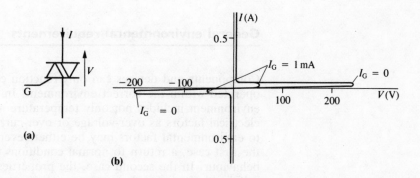

The diac

The **diac** is a two-terminal device used to provide the gate current pulses required to turn the triac on. The characteristic is similar to that of the triac but the breakdown voltage is much lower (about 30 V) and there is no gate control. The symbol and VI characteristic are shown in Fig. 2.58.

Fig. 2.58 The diac: (a)
symbol; (b) d.c.
characteristics.

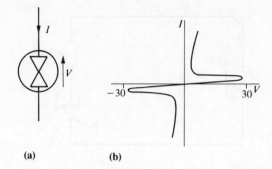

(a)   (b)

**The unijunction
transistor (double-bias
diode)**

The **unijunction transistor** is a three-terminal device mainly used to
control the turn-on of thyristors. The symbol and characteristics are
shown in Fig. 2.59. In this case, the controlled turn-on voltage is
V_{EB} and the control voltage is V_{BB}. The larger the value of V_{BB}, the
higher will be the turn-on value of V_{EB}; the ratio $\eta = V_{EB}/V_{BB}$ is
typically 0.5. When the turn-on value of V_{EB} is exceeded, V_{EB} falls
to about 1 V and the current I_E is determined by the external
circuitry. It is this current that is then used to turn on a thyristor.
Thus the unijunction transistor provides a voltage control for a
thyristor circuit.

Fig. 2.59 The unijunction
transistor: (a) symbol; (b)
d.c. characteristics.

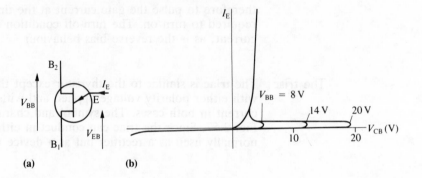

(a)   (b)

General environmental requirements

Components and devices can only function correctly if they are
operated within the correct environment. In this context, the
environment includes not only temperature and humidity but such
electrical factors as over-voltage or over-current. A malfunction due
to environmental factors may be either reversible or irreversible. In
the first case, a return to normal conditions will result in normal
behaviour. In the second case, the properties will have changed
sufficiently for normal operation to be impossible. This may range
from an unacceptable change in a particular parameter to total
destruction of the device.

If a device or component is likely to be exposed to environmental
extremes, it is often possible to protect it (or the complete system)
by a variety of methods, each appropriate to a particular
environmental problem.

System environmental factors

Temperature If the system temperature is excessive, resistor values will change, insulation may fail and the increased conductivity of semiconductors will cause devices to cease to function correctly. At very low temperatures, the reverse effect occurs and semiconductors become insulators. System temperature problems can be overcome by the use of heat sinks, forced air cooling, refrigeration, or heaters as required.

Humidity Excessive humidity can cause breakdown of insulation due to current leakage paths across damp surfaces and, in the long term, corrosion resulting in high contact resistance and other problems. Systems that are required to operate in high humidity are often protected by encapsulation in a suitable insulating resin.

Very low humidity can also cause problems as electrostatic charge can build up to provide local voltages of many kilovolts. These can cause over-voltage damage or, when there is electrical breakdown, a spurious signal may have undesirable effects on system operation.

Radiation High levels of radiation will adversely effect the operation of semiconductor devices. Screening can be provided if this is a likely environmental hazard.

Excessive electric or magnetic fields Excessive alternating fields are most likely to cause problems with signal interference. Steady fields can cause disturbances, particularly in display devices like cathode ray tubes, gas-filled devices and some instruments. If such fields are likely, the system can be protected by screening.

Factors relating to individual devices or components

Temperature rise due to excessive power dissipation All components and devices will have electrical voltages applied to them and pass currents. The resulting power input will be partly due to the d.c. bias conditions necessary for correct device operation and partly due to any signal power input. Some signal power will be transferred to external loads or other parts of the system, but any remaining power must be dissipated as heat. This, in turn, results in a rise in temperature of the device or component with an equilibrium condition occurring at a particular temperature when the power lost by dissipation is equal to the excess electrical power input. For small signal devices and most digital circuits, the equilibrium temperature is only slightly greater than the surrounding ambient temperature and no environmental problem arises. In devices used to amplify signals at higher power levels, the temperature rise may be sufficient to cause further increase in current and hence temperature (due to changes in semiconductor properties) leading to **thermal runaway** and the resulting destruction of the device. To avoid this problem, manufacturers quote the maximum dissipation for a particular device, with or without a heatsink which increases the heat loss and so reduces the equilibrium temperature. The **maximum dissipation** for a device is expressed in watts and this figure may be used with the *VI* characteristic to ensure that no problem occurs. Figure 2.60 shows the output characteristic for a BJT for which the manufacturer quotes maximum dissipation of 1.5 W and 500 mW, with and without heatsink respectively. The two additional curves on the graph are the loci for $V \times I = 1.5$ W and $V \times I = 500$ mW. For example, if I is 1 A, then a V of 1.5 V will give 1.5 W; so will 0.5 A with 3 V, 0.25 A and 6 V etc.

Fig. 2.60 Power dissipation in a transistor.

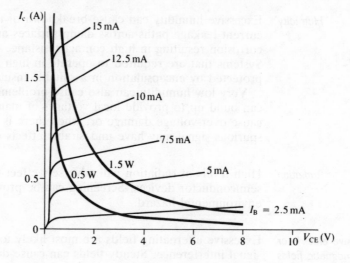

If, for a particular circuit, the Q point (the d.c. operating conditions) lies to the left of these maximum dissipation curves, the device will operate satisfactorily without a heatsink. If operation between the two curves is required. ($I_C = 0.5$ A, $V_{CE} = 2$ V, for example) then a heatsink will be required. Operation at still higher power levels would require a larger heatsink and perhaps forced air cooling.

Over-voltage Individual devices and components can suffer permanent damage after instantaneous application of excess voltage. The most important examples of this are capacitors and most semiconductor devices. For capacitors, the maximum voltage may be as low as a few volts for high capacitance electrolytics or as high as 10 kV for disc ceramic types. There are many other types with working voltges between 50 V and 500 V.

With semiconductor devices, reverse bias breakdown is the main problem; this need not cause permanent damage unless the resulting current rise causes excess dissipation. Breakdown values for junction devices range from 10 V to 2000 V but, for most types, figures between 30 V and 300 V are more typical. MOS devices are particularly susceptible to damage due to over-voltage on the gate source junction. The very high impedance here can allow the build up of high electrostatic charge with sufficient energy to cause permanent damage due to breakdown.

Protection due to over-voltage problems can be provided by zener diodes having a zener voltage less then the maximum value for the device being protected. Integrated circuit devices with FET inputs are commonly protected in this way.

Over-current   Excess current is unlikely to cause damage inherently but a maximum current may be quoted for a device operating at a relatively constant voltage. Examples include rectifying devices and saturated BJTs. In such cases, exceeding the maximum current rating will result in temperature rise and possible damage to the device. Quoted values vary widely with different device types, ranging from a few milliamps for some signal diodes to more than 100 A for some thyristors.

Summary

This chapter has been devoted to the description of the behaviour of components and devices used in electronic circuits and systems. In no instance has any physical explanation of the behaviour been given. The descriptions that have been given in each case have been those most suited to the understanding and analysis of the complete circuit operation. Some devices have much wider application than others; these include the BJT, the op-amp, logic gates and flip-flops. As a result, more attention has been paid to these devices. Transducers and display devices have not been included as these can be taken as signal sources and loads in any general analysis. The chapter concluded with a review of some environmental problems that provide limitations on the use of the components and devices discussed in the chapter.

3 Techniques of Analysis

The principal learning objectives of this chapter are to:

	Pages	Exercises
• Apply graphical analysis to circuits including devices, components, power supplies and signals	78–91	3.1–3.9
• Analyse d.c. operating conditions for circuits including devices, resistors and power supplies	91–108	3.10–3.19
• Analyse circuits using equivalent circuit techniques	108–16	3.20–3,24
• Analyse a.c. circuits using phasor diagrams and j notation	116–27	3.25–3.31
• Investigate frequency response analysis	127–36	3.32–3.34
• Analyse the transient response for passive circuits	136–45	3.35–3.37
• Analyse digital circuits of basic gates using Boolean algebra and Karnaugh mapping for circuit simplification	148–70	3.39–3.50
• Apply block diagram analysis	146–8	3.38

Analysis of electronic circuits provides a better understanding of the function and operation of such circuits and provides the essential background for the design of new circuits and fault finding on existing circuits. In Chapter 1, various levels of analysis are described; these range from verbal analysis to statistical analysis. In this chapter, the necessary techniques at intermediate levels are discussed. These include graphical and equivalent circuit methods for both d.c. conditions and a.c. signals, frequency response analysis, transient analysis, block diagram representation and analysis for digital circuits.

In each case, the techniques that are described are those that are specifically useful in the analysis of the electronic circuits described in the remainder of the book. Any electrical circuit techniques that are necessary are revised, and all techniques are illustrated by relevant examples.

Graphical analysis

One important 'description' of many electronic devices is the graphical set of characteristics. Since devices are normally combined with components, d.c supplies and signal sources, analysis of the combined electronic circuit requires a graphical description for the combination of components, d.c. supplies and signal sources. In its simplest form, such a description is known as a **load line** as it represents the effect of a d.c. supply in series with a load resistor and the device in question. When such a load line is superimposed on the device characteristics, interaction between the two graphs shows the operating point or Q point for the combination. This technique can be extended to combined d.c. supplies and a.c. signals and to more involved circuits of passive components.

D.C. load lines

Figure 3.1(a) shows a d.c. supply of E volts in series with a device whose characteristics are shown in Fig. 3.1(b).

Fig. 3.1 D.C. circuit with a device: (a) circuit; (b) device characteristics.

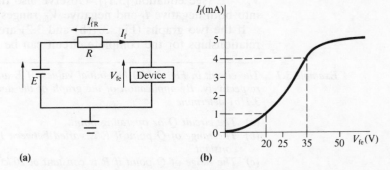

The device on its own can have any Q point along the characteristic. If, for example, V_{fe} is set to 20 V, I_f is 1 mA, or if I_f is set at 4 mA, V_{fe} is 35 V, etc. Once the device is connected in the circuit of Fig. 3.1(a), the two variables I_f, V_{fe} are no longer independent as they are further related by the d.c. voltage E and the resistor R. This relationship can be written

$$E = V_{fe} + I_f R$$

or

$$V_{fe} = E - I_f R \qquad [3.1]$$

or

$$I_f = \frac{E}{R} - \frac{V_{fe}}{R} \qquad [3.2]$$

These relationships are obtained by observing the arrow notation (see p. 00) which allows us to *add* V_{fe} and $I_f R$ to give E; $I_f R$ is positive in the direction shown since the voltage arrow direction is opposite to the I_f arrow direction. Any of the above relationships can be represented by a graph, but the device characteristics show how I_f depends upon V_{fe}, so equation [3.2] is in the correct form. The equation for a straight line is

$$y = mx + c$$

where y is I_f, x is V_{fe}, c is E/R and m is $-1/R$. Thus the required graph is a straight line with slope $-1/R$ passing through $I_f = E/R$ where V_{fe} is zero. This is shown in Fig. 3.2. The slope is given by $\Delta V_{fe}/\Delta I_f = R$ and is negative since an increase in V_{fe} produces a reduction in I_f. Note that the point where I_f is 0 corresponds to

Fig. 3.2 A d.c. circuit load line.

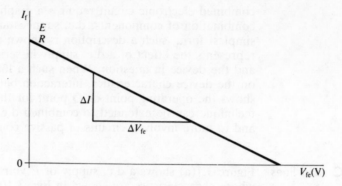

$V_{fe} = E$ (see equation [3.1]). Observe also that the load line extends into both negative I_f and negative V_{fe} ranges.

If the two graphs (Fig. 3.1(b) and 3.2) are superimposed, relationships for the complete circuit can be investigated.

Example 3.1   *The circuit in Fig. 3.1(a) has initial values of E and R of 40 V and 20 kΩ respectively. By application of the graph of the device characteristics in Fig. 3.1(b) determine*

(a) The circuit Q or operating point.
(b) The range of Q point if R is varied between 100 kΩ and 2 kΩ with E constant.
(c) The range of Q point if R is constant at 20 kΩ and E is varied between 25 V and 100 V.
(d) The required value of E if the Q point is to be 30 V, 4 mA when R is 1 kΩ.
(e) The required value of R if the Q point is to be 21 V, 2 mA when E is 50 V.

(a) Figure 3.3(a) shows the device characteristics redrawn with the superimposed load line passing through the points $I_f = 40/20 = 2$ mA and $V_f = E = 40$ V. This only intersects with the characteristic at one point, V_{fe} 17.5 V, I_f 1.13 mA and this point is therefore the circuit Q point.

(b) As *R* is varied, the lower end of the load line does not move $(V_{fe} = E)$ but the slope of the line does change. The required range is shown on the diagram by the two broken load lines. For the high resistance (100 kΩ), the load line passes through the point $40/100 = 0.4$ mA: the slope is less (slope $= -1/R$) and the resulting Q point (b$_1$) is 7.5 V, 0.32 mA. For the other extreme with *R* 2 kΩ, the slope is much greater. Construction now presents a problem since, when $V_{fe} = 0$, $I_f = 40/2 = 20$ mA which is outside the range of the characteristics. The slope however can be drawn by taking a change of I_f of some convenient value, e.g. 5 mA, and finding that the change in $V_{fe} = 5$ mA $\times$ 2 kΩ $= 10$ V. This allows the point 30 V (40 V $-$ 10 V), 5 mA to be drawn and the load line to be completed. The resulting Q point (b$_2$) is 31.7 V, 4.15 mA.

(c) The redrawn characteristics are shown again in Fig. 3.3(b). In this case, the slope of the load line does not change but it is shifted along the V_{fe} axis as *E* is changed. Two additional load lines pass through $V_{fe} = E = 25$ V and 100 V respectively. In the first case, the Q point (c$_1$) is 13 V, 0.6 mA. For the second load line, the slope method must be used. In this case the starting point is the $V_{fe} = 0$ axis, and since *E* and *R* are known, $I_f = 100/20 = 5$ mA. The load line can now be drawn parallel to the first or by taking a convenient change in I_f and calculating the change in $V_{fe}(I_f = 4$ mA, $V_{fe} = 20$ V). The resulting Q point (c$_2$) is 29 V, 3.6 mA.

(d) This requires a load line to pass through the Q point 30 V, 4 mA

Fig. 3.3 Graphical solution for Ex. 3.1.

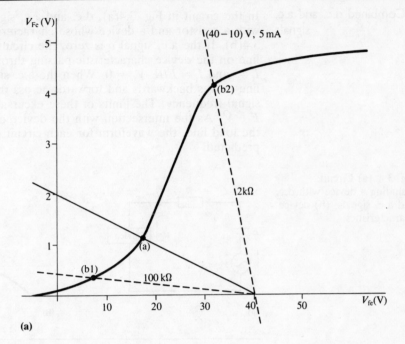

(a)

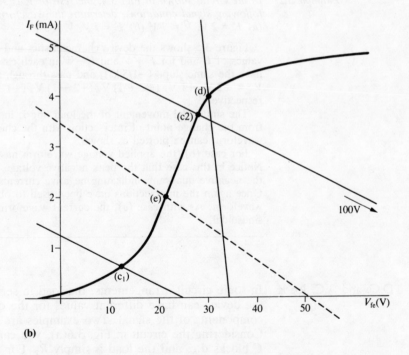

(b)

with a slope of $-1/1$ kΩ. A change of I_f of 4 mA (to zero) results in a change of 4 V, indicating the required E value to be 30 V + 4 V or 34 V.

(e) The required value of R is found in a similar manner by plotting the load line through the required Q point and the specified E. Since with $V_{fe} = 0$, $I_f = 3.5$ mA on the load line, $R = 50/3.5 = 14.28$ kΩ.

Combined d.c. and a.c. signals

In the circuit in Fig. 3.4(a), d.c. and a.c. signals are shown in series with a resistor and a device whose characteristics are shown in Fig. 3.4(b). If the a.c. signal v is zero, the circuit imposes a normal load line on the device characteristic passing through the points $V_{fe} = E$, $I_f =$ and $I_f = E/R$, $V_{fe} = 0$. When the a.c. signal is added, the load line moves backwards and forwards across the characteristic at the signal frequency. The limits of these excursions will be $E + \hat{V}$ and $E - \hat{V}$. As the intersection with the device characteristic moves with the load line, the waveform for each circuit current can be predicted.

Fig. 3.4 (a) Circuit including a device with d.c. and a.c. signals; (b) device characteristics.

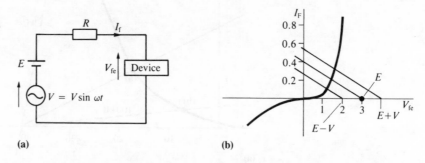

(a)　　　　　　　　　　　　　　(b)

Example 3.2   *In the circuit shown in Fig. 3.4, the resistor R is 5 kΩ. For each of the following signal conditions, determine the waveform of the current I_f:*
(a) E = 2 V, $\hat{V}$ = 1 V (b) E = 1 V, $\hat{V}$ = 3 V.

Figure 3.5 shows the device characteristics and the load lines for both values of E and for $E + V$ and $E - V$ in each case. These load lines all have the same slope $(-1/5\ k\Omega)$ and pass through the points $I_f = 0$, and $V_{fe} = +2\ V$, $+1\ V$, $(+2 + 1)\ V$, $(+2 - 1)\ V$, $(+1 + 3)\ V$ and $(+1 - 3)\ V$ respectively.

The sinusoidal movement of the load line is indicated for case (a) and from the change point of intersection with the characteristic, the current waveform can be plotted as shown.

For case (b), the applied voltage waveform has been drawn in Fig. 3.5. Notice in this case that the peak negative voltage takes the load line into the negative quadrant, indicating negative currents with negative voltages. Once again the intersections have been used to obtain the current waveform. As with case (a), the current waveform is distorted or non-sinusoidal.

D.C. and A.C. loads

In some circuit arrangements, the load in series with the signal and the device can have different values for the d.c. and a.c. components of the signal. Two examples are shown in Fig. 3.6. Considering the circuit in Fig. 3.6(a), you can see that the capacitor C blocks d.c. and the load is simply R_1. For the a.c. signal, if the frequency is high enough to make X_c very much less than R_2, the effective load is the parallel combination of R_1 and R_2. This results in an a.c. load line having a slope $-1/R_p$ and passing through the d.c. operating point given by the intersection between the d.c. load line and the device characteristic. The a.c. load line will then move across the graph according to the a.c. voltage input.

Fig. 3.5 Characteristics and
solution for Ex. 3.2.

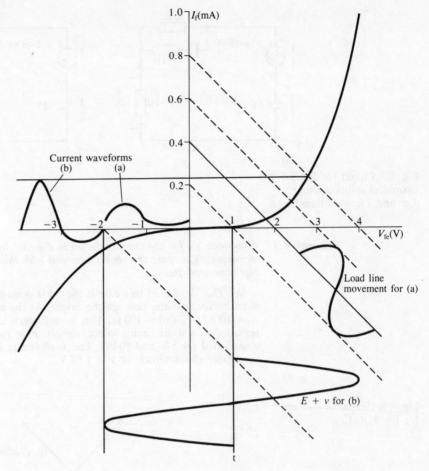

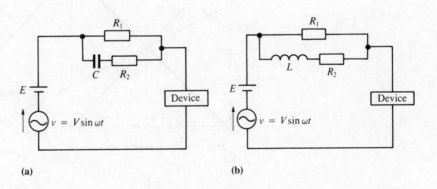

Fig. 3.6 Circuits having
different a.c. and d.c. load
lines.

(a)

(b)

A similar effect occurs in the circuit in Fig. 3.6(b), but in this
case the d.c load is R_p (as $X_L = 0$) and the high frequency load is
$R_1(X_L \gg R_2)$. These techniques are illustrated in example 3.3.

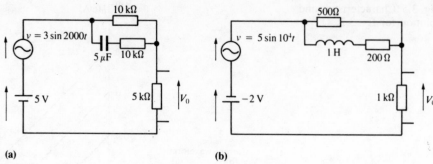

(a)   **(b)**

Fig. 3.7 Circuit for Ex. 3.3.
Graphical solution using
d.c. and a.c. load lines.

Example 3.3   *Determine V_O for the circuits shown in Fig. 3.7 by application of graphical techniques. Compare the results obtained with those given by conventional algebraic analysis.*

(a) The 'device' in this case is the 5 kΩ resistor having a straight line characteristic passing through the origin. At the signal frequency, $X_c = 10^6/(5 \times 2000) = 100$ Ω. This is very much less than 10 kΩ and will be regarded as a short circuit to a.c. signals. First the d.c. load line is constructed for 5 V and 10 kΩ. This is shown in Fig. 3.8 and intersects with the device characteristic at $V = 1.67$ V.

Fig. 3.8 Graphical solution
for Ex. 3.3(a).

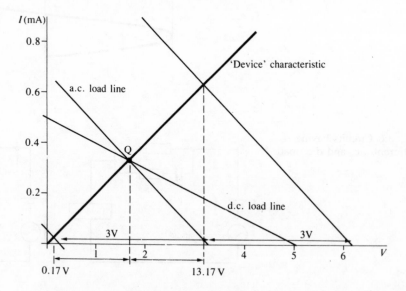

The a.c. load line has a slope $-1/5$ kΩ resulting from the two 10 kΩ resistors in parallel. This must pass through the d.c. operating point as, if the a.c. is reduced to zero, this point gives the correct V_O. This now moves horizontally by plus and minus 3 V (the peak a.c. voltage). This is shown in Fig. 3.8 where the moving a.c. load line intersects with the horizontal axis. The intersection with the device characteristic then gives the required V_O, which varies between 3.17 V and 0.17 V.

To obtain the answer algebraically, the superposition theorem can be applied

$$V_O \text{ due to d.c. alone} = \frac{5 \times 5}{10 + 5} = 1.67 \text{ V}$$

$$V_O \text{ due to a.c. alone} = 3 \sin 2000t \times \frac{5}{5 + 5} = 1.5 \sin 2000t$$

Total $V_O = 1.67 + 1.5 \sin 2000t$, which has maximum and minimum values of 3.17 V and 0.17 V.

(b) The graphical solution for the circuit in Fig. 3.7(b) is given in Fig. 3.9. At the signal frequency X_L is 10 kΩ, effectively making the 200 Ω branch open circuit. R_p for the d.c. load line is 500 Ω in parallel with 200 Ω or 143 Ω. The negative d.c. voltage moves the load line into the negative V, negative I quadrant. The device characteristic for 1 kΩ passes through the origin into both positive and negative quadrants. The a.c. load line is for 500 Ω and passes through the d.c operating point. This is moved horizontally by ±5 V to give the maximum and minimum intersections at +1.6 V and −5.1 V. Confirm for yourself that calculation by superposition gives the same results.

Fig. 3.9 Graphical solution for Ex. 3.3(b).

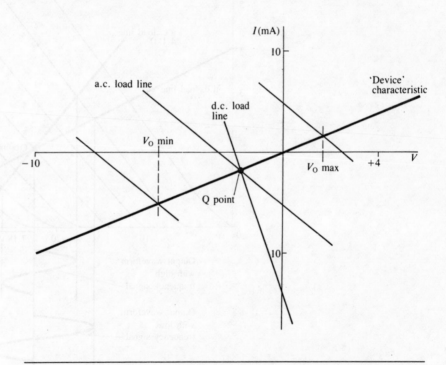

Load line analysis for three-terminal devices

With three-terminal devices such as BJTs, the device characteristics show a set of curves with one for each level of a control parameter. As the control parameter is varied, the Q point will move along the load line (imposed by the circuit) according to the variation of the control parameter. The resulting output voltage or current can then be predicted. The d.c. and a.c effects described above must still be applied.

Example 3.4   *The device shown in Fig. 3.10 has the characteristics given in Fig. 3.11. Determine (a) the d.c. operating or Q point, (b) the output V_{fe} for a low frequency signal input, and (c) the output V_{fe} for a high frequency signal input.*

Fig. 3.10 Circuit with three-terminal device for graphical analysis.

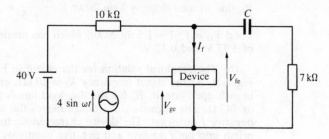

Fig. 3.11 Characteristics and load lines for Ex. 3.4.

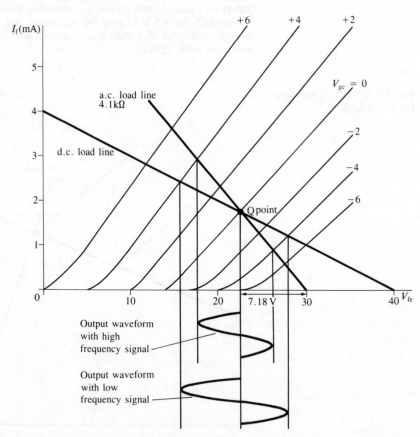

(a) For d.c, the device is in series with 40 V and 10 kΩ and the normal load line is drawn. The control voltage V_{ge} has no d.c. component, so the characteristic for d.c. is that indicated: $V_{ge} = 0$. The intersection with the load line gives the Q point at 22.5 V, 1.75 mA.

(b) With a low frequency signal input, the 7 kΩ resistor can be disregarded, as X_c will be approximately open circuit. As V_{ge} varies between +4 V and −4 V, the Q point will move along the load line between the two relevant characteristics. The extreme intersections show the output, V_{fe}, to move between 15.7 V and 27.7 V. Note that the output waveform is not strictly sinusoidal as the positive and negative peaks are unequal.

(c) For high frequency signals, X_c is effectively a short circuit. 'Looking' from the device back to earth, the 10 kΩ and 7 kΩ resistors are in parallel. The required a.c. load line will have a slope $-1/R_p$ where R_p is $(10 \times 7)/17$ or 4.1 kΩ. This line must pass through the Q point since, if the signal is reduced to zero, the device operating point must lie there. This a.c. load line has been constructed by taking a change from the Q point of 1.75 mA (to 0) and finding a voltage change of 1.75×4.1 or 7.18 V. The new intersections with the characteristics show maximum and minimum values V_{fe} of 26.2 V and 17.7 V respectively.

Example 3.5   *The circuit shown in Fig. 3.12 is a typical BJT amplifier arrangement, the characteristics of the device being given in Fig. 3.13. Using graphical methods, determine the d.c. Q point and the output waveform. It may be assumed that the capacitive reactances are negligible at the signal frequency.*

Fig. 3.12 BJT amplifier circuit for Ex. 3.5.

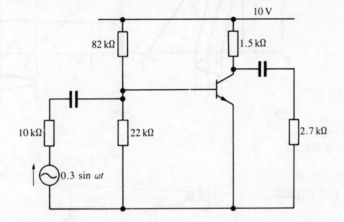

Although only a single d.c. supply is shown, this can be regarded as one 10 V supply in series with 1.5 kΩ and the output characteristic, and a second 10 V supply connected through the 82 kΩ, 22 kΩ potential divider to the base and thus to the input characteristic. The d.c circuits for these arrangements are shown in Figs 3.14(a) and (b).

The d.c. output load line is drawn directly on the output characteristic from Fig. 3.14(a) as shown on Fig. 3.13(b). The input circuit needs simplifying by Thévenin's theorem to give the circuit shown in Fig. 3.24(c).

$$V' = \frac{10 \times 22}{18 + 22} = 2.1 \text{ V}$$

$$R' = \frac{82 \times 22}{82 + 22} = 17.3 \text{ k}\Omega$$

This now allows the input d.c. load line to be drawn on Fig. 3.13(a) from $V_{BE} = 2.1$ V, $I_B = 0$ to $I_B = 2.1/17.3 = 121$ μA, $V_{BE} = 0$.

The intersection of the input load line with the input characteristic gives the input characteristic Q point at V_{BE} 0.65 V, I_B 2 μA. I_B is the parameter of the output characteristic and, although a curve for I_B 2 μA is not given, its position can easily be estimated as shown on Fig. 3.13(b). The intersection of this estimated curve with the output d.c load line gives the output Q point at I_C 4.1 mA, V_{CE} 3.83 V.

For the transistor alone, the Q point could lie anywhere in the range shown by the output characteristics; the d.c. load line (due to the d.c. circuit) restricts the range to anywhere along the load line: the value of the base current I_B fixes the Q point on that d.c. load line. If I_B is varied, the Q point will move up and down the load line showing a corresponding variation in V_{CE} and I_C.

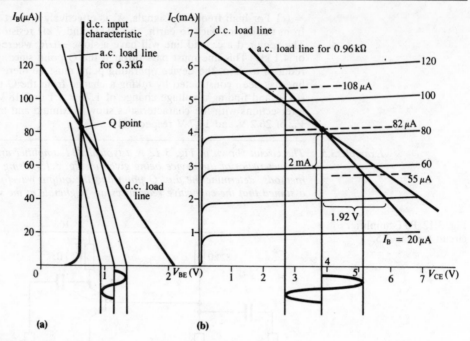

Fig. 3.13 Characteristics
and load lines for Ex. 3.5.

Fig. 3.14 D.C. part circuits
for Ex. 3.5.

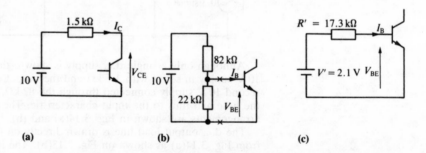

For a.c. signals, both load lines are changed; in the output circuit, the
1.5 kΩ and 2.7 kΩ resistors are effectively in parallel as the d.c. supply
appears to be a short circuit to a.c. signals (as in examples 3.2 and 3.3).
Thus the a.c. output load is $(1.5 \times 2.7)/(1.5 \times 2.7)$ or 0.96 kΩ. The a.c.
load line then has a slope $-1/0.96$ kΩ and is drawn through the Q point by
taking a change in current from the Q point and calculating the resulting
change in voltage. In this case a change of 2 mA from 4.1 mA causes a
change of 2×0.96 or 1.92 V.

Returning to the input circuit, the effective a.c. circuit is shown in
Fig. 3.15(a) and this can be simplified by Thévenin's theorem to the circuit
shown in Fig. 3.15(b). The resistance R'_s is the total resistance 'seen' looking
back from the points marked XX; this is the parallel combination of the
three resistors, which is 6.3 kΩ. The effective voltage e'_s is the voltage
measured across XX if the remainder of the circuit to the right is removed.
Therefore, since 82 kΩ and 22 kΩ in parallel are 17.3 kΩ,

$$e'_s = 0.3 \sin \omega t \times \frac{17.3}{17.3 + 10} = 0.19 \sin \omega t$$

The a.c. load line for 6.3 kΩ is now drawn on the input characteristic

Fig. 3.15 A.C. input circuit for Ex. 3.5.

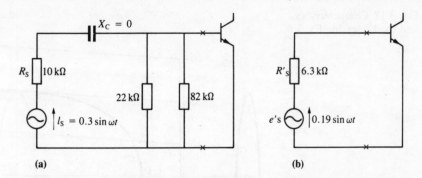

(a)   (b)

through the Q point as shown. This will now move horizontally by plus and minus 0.19 V due to the a.c. signal. Reading from the graph, the corresponding maximum and minimum values of I_B are 108 μA and 55 μA. The estimated characteristics for these two values of I_B are sketched on the output characteristics and their intersections with the a.c. output load line give the extremes of the output voltage and current waveforms. In this case, the voltage swings between 2.7 V and 5.15 V, as shown on Fig. 3.13(b).

Load lines due to reactive loads

With three-terminal devices, graphical methods can be extended to predict the behaviour resulting from reactive loads. The resulting a.c. load lines will be circular or elliptical. The construction of these load lines can be very complicated, particularly for complex loads. The application of such load lines is limited but the concept can be useful in the appreciation of the behaviour of certain circuits as illustrated in example 3.6.

Example 3.6

The device shown in the circuit in Fig. 3.16 has the characteristics given in Fig. 3.17. Assuming that the a.c. signal at the input results in a sinusoidal output current I_f of peak value 20 mA, construct the a.c. load line and determine the a.c. output voltage.

Referring to the circuit and characteristics, the d.c. load line will be vertical ($R = 0$) and pass through the point $V_{FE} = 20$ V, $I_F = 0$. There is no d.c. component in V_{GE} so the Q point is given by the intersection between the $V_{GE} = 0$ characteristic and the d.c. load line. This results in $I_F = 33$ mA. The a.c. output current has a specified peak value of 20 mA, causing the sinusoidal output current to swing between 53 mA and 13 mA as shown. The peak value of the alternating output voltage will be 0.02×750 or 15 V ($X_L = 3750 \times 0.2 = 750 \, \Omega$) but this will be 90° out of phase with the current. Thus when the current has its maximum and minimum values, the a.c. component of voltage will be zero (points A and B on the graph). On the other hand, when the current is passing through zero, the voltage will

Fig. 3.16 Circuit with reactive load for Ex. 3.6.

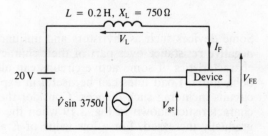

Fig. 3.17 Characteristics
and load lines for Ex. 3.6.

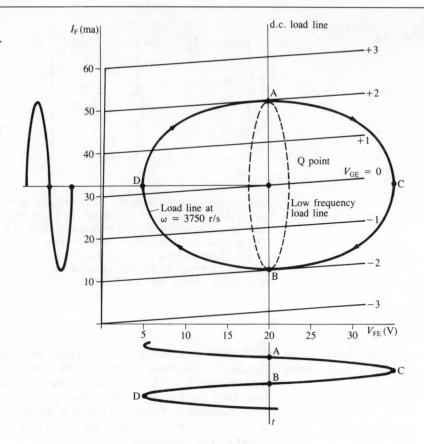

have its maximum values of +15 V and −15 V compared with the d.c. level
(points C and D on the graph). The path between points A, B, C and D,
the load line, forms an ellipse. If the frequency were reduced, the width of
the ellipse would also reduce until, at zero frequency, the d.c. load line
would remain.

Since the operating point moves around the ellipse, the direction of
movement can be determined. Referring to the circuit of Fig. 3.16,
$V_{fe} = 20 - V_L$ but leads the current. At point C (Fig. 3.17), the current is
falling and V_L will have its maximum negative value of −15 V. V_{FE} is
therefore 20 −(−15) or 35 V. The direction of rotation is thus that shown
by the arrows.

If the load is complex, the d.c. load line will be that due to the resistive
component of the load. The axis of the a.c. elliptical load line will no
longer be vertical but has the form shown in Fig. 3.18. The difficulty in the
construction of such load lines makes the practical uses very limited.

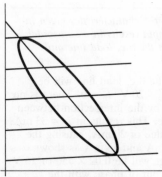

Fig. 3.18 An elliptical load
line for a complex load.

Load lines with negative resistance devices

Some devices such as thyristors and unijunction transistors exhibit
negative resistance over part of their characteristics (the
characteristics of some active circuits can also exhibit similar
behaviour). Load lines can be useful in explaining the behaviour of
circuits including such devices. Consider the circuit and device
characteristics shown in Fig. 3.19 when the d.c. voltage E is
gradually increased. For a low value of E at E_1, the load line cuts

Fig. 3.19 A negative
resistance device and load
lines.

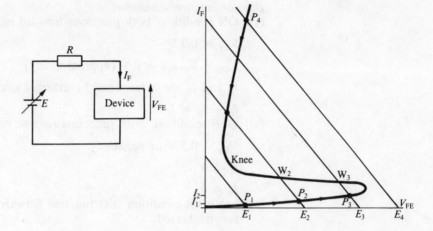

the characteristic at point P_1 only with $I_F = I_1$. As E is increased to E_2 and E_3, I_F rises to I_2 and I_3, but at E_4 the load line no longer cuts the low current part of the characteristic. The Q point therefore jumps to P_4 with a high current and low voltage V_{FE}. If E is now reduced, the high current intersection will be followed down the characteristic until, at the 'knee', the Q point jumps once again to the low current mode. Note that the third possible intersection at W_2 or W_3 etc. does not occur as the negative resistance (negative slope of characteristic) represents an unstable condition. If the circuit is forced into this state by an external voltage, then on removal of this voltage the circuit will switch to one of the stable states.

You should now be able to attempt exercises 3.1 to 3.9.

D.C. analysis of circuits including devices

The properties of electronic devices usually depend upon the d.c. bias conditions applied to the devices. This section is concerned with the techniques for analysing the d.c. conditions using approximate relationships for the devices in question. These include signal and zener diodes and both types of transistor; the following relationships are used, as required, throughout the section.

Signal and rectifying diodes
(i) Forward biased: if V is positive and greater than 0.5 V; assume that V is constant at 0.7 V (0.2 V for germanium diodes) and the current is determined by external circuit conditions
(ii) Reverse biased: if V is less then 0.5 V or is negative and the current is zero.
Zener diodes
(iii) Forward biased: as for signal diodes.
(iv) Reverse biased: as for signal diodes until $V = -V_Z$ after which the voltage is constant at V_Z and the current is determined by external conditions.

Bipolar junction transistors
(v) ON condition: both junctions forward biased.

$$V_{BE} = 0.7 \text{ V} \qquad\qquad\qquad\qquad\qquad\qquad\quad [3.3]$$

$$V_{CE} = V_{CESAT} = 0 \text{ V (approximately)} \qquad\qquad [3.4]$$

I_C and I_B are determined by external circuitry.

$$I_C + I_B = I_E \qquad\qquad\qquad\qquad\qquad\qquad\qquad [3.5]$$

(vi) OFF condition: both junctions reverse biased.

$$V_{BE} < 0.5 \text{ V or negative} \qquad\qquad\qquad\qquad\quad [3.6]$$

$$I_C \simeq 0 \qquad\qquad\qquad\qquad\qquad\qquad\qquad\qquad\quad [3.7]$$

$$I_B \simeq 0 \qquad\qquad\qquad\qquad\qquad\qquad\qquad\qquad\quad [3.8]$$

(vii) LINEAR condition: EB junction forward biased, CB junction
reverse biased.

$$V_{BE} = 0.7 \text{ V} \qquad\qquad\qquad\qquad\qquad\qquad\qquad [3.9]$$

I_B is determined by external conditions.

$$I_B + I_C = I_E \qquad\qquad\qquad\qquad\qquad\qquad\qquad [3.10]$$

or

$$I_C \simeq I_E \qquad\qquad\qquad\qquad\qquad\qquad\qquad\qquad [3.11]$$

$$I_C = h_{FE} I_B \qquad\qquad\qquad\qquad\qquad\qquad\qquad [3.12]$$

V_{CE} is determined by external circuit conditions and by I_C.

Field effect transistors
(viii) Pinch-off condition:

$$I_D = I_{DSS}\left(1 - \frac{V_{GS}}{V_P}\right)^2 \qquad\qquad\qquad\qquad [3.13]$$

(ix) Triode region:

$$I_D = I_{DSS}\left[2\left(1 - \frac{V_{GS}}{V_P}\right)\frac{V_{DS}}{-V_P} - \left(\frac{V_{DS}}{V_P}\right)^2\right] \qquad [3.14]$$

Diode circuits   In the analysis of circuits including d.c. supplies, diodes and
resistors, the first step is to ascertain the polarity of the diode bias
conditions after which relationships (i) and (ii) above can be applied
directly. This is now illustrated by some examples.

Example 3.7   *In the circuit shown in Fig. 3.20 the applied d.c. voltage may be either
positive or negative. In each case, calculate the output voltage V_O shown.*

(a) Positive applied voltage: with this polarity, D_2 is forward biased,
while D_1 and D_3 are reverse biased and can therefore be regarded as open
circuit. This leaves the 6 V battery in series with D_2 (0.7 V) and three
resistors. The current flowing is thus given by

$$I = \frac{6 - 0.7}{50 + 100 + 200} = 15.1 \text{ mA}$$

The output voltage is given by the *IR* voltage drop for the 100 Ω resistor
plus the diode volt drop. Working in kΩ and mA

$$V_O = 15.1 \times 0.1 + 0.7 = 2.21 \text{ V}.$$

Fig. 3.20 Diode resistor circuit for Ex. 3.7.

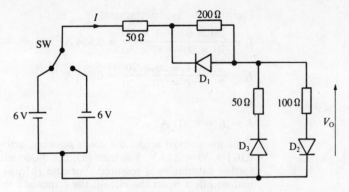

This result could have been written directly as

$$V_O = 0.7 + (6 - 0.7)\frac{100}{350} = 2.21 \text{ V}.$$

(b) Negative applied voltage: with this polarity D_2 is off and therefore open circuit, and D_1 and D_3 are on (0.7 V). The current I is now given by

$$I = \frac{-(6 - 0.7 - 0.7)}{50 + 50} = 46 \text{ mA}$$

and

$$V_O = -0.7 + 501 = -3 \text{ V}.$$

This is half the input voltage since it is the result of division between two equal resistors and two diodes. Note that the 200 Ω resistor does not come into the calculation but it will carry a current of 0.7/200 or 35 mA. This means that the two diodes pass slightly different currents.

In example 3.7, the bias states of the diodes could be determined by inspection. In some cases this is not possible without some preliminary calculation made while assuming that the diode in question is removed from the circuit. The polarity of the connection points is then found and hence the device condition is obtained. Analysis of the complete circuit can then be carried out. This is illustrated in the next example.

Example 3.8 *For the circuit shown in Fig. 3.21 calculate the current flowing in the 30 Ω resistor (a) with the diode as shown, and (b) with the diode reversed.*

Fig. 3.21 Diode resistor circuit for Ex. 3.8.

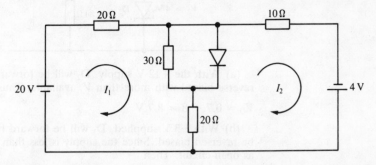

Since the polarity of the voltage across the diode is not obvious, assume the diode to be missing, insert unknown mesh currents I_1 and I_2 and solve by mesh analysis.

$$20 = 70I_1 - 50I_2$$
$$4 = -50I_1 + 60I_2$$

From which, by determinants,

$$I_1 = \frac{20 \times 60 + 4 \times 50}{70 \times 60 - 50 \times 50} = 0.824 \text{ A}$$

$$I_2 = \frac{70 \times 4 + 50 \times 20}{70 \times 60 - 50 \times 50} = 0.753 \text{ A}$$

and

$$I_1 - I_2 = 0.071 \text{ A}.$$

Thus the voltage across the *diode position*, across the 30 Ω resistor, is $0.071 \times 30 = 2.13$ V. For case (a), the diode will be forward biased and further calculation is required. For case (b), the diode is reverse biased and had no effect upon the circuit; the required current is simply I_1 or 0.824 A. Returning to case (a), the 0.7 V diode volt drop must be included in the equations and the 30 Ω resistor is left out.

$$20 - 0.7 = 40I_1 - 20I_2$$

$$4 + 0.7 = -20I_1 + 30I_2$$

Hence

$$I_1 = \frac{19.3 \times 30 + 4.7 \times 20}{40 \times 30 - 20 \times 20} = 0.84 \text{ A}.$$

In circuits involving zener diodes, there are three possible conditions, forward bias ($V = 0.7$ V), reverse bias less than V_z ($I = 0$) and reverse bias greater than V_z ($V = V_z$). It may be necessary to apply all three conditions to determine the behaviour of a particular circuit.

Example 3.9   *The input to the circuit shown in Fig. 3.22 will be, on different occasions, (a) +12 V, (b) −3 V or (c) −12 V. In each case determine the resulting output voltage V_O.*

Fig. 3.22 Zener diode circuit for Ex. 3.9.

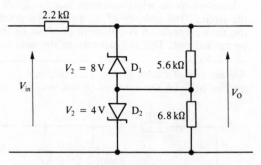

(a) With the +12 V supply, D_2 will be forward biased (0.7 V) while D_1 is reverse biased with more than V_z available, Thus

$$V_O = 0.7 + 8 = 8.7 \text{ V}$$

(b) With −3 V supplied, D_1 will be forward biased (0.7 V) and D_2 will be reversed biased. Since the supply is less than the V_z for D_2, D_2 is taken as open circuit. Then

$$V_O = -\left(0.7 + (3 - 0.7)\frac{6.8}{(6.8 + 2.2)}\right) = -2.44 \text{ V}$$

(c) With −12 V supplied, D_1 is forward biased and D_2 is reverse biased. The supply is greater than V_z; hence

$$V_O = -0.7 - 4 = -4.7 \text{ V}$$

BJT circuits

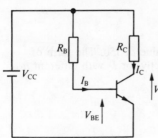

Fig. 3.23 BJT circuit for d.c. calculations.

If active devices such as BJTs are to operate correctly, circuitry involving d.c. supplies and resistors are used to provide the appropriate bias conditions and Q point. For BJTs, this will usually lie within the linear operating range (collector–base reverse bias, emitter–base forward bias) but in certain cases, a cut-off or saturated Q point can be used. The basic circuit arrangement for an npn transistor is shown in Fig. 3.23. Separate supplies could be provided for base and collector but no advantage is obtained by doing this. Both V_{BE} and V_{CE} are positive with respect to earth but if V_{CE} is greater than V_{BE} the required bias conditions will be satisfied. For example, as V_{BE} will be at +0.7 V, so if V_{CE} is at +3 V, V_{CB} is 3 − 0.7 or 2.3 V with the n collector positive with respect to the p base; this is the required reverse bias condition for linear operation.

In the base circuit, the current I_B is fixed by

$$I_B = \frac{(V_{CC} - V_{BE})}{R_B}$$

then from equation [3.12]

$$I_C = h_{FE}I_B$$

and

$$V_{CE} = V_{CC} - I_C R_C$$

In practice, the simple arrangement shown in Fig. 3.23 suffers from various disadvantages due to device properties. In particular, the sample-to-sample variation in h_{FE} can result in V_{CE} values anywhere between saturation and near cut-off for the same circuit and transistor type. A second-order problem is the sensitivity of the V_{BE}, I_B characteristic to operating temperature. If V_{BE} is constant (from a constant voltage supply), I_B will vary considerably with temperature. In the basic circuit shown, this problem is minimised since V_{CC} is much greater than V_{BE}, making I_B approximately V_{CC}/R_B.

These problems are overcome in various ways, which in simple circuits involve a resistor R_E connected in the emitter lead. Various arrangements are illustrated in the following examples which demonstrate the required techniques of analysis.

Example 3.10

The circuit shown in Fig. 3.24 is designed to operate at a nominal 2 mA collector current when h_{FE} is 150. The practical sample-to-sample range of h_{FE} is 50 to 300. Determine the range of Q points that will result from this variation.

Referring to the circuit diagram, the 12 V V_{CC} can be equated to the sum of the series volt drops between the V_{CC} line and earth.

$$V_{CC} = I_B R_B + V_{BE} + V_R \qquad [3.15]$$

and

$$V_{CC} = I_C R_C + V_{CE} + V_E \qquad [3.16]$$

In equation [3.15], V_{BE} is 0.7 V and V_E is $I_C R_E$ ($I_C \simeq I_E$). Substituting these and numerical values into equation [3.15]

$$12 = 690 I_B + 0.7 + 1 I_C$$

but $I_B = I_C/h_{FE}$, thus

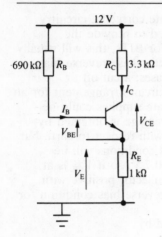

12 V

690 kΩ R_B R_C 3.3 kΩ

I_C

I_B

V_{BE} V_{CE}

R_E
V_E 1 kΩ

Fig. 3.24 BJT resistor circuit for Ex. 3.10.

$$12 - 0.7 = \frac{690}{150}I_C + 1I_C \qquad [3.17]$$

or

$$I_C = 2.02 \text{ mA}$$

This result is obtained using the nominal h_{FE} value of 150. The form of equation [3.17] can be rearranged to solve directly for I_C with different h_{FE} values. With $h_{FE} = 300$

$$I_C = \frac{11.3}{\frac{690}{300} + 1} = 3.42 \text{ mA}$$

or with $h_{FE} = 50$

$$I_C = \frac{11.3}{\frac{690}{50} + 1} = 0.76 \text{ mA}$$

The Q point is given by I_C, V_{CE} and for each value of I_O, equation [3.16] can be rearranged to give

$$V_{CE} = V_{CC} - I_C R_C - I_E R_E$$

where $I_E \simeq I_C$.

For the nominal values, this result is

$$V_{CE} = 12 - 2.02 \times 3.3 - 2.02 \times 1 = 3.3 \text{ V}$$

which is greater than the 0.7 V_{BE}, thus providing linear mode operation. The low h_{FE} gives

$$V_{CE} = 12 - 0.76 \times 3.3 - 0.76 \times 1 = 2.7 \text{ V}$$

which is also satisfactory. The high h_{FE} gives

$$V_{CE} = 12 - 3.42 \times 3.3 - 3.42 \times 1 = -2.7 \text{ V}$$

This indicates an $I_C(R_C + R_E)$ greater than V_{CC}, which is impossible. This means that the high I_C has reduced V_{CE} to the point of saturation and the equations based on h_{FE} are not valid. Equations [3.15] and [3.16] are still correct and V_{CE} is now V_{CESAT} (approximately 0) and $V_E = (I_C + I_B)R_E$.

Inserting values and rewriting these equations

$$12 = 690I_B + 0.7 + 1(I_C + I_B)$$

$$12 = 3.3I_C + 0 + 1(I_C + I_B)$$

Collecting terms

$$11.3 = 691I_B + I_C$$

$$12 = I_B + 4.31I_C$$

Solving by determinants

$$I_C = \frac{697 \times 12 - 11.3}{691 \times 4.3 - 1} = 2.79 \text{ mA}$$

$$I_B = \frac{11.3 \times 4.3 - 12}{691 \times 4.3 - 1} = 0.012 \text{ mA}$$

These results can be confirmed by noting that

$$1(2.79 + 0.012) + 3.3 \times 2.79 = 12 \text{ V}$$

In this case, the $I_C = I_E$ approximation would still be valid but the current ratio $I_C/I_B = 2.79/0.012 = 233$, which is less than the stated h_{FE} value of 300.

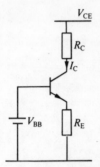

Fig. 3.25 A BJT bias arrangement with a constant V_{BB} supply.

From the above example, it may be seen that this circuit arrangement cannot provide for a constant I_C regardless of typical h_{FE} variation. This problem could be completely eliminated by the use of a constant V_{BB} supply as shown in Fig. 3.25. This would result in $I_C = (V_{BB} - 0.7)/R_E$ which is independent of h_{FE}. The problem with this arrangement is that no a.c. signals can be supplied to the base of the transistor as the d.c. V_{BB} supply would be an effective short circuit to a.c.

A compromise between the circuits of Fig. 3.24 and Fig. 3.25 minimises both problems. Effectively, the V_{BB} supply has a 'medium' value of resistance R_B in series with it. This avoids the low a.c. input impedance difficulty and reduces the $I_B R_B$ volt drop which caused the variation in the last example. In practice the V_{BB} supply is provided by a potential divider circuit across the V_{CC} supply. This circuit and the effect it has on I_C are shown in the next example.

Example 3.11

The circuit shown in Fig. 3.26(a) is designed to operate at a nominal 2 mA collector current when h_{FE} is 150. The practical sample-to-sample variation in h_{FE} is 50 to 300 (the same as in example 3.10). Determine the range of Q points that will result from this variation. Compare these results with those for example 3.10.

Fig. 3.26 Circuits for Ex. 3.11: (a) full circuit; (b) equivalent base circuit.

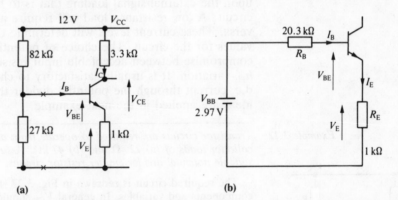

(a)   (b)

The 12 V V_{CC} and the potential divider resistors 82 kΩ and 27 Ω provide the effective V_{BB} supply. The equivalent circuit for this is obtained by applying Thévenin's theorem to the circuit at points XX, looking to the left. This results in the following values

$$V_{BB} = \frac{12 \times 27}{82 + 27} = 2.97 \text{ V}$$

$$R_B = \frac{82 \times 27}{82 + 27} = 20.3 \text{ k}\Omega$$

The resulting base–emitter part circuit is shown in Fig. 3.26(b). Equating the volt drops around this circuit

$$V_{BB} = I_B R_B + V_{BE} + I_E R_E$$

or since $I_E \simeq I_C$

$$V_{BB} - V_{BE} = I_B R_B + I_C R_E$$

and $I_B = I_C/h_{FE}$

$$V_{BB} - V_E = \frac{I_C R_B}{h_{FE}} + I_C R_E = I_C \left(\frac{R_B}{h_{FE}} + R_E \right)$$

Finally $I_C = \dfrac{V_{BB} - V_{BE}}{R_E + R_B/h_{FE}}$

Inserting numerical values

$I_C = \dfrac{2.97 - 0.7}{1 + 20.3/150} = 2.0 \text{ mA}$

and for the other values of h_{FE}

$I_C = \dfrac{2.97 - 0.7}{1 + 20.3/50} = 1.61 \text{ mA}$

$I_C = \dfrac{2.97 - 0.7}{1 + 20.3/300} = 2.13 \text{ mA}$

These results represent a variation compared with the nominal value of −20% and +5%. The equivalent results for the previous circuit in example 3.10 are −62% and +38%, the upper figure taking the transistor into saturation. The range of V_{CE} values is now 2.8 V to 5.1 V (from $V_{CE} = V_{CC} - I_C(R_E + R_C)$), which lie well within the linear operating range.

The design of a suitable bias circuit must start with the desired Q point, in particular the operating current I_C. This in turn depends upon the external signal loading that is to be connected to the circuit. A low resistance load will require a high current and vice versa. These current levels will determine the remaining resistor values for the circuit. The choice of potential divider values is a compromise between acceptable input resistance and sensitivity to h_{FE} variation. It is usually satisfactory to choose values that give a d.c. current through the potential divider that is about $0.1I_C$. These ideas are applied in the next example.

Example 3.12   *Transistor circuits are required to operate from a 10 V d.c. supply with final collector loads of (a) 220 Ω and (b) 47 kΩ. Assuming an h_{FE} of 120, design suitable potential divider emitter resistor circuits.*

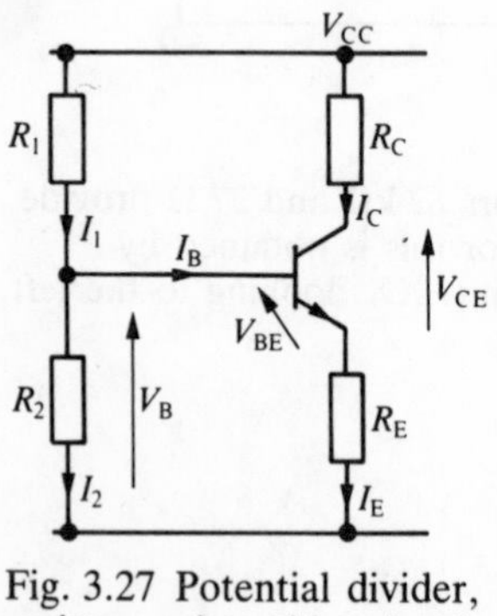

The required circuit is redrawn in Fig. 3.27 showing the various components and variables. In general V_{CE} should be about one-third of V_{CC} (this allows for maximum undistorted voltage swing when there is a capacitively coupled a.c. load) and R_E should be at least one-quarter of R_C to give adequate stability for h_{FE} variation. Thus

$(1 + 0.25)R_C \times I_C = 2/3\ V_{CC}$

or

$I_C = \dfrac{2/3 V_{CC}}{1.25R_C}$

In this example, for (a)

$I_C = \dfrac{2/3 \times 10}{1.25 \times 0.22} = 24.2 \text{ mA}$

Note that R_C is expressed in kΩ to give I_C in mA.
For (b)

$I_C = \dfrac{2/3 \times 10}{1.25 \times 47} = 0.114 \text{ mA}$

These figures are only nominal values and the final choice of components will be limited by preferred values of resistors. Thus suitable R_E values for one-quarter of R_C are 220/4 or 55 Ω (preferred value 56 Ω) and 47/4 or 11.75 kΩ (preferred value 12 kΩ).

Fig. 3.27 Potential divider, emitter resistor bias circuit for Ex. 3.12.

To determine R_1 and R_2 values, take I_2 as $0.1I_C$ or 2.42 mA and 0.0114 mA respectively. Now

$$V_C = I_C R_E + 0.7$$

For (a)

$$V_B = 24.2 \times 0.056 + 0.7 = 2.055 \text{ V}$$

For (b)

$$V_B = 0.114 \times 12 + 0.7 = 2.07 \text{ V}$$

$$R_2 = \frac{V_B}{0.1I_C}$$

For (a)

$$R_2 = \frac{2.055}{2.42} = 0.85 \text{ k}\Omega$$

(preferred value 820 Ω). For (b)

$$R_2 = \frac{2.07}{0.0114} = 182 \text{ k}\Omega$$

(preferred value 180 kΩ). Finally, the current I_1 is the sum of I_2 and the base current I_B. Hence

$$I_1 = I_2 + \frac{I_C}{h_{FE}}$$

and

$$R_1 = \frac{V_{CC} - V_B}{I_1}$$

For (a)

$$R_1 = \frac{10 - 2.055}{2.42 + 24.2/120} = 3.03 \text{ k}\Omega$$

(preferred value 3.3 kΩ). For (b)

$$R_1 = \frac{10 - 2.07}{0.0114 + 0.114/120} = 642 \text{ k}\Omega$$

(preferred value 630 kΩ).

The resulting designs are shown in Fig. 3.28. These circuits can be analysed using the techniques shown in example 3.12 and the resulting collector currents are found to be 21 mA and 0.15 mA respectively.

Fig. 3.28 Circuits found in the solution for Ex. 3.12.

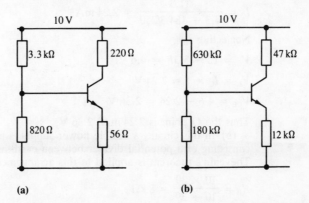

(a)   (b)

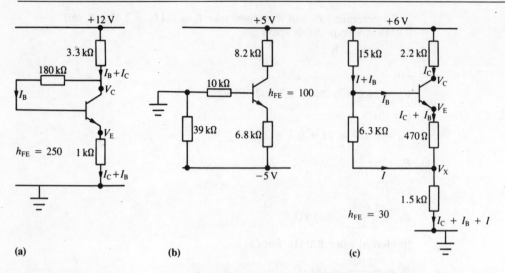

Fig. 3.29 A range of d.c.
circuits for BJTs for
Ex. 3.13.

Although the bias arrangement illustrated in the last two examples
is the most widely used for simple BJT circuits, other arrangements
are used in some circumstances. The next example illustrates the
techniques of d.c. analysis on three different circuit arrangements.

Example 3.13   *For each of the three circuits shown in Fig. 3.29, determine the transistor Q
point and all other node voltages and branch currents. In each case, nominal
h_{FE} is quoted on the circuit diagram.*

(a) For this circuit, the equation for the volt drops making up V_{CC}
include that due to R_C in which I_C and I_B flow. Thus

$$12 = 3.3(I_C + I_B) + 180I_B + 0.7 + 1(I_C + I_B)$$

Rearranging

$$11.3 = 4.3I_C + 184.3I_B$$

but

$$I_B = I_C/h_{FE}$$

$$11.3 = 4.3I_C + \frac{184.3}{250}I_C$$

and

$$I_C = \frac{11.3}{4.3 + 184.3/250} = 2.24 \text{ mA}$$

Neglecting I_B

$$V_C = 12 - 3.3I_C = 4.6 \text{ V.}$$

$$V_E = I_C \times 1 = 2.24 \text{ V}$$

$$V_{CE} = 4.6 - 2.24 = 2.36 \text{ V}$$

Thus the Q point is 2.24 mA, 2.36 V.

(b) In this circuit, a double power supply is used with the base circuit
consisting of a potential divider between earth and the negative supply.
Thévenin's theorem is applied to this arrangement to find

$$R_B = \frac{10 \times 39}{10 + 39} = 8 \text{ k}\Omega$$

$$V_B = \frac{5 \times 39}{10 + 39} = 4 \text{ V}$$

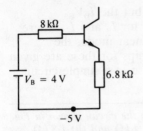

Fig. 3.30 Equivalent base circuit for Ex. 3.13 circuit (b).

This V_B is relative to the negative 5 V line and the equivalent d.c. circuit is shown in Fig. 3.30. The usual input equation can then be written

$$4 - 0.7 = 6.8 I_C + \frac{14.8}{100} I_C$$

hence

$$I_C = 0.47 \text{ mA}$$

and

$$V_C = +5 - 0.47 \times 8.2 = 1.1 \text{ V}$$

$$V_E = -5 + 0.47 \times 6.8 = 1.80 \text{ V}$$

Therefore

$$V_{CE} = 1.1 - (-1.80) = 2.90 \text{ V}$$

Thus the transistor Q point is 0.47 mA, 2.90 V.

(c) To analyse this circuit, we have to include not only I_C and I_B but also a current I flowing in the 6.3 kΩ resistor as shown in the circuit diagram. There are now two paths from the 6 V line past the base to earth, each of which provides an equation.

$$6 = 15(I + I_B) + 0.7 + 0.47(I_C + I_B) + 1.5(I + I_C + I_B)$$

$$6 = 15(I + I_B) + 6.3 I + 1.5(I + I_C + I_B)$$

Collecting terms and rearranging

$$5.3 = 16.5 I + 16.97 I_B + 1.97 I_C$$

$$6 = 22.8 I + 16.5 I_B + 1.5 I_C$$

Substituting for $I_B = I_C / h_{FE}$ and collecting terms

$$5.3 = 16.5 I + 2.54 I_C$$

$$6 = 22.8 I + 2.05 I_C$$

Solving by determinants

$$I_C = \frac{6 \times 16.5 - 22.8 \times 5.3}{16.5 \times 2.05 - 22.8 \times 2.53} = 0.906 \text{ mA}$$

and

$$I = \frac{5.3 \times 2.05 - 6 \times 2.54}{16.5 \times 2.05 - 22.8 \times 2.54} = 0.18 \text{ mA}$$

The voltage

$$V_x = 1.5(0.906 + 0.18) = 1.63 \text{ V}$$

The emitter voltage

$$V_E = 1.63 + 0.906 \times 0.47 = 2.05 \text{ V}$$

The collector voltage

$$V_C = 6 - 0.906 \times 2.2 = 4.01 \text{ V}$$

thus

$$V_{CE} = 4.01 - 2.05 = 1.96 \text{ V}$$

Thus the Q point is 0.906 mA, 1.96 V.

JFET bias circuits   The biasing of JFETs provides similar problems of correct polarity and thermal stability to the BJT and a similar circuit arrangement is used. Input current is, however, not involved but the $I_D V_{GS}$ relationships are more complicated than those for I_B and I_C. There are also two regions of operation within the linear mode, the pinch-of and the triode regions. The relationships for these are given by equations [3.13] and [3.14]. The use of these in a simple circuit arrangement is illustrated in the next example.

Example 3.14

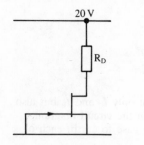

Fig. 3.31 JFET circuit for Ex. 3.14.

An n-channel FET has I_{DSS} 10 mA and V_P −4 V. In the circuit shown in Fig. 3.31, determine the operating conditions if R_D is (a) 1 kΩ and (b) 1.8 kΩ.

V_{GS} is 0 V, so if the JFET is in pinch-off, $I_D = I_{DSS} = 10$ mA. Assuming this to be correct, we test for the given values to find V_{DG} using equation [3.13]. If V_{DG} is less than V_P, the assumption is not correct and the solution is obtained using equation [3.14].

(a) R_D is 1 kΩ, therefore

$$V_{DS} = 20 - I_D R_D = 20 - 10 = 10 \text{ V}$$

This is greater than V_P so the assumption is correct. The required Q point is I_D 10 mA, $V_{DS} = 10$ V.

(b) $R_D = 1.8$ kΩ, therefore

$$V_{DS} = 20 - 18 = 2 \text{ V}.$$

This is less than V_P so the assumption is incorrect and we must therefore use equation [3.14].

$$I_D = 10\left[2(1 - 0)\frac{V_{DS}}{4} - \left(\frac{V_{DS}}{4}\right)^2\right]$$

rearranging

$$I_D = 5V_{DS}\left(1 - \frac{V_{DS}}{8}\right)$$

but $V_{DS} = 20 - 1.8I_D$. Substituting

$$I_D = 5(20 - 1.8I_D)\left(1 - \frac{(20 - 1.8I_D)}{8}\right)$$

Multiplying out and rearranging

$$I_D^2 - 17.28I_D + 74 = 0$$

from which

$$I_D = 9.42 \text{ mA and } V_{DS} = 3.04 \text{ V}$$

or

$$I_D = 7.86 \text{ mA and } V_{DS} = 5.85 \text{ V}$$

The second result is incorrect since V_{DS} is greater than V_P, putting the device in pinch-off and this condition has already been eliminated. Thus the required operating conditions are I_D 9.42 mA and V_{DS} 3.04 V.

Practical devices show considerable sample-to-sample variation in the parameters I_{DSS} and V_p. Bias circuitry simlar to that used for BJTs can stabilise the operating current. A potential divider source resistor arrangement is illustrated in the next example.

Example 3.15

An n-channel JFET has I_{DSS} 10 mA and V_P −3 V. Determine, for the circuit shown in Fig. 3.32, the d.c. conditions for the device.

$$V_G = \frac{16 \times 0.5}{1.5 + 0.5} = 4 \text{ V}$$

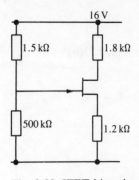

Fig. 3.32 JFET bias circuit for Ex. 3.15.

From equation [3.13]

$$I_D = 10\left(1 - \frac{4 - 1.2I_D}{-3}\right)^2$$

$$= 10(1 + 1.33 - 0.4I_D)^2$$

$$= 10(5.43 - 1.86I_D + 0.61I_D^2)$$

Rearranging and solving the quadratic gives

$$I_D = 8.0 \text{ mA or } 4.22 \text{ mA}$$

For the 8.0 mA result

$$V_D = 16 - 1.8 \times 8$$

and is not negative so this result is invalid. For the 4.22 mA result

$$V_D = 16 - 4.22 \times 1.8 = 8.4 \text{ V}$$

and

$$V_{DG} = 8.4 - 4 = 4.4 \text{ V}$$

which is greater than V_P thus confirming the pinch-off condition. Hence the required Q point is I_D 4.22 mA, V_{GS} 3.34 V.

D. C. for circuits involving more than one device

Circuit arrangements often include two transistors or transistors and diodes. A selection of typical circuits is shown in this section, Analytical techniques are similar to those covered in the previous sections.

Example 3.16

In the cascode amplifier circuit shown in Fig. 3.33(a) both transistors have an h_{FE} of 100. Calculate the Q point for both transistors.

Fig. 3.33 Cascode amplifier circuit for Ex. 3.16: (a) full circuit; (b) equivalent base circuit for Tr_2.

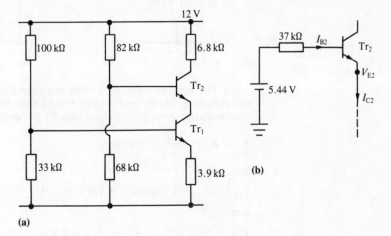

Starting with Tr_1, the base circuit is simplified by Thévenin's theorem to give $V_{B1} = 3$ V and $R_{B1} = 24.8$ kΩ. Hence

$$I_{C1} = \frac{3 - 0.7}{\dfrac{28.7}{100} + 3.9} = 0.55 \text{ mA}$$

Since Tr_1 and Tr_2 are in series, $I_{C2} = I_{C1}$ (neglecting I_{B2}). The base circuit for Tr_2 can also be found, using Thévenin's theorem, to be $V_{B2} = 5.44$ V and $R_{B2} = 37$ kΩ. In this case, I_C is already known but I_B is required to find the voltage at the base and emitter of Tr_2.

$$I_{B2} = \frac{I_{C2}}{100} = 0.0055 \text{ mA}$$

Referring to the base circuit of Tr$_2$ which is shown in Fig, 3.33(b)

$$V_{E2} = 5.44 - 0.0055 \times 37 - 0.7 = 2.7 \text{ V}$$

which is also V_{C1}. The remaining voltages are

$$V_{C2} = 12 - 0.55 \times 6.8 = 8.3 \text{ V}$$

$$V_{EI} = 0.55 \times 3.9 = 2.1 \text{ V}.$$

Hence, for Tr$_1$,

$$I_C = 0.55 \text{ mA}, \quad V_{CE} = 2.7 - 2.1 = 1.7 \text{ V}$$

for Tr$_2$,

$$I_C = 0.55 \text{ mA}, \quad V_{CE} = 8.3 - 2.7 = 5.6 \text{ V}.$$

Example 3.17   *The long-tailed pair amplifier circuit in Fig. 3.34(a) has identical transistors. Calculate the Q point (a) with the circuit as shown and (b) with the 10 kΩ resistor replaced by the constant current 'tail' circuit shown in Fig. 3.34(b).*

Fig. 3.34 Long-tailed pair amplifier circuit for Ex. 3.17: (a) basic circuit; (b) alternative 'constant current tail' circuit.

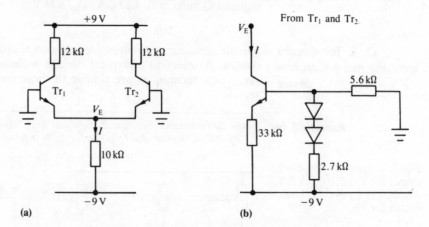

(a)   The emitter voltage for both transistors is simply -0.7 V. Thus the tail current $I = [-0.7 - (-9)]/10 = 0.83$ mA. Since the transistors are identical, this current is divided equally between them.

$$I_{C1} = I_{C2} = \frac{0.83}{2} = 0.45 \text{ mA}$$

and

$$V_{C1} = V_{C2} = 9 - (0.425 \times 12) = 4.02 \text{ V}$$

and

$$V_{CE1} = V_{CE2} = 4.02 - (-0.7) = 4.2 \text{ V}$$

(b)   Referring to Fig. 3.4(b), first V_B is found by neglecting the base current and taking the diode volt drops as 0.7 V each.

$$V_B = -9 + 1.4 + (9 - 1.4)\frac{2.7}{5.6 + 2.7} = -5.12 \text{ V}.$$

Hence

$$V_E = -5.12 - 0.7 = -5.82 \text{ V}$$

$$I_E = \frac{-5.82 - (-9)}{3.3} = 0.96 \text{ mA}$$

As with the original circuit in Fig. 3.4(a), this current is divided between the two identical transistors, making $I_{C1} = I_{C2} = 0.48$ mA.

$$V_{C1} = V_{C2} = 9 - 0.48 \times 12 = 3.24 \text{ V}$$

$$V_{CE1} = V_{CE2} = 3.24 - (-7) = 3.94 \text{ V}$$

Example 3.18   *The direct coupled amplifier shown in Fig. 3.35 employs an npn transistor and a pnp transistor. Both have h_{FE} of 80. Calculate all relevant voltages and currents.*

Fig. 3.35 Circuits using complementary transistors for Ex. 3.18: (a) full circuit; (b) equivalent base circuit for Tr_2.

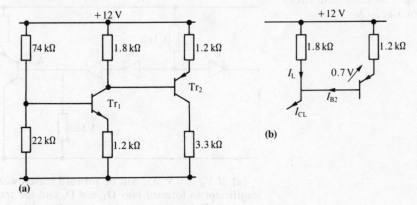

Calculation of I_{C1} follows the usual method.

$V = 2.75$ V, $R_{B1} = 17$ kΩ

Therefore

$$I_{C1} = \frac{2.05}{\dfrac{18.2}{80} + 1.2} = 1.43 \text{ mA}$$

This current is the sum of the load current and I_{B2} (see Fig. 3.35(b))

$$1.43 = I_L + I_{B2} = I_L + \frac{I_{C2}}{80}$$

or

$$114.4 = 80I_L + I_{C2} \tag{3.18}$$

Equating the voltages for Tr_2 emitter–base circuit

$$1.8I_L = 1.2I_{C2} + 0.7$$

or

$$0.7 = 1.81I_L - 1.2I_{C2} \tag{3.19}$$

Rewriting equations [3.18] and [3.19]

$$114.4 = 80I_L + I_{C2}$$

$$0.7 = 1.8I_L - 1.2I_{C2}$$

Solving by determinants

$$I_L = \frac{114.4 \times 1.2 - 0.7}{-80 \times 1.2 - 1.8} = 1.41 \text{ mA}$$

Hence

$$V_{C1} = 12 - 1.8 \times 1.41 = 9.46 \text{ V}$$

$$V_{E2} = 9.46 + 0.7 = 10.16 \text{ V}$$

$$I_{C2} = \frac{12 - 10.16}{1.2} = 1.53 \text{ mA (or by determinants from above)}$$

$$V_{CE2} = 12 - 1.53(1.2 + 3.3) = 5.1 \text{ V}$$

$$V_{CE1} = 12 - 1.41 \times 1.8 - 1.2 \times 1.43 = 7.75 \text{ V}$$

Example 3.19   *Fig.3.36 shows the circuit of a DTL logic gate. Determine V_O for V_{in} of (a) 0 V and (b) 5 V. Assume a transistor h_{FE} of 30.*

Fig. 3.36 Logic gate circuit for Ex. 3.19.

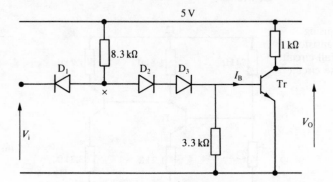

(a) If V_{in} is 0 V, D_1 will be forward biased with 0.7 V at point X. This is insufficient to forward bias D_2 *and* D_3 *and* the transistor base–emitter junction. The transistor is therefore off and $V_0 = 5$ V.

(b) If V_{in} is 5 V, D_1 is off and there are three pn junctions in series with the 5 V supply and the 8.3 kΩ resistor. From the circuit

$$I = \frac{5 - 3 \times 0.7}{8.3} = 0.35 \text{ mA}.$$

The current in the 33 kΩ resistor is $\dfrac{0.7}{3.3}$, so

$$I_B = 0.35 - \frac{0.7}{3.3} = 0.137 \text{ mA}$$

$$I_C = 30 \times 0.137 = 4.14 \text{ mA}$$

$$V_0 = 5 - 4.14 = 0.86 \text{ V}$$

Example 3.20   *The circuit shown in Fig. 3.37 is a discrete component Schmitt trigger bistable. The transistors have h_{FE} 100 and V_{CESAT} 0.1 V. Calculate collector and emitter voltages for both transistors if V_{in} is (a) 0 V and (b) 4 V.*

Fig. 3.37 Schmitt trigger bistable circuit for Ex. 3.20.

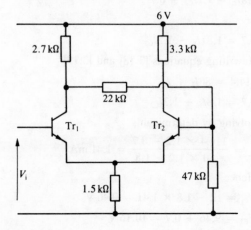

(a) If V_{in} is 0 V, the base of Tr_1 cannot be positive with respect to the emitter and we can assume that the device is off. The base–emitter circuit for Tr_2 consists of the potential divider $(2.7 + 22)$ kΩ, 47 kΩ and the 6 V V_{CC} together with the 1.5 kΩ R_E. The Thévenin equivalents are V_B 3.93 V and R_B 16.2 kΩ. Thus if Tr_2 is in the linear mode

$$I_C = \frac{3.23}{\dfrac{17.7}{100} + 1.5} = 1.93 \text{ mA}.$$

and

$$V_{CE} = 6 - 1.93(3.3 + 1.5) = -3.26 \text{ V}$$

As this is less than V_{CESAT}, the assumption of linear mode operation was incorrect. Equations assuming saturation can now be written. For the base circuit

$$3.93 - 0.7 = I_B \times 16.2 + (I_B + I_C)1.5$$

For the collector circuit

$$6 - V_{CESAT} = 3.3I_C + (I_B + I_C)1.5$$

The term h_{FE} can no longer be used, but $V_{CE} = V_{CESAT} = 0.1$ V. Rearranging

$$3.23 = 17.7I_B + 1.51I_C$$

$$5.9 = 1.5I_B + 4.81I_C$$

Solving by determinants,

$$I_B = \frac{3.23 \times 4.8 - 5.9 \times 1.5}{17.7 \times 4.8 - 1.5 \times 1.5} = 0.0805 \text{ mA}$$

$$I_C = \frac{17.7 \times 5.9 - 1.5 \times 3.23}{17.7 \times 4.8 - 1.5 \times 1.5} = 1.20 \text{ mA}$$

These may be checked by

$$V_{CE} = 6 - 1.204 \times 3.3 - 1.5(1.204 + 0.0805) = 0.1 \text{ or } V_{CESAT}$$

Now V_E for both transistors is $1.5(1.2 + 0.08) = 1.92$ V. For Tr_2

$$V_B = 1.92 + 0.7 = 2.62 \text{ V}$$

and therefore

$$V_{C1} = 2.62 + (6 - 2.62)\frac{22}{22 + 2.7} = 5.63 \text{ V}$$

Thus for Tr_1

$V_{CE} = 5.63 - 1.92 = 3.71$ V and $I_C = 0$

(b) When V_{in} is 4 V, Tr_1 is probably on and the resulting low V_{C1} will hold Tr_2 off. For Tr_1

$$I_E = \frac{4 - 0.7}{1.5} = 2.2 \text{ mA}$$

Assuming $I_C = I_E$

$$V_{C1} = 6 - 2.2 \times 2.7 = 0.06 \text{ V}.$$

This is negative with respect to the emitter voltage, thus Tr_1 is saturated and

$$V_{C1} = (4 + 0.7) + 0.1 = 3.4 \text{ V}$$

The part circuit in Fig. 3.38 shows the relevant currents and voltages for this part of the solution.

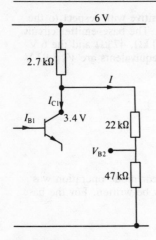

Fig. 3.38 Part circuit for solution of Ex. 3.20.

The current I is given by

$$I = \frac{3.4}{22 + 47} = 0.049 \text{ mA}$$

Therefore

$$V_{B2} = 0.049 \times 47 = 2.32 \text{ V}$$

This is negative with respect to the V_E of $(4 - 0.7)$ or 3.3 V so the assumption that Tr_2 is off was correct. Now

$$I_{C1} = \frac{6 - 3.4}{2.7} - I = 0.963 - 0.049 = 0.914 \text{ mA}$$

Thus

$$I_{B1} = I_E - I_C = 2.2 - 0.914 = 1.29 \text{ mA}$$

This figure may seem rather large for a base current, but the 4 V input has driven Tr_1 hard into saturation.

In this part of the book, the techniques of analysis for common electronic circuit arrangements have been examined. The many worked examples have illustrated a wide range of common device component circuits which are to be found in electronic systems. These techniques are applied where necessary in later chapters of this book.

You should now be able to attempt exercises 3.10 to 3.19.

Analysis by equivalent circuit techniques

Equivalent circuits, or models, are used in various situations to simplify analysis. In most cases, the resulting simplification is only applicable for some limited range of conditions. For example, the circuit shown in Fig. 3.39 shows some Thévenin equivalents for a more complicated circuit.

With very low frequency signals (less than 0.1 Hz), the 25 μF capacitor and the 1 nF capacitor are effectively open circuit. The resulting equivalent in (b) can be further reduced by Thévenin to a generator $e' = 0.154e$ and $R' = 1.69$ kΩ. For frequencies between 10 Hz and 1600 Hz, the 25 μF capacitor is effectively short circuit while the 1 nF capacitor is still effectively open circuit. The resulting equivalent in Fig. 3.39(c) reduces to $e' = 0.129e$ and $R' = 1.69$ kΩ. At high frequencies, greater than 160 kHz, all that remains is (d) which reduces to $0.59e$ and 0.59 kΩ. The final circuit in Fig. 3.39(e) is the equivalent for all these cases but its components depend upon the frequency range of the input signal e.

When equivalent circuits are found for non-linear devices or circuits, the signal range is limited not only by frequency but also by signal amplitude and d.c. level. This is because, over a limited range, the characteristics can be approximated to straight lines. The

Fig. 3.39 A circuit and
equivalents at different
signal frequencies.

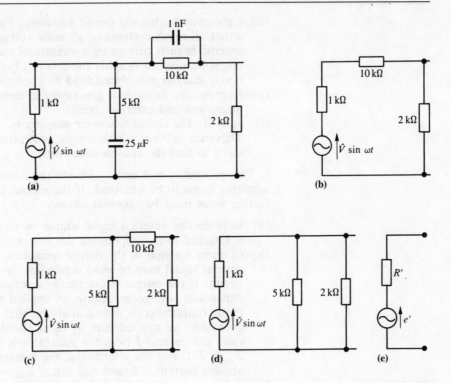

(a)

(b)

(c)

(d)

(e)

slope of such straight lines depends upon the operating point or the
d.c. bias conditions. In Chapter 2, equivalent circuits for each device
are discussed and typical values are given. In this section, the
techniques of analysis of circuits including these devices is explained
and demonstrated. The two stages in the procedure are first to draw
an equivalent for the circuit including the device and then carry out
the analysis using linear network methods.

**The equivalent circuit
method**

A complex circuit involving d.c. supplies, components, signals and
one or more devices can be reduced to the small-signal equivalent
by the following procedures.

(a) For each device shown in the full circuit, draw the appropriate
equivalent. The choice of equivalent will be determined by the
available information.

(b) Taking each terminal of each device in turn, connect it to earth
(or to other device terminals) through any component or signal
sources shown in the full circuit. Remember that, to signals,
d.c. supplies can be regarded as short circuit and that capacitors
and inductors can be regarded as short circuit or open circuit
over certain frequency ranges.

(c) With complicated circuits, it may be helpful to redraw this
result so that connections do not cross etc.

To analyse the circuit obtained by this procedure, the following
steps can be taken.

(d) Choose and label the circuit variables. These will usually be either all mesh currents or all node voltages. Occasionally a specific branch current or a mixture of current and voltages might be used to simplify the analysis but the basic mesh or nodal analysis will always lead to a solution.

(e) Express any dependent generators in terms of the chosen unknowns and external signals.

(f) Simplify the circuit wherever possible by application of Thévenin or Norton's theorems: write circuit equations and solve to find the chosen unknowns.

This procedure will usually enable the input impedance and amplifier gains to be obtained. If the output impedance is required, further steps must be taken as follows.

(g) Replace any external signal source by its internal resistance (this is referred to as suppressing the source).

(h) Connect a signal to the output terminals; either a voltage or a current signal may be used depending on the form of the output circuit. If the output is essentially a series circuit with a dependent voltage generator, an applied votage should be used. If the output circuit is essentially parallel in form with a current generator, an applied current signal should be used. In the first case, the current I from the generator is calculated to find $Z_{out} = E/I$. For the alternative, the voltage V resulting from the applied current is found and hence $Z_{out} = V/I$.

These techniques will now be illustrated by some examples.

Example 3.21   *A BJT is specified as having h_{fe} 75, h_{oe} 10^{-5} S and h_{ie} 3.3 kΩ (see Chapter 2). It is to be used in the complete circuit shown in Fig. 3.40. Assuming a signal frequency such that all capacitors are effectively short circuit, draw an equivalent circuit and calculate the output voltage across the 10 kΩ resistive load.*

Fig. 3.40 Circuit for Ex. 3.21.

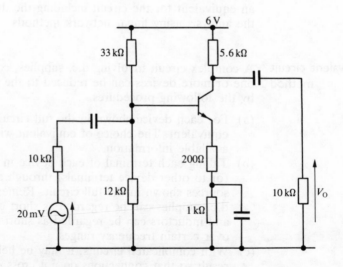

The three separate parts of the small-signal equivalent circuit are shown in Fig. 3.41. Part (a) shows the device equivalent drawn and labelled; (b) shows the emitter connection to earth; the 1 kΩ resistor is short-circuited by

Fig. 3.41 Stages in drawing the equivalent circuit in Ex. 3.21: (a) device equivalent; (b) connecting the emitter; (c) connecting the base; (d) connecting the collector; (e) the complete equivalent circuit; (f) the circuit simplified and labelled for analysis.

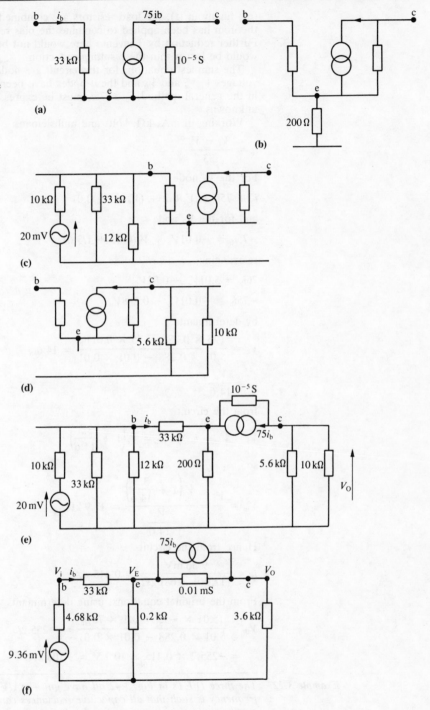

the capacitor; (c) shows the base connection to earth with the capacitor short circuit and the 6 V d.c. supply short circuit for a.c. signals; (d) shows the collector connection to earth through the 6 V supply and the capacitor to earth; all three of the parts are included in the full equivalent circuit in (e); finally in (f), the load resistors are combined in parallel and Thévenin's theorem has been applied to combine the bias resistors and the signal. Further reduction by Thévenin here would not be helpful as the variable i_b would be lost within the resulting reduction.

The simplest equations for this circuit are nodal equations written for the voltages V_i V_e and V_0 and these nodes have been so labelled. From step (e) in the general method, i_b or $75i_b$ must be expressed in terms of these chosen unknowns.

Working in mA, kΩ, volts and millisiemans:

$$i_b = \frac{V_i - V_e}{33}$$

For the V_e node

$$i_b + 75i_b = V_e/0.2 + (V_e - V_0)0.01$$

and for the V_0 node

$$-75i_b = -0.01V_e + V_0(0.01 + 1/3.6)$$

rearranging

$$76i_b = 5.01V_e - 0.01V_0$$

$$-75i_b = -0.01V_e + 0.288V_0$$

by determinants

$$V_e = \frac{(76 \times 0.288 - 75 \times 0.01)i_b}{5.01 \times 0.288 - 0.01 \times 0.01} = 14.6i_b$$

$$i_b = \frac{V_e}{14.6}$$

from the circuit

$$V_i = V_e + i_b \times 3.3 = V_e\left(1 + \frac{3.3}{14.6}\right)$$

so

$$V_{in} = \frac{V_i}{i_b} = \frac{V_e\left(1 + \frac{3.3}{14.6}\right)}{\frac{V_e}{14.6}} = 17.9 \text{ k}\Omega$$

Hence from the circuit

$$i_b = \frac{9.36 \text{ mV}}{(17.9 + 4.68) \text{ k}\Omega} = 0.415 \text{ }\mu\text{A}$$

From the original equations, using determinants

$$V_0 = \frac{(5.01 \times -75 + 0.1 \times 76)i_b}{5.01 \times 0.288 - 0.01 \times 0.01} = -255.2i_b$$

$$= -255.2 \times 0.415 \times 10^{-3} \text{ V} = -106 \text{ mV}$$

Example 3.22   *The three JFETs in Fig. 3.42 all have gm 4 mA/V and r_0 100 kΩ. The signal frequency is such that all capacitive reactances can be neglected. Draw a small-signal equivalent circuit and calculate the overall voltage gain for the amplifier.*

The equivalent circuit is shown in Fig. 3.43. This circuit consists of four isolated parts having gains A_1, A_2, A_3 and A_4. The parallel resistors in each case have the combined value shown in the circuit.

By inspection

Fig. 3.42 JFET circuit for
Ex. 3.22.

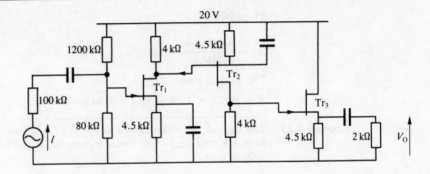

Fig. 3.43 Small-signal
equivalent circuit for
Ex. 3.22.

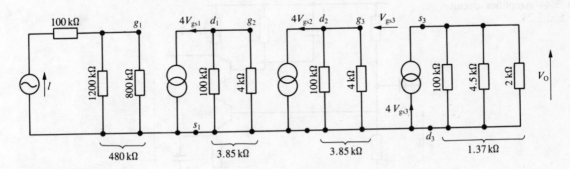

$$A_1 = V_{g1}/e = \frac{480}{580} = 0.827$$

$$A_2 = V_{g1}/V_{g1} = \frac{-4V_{gs1} \times 3.85}{V_{g1}} = -15.4$$

since $V_{gs1} = V_{g1}$

$$A_3 = \frac{V_{g3}}{V_{g2}} = \frac{-4V_{gs2} \times 3.85}{V_{gs}} = -15.4$$

since $V_{gs2} = V_{g2}$

$$A_4 = \frac{V_O}{V_{g3}}$$

but

$$V_{gs3} = V_{g3} - V_O$$

and

$$V_O = 4V_{gs3} \times 1.37 + 5.46V_{gs3}$$

Therefore

$$V_{gs3} = V_{g3} - 5.48V_{gs3}$$

or

$$V_{g3} = V_{gs3}(1 + 5.48)$$

and

$$V_{gs3} = \frac{V_{g3}}{6.48}$$

Therefore

$$A_4 = \frac{5.48V_{gs3}}{6.48V_{gs3}} = 0.845$$

The overall gain

$$A = 0.827 \times 15.4 \times 15.4 \times 0.845 = 166$$

Example 3.23   *The cascode amplifier shown in Fig. 3.44 employs BJTs with h_{ie} 1.5 kΩ, gm 80 mA/V and h_{oe} 10^{-4} S. Draw a complete equivalent circuit and calculate the voltage gain.*

Fig. 3.44 Amplifier circuit for Ex. 3.23.

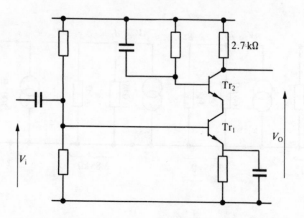

The equivalent circuit is drawn in Fig. 3.45. All the device terminals are labelled. The input voltage V_i is V_{be1}. Node voltages V_x and V_O are also labelled and V_{be2} is V_x. Writing nodal equations

at V_x: $-80V_i + 80(-V_x) = V_x\left(\frac{1}{10} + \frac{1}{1.5} + \frac{1}{10}\right) - \frac{V_O}{10}$

at V_O: $-80(-V_x) = \frac{V_x}{10} + V_0\left(\frac{1}{10} + \frac{1}{2.7}\right)$

Rearranging and collecting terms

$$-80V_i = 80.87V_x - 0.1V_0$$

$$0 = -80.1V_x + 0.47V_0$$

from the second equation

Fig.. 3.45 Equivalent circuit for Ex. 3.23.

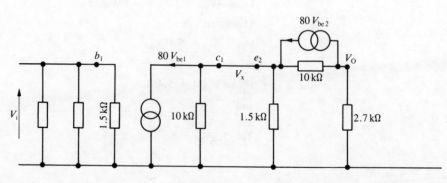

$$V_x = \frac{0.47V_0}{80.1}$$

and

$$-80V_i = 80.87 \times \frac{0.47}{80.1}V_0 - 0.1V_0$$

or

$$V_0 = \frac{-80}{0.375}V_i = -213.6V_i$$

Hence the voltage gain is 213.6 with a phase shift of 180°.

Example 3.24

An amplifier is specified as having a gain of −500 with input and output impedances of 5 kΩ and 2 kΩ respectively. It is connected to an external load of 4 kΩ and the input is obtained from a source of 1 mV with internal resistances 2 kΩ. A 20 kΩ resistor is connected between input and output of the amplifier. Draw an equivalent circuit for the arrangement and derive the Thévenin equivalent generator at the output terminals.

The equivalent circuit for the amplifier is shown (from the description in the question) in Fig. 3.46(a) and that for the complete arrangement in Fig. 3.46(b).

Fig. 3.46 Circuits for Ex. 3.24: (a) amplifier equivalent circuit (b) full equivalent circuit.

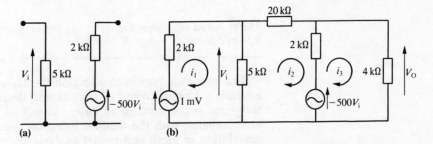

The full circuit can be solved by mesh analysis to find V_O. To find the circuit output impedance, it must be redrawn with the signal source suppressed and a signal V applied to the output.

First, for V_O, mesh currents i_1, i_2 and i_3 are chosen as shown. Now, for the dependent generator

$$V_1 = 5(i_1 - i_2)$$

The mesh generators are

$$e = 7i_1 - 5i_2$$

$$2500(i_1 - i_2) = -5i_1 + 27i_2 - 2i_3$$

$$-2500(i_1 - i_2) = -2i_2 + 6i_3$$

Rearranging and solving by determinants gives $V_O = 4i_3 = -48$ mV.

The redrawn circuit for the output impedance calculation is shown in Fig. 3.47. From this circuit,

$$I = I_1 + I_2 + I_3$$

and

$$V_i = V \times \frac{1.43}{20 + 1.43} = \frac{V}{15}$$

$$I_2 = \frac{V}{21.43}, \qquad I_3 = \frac{V}{4}$$

$$I_1 = \frac{V - (-500)\frac{V}{15}}{2} = \frac{V}{2}\left(1 + \frac{500}{15}\right) = 17.17 \text{ V}$$

therefore

$$I = \left(\frac{1}{21.43} + \frac{1}{4} + 17.17\right)V = 17.47 \text{ V}$$

and

$$Z_{out} = \frac{V}{I} = 0.057 \text{ k}\Omega \text{ or } 57 \text{ }\Omega$$

Fig. 3.47 Circuit for output impedance calculation in Ex. 3.43.

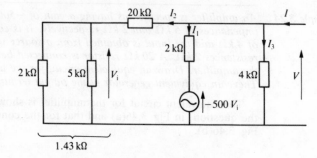

The required equivalent circuit at the output terminals is a generator of 9.54 mV in series with a 57 Ω resistor.

The four examples in this section illustrate the drawing of equivalent circuits from full circuits and descriptive information. The technique is the same in all cases, as will be seen if these examples are compared with the method listed on page 109. Further description of mesh and model analysis and of solution by determinants is given in the appendix at the end of the book. Many further examples will be found in later chapters, including various practical short-cuts and approximations.

You should now be able to attempt exercises 3.20 to 3.24.

A.C. analysis

The majority of analogue electronic systems are concerned with the processing of sinusoidal or a.c. signals. In many cases normal or mid-band operation is independent of signal frequency and waveform and the only interest is in the limiting conditions at extremes of signal frequency. In other circuits, selection of signal by frequency is used, for example in filters. In such cases, a.c. analysis is required to show variation of performance as the signal frequency is changed. This analysis is usually applied to investigate the variation of gain and phase shift but it may also be used to indicate the variation of an impedance level. The general term used for this form of information is frequency response information. Analysis at a spot or single frequency is required for some circuits, particularly those operating at a single (mains) frequency or those for which a resonant condition is used, such as oscillators.

The techniques for a.c. analysis of circuits include phasor diagram analysis and analysis using j notation. Phasor diagrams are useful for the explanation of a specific behaviour at spot frequencies but for anything more than simple circuit arrangements, the technique becomes most unwieldy. Complex algebra or j notation methods allow 'd.c.' relationships to be applied to a.c. circuits. These methods include series parallel combinations, mesh and nodal analysis. Thévenin and Norton's theorems and other simplifying theorems. These methods also allow an analysis leading to expressions in frequency response form which may then be used to provide a graphical representation of the frequency response. In this section, each of these techniques is examined and illustrated by example.

Phasor diagram analysis The voltage and current phasor relationships for individual components are discussed in Chapter 2. Such components may initially be connected in series or in parallel and then, with further connections, series parallel combinations result. For any combination, a reference phasor must be used: for series circuits, this will be the current that is common to all the series components; for parallel circuits, the common voltage is used as the reference. With further interconnections (e.g. two parallel circuits in series), the resultant of an initial combination may become the reference for further analysis. As phasors for both currents and voltages will be used, separate scales for the length of voltage and current phasors must be selected. The practice of phasor diagram analysis is best demonstrated by examples.

Example 3.25   *The input current I_{in} to the circuit shown in Fig. 3.48(a) is 8 mA at a frequency of 100 r/s. Use phasor diagram analysis to find the current in the 3.3 kΩ resistor.*

Fig. 3.48 Circuit and phasor diagram for Ex. 3.25.

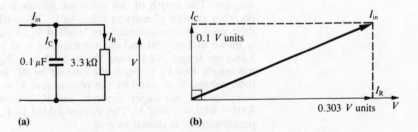

Since the two components are in parallel, the common reference must be the voltage across the circuit. This is labelled V, with the sense of measurement shown. The input current I_{in} and the component currents are also labelled. We also require the value of the capacitive reactance X_c at the signal frequency. This is given by $X_c = 1/\omega C = 1/(10^3 \times 10^{-7}) = 10 \text{ k}\Omega$. The required phasor diagram is shown in Fig. 3.48(b). The reference phasor V is not scaled for length as it is the only voltage phasor. The I_R phasor is in phase with the voltage and of length $V/3.3$ units or $0.303 V$ units. The I_C phasor *leads* V by 90° and has length $V/10$ or $0.1 V$ units (working in mA, V and kΩ). The phasor sum of these two currents will give the input or total current I_{in}. This is obtained by completing the parallelogram as shown.

The resultant can be measured or calculated by

$$I_{in} = \sqrt{(0.1)^2 + (0.303)^2} \; V$$

and since the current is given as 8 mA

$$8 = V\sqrt{(0.01 + 0.092)} = 0.319V$$

Hence

$$V = \frac{8}{0.319} = 25.1 \; V$$

The required current in the resistor is given by $I_R = 25.1/3.3 = 7.6$ mA and it lags I_{in} by $\tan^{-1} 0.1/0.303 = 18.3°$.

Example 3.26   *The voltage signal applied to the circuit shown in Fig.3.49 is given by 2 sin 10⁵t. By means of a phasor diagram, determine the input impedance and the voltage ratio V_O/V_i. Without calculation, deduce how these properties would vary with signal frequency.*

Fig. 3.49 *RLC* circuit for Ex. 3.26.

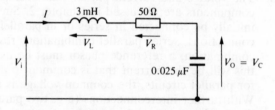

The inductive and capacitive reactances are obtained from the signal frequency and the component values.

$$X_1 = 3 \times 10^{-3} \times 10^5 = 300 \; \Omega$$

$$X_C = \frac{10^6}{0.025 \times 10^5} = 400\Omega$$

Since we have a series circuit, the current I is common to all parts of the circuit and will be the reference phasor. The sense of I is shown in the diagram. The length of this reference phasor is unimportant as there will be no other current phasors to relate to it. The various circuit voltages and their sense of measurement are labelled V_L, V_C and V_R. The phasor diagram is shown in Fig. 3.50 and the construction is as follows: V_R is in phase with I and has length $50I$ V using the scale shown. V_C lags the current by 90° and has length $400I$ V. V_L leads the current by 90° and has length $300I$ V. The total voltage V_i is given by the phasor sum $V_i = V_R + V_C + V_L$. This is completed in two stages; since V_L and V_C are 180° out of phase, $V_L + V_C = 300I - 400I = -100I$ V. This is then added to V_R by completing the parallelogram as shown to give

$$V_i = \sqrt{(100I)^2 + (500I)^2} \; \angle -\tan^{-1}\frac{100}{50} = 112I \; \angle -63.4°$$

Fig. 3.50 Phasor diagram solution to Ex. 3.26.

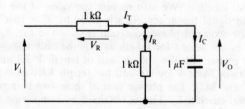

This result could have been obtained by direct measurement on the diagram. The required properties can now be found. Input impedance

$$Z_{in} = \frac{V_i}{I} = \frac{112I \angle -63.4°}{I} = 112 \angle -63.4° \; \Omega.$$

Voltage ratio (since $V_o = V_C$)

$$\frac{V_o}{V_i} = \frac{400 \angle -90°}{112I \angle -63.4°} = 3.57 \angle -26.5°$$

The current I is

$$\frac{V_i}{Z_{in}} = \frac{2 \sin 10^5 t}{112 \angle -63.4°}$$

or taking rms values, $I = 12.7 \angle 63.4°$ mA

As the angular frequency is increased, V_L will increase and V_C will reduce. At the resonant frequency ω_0, V_L and V_C will be equal and cancel leaving $Z_{in} = 50 \angle 0° \; \Omega$. This occurs at a frequency of 1.15×10^5 r/s at which $X_L = X_C = 346 \; \Omega$. Thus, although $V_i = 50I$, $V_o = 346I$, and $V_o/V_i = 346/50 = 6.92$. This is the maximum ratio and further increase in frequency makes the inductive reactance dominant and in the limit, $V_i = \omega LI$ while $V_o = 1/\omega CI$ so the ratio V_o/V_i tends to $1/\omega^2 LC$ which becomes very small at high frequencies.

At frequencies below resonance, X_C becomes large and X_L becomes small. In the limit here, at d.c., X_C is effectively open circuit, so $V_o = V_i$

Example 3.27   *The circuit shown in Fig. 3.51 has an a.c. applied voltage V_i at an angular frequency of 2000 r/s. By means of a phasor diagram, calculate the input output voltage ratio V_o/V_i and the input impedance. Without calculation state how these properties would change if the signal frequency was increased or decreased.*

Fig. 3.51 *RC* circuit for Ex. 3.27.

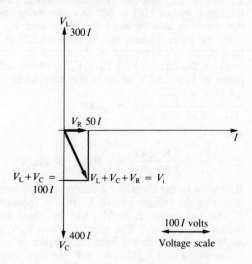

At the stated signal frequency, the capacitive reactance X_C is given by

$$X_C = \frac{1}{2000 \times 10^{-6}} = 500 \; \Omega$$

It is essential that the variables which are to be used are clearly labelled on the phasor diagram. These are the currents in the parallel circuit I_C and I_R and the total current I_T. V_o and V_i are specified in the problem but the voltage across the series resistor V_R is also required. These are all shown using the arrow notation.

The phasor diagram shown in Fig. 3.52 is constructed as follows. Starting with the parallel combination, choose a convenient length for the reference voltage phasor V_o. This may be referred to as a voltage unit, or a specific value of 1 V can be chosen if preferred. The current phasor I_R is now drawn in phase with V_o; the length relationship is arbitrary, but using the 1 V V_o, I_R will be 1 mA and the length chosen represents 1 mA. Since X_C is 0.5 kΩ, I_C will be 2 mA and it will lead I_R by 90°, with length twice as long as that for I_R.

Fig. 3.52 Phasor diagram solution to Ex. 3.27.

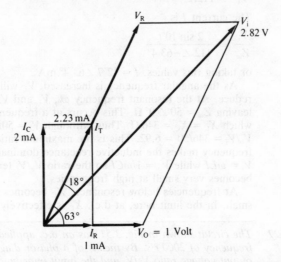

The two parallel currents I_R and I_C are now added on the phasor diagram by completing the parallelogram to give I_T, the length of which is $\sqrt{2^2 + 1^2} = 2.23$ mA (or by measurement). This current leads the original reference by $\tan^{-1} 2/1$ or 63°. For the series combination, this current I_T is the reference and V_R will be in phase with I_T. Since the voltage scale is twice the current scale, the V_R phasor (2.23 mA × 1 kΩ) must be twice as long as the I_T phasor and in phase with it.

From the circuit, $V_i = V_R + V_o$ by phasor addition and this is shown to result in $V_i = 2.82$ V which in this case leads V_o by 45°. Thus

$$\frac{V_o}{V_i} = \frac{1 \angle 0°}{2.82 \angle 45°} = 0.35 \angle -45°$$

and

$$Z_{in} = \frac{V_i}{I_T} = \frac{2.82 \angle 45°}{2.23 \angle 63°} = 1.26 \angle -18° \text{ kΩ}$$

Referring to the diagram, a change in frequency will initially only alter the length of the current phasor I_C. If ω is reduced, X_C is increased and I_C falls. This will reduce I_T and V_R and the phase angle between then and V_o. V_i is thus also reduced. The resulting ratio V_o/V_i is increased and its phase angle is reduced. In the limit, as ω tends to zero, all currents and voltages are in phase and V_o/V_i is 0.5 $\angle 0°$. At high frequencies, the currents I_C and I_T increase and, with them, V_i. The limiting case here is V_o tends to 0 $\angle 90°$.

The above examples using phasor diagram analysis do show a.c. voltage current relationships very clearly but they also show how complicated these techniques become when circuits involve more than two or three components. The solution by a phasor diagram is in general at a single frequency, although it is possible in some cases to construct locus diagrams to show phasors varying with frequency.

An alternative is to use complex algrebra or j notation which is the subject of the next section.

Analysis by j notation   Analysis of a.c. circuits by the use of j notation allows for the phasor relationship resulting from the combination of resistive and reactive elements. The analysis takes the form of writing normal circuit equations and solving by algebraic methods. The circuit equations can be written in the same way as those for d.c. circuits using Kirchhoff's law methods together with any of the simplifying theorems such as Thévenin or Norton. The algebraic methods follow normal rules with the following exceptions:

(i) 'Real' and 'j' terms cannot be directly added or subtracted; e.g. $3 + j2$ cannot be reduced further by addition.

(ii) Real and j terms can be multiplied and divided; e.g. $3 \times j2 = j6$ or $j10/2 = j5$.

(iii) j terms can be added or subtracted alone or in combination with 'real' terms; e.g. $3 + j2 - 2 + j4 = 1 + j6$.

(iv) If j terms are multiplied, the relationship $j^2 = -1$ is used:

$$(3 + j2)(-2 + j4) = -6 + j12 - j4 + j^2 8 = -6 + j8 - 8$$
$$= -14 + j8.$$

(v) Complex terms containing real and j components can be converted from **rectangular** to **polar** form or vice versa using the following relationships

$$a + jb = \sqrt{a^2 + b^2} \; \angle\tan^{-1} b/a$$

$$R \angle\theta = R \cos\theta + jR \sin\theta$$

(vi) Polar form complex terms cannot be added or subtracted directly (unless they have the same angles) but they can be multiplied or divided by multiplying or dividing the modulus and adding or subtracting the angle. For example

$$10 \angle 20° \times 3 \angle 90° = 30 \angle 110°$$

and

$$\frac{20 \angle 120°}{4 \angle 70°} = 5 \angle 50°$$

and

$$\frac{20 \angle -10°}{4 \angle -55°} = 5 \angle(-10° + 55°) = 5 \angle 45°$$

(vii) Division of complex terms in rectangular form is usually accomplished by converting to polar form and then, if necessary, reconverting the result for further manipulation. The alternative procedure is to multiply both numerator and denominator by the conjugate of the denominator. For example

$$\frac{3 + j6}{1 - j2} = \frac{(3 + j6)(1 + j2)}{(1 - j2)(1 + j2)} = \frac{3 + j6 + j6 - 12}{1 - j2 + j2 + 4}$$

$$= \frac{-9 + j12}{5} = -1.8 + j2.4 = 3 \angle 127°.$$

or by polar methods

$$\frac{3 + j6}{1 - j2} = \frac{6.708 \angle 63.5°}{2.236 \angle -63.5°} = 3 \angle 127°$$

As mentioned above, the writing of circuit equations follows identical methods to those used for d.c. In particular, the arrow notation for sense of voltages and currents must be strictly applied as this will not only relate the polarity of relationships but also determine the correct phase relationships (as was shown in phasor diagram analysis). It is also essential that if two or more signal sources are included, the phase relationship between them must be known and indicated using the complex notation ($a + jb$ or $R \angle \theta$). If this is not known, the problem is meaningless.

Analysis will be used for two general areas: solution of a circuit at one or more spot frequencies to otain a numerical answer for impedance, output voltage etc., or a more general solution in terms of ω which is then used to obtain the frequency response of the voltage ratio or impedance etc. The general solution can of course be used to obtain spot frequency answers if required.

We shall start by examining some examples of the first type, starting with the last problem solved by phasor diagram methods.

Example 3.28   *In the circuit shown in Fig. 3.53 the applied voltage V_i is at an angular frequency of 2000 r/s. Using the j notation methods, calculate the input impedance and the voltage ratio V_o/V_i.*

Fig. 3.53 *RC* circuit for Ex. 3.28.

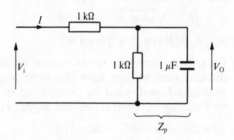

The capacitive reactance of the $1 \, \mu F$ capacitor is given by

$$X_C = \frac{1}{\omega C} = \frac{1}{2000 \times 10^{-6}} = 0.5 \, k\Omega$$

The impedance Z_P of the parallel combination is given by

$$Z_P = \frac{1 \times -j0.5}{1 - j0.5} = \frac{-j0.5(1 + j0.5)}{(1 - j0.5)(1 + j0.5)}$$

$$= \frac{0.25 - j0.5}{1.25} = 0.2 - j0.4 \, k\Omega$$

Total imput impedance $Z_{in} = 1 + 0.2 - j0.4 = 1.2 - j0.4$

$$= 1.26 \angle -18.4° \, k\Omega$$

By potential divider methods, the voltage ratio is given by

$$\frac{V_o}{V_1} = \frac{0.2 - j0.4}{1.2 - j0.4} = \frac{0.447 \angle -63.4°}{1.26 \angle -18.4°} = 0.35 \angle -45°$$

These answers and the method should be compared with the solution of example 3.26.

Example 3.29   *In the circuit in Fig. 3.54, the common signal frequency is such that the capacitive reactances have the values shown. Calculate the current flowing in the 5 kΩ resistor and hence the voltage V_x shown.*

Fig. 3.54 Circuit for
Ex. 3.29.

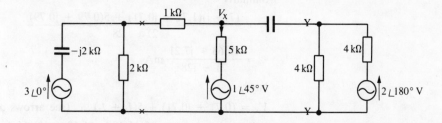

This circuit could be solved directly by using mesh analysis but, as this would result in four equations and four unknowns requiring the solution of a fourth-order determinant, some initial simplification is desirable.
Thévenin's theorem may be used to reduce those parts of the circuit to the left of XX and to the right of YY as follows:
To the left of XX

$$Z' = \frac{-j2 \times 2}{2 - j2} = \frac{-j4(2 + j2)}{2^2 + 2^2} = (1 - j) \text{ k}\Omega$$

$$V' = \frac{3 \times 2}{2 - j2} = \frac{6(2 + j2)}{2^2 + 2^2} = 1.5 + j1.5$$

(note that 3 ∠0° in j form is simply 3)
To the right of YY, we have a simple resistive voltage divider, and by inspection

$Z'' = 2 \text{ k}\Omega$ and $V'' = 1 \angle 180°$ or $1 \angle 0°$ with the sense arrow reversed.

Fig. 3.55 Simplified circuit
for Ex. 3.29 after
application of Thévenin's
theorem.

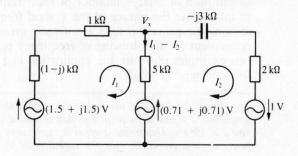

The simplified circuit is shown in Fig. 3.55. All three generators are shown in their rectangular or j form (1 ∠45° = 0.71 + j0.71). Circulating mesh currents are indicated and the circuit equations are

$$(1.5 + j1.5) - (0.71 + j0.71) = I_1(7 - j) - 5I_2$$

$$(0.71 + j0.71) + 1 = -5I_1 + (7 - j3)I_2$$

Simplifying

$$0.79 + j0.79 = I_1(7 - j) - 5I_2$$

$$1.71 + j0.71 = -5I_1 + (7 - j3)I_2$$

By determinants

$$I_1 = \frac{(0.79 + j0.79)(7 - j3) - (1.71 + j0.71)(-5)}{(7 - j)(7 - j3) - (-5)(-5)}$$

Multiplying out and collecting terms

$$I_1 = \frac{16.45 + j6.71}{21 - j28} \text{ mA}$$

Similarly

$$I_2 = \frac{(7 - j)(1.71 + j0.71) + 5(0.79 + j0.79)}{21 - j28}$$

$$= \frac{16.63 + j7.21}{21 - j28} \text{ mA}$$

Now,

$$V_X = (0.71 + j0.71) + 5(I_1 + I_2) \qquad \text{(see arrows on the circuit)}$$

$$= 0.71 + j0.71 + \frac{5(16.45 - 16.63 + j16.65 - j7.21)}{21 - j28}$$

which simplifies to $V_X = 0.769 + j0.67$ V.

Frequency response analysis

The last two examples demonstrate the use of j notation methods in the solution of a.c. problems at a single frequency. In electronic systems, we are usually concerned with a wide range of single frequencies or waveforms consisting of sums of single frequencies. Thus any circuit analysis that is used should be applicable to such input information. In other words, we should be looking for a general result into which values of frequency (ω or f) can be substituted to find the result for particular conditions. This is not sufficient though to appreciate the performance of the system, as substitution of a large number of spot frequencies would be required to investigate the nature of the system frequency response. For this reason, the general results are usually arranged in a form that is convenient for the drawing of frequency response curves. The next two examples illustrate the application of j notation analysis to this form of general result.

Example 3.30   *The sinusoidal signal V_i applied to the circuit shown in Fig. 5.56 has amplitude 2 V and a frequency ω that is varied between 200 r/s and 2×10^4 r/s. Determine a general expression for the output voltage and hence draw a graph showing the variation of V_o with frequency.*

Fig. 3.56 *RLC* circuit for Ex. 3.30.

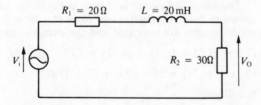

A general expression for the output voltage can be obtained by the simple potential divider relationship

$$V_o = V_i \times \frac{R_2}{R_1 + R_2 + j\omega L} \qquad \qquad [3.20]$$

The modulus of the ratio V_o/V_i is found,

$$\frac{V_o}{V_i} = \frac{R_2}{\sqrt{(R_1 + R_2)^2 + \omega^2 L^2}}$$

Values of R_1, R_2, L and ω could now be substituted into the result to give the required frequency response. The information can be more readily obtained if equation [3.20] is rearranged into a more convenient form. The numerator and denominator are divided by $(R_1 + R_2)$ and the required ratio is given by

$$\frac{V_o}{V_i} = \frac{\dfrac{R_2}{R_1 + R_2}}{1 + j\dfrac{\omega L}{R_1 + R_2}}$$

This expression has two constants, a ratio $R_2/(R_1 + R_2)$ and a **time constant** $L/(R_1 + R_2)$. In this application, we are concerned with frequency rather than time and it is usual to express the result as

$$\frac{V_o}{V_i} = \frac{K_o}{1 + j\omega/\omega_1}$$

where $\omega_1 = (R_1 + R_2)/L$. $K_o = R_2/(R_1 + R_2)$ is the basic circuit gain and ω_1 is a critical frequency associated with the frequency response of the circuit. Inserting values

$$K_o = \frac{30}{20 + 30} = 0.6 \quad \text{and} \quad \omega_1 = \frac{30 + 20}{0.02} = 2500 \text{ r/s}$$

Thus at a sample signal frequency of $= 3000$ r/s

$$\left|\frac{V_o}{V_i}\right| = \left|\frac{0.6}{1 + j\frac{3000}{2500}}\right| = \frac{0.6}{\sqrt{1 + (\frac{3}{2.5})^2}} = 0.384$$

also

$$\angle V_o/V_i = \tan^{-1}\left(\frac{3/2.5}{1}\right) = -50.2°$$

Results for a range of frequencies can be found in the same way to give the following values

ω r/s	2000	1000	3000	10^4	2×10^4
$|V_o/V_i|$	0.598	0.56	0.384	0.146	0.074
$\angle V_o/V_i$	$-4.6°$	$-21.8°$	$-50.2°$	-76	$-82.9°$

Since V_i is specified as 2 V, V_o is obtained by multiplying the V_o/V_i figures by 2. The required graph is shown in Fig. 3.57.

Fig. 3.57 Frequency response graph for Ex. 3.30.

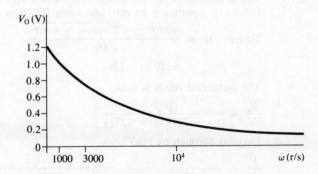

Example 3.31 *The frequency of the input V_i to the circuit shown in Fig. 3.58 varies over a wide range. Derive an expression for the gain V_o/V_i and illustrate your answer with graphs showing the gain and phase response for the circuit.*

Fig. 3.58 *RC* circuit for Ex. 3.31.

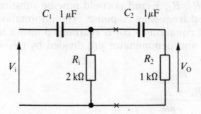

This network could be solved using mesh analysis but it is probably quicker to simplify the circuit using Thévenin's theorem and then obtain the required answer using a potential divider relationship.

Simplifying the circuit to the left of XX

$$V' = \frac{R_1}{R_1 + 1/j\omega C_1} \times V_i = \frac{j\omega C_1 R_1 V_i}{1 + j\omega C_1 R_1} = \frac{j\omega 2 \times 10^{-3} V_i}{1 + j\omega 2 \times 10^{-3}}$$

$$Z' = \frac{R_1/j\omega C_1}{R_1 + 1/j\omega C_1} = \frac{R_1}{1 + j\omega C_1 R_1} = \frac{2000}{1 + j\omega 2 \times 10^{-3}}$$

We now have a simple series circuit with V' applied to Z', C_2 and R_2 in series and V_o taken across R_2. Therefore

$$V_o = \frac{V' \times R_2}{Z' + 1/j\omega C_2 + R_2} = \frac{\dfrac{j\omega 2 \times 10^{-3}}{1 + j\omega 2 \times 10^{-3}} \times 1000 V_i}{1000 + \dfrac{1}{j\omega 10^{-6}} + \dfrac{2000}{1 + j\omega 2 \times 10^{-3}}}$$

Numerator and denominator are now multiplied by $j\omega 10^{-6}(1 + j\omega 2 \times 10^{-3})$ to clear the fractions

$$\frac{V_o}{V_i} = \frac{(j\omega)^2 \times 2 \times 10^{-6}}{(1 + j\omega 2 \times 10^{-3}) + 10^{-3} j\omega (1 + j\omega 2 \times 10^{-3}) + j\omega 2 \times 10^{-3}}$$

$$= \frac{(j\omega)^2 \times 2 \times 10^{-6}}{2 \times 10^{-6}(j\omega)^2 + 5 \times 10^{-3} j\omega + 1}$$

At this stage, $(j\omega)^2$ is not written $(-\omega^2)$ as the objective is a general expression for frequency response with a form similar to that used in the last example. Dividing through by 2×10^{-6}

$$\frac{V_o}{V_i} = \frac{(j\omega)^2}{(j\omega)^2 + 2500 j\omega + 5 \times 10^5}$$

The denominator is a quadratic in $j\omega$ which may be solved to find two factors by equating it to zero and using the 'formula'.

$$\text{Hence} \quad j\omega = \frac{-2500 \pm \sqrt{2500^2 - 2 \times 10^6}}{2000}$$

$$= -219 \text{ or } -2281$$

The factorised result is thus

$$\frac{V_o}{V_i} = \frac{(j\omega)^2}{(j\omega + 219)(j\omega + 2281)}$$

Dividing through by $(j\omega)^2$

$$\frac{V_o}{V_i} = \frac{1}{\left(1 + \dfrac{219}{j\omega}\right)\left(1 + \dfrac{2281}{j\omega}\right)} = \frac{1}{\left(1 - j\dfrac{219}{\omega}\right)\left(1 - j\dfrac{2281}{\omega}\right)}$$

which is the result for the gain in a convenient form for investigation of the frequency response. A sample calculation at $\omega = 500$ r/s gives

$$\frac{V_o}{V_i} = \frac{1}{(1 - j0.438)(1 - j4.562)} = \frac{1}{1.092 \angle -23.7° \times 4.67 \angle -77.6°}$$

$$= 0.196 \angle 101.3°$$

Other points are calculated in the same way to give the following results.

ω r/s	100	200	500	1000	2000	5000	10 000
$\left\lvert \dfrac{V_o}{V_i} \right\rvert$	0.0182	0.059	0.196	0.39	0.654	0.91	0.975
$\angle V_o/V_i$	+153°	+133°	+101°	+78.7°	+55°	+27°	+13°

The graph for these responses is shown in Fig. 3.59.

Fig. 3.59 Frequency responses for Ex. 3.31: (a) gain response; (b) angle response.

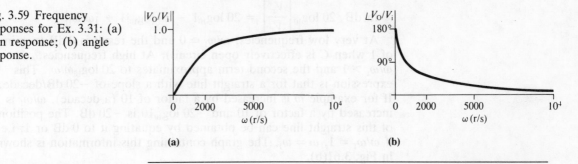

Bode plots for frequency response curves

In both the last two examples, further difficulties concerning the obtaining and the plotting of frequency response information have appeared. Obtaining points from general results is quite a tedious process and then, having got them, the graphs tend to compress the information at low frequencies and, in any case, can only conveniently represent a narrow range of gains or frequencies without loss of detail at the lower end. The problem of scales is overcome by the use of logarithmic scales for both frequency and gain. For frequency, a direct $\log f$ or $\log \omega$ is used. For voltage or current ratios, the dB scale is used where

$$\text{Number of dB} = 20 \log_{10} \left\lvert \frac{V_o}{V_i} \right\rvert$$

The resulting combined frequency responses are known as the **Bode plot** for the circuit or system.

A further advantage obtained by the use of log scales is that if the response due to individual terms is found, the overall reponse is given by the addition of the individual terms. These points may be illustrated by taking the result from the last example.

$$\frac{V_o}{V_i} = \frac{1}{\left(1 - j\dfrac{219}{\omega}\right)\left(1 - j\dfrac{2281}{\omega}\right)}$$

In dB, $20 \log_{10} \left\lvert \dfrac{V_o}{V_i} \right\rvert$

$$= 20 \log_{10} 1 - 20 \log_{10} \left\lvert 1 - j\frac{219}{\omega} \right\rvert - 20 \log_{10} \left\lvert 1 - j\frac{2281}{\omega} \right\rvert$$

Figure 3.60 shows the individual terms (broken lines) and the total Bode gain response.

Another aspect of the Bode plot is that a straight line approximation can be drawn very quickly without calculating a large number of points. All the general expressions arranged for frequency response investigation consist of terms such as $(1 + j\omega/\omega_1)$, $(-j\omega_2/\omega)$, $K/j\omega$ and constants. If straight line graphs can be drawn for each of these terms, then the total is obtained by addition of the individual graphs.

Consider the simple case shown in Fig. 3.61.

$$\frac{V_o}{V_i} = \frac{1/j\omega C}{R + 1/j\omega C} = \frac{1}{1 + j\omega CR} = \frac{1}{1 + j\omega/\omega_1} \qquad [3.21]$$

where $\omega_1 = 1/CR$.

In dB, $20 \log_{10}\left|\frac{V_o}{V_i}\right| = 20 \log_{10}1 - 20 \log_{10}|1 + j\omega/\omega_1|$

At very low frequencies, $\omega/\omega_1 \simeq 0$ and the result is 0 dB (a gain of 1 when C is effectively open circuit). At high frequencies, $\omega/\omega_1 > 1$ and the second term approximates to $20 \log_{10}\omega/\omega_1$. This expression is that for a straight line with a slope of -20 dB/decade. If for example ω is increased by a factor of 10 (a decade), ω/ω_1 is increased by a factor of 10 and $-20 \log_{10}10$ is -20 dB. The position of this straight line can be obtained by equating it to 0 dB or 1; i.e. if $\omega/\omega_1 = 1$, $\omega = \omega_1$. The graph containing this information is shown in Fig. 3.61(b).

This shows (as unbroken lines) the low and high frequency asymptotes that the true graph approaches when ω is much less or much greater than ω_1.

When $\omega = \omega_1$,

$$20 \log_{10}\left|\frac{V_o}{V_i}\right| = 20 \log_{10}1 - 20 \log_{10}|1 + j\omega/\omega_1|$$

$$0 \text{ dB} - 20 \log_{10}\sqrt{2} \text{ dB} = -3 \text{ dB}$$

When this point has been added to the graph, the remainder of the curve can be sketched in as shown with a broken line. The error between the straight line approximation and the true curve has a maximum error of 3 dB at the **break frequency** (corner frequency or 3 dB frequency) and the error is less than 1 dB, a ratio of 0.9, at half or double the break frequency. This means that for most applications the straight line approximation may be used with confidence instead of the true curve.

A similar approach may be made to the phase plot. From [3.21],

$$\frac{V_o}{V_i} = \tan^{-1}\omega/\omega_1$$

This may be plotted to a base of log ω with the result shown in Fig. 3.62. The angle approaches zero at very low frequencies, $-90°$ at very high frequencies, and is $-45°$ at the break frequency ω_1. The best straight line approximation to this S-shaped curve may be drawn between $0°$ at $\omega = 0.1\omega_1$ and $-90°$ at $\omega = 10\omega_1$. For lower and higher frequencies respectively, the angle is taken as $0°$ and $-90°$. Comparison between the approximation and the true curve in Fig. 3.62 shows maximum errors of about $6°$ in four places.

Fig. 3.60 The Bode gain frequency response for Ex. 3.31.

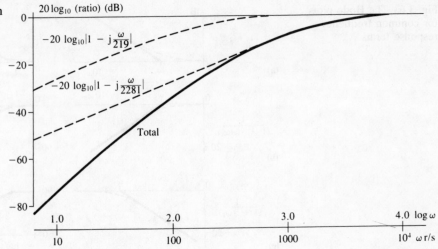

Fig. 3.61 The Bode gain plot for a simple *CR* circuit.

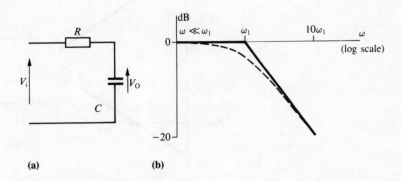

(a) (b)

Fig. 3.62 The Bode phase plot for a simple *CR* circuit.

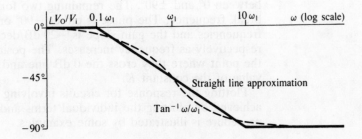

The analysis and description of other terms found in general frequency response expression follow the same pattern. The results are summarised in Fig. 3.63. The first three are similar in that they each have a frequency range with unity gain (0 dB) and a range of slope 20 dB/decade. These two ranges are separated by the break frequency ω_1. In (a) and (b), the gain is constant below ω_1 and then rises and falls respectively by 20 dB/decade for higher frequencies. Example (c) operates in the reverse manner with constant gain at

Fig. 3.63 The Bode plots
for common frequency
response terms.

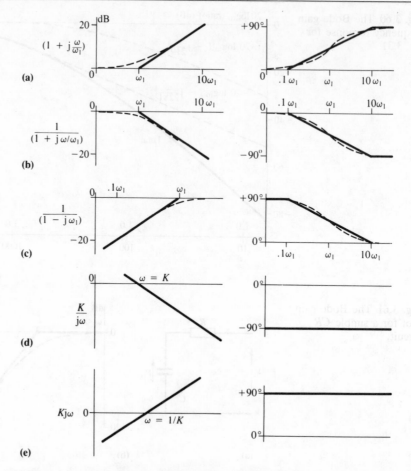

(a)

(b)

(c)

(d)

(e)

high frequencies and −20 dB/decade as the frequency falls below ω_1.
The pattern for the phase shift is similar for these three, varying
between 0° and ±90°. The remaining two forms (d) and (e) have no
break frequency. The phase is either −90° or +90° for all
frequencies and the gain slope is −20 dB/decade and +20 dB/decade
respectively as frequency increases. The position of these is fixed by
the point where they cross the 0 dB line and this depends upon the
value of the constant K.

Plotting the response for circuits involving several terms is
achieved by plotting the individual terms and then adding. This
procedure is illustrated by some examples.

Example 3.32   *Draw the Bode plots for each of the following voltage ratio expressions.
Hence, in each case, find the unity gain frequencies and the frequency at
which the phase shift is 0°.*

(a) $$\dfrac{700}{\left(1 - j\dfrac{50}{\omega}\right)\left(1 - j\dfrac{200}{\omega}\right)\left(1 + j\dfrac{\omega}{1400}\right)^2}$$

(b) $$\dfrac{0.2(1 + j\omega/100)}{(1 + j\omega/10^4)(1 + j\omega/2 \times 10^5)(1 + j\omega/5 \times 10^5)}$$

(c) $$\dfrac{100(1 + j\omega/500)}{(j\omega)(1 + j\omega/4000)(1 + j\omega/10^4)}$$

(d) $\dfrac{0.01j\omega}{(1 + j\omega/800)^3}$

The first step in drawing Bode plots is to choose suitable scales. For the frequency scale, it is sufficient to range from one-tenth of the lowest break frequency to ten times the highest break frequency. The gain scale is more difficult to estimate but a starting range of 20 dB per term is usually sufficient. For the phase plot, each term requires 90°, +90° for terms (a), (c) and (e) in Fig. 3.63 and −90° for terms (b) and (d).

(a) Frequency scale from 5 r/s to 1.4×10^4 r/s but it is more convenient to use complete decades, i.e. from 1 r/s to 10^5 r/s. Gain scale from +60 dB (allowing for the 700 gain which is equivalent to 56.9 dB). The four terms suggest a range of 80 dB so a minimum of −20 dB should be satisfactory. For the phase plot, there are two +90° terms and two −90° terms ((1 + $j\omega/1400)^2$); the required phase scale is thus from +180° to −180°.

The construction is shown in Fig. 3.64. For the gain plot, the individual terms have been drawn separately (as labelled) and then added to give the total. This could have been drawn more easily by starting at mid-band with a constant 56.9 dB and then by increasing the slope by 20 dB/decade at each break frequency. Thus, from 200 r/s to 50 r/s the slope is 20 dB/decade and below 50 r/s it is 40 dB/decade. Above 1400 r/s, the slope is 40 dB/decade for the double term.

The phase plot is constructed in a similar manner, with the individual terms drawn separately and then added to obtain the total. If necessary, this may be done on a point frequency basis; this is illustrated at 100 r/s as

Fig. 3.64 Bode plots for Ex. 3.32(a).

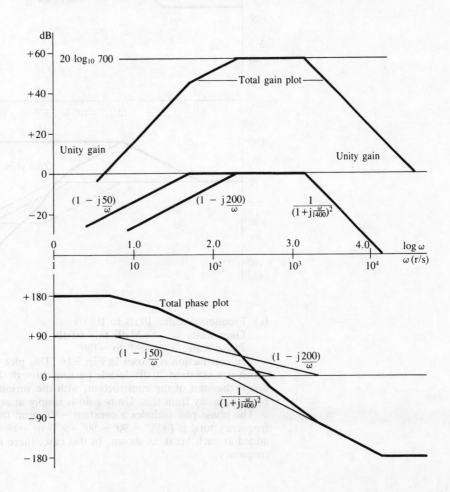

follows. The lowest frequency term provides $+32°$, the next term gives $+59°$, the high frequency term gives zero, here making the total $+91°$.

The unity gain frequencies are antilog 0.64 or 4.3 r/s and antilog 4.56 which is 3.6×10^4 r/s. These are taken from the points where the total gain crosses the 0 dB line. The log of ω is read from the graph and then converted. The zero phase frequency is taken from the phase plot where $\log \omega$ is 2.56 and ω is therefore 36.3 r/s.

(b) Frequency scale: 10 r/s to 10^7 r/s
     Gain scale: $+40$ dB to -20 dB
     Phase scale: $+90°$ to $-180°$

The construction is shown in Fig. 3.65 and follows the same steps as in part (a). From the graphs, unity gain occurs at antilog 2.7 or 500 r/s and antilog 6.15 or 1.4×10^6 r/s; zero phase occurs at antilog 4.6 or 4×10^4 r/s.

Fig. 3.65 Bode plots for Ex. 3.32(b).

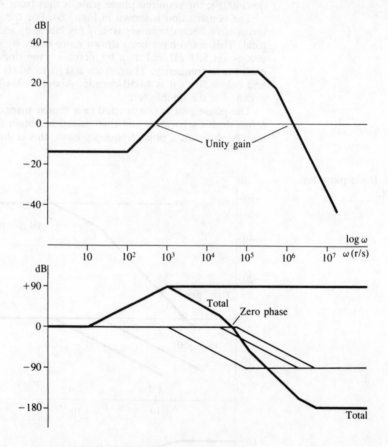

(c) Frequency scale: 10 r/s to 10^5 r/s
     Gain scale: $+20$ dB to -60 dB
     Phase scale: $+90°$ to $-180°$

The constuction is shown in Fig. 3.66. This plot includes the term $100/j\omega$ which is a constant 20 dB/decade passing through 0 dB at $\omega = 100$ r/s. This line is the start of the construction, with the various changes in slope breaking away from this. Unity gain is simply at $\omega = 100$ r/s.

The phase plot includes a constant $-90°$ term from $100/j\omega$. The high frequency total is $(+90° - 90° - 90° - 90°)$ or $-180°$, and the terms are added at each break as shown. In this case, there is no zero phase frequency.

Fig. 3.66 Bode plots for
Ex. 3.32(c).

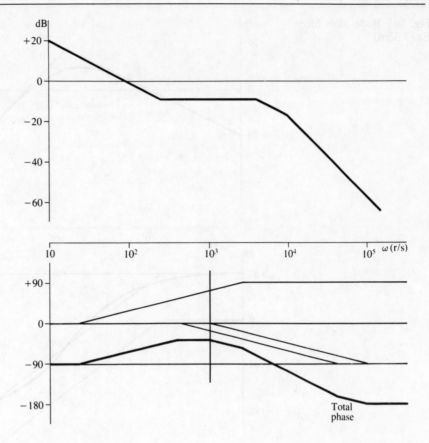

(d) Frequency scale: 10 r/s to 10^4 r/s
     Gain scale: +20 dB to −40 dB
     Phase scale: +90° to −180°
   There are only two straight lines in the approximate gain plot here; the
0.01jω with a constant of 20 dB per decade passing through 0 dB at
$\omega = 100$ r/s and −60 dB per decade for frequencies greater than 800 r/s.
Thus the total falls by 40 dB per decade above 800 r/s. In this case, the
error of the straight line approximation will be a maximum of 3×3 dB at
800 r/s. This point and the resulting true plot have been included in
Fig. 3.67. The unity gain frequencies are 100 r/s and 2187 r/s (from the true
curve).
   The phase plot changes from +90° to −180° and, once again, the cubed
term results in a larger maximum error of about 16°.
   The true curve is again sketched using this figure. The zero phase
frequency shows the maximum discrepancy between the approximate and
true plots with 355 r/s and 457 r/s respectively. In practical systems such
high order terms rarely occur but, where they do or where other
accumulation of errors occur, it is a simple matter to adjust from straight
line approximations as has been done in this example.

Fig. 3.67 Bode plots for
Ex. 3.32(d).

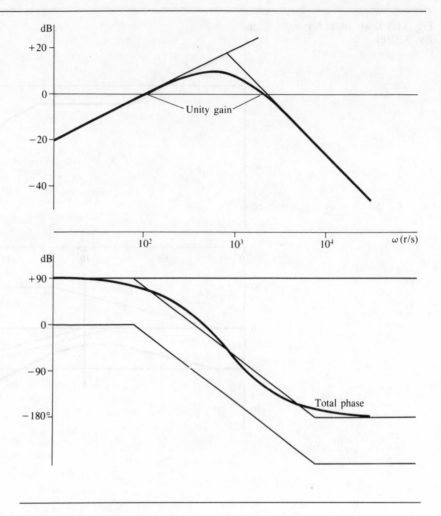

Polar plots   An alternative display of frequency response information is to
include both gain and phase on the same plot by drawing the locus
of the gain phasor. The resulting diagram is known as a **polar plot**
(or **Nyquist diagram** when applied to feedback systems). It has the
advantage of containing all the information on a single graph but
there are several disadvantages. The gain scale is linear, limiting the
range that can be displayed; points must be calculated or measured
as no straight line approximation technique is possible; the
frequency scale around the locus is non-linear. As an example, the
results from example 3.31 may be used. For convenience, these are
repeated here.

ω r/s	100	200	500	1000	2000	5000	10 000
$|A|$	0.018	0.059	0.196	0.39	0.654	0.91	0.975
$\angle A$	+153°	+133°	+101°	+78.7°	+55°	+27°	+13°

In Fig. 3.68, the gain phasors have been drawn for ω values of 10^4, 5000 and 2000 r/s. For the lower frequencies, only the end points are shown. Usual practice only shows the locus of the end points with a frequency scale as shown here. This polar plot has been extended to very high frequencies where the gain tends to $1 \angle 0°$. It can be seen that detail is lost at very low frequencies.

Information for polar plots can be taken from Bode plots by preparing a table of results as above and using antilog(Bode gain/20) for A and reading $\angle A$ directly.

Fig. 3.68 Polar plot for Ex. 3.31.

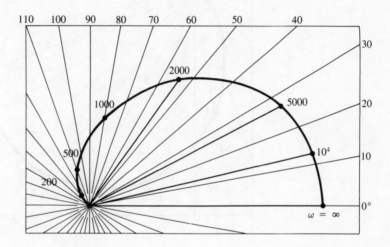

Working from Fig. 3.65 for example 3.32(b), we obtain (allowing for 3 dB at the break frequencies):

ω r/s	10	100	10^3	10^4	10^5	4×10^5	10^6	1.4×10^6	10^7
$\|A\|$	0.2	0.63	2	14.1	20	10	2	1	0.02
$\angle A$	0°	+45°	+80°	+45°	−45°	−99°	−135°	−147°	−180°

The polar plot for these results is shown in Fig. 3.69.

In this section, we have considered the anlysis of circuits with a.c. or steady-state sinusoidal signals. The alternatives of phasor diagram and j notation analysis have been demonstrated for single frequency problems. For variable frequency signals, frequency response methods have been compared using Bode and polar plotting methods.

You should now be able to attempt exercises 3.25 to 3.44.

Fig. 3.69 Polar plot for
Ex. 3.32(b).

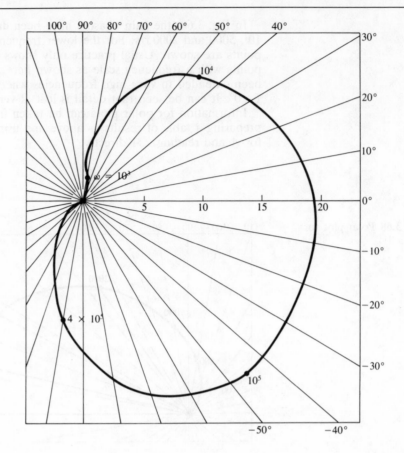

Transient analysis

In the analysis of electronic circuits, it is often convenient to
consider steady-state conditions. This means that the analysis
assumes constant a.c. or d.c. levels. In practice, such signals must
be switched ON or the level may be switched during normal
operation. Such sudden changes will cause transient signals which
die away to nothing, leaving only the new steady-state conditions. In
this section we will consider the analysis of the transient response
for some simple circuit arrangements. The technique depends upon
the instantaneous relationships for v and i in electrical components
(see Chapter 2). Since these are differential relationships, the
equations used will be differential equations. Full solution of such
equations would not be appropriate in this book but the results can
be examined and applied to a range of situations commonly
encountered in electronic circuits.

**Transients for series L
and R**
The circuit shown in Fig. 3.70 shows a series inductance and
resistance which can be switched to 0 or to a battery of E volts. At
the starting point of the analysis, the time is referred to as t_0 and in
this case, at $t = t_0$, the switch is moved to the battery connection.

Fig. 3.70 Circuit for *LR* transients.

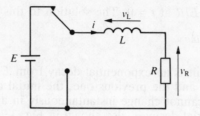

The equation for the circuit is

$$E = v_L + v_R$$

Expressing this in terms of the unknown current i

$$E = L\frac{di}{dt} + iR$$

which can be solved to give

$$i = \frac{E}{R}(1 - e^{-(R/L)t})$$

This shows that, at $t = 0$,

$$i = \frac{E}{R}(1 - 1) = 0$$

and at $t = \infty$,

$$i = \frac{E}{R}(1 - 0) = \frac{E}{R}$$

From the time at which the switch is closed, the current grows exponentially towards E/R. This is shown in Fig. 3.71. The time factor in this exponential growth of current is given by the circuit constant R/L or if taken as L/R, the circuit **time constant**. Now when $t = L/R$, the time constant

$$i = \frac{E}{R}(1 - e^{-1}) = 0.63\frac{E}{R}$$

Fig. 3.71 Transient waveforms for a series *LR* circuit.

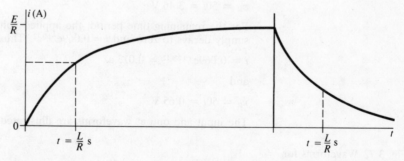

This very useful result shows that in an exponential growth, the value reached after one time constant is 0.63 of the final value. (In practice this can often be approximated to 2/3.)

If, after a long time, the switch is returned to zero, the equation can be rewritten

$$0 = L\frac{di}{dt} + iR$$

with $i = E/R$ at $t = 0$. The solution to this equation is

$$i = \frac{E}{R}e^{-(R/L)t}$$

This result is an exponential decay from E/R to zero. In both this situation and the previous one, the initial condition shows that the current cannot change instantaneously in an inductor. The exponential decay is also shown in Fig. 3.71; in this case, the current has fallen to $E/R(e^{-1})$ after one time constant, which is the initial value multiplied by 0.37 or approximately one-third.

The remaining problem is the situation if there is an initial current prior to the application of the battery E. This could occur if the switch was returned to E before i had decayed to zero. This may be illustrated by an example.

Example 3.33   *A circuit consists of a 6 mH inductance in series with a 50 Ω resistor. An applied d.c. signal is switched from 0 to +2 V at $t = 0$, from 2 V to 4 V at $t = 0.1$ ms and from 4 V to 0 at $t = 0.3$ ms. Calculate the variation in current and hence sketch the voltage waveform across the 50 Ω resistor.*

The circuit time constant is given by

$$\tau = \frac{L}{R} = \frac{6 \times 10^{-3}}{50} = 0.12 \text{ ms}$$

After the first time period, from 0 to 0.1 m s,

$$i = \tfrac{2}{50}(1 - e^{-0.1/0.12}) = 0.0226 \text{ A}$$

and $v_R = 50\,i = 1.13$ V. For the next time period between 0.1 and 0.3 ms, the possible change in current is from 0.0226 to 4/50 or 0.08 A. The exponential change is then given by

$$i = 0.0226 + (0.08 - 0.0226)(1 - e^{-t/0.12})$$

The next change occurs at $t = 0.3$ ms, but this is 0.2 ms after the beginning of this section. Therefore at $t = 0.3$ ms

$$i = 0.0226 + 0.0575(1 - e^{-0.2/0.12}) = 0.069 \text{ A}$$

and

$$v_R = 50i = 3.46 \text{ V}$$

For the remaining time period, the applied voltage is zero; the current now simply decays to zero and $i = 0.069e^{-t/0.12}$. Thus, after a further 0.2 ms,

$$i = 0.069e^{-0.2/0.12} = 0.013 \text{ a}$$

and

$$v_R = 50i = 0.65 \text{ V}$$

The input and output waveforms are illustrated in Fig. 3.72.

Fig. 3.72 Waveforms for the solution of Ex. 3.33.

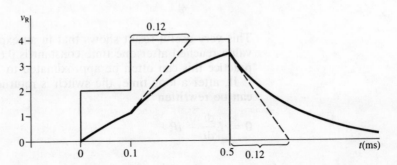

A useful aid in sketching results such as these is to note that if the initial slope after a change is continued in a straight line, it will reach the final value after one time constant. This is illustrated in Fig. 3.72 for each of the exponential transients.

CR transients

Figure 3.73 shows two arrangements with a series *CR* circuit switched to a d.c. voltage. In each case, the equation is the same for the time period following operation of the switch

$$E = v_R + v_C$$

$$= iR + \frac{1}{C}\int i \, dt$$

If this equation is solved, the following results are obtained,

$$i = \frac{E}{Re^{-t/CR}}$$

$$v_R = Ee^{-t/CR}$$

$$v_C = E(1 - e^{-t/CR})$$

If the switch is now returned to the zero position, the equation is

$$0 = v_R + v_C$$

$$= iR + \frac{1}{C}\int i \, dt \qquad\qquad [3.22]$$

Fig. 3.73 Circuits for series *CR* transients.

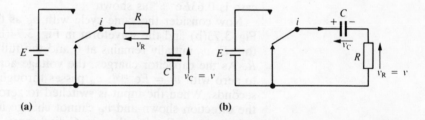

(a)   (b)

Under this condition the circuit diagram shows that the voltages across *C* and *R* are identical since the components are in parallel as viewed from the output terminals; however, v_R and v_C having appropriate senses of measurement making $v_R = -v_C$, solution of equation [3.22] gives

$$v = V_0 e^{-t/CR}$$

where v is the common output voltage with respect to ground and V_0 is the value of the common output voltage at the moment the switch has been changed. These relationships can be better appreciated by examining a cycle of events for the two circuits in Fig. 3.74.

The waveforms for the circuit in Fig. 3.73(a) are shown in Fig. 3.74(a). When the input is switched to *E*, the capacitor is uncharged and, since v_C cannot change instantaneously, the full *E* volts appears across *R* giving an initial charging current of *E/R*. As *C* charges, v_C increases and v_R and *i* get smaller. As the rate of charge reduces, v_C follows the exponential curve eventually to reach

Fig. 3.74 Transient
waveforms for series *CR*
circuits.

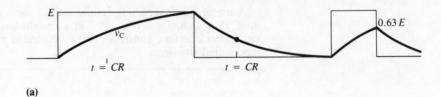

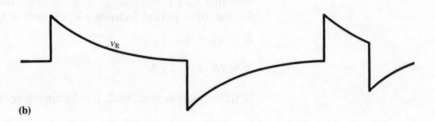

(a)

(b)

E. As with the inductor circuit the timing is determined by the **time
constant**, in this case *CR* seconds. Once again, v_C passes 0.63 of the
final value when *t* is equal to the time constant. In this example, the
switch remains at *E* for a time much longer than the time constant.
For practical purposes it may be considered that v_C reaches *E* in
about 5× the time constant.

When the switch is returned to zero, *C* discharges through *R*
following the simple exponential curve $Ee^{-t/CR}$. For the second cycle,
the switch remains at *E* for only about one time constant and *C*
charges to about $E(1 - e^{-1})$ or 0.63*E*. The discharge voltage in this
case is $0.63Ee^{-t/CR}$ as shown.

Now consider the same cycle with v_R as the output, as shown in
Fig. 3.73(b) and the waveform in Fig. 3.74(b). When the input rises
to +*E*, v_C initially remains at 0 and the full *E* volts appears across
R. As the capacitor charges, the voltage across *R* falls exponentially
to zero with $v_R = Ee^{-t/CR}$. v_R passes through 0.37*E* at *t* = *CR*
seconds. When the input is switched to zero, *C* is charged + − in
the direction shown and v_C cannot change instantaneously. v_R thus
switches to −*E* volts; then as *C* discharges, v_R returns exponentially
to zero by $-Ee^{-t/CR}$. This situation is best thought of as: 'any
switched voltage change at one side of a capacitor must appear
unchanged at the other side'. In this case, one side is switched from
+*E* to 0, a change of −*E* volts. The other side of the capacitor
(which is v_R) also changes initially by the same −*E* volts. During
the shorter pulse, v_R does not have time to fall to zero but the step
of −*E* still applies at the end of the pulse.

The transient changes discussed in this section can be applied to
other changes in voltage and to other circuit arrangements, as can
be illustrated by an example.

Example 3.34   *If the voltage waveform given in Fig. 3.75(a) is derived from the switched
circuit shown in (b), determine the waveform across the 2 kΩ resistor.*

When the switch is moved to the 3 V line, there will be no instantaneous
change in the capacitor voltage; the initial current will be 3/2.5 mA and v_o
will rise to (3 × 2)/2.5. The time constant is $2500 × 2 × 10^{-7}$ s or 0.5 ms, so
v_o will decay exponentially to zero by

$$v_o = \frac{3 \times 2}{2.5} e^{-t/5 \times 10^{-4}}$$

$$= 2.4 e^{-t/0.5}$$

where t is in milliseconds.

Fig. 3.75 Circuit and
waveforms for Ex. 3.34.

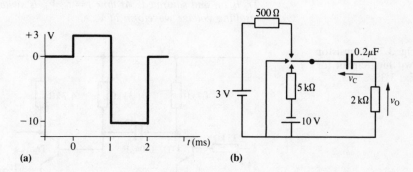

Fig. 3.75 Circuit and
waveforms for Ex. 3.34.

(a)   (b)

After 1 ms, $v_o = 2.4 e^{-2} = 0.325$ V. The voltage across the 500 Ω resistor will
have fallen in the same ratio to $0.325 \times 0.5/2 = 0.081$ V. Thus
$v_C = 3 - 0.325 - 0.081 = 2.59$ V, and the capacitor has nearly charged up to
the 3 V supply. The input now changes to -10 V but v_C cannot change
instantaneously so there will be a total of $-(10 + 2.59)$ V across the two
resistors with a new initial current of $12.59/(2 + 5)$ mA or 1.8 mA. The time
constant is now $7000 \times 2 \times 10^{-7}$ or 1.4 ms, so v_o is given by

$$v_o = -1.8 \times 2 e^{-t/1.4}$$

with t in milliseconds.

After 1 ms, $v_o = -1.76$ V

This time,

$$v_C = -10 + 1.76 \times 7/2 = -3.83 \text{ V}$$

For the last time period, when the switch is returned to zero, the capacitor
discharges from -3.83 V to zero with a time constant of $2000 \times 2 \times 10^{-7}$ or
0.4 ms.

$v_o = +3.83 e^{-t/0.4}$ with t in milliseconds.
The output waveform is shown in Fig. 3.76.

Fig. 3.76 Output waveforms
for Ex. 3.34.

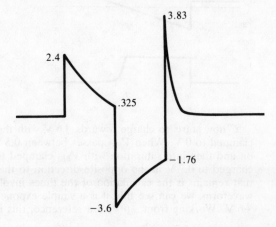

In electronic circuits, simple *CR* circuits are often associated with active devices such as transistors or op-amps. The *CR* circuit then controls the time for which the device is switched off. This is illustrated in the next example.

Example 3.35   *In the circuit shown in Fig. 3.77 the transistor Tr_1 is held off with V_{in} at 0 V. Tr_2 is on and saturated. At time $t = t_0$, V_{in} is switched to +6 V. Calculate the resulting voltage waveform at V_o.*

Fig. 3.77 Transistor switching circuit for Ex. 3.35.

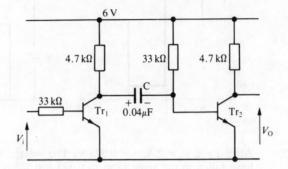

Before V_{in} is switched, C will be charged in the direction shown with the positive side at +6 V (no I_C, therefore no volt drop across the 47 kΩ resistor) and the negative side at 0.7 V (V_{BE2}). When V_{in} is switched, Tr_1 turns on and the collector falls to V_{CESAT} or approximately 0 V. The capacitor voltage cannot change instantaneously, so the *same change in voltage* appears at the other side, taking V_{BE} from +0.7 V to −5.3 V. This turns Tr_2 off with V_o rising to +6 V. This starting point is shown in Fig. 3.78

Fig. 3.78 Waveforms for Ex. 3.35.

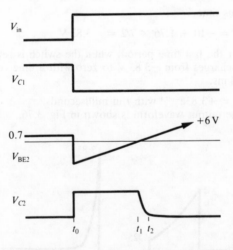

C now starts to charge towards +6 V with the original positive side clamped to 0 V. When V_{BE2} passes between 0.5 V and 0.7 V, Tr_2 is turned on and then into saturation with V_{BE2} clamped to 0.7 V. This leaves C charged to 0.7 V in the opposite direction to that shown in Fig. 3.95. All that remains is the calculation of the times involved. Re-examining the waveform, we can see that it is a simple exponential from −5.3 V towards +6 V. Working from +6 V as a reference, this is given by

$$v = -(6 + 5.3)e^{-t/CR}$$

where CR is the time constant $4 \times 10^{-8} \times 33 \times 10^3$ ms or 1.32 ms. Starting from t_0, at t_1 and t_2 respectively, V_{BE2} is $+0.5$ V and $+0.7$ V. But with reference to $+6$ V, these levels are -5.5 V and -5.3 V. Substituting in the above equation

$$-5.5 = -11.3e^{-t_1/1.32} \text{ and } -5.3 = -11.3e^{-t_2/1.32}$$

Rearranging

$$e^{+t_1/1.32} = \frac{11.3}{5.5} = 2.05 \qquad e^{+t_2/1.32} = \frac{11.3}{5.3} = 2.132$$

taking $\log_e$ of both sides

$$\frac{t_1}{1.32} = 0.72 \qquad \frac{t_2}{1.32} = 0.757$$

and

$$t_1 = 0.95 \text{ ms} \quad \text{and} \quad t_2 = 1.0 \text{ ms}$$

This analysis is not accurate for t_2 since as the transistor turns on the time constant will be shortened, reducing the time between t_1 and t_2.

Double time constant systems

In circuits including two capacitors and two resistors (or two inductors and two resistors), there will usually be a double time constant effect. Consider the circuit shown in Fig. 3.79 when the input voltage is switched from 0 V to $+E$ V.

Fig. 3.79 A double time constant RC circuit.

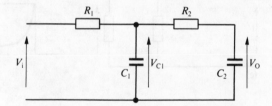

Initially, both capacitors are uncharged and the current in R_1 will be E/R_1. As V_{C1} starts to rise, the current through R_1 will get less and only part of this current charges C_1. The remainder flows through R_2 charging C_2. Finally, both capacitors will be fully charged to $+E$ V. A typical waveform for v_0 is shown in Fig. 3.80(b). The equation for the circuit has the form

$$v_0 = E(1 + Ae^{-t/\tau_1} - Be^{-t/\tau_2})$$

Fig. 3.80 The transient response for a double time constant circuit.

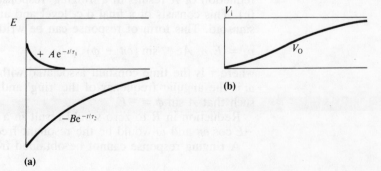

where τ_1 and τ_2 are the two circuit time constants. The two amplitude constants A and B will be such that $A = B = -1$. The time constants will be determined by C_1, C_2, R_1 and R_2 but they are not the simple C_1R_1 and C_2R_2 from the two parts of the circuit. The three components which make up the output waveform are shown in Fig. 3.80(a) and examination shows that they add up to make the waveform in (b). Other arrangements can lead to different combinations of the two time constant terms, with or without a final d.c. term depending if there is a series capacitor. Some example are shown in Fig. 3.81.

Fig. 3.81 Examples of double time constant circuits and their transient step responses.

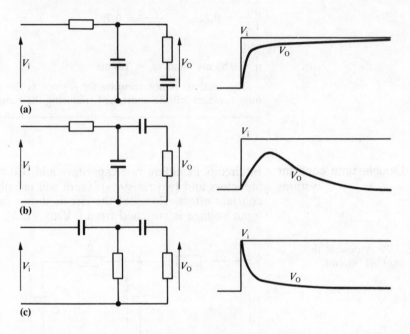

In systems involving inductance, capacitance and resistance, there will again be two time constants but if the loss is reduced (series R made smaller or parallel R made larger), another transient effect can be observed. This is illustrated in Fig. 3.82.

Each waveform is the voltage across the capacitor after the switch is closed. The first, (b), is a typical double time constant response with one short and one long time constant and a final d.c. level of E. As R is reduced, the time constants come closer together until they are equal as in (c) (this is said to be critical damping). Further reduction of R results in a 'ringing' response as shown in (d) and (e). This consists of a final d.c. level and an exponentially decaying sinusoid. This form of response can be written

$$v_o = E + Ae^{-t/\tau} \sin(\omega t + \phi)$$

where τ is the time constant associated with the exponential decay, ω is the angular frequency of the 'ring' and A and ϕ are constants such that $A \sin \phi = -E$.

Reduction in R to zero would result in a steady cosine $-E \cos \omega t$ and ω would be the resonant frequency $\omega_o = 1/\sqrt{LC}$.

A ringing response cannot be obtained from simple circuits

Fig. 3.82 (a) A double time constant circuit involving L and C; (b) overdamped response; (c) critically damped response; (d) and (e) underdamped responses.

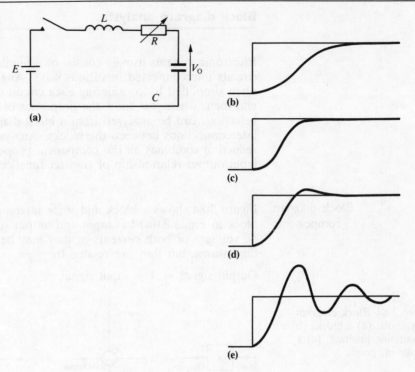

involving only capacitors and resistors or only inductors and resistors but, if such circuits are combined with an amplifier in a feedback arrangement, responses identical to those in Fig. 3.82 can be obtained. The circuit shown in Fig. 3.83 is an example of this, which will be used in Chapter 5. Its behaviour will be like that of Fig. 3.82(b) if the amplifier gain A is less than 1, but for A values between 1 and 3 there will be a ringing response like that in (d) and (e). For A greater than 3, the amplitude of the ring will grow until it is limited by the amplifiers saturating. The system is then said to be oscillating or unstable. This situation is discussed in detail in Chapter 5.

Fig. 3.83 A double time constant circuit involving amplifiers and feedback.

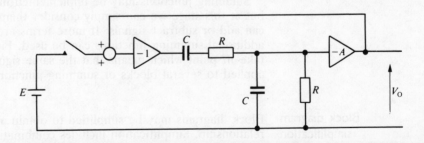

You should now be able to attempt exercises 3.35 to 3.37.

Block diagram analysis

Electronic systems usually consist of a number of sub-systems or circuits interconnected in various ways. Analysis of the system is often simplified by considering each circuit or sub-system as a block and then, when you know the properties of the blocks, the system behaviour can be analysed from a **block diagram** showing the interconnections between the blocks. Analysis by block diagram reduction combines all the component properties to give the system input/output relationship or transfer function.

Block diagram components

Figure 3.84 shows a block and some interconnecting elements. The block in Fig. 3.84(a) has input and output signals; these may both be voltages or both currents or they may be variables of different dimensions, but they are related by

Output signal $= A \times$ input signal

Fig. 3.84 Block diagram symbols: (a) a block; (b) a summing junction; (c) a take-off point.

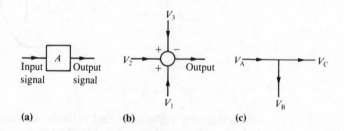

(a)   **(b)**   **(c)**

In this context, A can be referred to as a **transfer function** and it may be a simple gain ($+100$ or -0.2, for example) or it may be frequency dependent such as $20/(1 + j0.1\omega)$. Ideally, the value of A will be unaffected by any elements connected to the block but in practice any such loading effects can be included in the calculation of A. Fig 3.84(b) shows a **summing junction** which has the property of adding or subtracting signals. In the example shown, the output (the line with no $+$ or $-$) is given by

Output $= V_1 + V_2 - V_3$

Summing junctions may be implemented in a number of ways, but at this stage we can simply consider them to be devices which can add or subtract signals. If more terms are to be added, additional summing junctions can be used. Fig. 3.84(c) shows a **take-off point** which means that the same signal is taken off and applied to several blocks or summing junctions.

Block diagram simplification

Block diagrams may be simplified to obtain an overall input/output relationship. Simplification includes combination of cascaded or parallel blocks and elimination of feedback loops.

Cascaded blocks

Figure 3.85(a) shows three blocks cascaded. The output V_d is given by

$$V_d = V_c \times G_3 = (V_b \times G_2) \times G_3 = [(V_a \times G_1) \times G_2] \times G_3$$
$$= V_a \times G_1 G_2 G_3$$

Thus the gains of cascaded blocks can be multiplied together.

Fig. 3.85 Block diagram simplification: (a) cascaded blocks; (b) parallel blocks; (c) forward and feedback blocks.

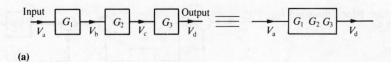

(a)

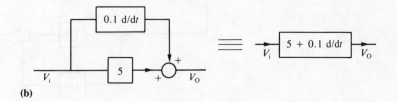

(b)

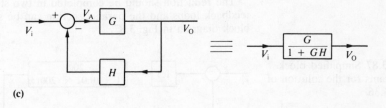

(c)

Parallel blocks   Figure 3.85(b) shows two blocks in parallel using a take-off point and a summing junction. In this example, one of the blocks involves differentiation and V_o is given by $V_o = 5V_i + 0.1 \, dV_i/dt$. By inspection, parallel blocks can be simply added.

Feedback loops   Figure 3.85(c) shows a system involving feedback. The input to block G is the sum of an external signal and a signal derived from the output signal. In this case, the output is inverted and multiplied by H before adding to the input V_i. The input to block G is given by

$$V_A = V_i - HV_o \text{ but } V_o = GV_A$$

Therefore

$$V_A = V_i - HGV_A$$

Collecting terms

$$V_A(1 + HG) = V_i$$

or $$V_A = \frac{V_i}{1 + HG}$$

and $$V_o = \frac{G}{1 + HG} \times V_i$$

Thus the arrangement can replaced by a single block as shown. If the summing junction had two + signs, the result would be $G/(1 - GH)$. The application of these reductions is best illustrated by an example.

Example 3.36   *Determine the overall voltage gain for the system shown in Fig. 3.86.*

Fig. 3.86 Block diagram for Ex. 3.36.

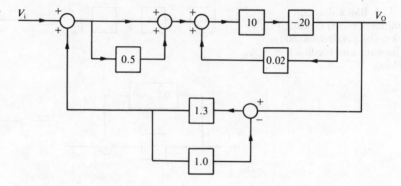

The reduction should be completed in two stages: first the two inner feedback loops and the parallel section can be simplified to result in the block diagram in Fig. 3.87.

Fig. 3.87 Simplified block diagrams for the solution of Ex. 3.36.

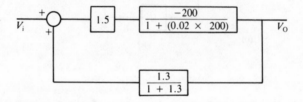

The simplified blocks are $\dfrac{-20 \times 10}{1 - (-20 \times 10 \times 0.02)} = -40$

$$\text{and } \frac{1.3}{1 + 1/3} = 0.565$$

The final ratio V_o/V_i is given by

$$\frac{V_o}{V_i} = \frac{-40 \times 1.5}{1 + 40 \times 1.5 \times 0.565} = -1.72.$$

You should now be able to attempt exercise 3.38.

Digital analysis

The concept of a logic gate has already been discussed. Gates may be grouped together in order to achieve a required switching function, and the arrangement is known as **combinational logic**. Rules as defined by George Boole (1815–1864) can be used to develop expression for a logical operation and these rules may also be applied for simplification of the logic expression and hence the logic circuit. The rules and their application will discussed in this section. The use of sum of products, products of sums and the AND-OR-INVERT (A-O-I) circuits will be investigated. The use of Karnaugh mapping techniques for simplifying logic circuits and the use of universal inverting logic will also be discussed.

Logic rules

Boolean algebra has evolved as a system where only two types of statement are allowable, namely **true** and **false**.

Using the concept of true and false allows the use of a truth table, an example of which is shown in Fig. 3.88(a) for a two-variable input.

If logic level 1 is used for true and logic level 0 for false, the table could be redrawn as shown in Fig. 3.88(b). This is in fact the truth table for a two-input OR gate using positive logic. The truth table in any row specifies the expected output logic state in terms of a particular input combination. Four rows are shown in Fig. 3.88 because two inputs are used, which give four possible combinations of those inputs, i.e. both inputs false, both inputs true, one input true and one false and vice versa. The table could be extended in terms of number of rows and columns to cater for a larger number of input variables. For example, three-input variables have eight possible combinations so the truth table in this case requires four columns and eight rows to allow for all input variations to give a particular output.

Using logic level 1 for true and logic level 0 for false allows the rules of Boolean algebra to be applied to the analysis of logic circuits. Some of the rules are as follows:

A	B	F
F	F	F
F	T	T
T	F	T
T	T	T

(a)

A	B	F
0	0	0
0	1	1
1	0	1
1	1	1

(b)

Fig. 3.88 Two-variable truth table for a logic function: (a) general table; (b) using positive logic.

AND and OR gates with an input of 1 or 0

A	B	F
0	0	0
0	1	0
1	0	0
1	1	1

(a)

A	B	F
0	0	0
0	1	1
1	0	1
1	1	1

(b)

Fig. 3.89 (a) truth table for a two-input AND gate; (b) truth table for a two-input OR gate.

$A.1 = A$ This is an ADD function with one of the inputs at logic 1. It follows from the truth table of the AND gate shown in Fig. 3.89(a) that the output is always the same as the A input.

$A.0 = 0$ For an AND circuit, if any input is at logic 0 the output must be at logic 0 regardless of the state of the other input (A).

$A + 1 = 1$ For an OR gate with one of the inputs always at logic 1, it follows that the output must always be a logic 1 regardless of the value of the other input (A). This can be verified from the truth table of the OR gate shown in Fig. 3.89(b).

$A + 0 = A$ The output of an OR gate is only a logic 1 when either input is a logic 1, hence the second input at logic 0 will have no effect.

Complementation laws

$A + \overline{A} = 1$ Because the function is a two-input OR gate with one input the complement of the other it follows that one input must always be 1 and hence the output must always be 1.

$A.\overline{A} = 0$ Because the function is a two-input AND gate with one input the complement of the other it follows that both inputs can never be simultaneously at 1, hence the output must always be 0.

Involution	$\overline{\overline{A}} = A$

Commutative law   This states that the order in which variables, or logic levels, appear in the equation is not important; i.e.

$$A + B = B + A$$
$$A.B = B.A$$

Distributive law $A + (B.C) = (A + B).(A + C)$
$$A.(B + C) = (A.B) + (A.C)$$

These rules are useful, as will be seen later, in giving expressions that are **product of sums** and **sum of products** respectively.

Associative law   This states that the order in which variables may be grouped in an equation is unimportant; i.e.

$$A + B + C = (A + B) + C = A + (B + C) = (A + C) + B$$
$$A.B.C = (A.B).C = (A.C).B = (B.C).A$$

Absorption law $A + A = A$

$$A + (A.B) = A$$

$$A.A = A$$

$$A.(A + B) = A$$

The absorption law is useful in complicated logic expressions since it allows some terms to be absorbed into others, thus simplifying the expression. The validity of the rules may be checked by reference to the truth tables for AND and OR gates or they may be checked by using some of the earlier rules. As an example, consider the expression

$$F = A + (A.B)$$

This may be written as

$F = A.1 + A.B$	since $A.1 = A$
then $F = A.(1 + B)$	from distributive law
and $F = A.1$	since $B + 1 = 1$ or $F = A$

so that

$$F = A + (A.B) = A$$

Negative absorption law $A + (\overline{A}.B) = A + B$

$$\overline{A} + (A.B) = \overline{A} + B$$

$$\overline{A}.(A + B) = \overline{A}.B$$

$$A.(\overline{A} + B) = A.B$$

Again, the validity of the rules may be checked using truth tables. As an example, consider the truth table shown in Fig. 3.90.

A	$\overline{A}$	B	$\overline{B}$	$\overline{A}.B$	$A+(\overline{A}.B)$	$A+B$
0	1	0	1	0	0	0
0	1	1	0	1	1	1
1	0	0	1	0	1	1
1	0	1	0	0	1	1

Fig. 3.90 Truth table to test the validity of the expression $A + (\overline{A}.B) = A + B$.

Since the last column of the truth table represents the function $A + B$ and the penultimate column the function $A + (\overline{A}.B)$ and the columns are identical it follows that

$$A + (\overline{A}.B) = A + B$$

Also, since $A.1 = A$ and $B + \overline{B} = 1$, then the expression $A + (\overline{A}.B)$ may be rewritten as

$$F = A.(B + \overline{B}) + (\overline{A}.B)$$

or $F = A.B + A.\overline{B} + \overline{A}.B$ by the distributive law

$F = A.B + A.\overline{B} + \overline{A}.B + A.B$ since $A.B + A.B = A.B$

$F = A.(B + \overline{B}) + B.(A + \overline{A})$ by the distributive law

$F = A.1 + B.1$ since $A + \overline{A} = 1$ and $B + \overline{B} = 1$

$F = A + B$ since $A.1 = 1$ and $B.1 = 1$

so that

$$A + (\overline{A}.B) = A + B$$

The other rules could be proved in the same way.

De Morgan's theorem   Used with inverting gates, De Morgan's theorem states that:

1. For a NAND function the result is the same as the function where the inverted inputs are OR-ed; i.e.

$$\overline{A.B} = \overline{A} + \overline{B}$$

or $\overline{A.B.C.D} = \overline{A} + \overline{B} + \overline{C} + \overline{D}$ etc.

2. For a NOR function the result is the same as the function where the inverted inputs are AND-ed; i.e.

$$\overline{A + B} = \overline{A}.\overline{B}$$

or $\overline{A + B + C + D} = \overline{A}.\overline{B}.\overline{C}.\overline{D}$ etc.

Again the results may be proved using a truth. Refer to the truth table of Fig. 3.91 for $\overline{A.B} = \overline{A} + \overline{B}$. The final column represents $\overline{A} + \overline{B}$ while the penultimate column represents $\overline{A.B}$, and since the columns are identical it follows that

$$\overline{A.B} = \overline{A} + \overline{B}$$

A	B	$\overline{A}$	$\overline{B}$	$\overline{A.B}$	$\overline{A}+\overline{B}$
0	0	1	1	1	1
0	1	1	0	1	1
1	0	0	1	1	1
1	1	0	0	0	0

Fig. 3.91 Truth table to test the validity of the expression $\overline{A.B} = \overline{A} + \overline{B}$

Application of logic rules   The logic elements described earlier can be used as the basis of a logic equation which defines a logic system required to perform a particular task. It may be possible using the rules outlined earlier in this section to reduce the circuit to a simpler form – a process known as **minimisation**. This process produces an expression which can be implemented by using the fewest number of gates. However, the minimised circuit may not be the 'best' circuit from a practical application point of view. Consider the following expression, for example:

$$F = A.(B + \overline{B}.C)$$

This equation could be satisfied by the circuit of Fig. 3.92. The

Fig. 3.92 Possible circuit to fulfil the required logic function $F = A.(B + \bar{B}.C)$

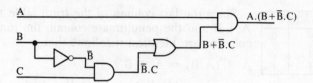

circuit would require an inverter, one two-input OR gate and two two-input AND gates to satisfy the logic expression *as it stands*. Using the logic rules the expression $F = A.(B + \bar{B}.C)$ could be minimised as follows:

$F = A.(B.1 + \bar{B}.C)$ since $B = B.1$

$F = A.(B.(C + \bar{C}) + \bar{B}.C)$ since $C + \bar{C} = 1$

$F = A.(B.C + B.\bar{C} + \bar{B}.C)$ by the distributive law

$F = A.(B.C + B.\bar{C} + \bar{B}.C + B.C)$ since $B.C + B.C = B.C$

$F = A.(B.(C + \bar{C}) + C.(B + \bar{B}))$ by the distributive law

$F = A.(B.1 + C.1)$ since $B + \bar{B} = 1, C + \bar{C} = 1$.

$F = A.(B + C)$ since $B.1 = B, C.1 = 1$

Thus the expression $F = A.(B + \bar{B}.C)$ can be reduced to $F = A.(B + C)$ and this circuit requires one two-input OR gate and one two-input AND gate as shown in Fig. 3.93.

Fig. 3.93 Minimised form of Fig. 3.92.

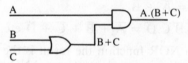

The circuit of Fig. 3.93 is obviously simpler and more economical in gates compared to the original circuit of Fig. 3.92. Yet it may be more convenient to use the original circuit, for example to utilise to the full spare circuits from other logic functions. Alternatively, there may be a requirement for an input of, say, $\bar{B}.C$ at another point in the system so that the circuit of Fig. 3.92 may be preferred for that reason. The selected curcuit is often a compromise between various factors. However, for the purposes of explanation of minimisation in this section, the minimal circuit will be developed regardless of other factors which could influence the final choice of circuit.

Example 3.37   *Using the logic algebra rules, simplify the expression:*

$F = (A + B).(A + C)$

Since $F = A.A + A.B + A.C + B.C$ using the distributive law

then $F = A + A.B + A.C + B.C$ since $A.A = A$

and $F = A.(1 + B) + A.C + B.C$ using the distributive law

$\quad\quad = A + A.C + B.C$ since $B + A = 1$ and $A.1 = A$

$\quad\quad = A.(1 + C) + B.C$ using the distributive law

$\quad\quad = A + B.C$ since $1 + C = 1$ and $A.1 = A$

Example 3.38   *Verify the results of example 3.37 using a truth table.*

Referring to the truth table of Fig. 3.94 it can be seen that the column for (A + B).(A + C) is identical to the column for A + B.C, so that the relationship established in example 3.37 is verified.

Fig. 3.94 Truth table to verify the expression (A + B).(A + C) = A + B.C

A	B	C	A+B	A+C	B.C	(A+B).(A+C)	A+B.C
0	0	0	0	0	0	0	0
0	0	1	0	1	0	0	0
0	1	0	1	0	0	0	0
0	1	1	1	1	1	1	1
1	0	0	1	1	0	1	1
1	0	1	1	1	0	1	1
1	1	0	1	1	0	1	1
1	1	1	1	1	1	1	1

The original circuit used in example 3.37 is shown in Fig. 3.95 while the simplified circuit is shown in Fig. 3.96. The simplified circuit shows a saving in the use of one of the original two-input OR gates.

Fig. 3.95 Logic circuit to produce the function F = (A + B).(A + C)

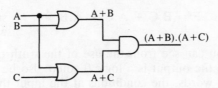

Fig. 3.96 Logic circuit to produce the function F = A + B.C

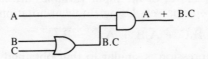

Sometimes circuits may be derived from the original design specification. If it is required, say, to design a circuit which sets an alarm if any two of three possible logic circuit inputs are present then, assuming the input and output states are specified by logic 1 level, the equation may be deduced by inspection as

F = A.B + A.C + B.C

(where A, B and C are inputs and F is the output) and the circuit may be arranged as shown in Fig. 3.97.

Fig. 3.97 Logic circuit for the function F = A.B + A.C + B.C

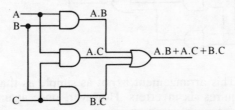

Alternatively the truth table could be used, and by writing down all the possible combinations of A, B and C, a logic 1 may be placed in the F output column if any two, or more, of the three input variables are at logic 1. This is shown in Fig. 3.98.

A	B	C	F	Circuit arrangement using AND gates
0	0	0	0	
0	0	1	0	
0	1	0	0	
0	1	1	1	B.C
1	0	0	0	
1	0	1	1	A.C
1	1	0	1	A.B
1	1	1	1	A.B.C

Fig. 3.98 Truth table to give logic 1 output if any two, or more, of the three input variables are at logic 1.

From Fig. 3.98 the required logic function could be written as a sum of products, in that the required function is obtained if any one *or* other of the listed arrangements is obtained; i.e.

$$F = B.C + A.C + A.B + A.B.C$$

The result seems more complicated than the one obtained by inspection. However, this is only so because of the extra term A.B.C which allows for *all* inputs being at logic level 1 to give a logic 1 output. The longer expression can be reduced to the shorter expression by using the Boolean algebra rules. Since $A + A = A$, there is no loss of generality to write

$$F = B.C + A.C + A.B + A.B.C + A.B.C + A.B.C$$

from which

$$F = B.C.(A + 1) + A.C.(B + 1) + A.B.(C + 1)$$

from the distributive law

and $F = B.C.1 + A.C.1 + A.B.1$ since a variable OR-ed with 1 = 1

$$F = B.C + A.C + A.B$$ since a variable AND-ed with 1 = the variable

You can see from the use of the truth table that the condition when the output is a logic 0 could also be used to set the alarm. In other words, the conditions at the input that occur such that any two, or more, of the input variables are *not* logic 1 then the alarm should *not* be set. Using Boolean notation this could be written as

$$\bar{F} = \bar{A}.\bar{B}.\bar{C} + \bar{A}.\bar{B} + \bar{A}.\bar{C} + \bar{B}.\bar{C}$$

The expression is similar to that obtained earlier, except that all terms are individually inverted, and it follows that the same use of Boolean algebra rules could be used to simplify the expression to

$$\bar{F} = \bar{A}.\bar{B} + \bar{A}.\bar{C} + \bar{B}.\bar{C}$$

or $F = \overline{\bar{A}.\bar{B} + \bar{A}.\bar{C} + \bar{B}.\bar{C}}$

The circuit could be built as shown in the equation to give the circuit of Fig. 3.99.

Fig. 3.99 One form of logic circuit to give the expression
$\bar{F} = \bar{A}.\bar{B} + \bar{A}.\bar{C} + \bar{B}.\bar{C}$

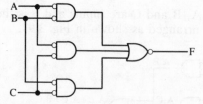

This arrangement is not as simple as that of Fig. 3.98 since it requires six inverters. However, once again the equation

$$F = \overline{\bar{A}.\bar{B} + \bar{A}.\bar{C} + \bar{B}.\bar{C}}$$

can be simplified, since

$$F = \overline{\overline{A + B} + \overline{A + C} + \overline{B + C}}$$ by De Morgan's theorem

and $F = (A + B).(A + C)(.(B + C)$ by De Morgan's theorem

This shows that the desired result can be obtained using a product of sums expression. The circuit for this is shown in Fig. 3.100. Note the absence of any inverters.

Fig. 3.100 Basic product of sums circuit for a given function. Compare with Fig. 3.99.

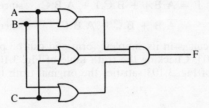

The truth table for the equation

$$F = (A + B).(A + C).(B + C)$$

is shown in Fig. 3.101 and shows that the circuit of Fig. 3.100 is completely equivalent to that of Fig. 3.97.

Fig. 3.101 Truth table to verify that the product of sums solution for a logic function is completely equivalent to the sum of products solution.

A	B	C	A+B	A+C	B+C	(A+B).(A+C).(B+C)
0	0	0	0	0	0	0
0	0	1	0	1	1	0
0	1	0	1	0	1	0
0	1	1	1	1	1	1
1	0	0	1	1	0	0
1	0	1	1	1	1	1
1	1	0	1	1	1	1
1	1	1	1	1	1	1

In fact the equation $F = (A + B).(A + C).(B + C)$ can be reduced to the original sum of products form by the use of the Boolean algebra rules; i.e.

$F = (A.A + A.B + A.C + B.C).(B + C)$ by distributive law

$\quad (A + A.B + A.C + B.C).(B + C)$ since $A.A = A$

$\quad (A.(1 + B) + A.C + B.C).(B + C)$ by distributive law

$\quad (A.1 + A.C + B.C).(B + C)$ since $1 + B = 1$

$\quad (A.(1 + C) + B.C).(B + C)$ by distributive law

$\quad (A.1 + B.C).(B + C)$ since $C + 1 = 1$

$\quad (A + B.C).(B + C)$ since $A.1 = A$

$\quad A.B + B.B.C + A.C + B.C.C$ by distributive law

$\quad A.B + B.C + A.C + B.C$ since $B.B = B$ and $C.C = C$

so that $F = A.B + A.C + B.C$ since $B.C + B.C = B.C$

Example 3.39   *A logic circuit is to be designed to satisfy the truth table shown in Fig. 3.102. Obtain the logic expression for the circuit from the truth table and simplify the expression if possible.*

From the truth table the sum of products solution that gives a logic 1 output is given by

$$F = \overline{A}.B.\overline{C} = \overline{A}.B.C + A.\overline{B}.C + A.B.\overline{C}$$

and this may be simplified, since

A	B	C	F
0	0	0	0
0	0	1	0
0	1	0	1
0	1	1	1
1	0	0	0
1	0	1	1
1	1	0	1
1	1	1	0

Fig. 3.102 Truth table for a logic function as defined in example 3.39.

Fig. 3.103 Logic circuit to satisfy the requirements of example 3.39.

$F = \overline{A}.B.(\overline{C} + C) + B.\overline{C}.(\overline{A} + A) + A.\overline{B}.C$ by distributive law

(The term $\overline{A}.B.\overline{C}$ has been used twice to simplify the equation. This is possible since $A + A = A$ so that $\overline{A}.B.\overline{C} = \overline{A}.B.\overline{C} + \overline{A}.B.\overline{C}$)

Hence $F = \overline{A}.B.1 + B.\overline{C}.1 + A.\overline{B}.C$ since $A + \overline{A} = 1$ and $C + \overline{C} = 1$

$\overline{A}.B + B.\overline{C} + A.\overline{B}.C$ since $A.1 = A$

This circuit is in its simplest form and can be produced as the circuit of Fig. 3.103. Checking the truth table of Fig. 3.104 shows that the simplified circuit of Fig. 3.103 satisfies the original truth table.

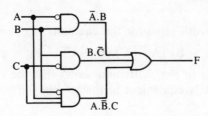

Fig. 3.104 Truth table comparison to show equivalence of original and minimised circuits of example 3.39.

A	B	C	A.B	B.C	A.B.C	A.B + B.C. + A.B.C
0	0	0	0	0	0	0
0	0	1	0	0	0	0
0	1	0	1	1	0	1
0	1	1	1	0	0	1
1	0	0	0	0	0	0
1	0	1	0	0	1	1
1	1	0	0	1	0	1
1	1	1	0	0	0	0

The alternative solution is to obtain an expression where the output is logic 0 or $\overline{F} = 1$; i.e.

$\overline{F} = \overline{A}.\overline{B}.\overline{C} + \overline{A}.\overline{B}.C + A.\overline{B}.\overline{C} + A.B.C$

Again the expression can be simplified

$\overline{F} = \overline{B}.\overline{C}.(\overline{A} + A) + \overline{A}.\overline{B}.(\overline{C} + C) + A.B.C$
    by distributive law

$= \overline{B}.\overline{C} + \overline{A}.\overline{B} + A.B.C$ since $\overline{A} + A = 1$ and $A.1 = A$

and $F = \overline{\overline{B}.\overline{C} + \overline{A}.\overline{B} + A.B.C}$

$= \overline{\overline{B} + \overline{C} + \overline{A} + \overline{B} + A.B.C}$
    by De Morgan's theorem

$= (\overline{\overline{B} + \overline{C}}).(\overline{\overline{A} + \overline{B}}).(\overline{A.B.C})$
    by De Morgan's theorem

$= (B + C).(A + B).(\overline{A} + \overline{B} + \overline{C})$ by De Morgan's theorem

$= (A.B + A.C + B.B + B.C).(\overline{A} + \overline{B} + \overline{C})$
    by distributive law

$= (A.B + A.C + B + B.C).(\overline{A} + \overline{B} + \overline{C})$
    since $B.B = B$

$= A.\overline{A}.B + A.\overline{A}.C + \overline{A}.B + \overline{A}.B.C$
$+ A.B.\overline{B} + A.\overline{B}.C + B.\overline{B} + \overline{B}.B.C$
$+ A.B.\overline{C} + A.C.\overline{C} + B.\overline{C} + B.\overline{C} + B.C.\overline{C}$
    by distributive law

$$= 0 + 0 + \overline{A}.B + \overline{A}.B.C + 0 + A.\overline{B}.C$$
$$+ 0 + 0 + A.B\overline{C} + 0 + B.\overline{C} + 0$$

$$= \overline{A}.B + \overline{A}.B.C + A.\overline{B}.C + A.B.\overline{C} + B.\overline{C}$$

$$= \overline{A}.B.(1 + C) + B.\overline{C}.(1 + A) + A.\overline{B}.C$$

by distributive law

$$F = \overline{A}.B + B.\overline{C} + A.\overline{B}.C \qquad \text{since } (1 + C) = 1 \text{ and } A.1 = A \text{ etc.}$$

Example 3.40

A	B	C	F
0	0	0	0
0	0	1	1
0	1	0	0
0	1	1	1
1	0	0	1
1	0	1	0
1	1	0	1
1	1	1	1

Fig. 3.105 Truth table for a logic function as defined in example 3.40.

Fig. 3.106 Solution for example 3.40: (a) Sum of products; (b) Product of sums.

From the truth table of Fig. 3.105 derive a logic circuit to satisfy the requirements of the truth table. (i) using the sum of products, (ii) using the product of sums.

Since $F = \overline{A}.\overline{B}.C + \overline{A}.B.C. + A.B.\overline{C} + A.B.C$

$$F = \overline{A}.C.(\overline{B} + B) + A.B.(\overline{C} + C) \text{by distributive law}$$

$$= \overline{A}.C + A.B \qquad \text{since } (B + B) = 1 \text{ and } A.1 = A \text{ etc.}$$

This the sum of products solution in its simplest form. The circuit is shown in Fig. 3.106(a)

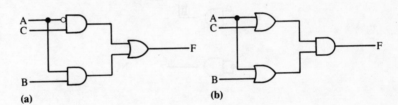

(a)   **(b)**

Alternatively

$$\overline{F} = \overline{A}.\overline{B}.\overline{C} + \overline{A}.B.\overline{C} + A.\overline{B}.\overline{C} + A.\overline{B}.C$$

$$= \overline{A}.\overline{C}.(\overline{B} + B) + A.\overline{B}.(C + \overline{C})$$

by distributive law

$$= \overline{A}.\overline{C} + A.\overline{B} \qquad \text{since } (B + B) = 1 \text{ and } A.1 = A \text{ etc}$$

and $\quad F = \overline{\overline{A}.\overline{C} + A.\overline{B}}$

$$= \overline{\overline{(A + C)} + \overline{(A + \overline{B})}} \qquad \text{by De Morgan's theorem}$$

$$= \overline{\overline{(A + C)}}.\overline{\overline{(A + \overline{B})}} \qquad \text{by De morgan's theorem}$$

$$F = (A + C).(A + \overline{B})$$

The product of sums circuit is shown in Figure 3.106(b)

You should now be able to attempt exercises 3.39 to 3.41.

The use of inverting logic

Example 3.38 shows that a logic function can be implemented using either the product of sums (OR gates into an AND gate) or the sum of products (AND gates into an OR gate) solution. There are advantages to be gained in using inverting gates since they can produce amplification of the logic signal. Such gates include NAND, NOR or simply the inverting NOT gate. The sum of products solution is most easily implemented using NAND gates while the product of sums solution is best achieved with NOR gates. An alternative is to use A-O-I (AND-OR-INVERT) which is AND gates followed by a NOR gate and for which integrated circuits are readily available. An example of such an arrangement was shown in Fig. 3.98. The circuit arrangements using NAND gates to implement the basic AND, OR and NOT functions are shown in Fig. 3.107 while the circuit arrangements using NOR gates to implement the basic AND, OR and NOT gates are shown in Fig. 3.108.

Fig. 3.107 Use of NAND gates to produce the basic AND, OR and NOT functions.

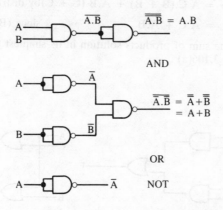

Fig. 3.108 Use of NOR gates to produce the basic AND, OR and NOT functions.

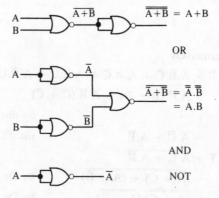

Conversion of any logic circuit to one using NAND and NOR gates only is easily done using the equivalent circuits of Fig. 3.107 and 3.108 respectively.

Example 3.41 *Convert the circuit of Fig. 3.92 to one using (i) NAND gates only, and (ii) NOR gates only.*

(i) The individual gates of Fig. 3.92 can be converted to NAND gates only using the equivalent circuits of Fig. 3.107, as shown in Fig. 3.109(a).

(ii) The original circuit of Fig. 3.92 can be converted to NOR gates only using the equivalent circuits of Fig. 3.108 as shown in Fig. 3.109(b).

Fig. 3.109 Solution of logic requirement of example 3.41: (a) using NAND gates only; (b) using NOR gates only; (c) minimised NOR gate circuit after removal of redundant gates.

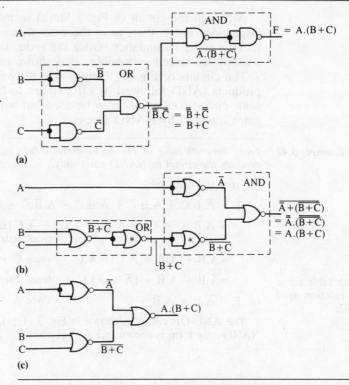

(a)

(b)

(c)

A check of Fig. 3.109(b) shows that two of the NOR gates are in fact redundant. Gates marked * could simply be replaced by a piece of wire without altering the logic equation at any point in the circuit. Thus the circuit could be simplified to the circuit of Fig. 3.109(c) which requires only four NOR gates.

It can be seen from any of the circuits in Fig. 3.109 that the original circuit of Fig. 3.92 can be replaced by one using inverting gates only while still giving the correct logic function requirement at the output. Although not economical in the use of gates there are advantages to be gained in the use of gates of only one type. Gates in integrated circuit form are available with several gates per chip, i.e. a quad two-input NOR gate chip would contain four of the two-input NOR gates on a single chip. Thus the circuit of Fig. 3.109(c) could be built using a single IC.

Some circuit arrangements of AND and OR gates do, however, lend themselves to conversion to NAND or NOR only forms without an increase in the number of gates required. Consider the circuit of Fig. 3.106(a) and convert it to a NAND-only circuit arrangement using the techniques of Fig. 3.107 for the conversion. The result is shown in Fig. 3.110(a).

Fig. 3.110 (a) NAND gates only solution to the circuit of Fig. 3.106(a); (b) Minimised version after removal of redundant gates.

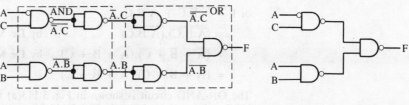

(a)

(b)

Although the circuit of Fig. 3.110(a) seems more complicated, inspection reveals that, as in the case of the circuit in Fig. 3.109(b), there is some redundancy. Once the redundant gates are eliminated a simplified circuit is-possible, as is shown in Fig. 3.110(b).

The circuits of Fig. 3.110 show that to convert the sum of products (AND followed by OR) circuit to NAND-only requires the same circuit configuration to be used but with the AND and OR gates replaced by NAND gates.

Example 3.42

From the truth table of Fig. 3.111 derive a sum of products logic circuit and produce the circuit in NAND gates only.

A	B	C	F
0	0	0	0
0	0	1	1
0	1	0	1
0	1	1	1
1	0	0	1
1	0	1	1
1	1	0	0
1	1	1	1

Fig. 3.111 Truth table for a required logic function used in example 3.42.

From the truth table

$$F = \overline{A}.\overline{B}.C + \overline{A}.B.\overline{C} + \overline{A}.B.C + A.\overline{B}.\overline{C} + A.\overline{B}.C + A.B.C$$

$$= \overline{A}.B.(\overline{C} + C) + A.\overline{B}.(\overline{C} + C) + \overline{A}.C.(\overline{B} + B) + A.C.(\overline{B} + B)$$
from distributive law

$$= \overline{A}.B + A.\overline{B} + \overline{A}.C + A.C \qquad \text{since } \overline{C} + C = 1 \text{ and A.1} = 1 \text{ etc.}$$

$$= \overline{A}.B + A.\overline{B} + (\overline{A} + A).C \qquad \text{from distributive law}$$

or $F = \overline{A}.B + A.\overline{B} + C \qquad \text{since A + A = 1 and A.1 = 1 etc.}$

The AND-OR circuit is shown in Fig. 3.112(a). The circuit converted to NAND-only form is shown in Fig. 3.112(b).

Fig. 3.112 Solution of example 3.42: (a) Sum of products; (b) Sum of products using NAND gates.

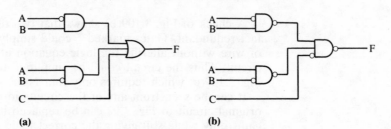

(a)   **(b)**

In this case, because there is a single-term variable input to the OR gate, this term must be inverted in the NAND-only circuit prior to its application to the output NAND gate. Use of the rules for changing AND-OR circuits to NAND-only form as shown in Fig. 3.107 will show this point.

Example 3.43

Using the truth table of Fig. 3.111 derive a product of sums logic circuit in NOR gates only.

From the truth table

$$\overline{F} = \overline{A}.\overline{B}.\overline{C} + A.B.\overline{C}$$

or $F = \overline{\overline{A}.\overline{B}.\overline{C} + A.B.\overline{C}}$

$$= (\overline{\overline{A}.\overline{B}.\overline{C}}).(\overline{A.B.\overline{C}}) \qquad \text{by De Morgan's theorem}$$

$$= (\overline{\overline{A}} + \overline{\overline{B}} + \overline{\overline{C}}).(\overline{A} + \overline{B} + \overline{\overline{C}}) \qquad \text{by De Morgan's theorem}$$

$$= (A + B + C).(\overline{A} + \overline{B} + C)$$

The OR-AND circuit is shown in Fig. 3.113(a) while the conversion to NOR-only gates is shown in Fig. 3.113(b).

Fig. 3.113 Solution of example 3.43: (a) Product of sums; (b) Product of sums using NOR gates.

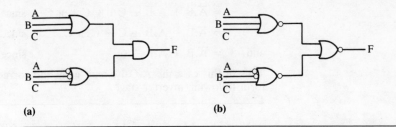

(a)   (b)

The A-O-I solution for a logic network can be based on the sum of products or the product of sums type of circuit. The following example should illustrate the point.

Example 3.44   *Using the truth table of Fig. 3.111 and the sum of products logic circuit of example 3.41 derive an A-O-I logic circuit to satisfy the truth table.*

The same simplified logic equation from example 3.41 can be used; i.e.

$$F = \overline{A}.B + A.\overline{B} + C$$

Using AND gates followed by a NOR gate gives the circuit of Fig. 3.114(a).

Fig. 3.114 (a) A-O-I solution of the logic function of example 3.44; (b) Alternative A-O-I solution.

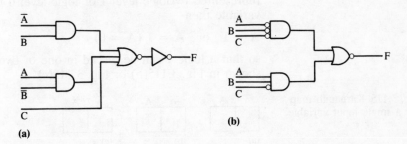

(a)   (b)

Since the output from the NOR gate is an inverted version of the required function, the A-O-I circuit must be followed by an inverter as shown.

Example 3.45   *Repeat the above example using the product of sums arrangement.*

Since in this case

$$\overline{F} = \overline{A}.\overline{B}.\overline{C} + A.B.\overline{C}$$

the circuit using A-O-I configuration can be achieved as shown in Fig. 3.114(b). Since in this case

$$
\begin{aligned}
F &= \overline{\overline{A}.\overline{B}.\overline{C} + A.B.\overline{C}} \\
&= (\overline{\overline{A}.\overline{B}.\overline{C}}).(\overline{A.B.\overline{C}}) && \text{by De Morgan's theorem} \\
&= (\overline{\overline{A}} + \overline{\overline{B}} + \overline{\overline{C}}).(\overline{A} + \overline{B} + \overline{\overline{C}}) && \text{by De Morgan's theorem} \\
&= (A + B + C).(\overline{A} + \overline{B} + C)
\end{aligned}
$$

Expanding the function for F gives

$$F = A.\overline{A} + \overline{A}.B + \overline{A}.C + A.\overline{B} + B.\overline{B} + \overline{B}.C + A.C + B.C + C.C$$

$$= 0 + \overline{A}.B + \overline{A}.C + A.\overline{B} + 0 + \overline{B}.C + A.C + B.C + C$$
$$\text{since } A.\overline{A} = 0 \text{ and } C.C = C$$

$$= \overline{A}.B + A.\overline{B} + C.(A + \overline{A}) + C.(B + \overline{B}) + C$$
$$\text{by distributive law}$$

$$= \overline{A}.B + A.\overline{B} + C.1 + C.1 + C \quad \text{since } A.1 = 1 \text{ etc.}$$
$$= \overline{A}.B + A.\overline{B} + C + C + C \quad \text{since } C.1 = C$$
$$\text{and} \quad F = \overline{A}.B + A.\overline{B} + C \quad \text{since } C + C = C \text{ etc.}$$

Thus in this case the A-O-I circuit gives the correct output without the need for a following inverter stage.

You should now be able to attempt exercises 3.42 to 3.46.

Karnaugh mapping

All circuit minimisation attempted so far has been done using the Boolean algebra rules. In some cases the steps taken to achieve minimisation have not always been obvious, and in most cases the process has been lengthy. **Karnaugh mapping** is a technique which allows circuit minimisation to be achieved simply and quickly, even for large numbers of variables. The Karnaugh map is a pictorial method of plotting the circuit truth table.

Single-variable map

Using binary representation a single variable is either true or false, represented by logic level 1 or logic level 0 respectively. If A is the variable then

$$A = 1 \quad \text{or} \quad \overline{A} = 1 \, (A = 0)$$

so that a logic 1 can be placed in one of two squares, or cells, as shown in Fig. 3.115(b) or Fig. 3.115(c).

Fig. 3.115 Karnaugh map for a single input variable.

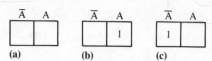

(a)   (b)   (c)

It is usual to represent true outputs only (by use of logic 1) but the technique is also valid if false outputs are recorded by logic 0.

Two-variable map

Fig. 3.116 Karnaugh map for two input variables.

Fig. 3.117 Use of the two-variable map to minimise a logic equation.

Each of two variables can have two possible states so that the total combinations possible for the two variables is four (2^2). The second variable map can be drawn superimposed on the first variable map after first rotating it by 90° to give the map of Fig. 3.116.

From this map it can be seen that A is represented by the right-hand segments while B is represented by the bottom half segments. Thus A and B together can be found in the bottom right-hand cell, i.e. this cell represents the function A.B. Similarly the remaining three cells represent the other possible combinations of $\overline{A}.\overline{B}$, $\overline{A}.B$ and $A.\overline{B}$.

Any function represented by a two-variable equation or truth table can be plotted on the map. As an example, consider the function

$$F = A.B + A.\overline{B}$$

Using the map shown in Fig. 3.117 the cells corresponding to the terms A.B and A.$\overline{B}$ are marked with a 1 since they represent the possible true states for that equation.

Using Boolean algebra rules the circuit can be minimised, since

$$F = A.(B + \overline{B}) \qquad \text{by distributive law}$$

$$= A.1 \qquad \text{since } B + \overline{B} = 1$$

$$= A \qquad \text{since } A.1 = A$$

From the map of Fig. 3.117 it can be seen that by considering the logic 1 states as a pair the function $F = A$ is produced. What the map is in fact representing by allowing the two adjacent cells containing logic 1 to be paired is that, apart from commonality with A, in this case, adjacent squares also represent B or $\overline{B}$ $(B + \overline{B})$ which by the Boolean rules can be neglected since $B + \overline{B} = 1$.

Example 3.46

Use the truth table of Fig. 3.118 to plot a two-variable Karnaugh map and hence produce a minimised logic expression.

A	B	F
0	0	0
0	1	1
1	0	1
1	1	1

Fig. 3.118 Truth table to satisfy the required logic function of example 3.46.

Fig. 3.119 Two-variable Karnaugh map to satisfy the required logic function of example 3.46.

From the truth table the sum of products expression for $F = 1$ can be obtained; i.e.

$$F = \overline{A}.B + A.\overline{B} + A.B$$

The map for this equation is shown in Fig. 3.119.

By vertically grouping two adjacent cells containing logic 1, A is represented and horizontally grouping two adjacent cells containing logic 1, B is represented, so that the function simplifies to

$$F = A + B$$

Grouping of cells $A.\overline{B}$ and $\overline{A}.B$ is not allowed since there is no commonality of one variable to allow the remaining variables to disappear; i.e. grouping $\overline{A}.B$ with A.B gives commonality in B and the $(A + \overline{A})$ term disappears. Adjacency in this case refers to adjoining cells where only one of the two variables changes as you move from square to square. Thus such movements must be done horizontally or vertically, *not* diagonally, for this reason.

In grouping $\overline{A}.B$ and A.B and A.$\overline{B}$ and A.B, the cell A.B has been used twice. This is quite in order since $A + A = A$. Using Boolean algebra rules for the above function

$$F = \overline{A}.B + A.\overline{B} + A.B + A.B \quad \text{since } A.B + A.B = A.B$$

$$= A.(\overline{B} + B) + B(\overline{A} + A) \quad \text{distributive law}$$

$$= A.1 + B.1 \quad \text{since } B + \overline{B} = 1 \text{ etc.}$$

$$\text{or} \quad F = A + B \quad \text{since } A.1 = A \text{ etc.}$$

Three-variable map

Since there are 2^3 (8) possible combinations of the variables there must be eight cells in the Karnaugh map. Letting the variable equal A, B and C, then once again A must be represented in half (4) of the cells while $\overline{A}$ must be represented in the other half (4) of the cells. Similarly B must be represented in four cells, two of which it shares with A and two with $\overline{A}$. Also variable C must be represented in four cells to link with all possible combinations of A and B. The arrangement is shown in Fig. 3.120(a).

The method of identifying individual cells in Fig. 3.120(a) is immaterial as long as the rule that adjacent cells, horizontally or vertically, only differ by one bit in their respective code groups is

Fig. 3.120 Karnaugh map
for three input variables.

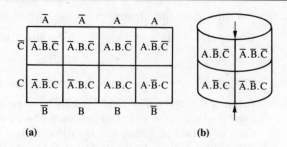

(a)   (b)

applied. Figure 3.120(a) can be seen to satisfy this criteria. It can be
seen from the three-variable map that the top right-hand corner cell
has the value $A.\overline{B}.\overline{C}$ while the top left-hand corner has the value
$\overline{A}.\overline{B}.\overline{C}$, i.e. only one variable has changed. Similarly for the bottom
left-hand corner $(\overline{A}.\overline{B}.C)$ and the bottom right-hand corner $(A.\overline{B}.C)$
only one bit has changed. In other words the edges of the left-hand
and right-hand sides may be considered coincident, i.e. the map can
be considered as the curved surface of a cylinder so that the four
cells of $\overline{B}$ also form a **block** of cells; see Fig. 3.120(b). This gives
what is known as end-to-end adjacency for the map.

In this arrangement simplification can be achieved by grouping
wherever possible into groups of four; if not possible, then groups
of two and finally single cells. Again, in the grouping only adjacent
cells may be collected together. Figure 3.121 shows examples where
cells may be grouped. Check that the values of grouped cells is as
stated in the diagram.

Fig. 3.121 Minimisation
examples for three-variable
Karnaugh maps.

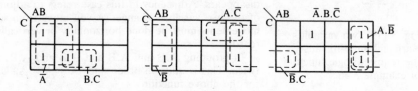

Example 3.47

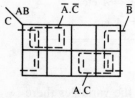

Fig. 3.122 Karnaugh three-
variable map to satisfy the
logic requirements of
example 3.47.

A function is to be represented using the expression

$$F = \overline{A}.\overline{B}.\overline{C} + \overline{A}.\overline{B}.C + \overline{A}.B.\overline{C} + A.\overline{B}.C + A.\overline{B}.\overline{C} + A.B.C$$

Determine, using the Karnaugh map, a simplified expression for the circuit.

Using the map of Fig. 3.122 and placing a 1 in each of the cells identified
by the above expression allows circuit equation simplification to

$$F = \overline{B} + A.C + \overline{A}.\overline{C}$$

and this could be checked as valid by simplification (by using Boolean
algebra rules or checking truth tables).

The solution of example 3.47 gives a sum of products equation, the
circuit for which is shown in Fig. 3.123. The solution for the logic circuit in
example 3.47 is required to fulfil six of the possible eight combinations as
logic 1, as the map if Fig. 3.122 shows. An alternative solution using cells
with logic 0 is valid, i.e.

$$\overline{F} = A.B.\overline{C} + \overline{A}.B.C$$

A glance at the map of Fig. 3.122 shows that to place 0 in the cells $A.B.\overline{C}$
and $\overline{A}.B.C$ will not allow simplification since the cells are not adjacent. The

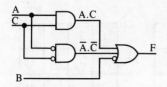

Fig. 3.123 Logic circuit solution to the requirements of example 3.47.

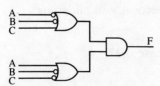

Fig. 3.124 Product of sums solution for example 3.47.

expression $\overline{F} = A.B.\overline{C} + \overline{A}.B.C$ can be modified by inverting both sides of the equation, so that

$$F = \overline{A.B.\overline{C} + \overline{A}.B.C}$$

$$= \overline{(A.B.\overline{C})}.\overline{(\overline{A}.B.C)} \text{ by De Morgan's theorem}$$

and $F = (\overline{A} + \overline{B} + C).(A + \overline{B} + \overline{C})$ also by De Morgan's theorem

so that the function may be represented by a product of sums equation, the circuit for which is shown in Fig. 3.124.

Conversion of the original function to NAND-only or NOR-only circuits could follow, if required, along the lines mentioned earlier in this section.

Four-variable map

There are 16 (2^4) possible combinations of the variables and hence 16 cells are required. Each of the variables must be included in eight of the cells and the layout of the cells must be such that again adjacent cells, horizontally and vertically, only differ by one bit in their respective code groups. A possible arrangement using variables A, B C and D is shown in Fig. 3.125.

Fig. 3.125 Karnaugh map for four input variables.

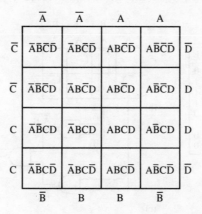

In this arrangement whether using 1 or 0 in each cell, grouping should be done, where possible, in groups of 8, 4, 2 or single cells. Possible arrangements are shown in Fig. 3.126.

As Fig. 3.126 shows, in this arrangement there is top-to-bottom adjacency as well as end-to-end adjacency, since the cells at the top and bottom of the map satisfy the rules for adjacency. Thus the map may be bent in the form of a cylinder with the top and bottom edges coincident. Since there is end-to-end adjacency as well, the four-variable map can be considered as **toroidal**, i.e. rather like an inner tube of a tyre.

Fig. 3.126 Minimisation examples for four-variable Karnaugh maps.

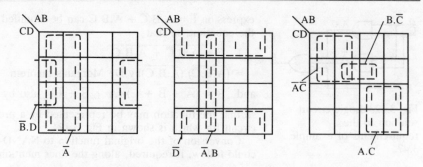

Karnaugh mapping can be extended as necessary to allow for more than four variables.

You should now be able to attempt exercise 3.47.

Conversion to logic circuits direct from Karnaugh maps

It is not really necessary when using Karnaugh maps to simplify equations and then convert the simplified equations to circuit diagrams. The derivation of a simplified circuit to universal inverting logic can be accomplished directly from the map. The following sections will deal with the process of conversion of a mapped circuit function to NAND-only, NOR-only and A-O-I gates.

Conversion directly to NAND gate circuit

Using a three-variable map as an example and assuming a required logic function is stated in the truth table of Fig. 3.127(a), the map may be drawn as shown in Fig. 3.127(b).

Fig. 3.127 Determination of NAND-only circuit directly from a Karnaugh map.

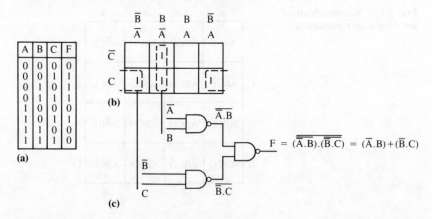

It follows from previous work that the result of grouping adjacent cells in Fig. 3.127(b) gives an expression

$$F = \overline{A}.B + \overline{B}.C$$

which could be represented by an AND-OR circuit since it is a sum of products equation. Since such a circuit is easily converted into a NAND-only circuit, the diagram can be drawn directly as shown in Fig. 3.127(c). If a group of four adjacent cells occurs so that a single variable is obtained for the logic equation, then that variable must be inverted before it is applied as an input to the final NAND gate. This point has already been brought out in example 3.42.

*Conversion directly to NOR
gate circuit*

Using the same truth table as for Fig. 3.127(a) the map and NOR circuit is derived as shown in Fig. 3.128(a) and 3.128(b) respectively.

Fig. 3.128 Determination of NOR-only circuit directly from a Karnaugh map.

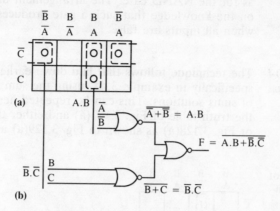

(a)

(b)

The 0 functions are grouped in this case, since the NOR circuit is derived from the product of sums circuit as discussed earlier. Since each group in Fig. 3.128(a) represents an AND function (i.e. $\overline{F} = A.B$) then it can be converted to an OR function by using De Morgan's theorem; i.e.

If $\overline{F} = A.B$

then $F = \overline{A.B} = \overline{A} + \overline{B}$ by De Morgan's theorem

Therefore all that is necessary is to 'separate' and invert the inputs to the first stage of NOR gates as shown in Fig. 3.128(a). Should a group of four adjacent cells have arisen in this case the single-variable input could be taken directly as an input to the final NOR gate *without* inversion (i.e. if a group of four 0 functions gave $\overline{F} = \overline{C}$, say then $F = C$ and this inversion would be achieved by the final NOR gate so that $\overline{C}$ as an input would go directly to the final NOR gate).

The output of Fig. 3.128(b) is given as

$$F = \overline{\overline{A.B} + \overline{B.\overline{C}}}$$

$$= (\overline{A.B}).(\overline{B.\overline{C}}) \qquad \text{by De Morgan's theorem}$$

$$= (\overline{A} + \overline{B}).(B + C) \qquad \text{by De Morgan's theorem}$$

$$= \overline{A}.B + \overline{B}.B + \overline{A}.C + \overline{B}.C \quad \text{by distributive law}$$

$$= \overline{A}.B + \overline{A}.C + \overline{B}.C \qquad \text{since } B.\overline{B} = 0$$

$$= \overline{A}.B + \overline{B}.C + \overline{A}.C.(B + \overline{B})$$

$$\qquad\qquad\qquad \text{since } B + \overline{B} = 1 \text{ and } \overline{A}.C.1 = \overline{A}.C$$

$$= \overline{A}.B + \overline{B}.C + \overline{A}.C.B + \overline{A}.C.\overline{B}$$

$$\qquad\qquad \text{by distributive law}$$

$$= \overline{A}.B.(1 + C) + \overline{B}.C.(1 + \overline{A})$$

$$\qquad\qquad \text{by distributive law}$$

and $F = \overline{A}.B + \overline{B}.C$ since $1 + C = 1$
and $\overline{A}.B.1 = \overline{A}.B$ etc.

This shows that when producing the NOR circuit directly from the truth table the output F is produced for the input 1, *not* input 0, even though the mapping has been based in the 0. In this case the true output is obtained for the input combination of $\overline{A}.B + \overline{B}.C$ just as for the NAND case. The arrangement using NOR gates is based on the knowledge that such a gate produces a true output only when all inputs are false.

Conversion directly to A-O-I circuit

The technique follows that laid down earlier in this section, and specifically in example 3.44, using the sum of products and product of sums solutions. This can be repeated here by again referring to the truth table of Fig. 3.127(a) and either the map of Fig. 3.127(b) or Fig. 3.128(a) as shown in Fig. 3.129(a) and 3.130(a) respectively.

Fig. 3.129 Determination of A-O-I circuit directly from a Karnaugh map.

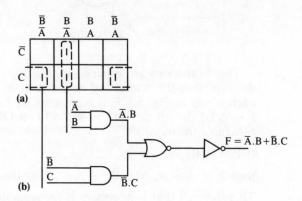

Fig. 3.130 Determination of an alternative A-O-I circuit directly from a Karnaugh map.

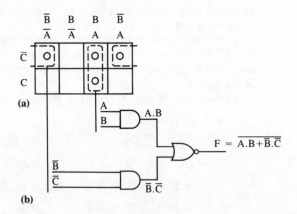

In Fig. 3.129(b), since the output of the NOR gate is the inverse of the required function, the NOR gate is followed by an inverter so that

$$F = \overline{A}.B + \overline{B}.C$$

As $F = \overline{A.B + B.\overline{C}}$ in Fig. 3.130(b), which is the same value as found for F from the all-NOR circuit of Fig. 3.128(b), it follows that the A-O-I circuit of Fig. 3.130(b) can achieve the required function for F without the need for a following inverter.

You should now be able to attempt exercises 3.48 to 3.50.

Exclusive-OR and Exclusive-NOR circuits

An interesting circuit that takes the form of a sum of products circuit is that part of a **half-adder** circuit which produces the **sum** function. This is produced by the requirement to add two bits (binary digits) to give a sum and possible carry, as the truth table of Fig. 3.131 shows.

Fig. 3.131 Circuit for the addition of two bits: (a) truth table; (b) the Karnaugh map; (c) logic circuit arrangement; (d) symbol for the Exclusive-OR function

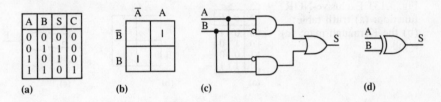

A	B	S	C
0	0	0	0
0	1	1	0
1	0	1	0
1	1	0	1

(a) (b) (c) (d)

Details of the half-adder circuit are dealt with fully in Chapter 4 but it is interesting to examine the circuit that would realise the sum function of Fig. 3.131 where

$$S = \overline{A}.B + A.\overline{B}$$

This gives what is known as an Exclusive-OR function, where the output is a logic 1 only when both inputs are complementary. It is often expressed as

$$S = A \oplus B$$

Examination of the two-variable map of Fig. 3.131 shows that the circuit is not capable of minimisation and should be constructed as in Fig. 3.131(c). The symbol is shown in Fig. 3.131(d). Conversion to universal inverting logic would indicate NAND gates and, because of an inverted input to each AND gate of the basic circuit, it would appear that five NAND gates would be required. However, manipulation of the original equation using Boolean algebra rules shows that only four NAND gates are required.

Since $S = \overline{A}.B + A.\overline{B}$

then $S = \overline{A}.B + A.\overline{B} + 0 + 0$

$\quad = \overline{A}.B + A.\overline{B} + A.\overline{A} + B.\overline{B}$ since $A.\overline{A} = 0$ etc.

$\quad = A.(\overline{A} + \overline{B}) + B.)\overline{A} + \overline{B})$ by distributive law

$\quad = A.(\overline{A.B}) + B.(\overline{A.B})$ by De Morgan's theorem

Hence $\overline{S} = \overline{A.(\overline{A.B}) + B.(\overline{A.B})}$ by De Morgan's theorem

$\quad = \overline{A.(\overline{A.B})}.\overline{B.(\overline{A.B})}$ by De Morgan's theorem

and $S = \overline{\overline{A.(\overline{A.B})}.\overline{B.(.\overline{B})}}$

The circuit is shown in Fig. 3.132.

Fig. 3.132 A minimised NAND only form of the Exclusive-OR function

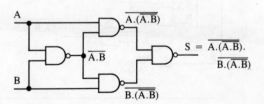

$A.(\overline{A.B})$

$\overline{A.B}$

$S = \overline{A.(\overline{A.B}).}$
$\overline{B.(\overline{A.B})}$

$B.(\overline{A.B})$

By inverting the output of an Exclusive-OR gate the truth table and Karnaugh map of Fig. 3.133(a) and (b) are obtained. Since $F = \overline{A}.\overline{B} + A.B$ form the truth table and map, it would appear that the circuit is not capable of minimisation but could be implemented in various forms such as all NANDs, A-O-I etc.

Fig. 3.133 Exclusive-NOR function: (a) truth table; (b) the Karnaugh map; (c) symbol.

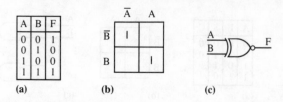

(a)   (b)   (c)

The function where $F = 1$ only when both A and B have the same value is known as an Exclusive-NOR. It is often expressed as

$$F = A \odot B$$

The symbol for the Exclusive-NOR function is shown in Fig. 3.133(c).

Example 3.48   *Convert the Exclusive-OR function to A-O-I directly from the Karnaugh map.*

The 0 functions of the map have been used since it means that an inverter after the NOR gate is not required. See the circuit of Fig. 3.134.

Fig. 3.134 Exclusive-OR circuit using A-O-I form.

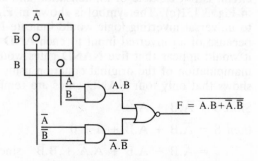

Example 3.49   *Convert the Exclusive-NOR function to NAND form directly from the Karnaugh map.*

The 1 functions have been used to give the circuit of Fig. 3.135.

Fig. 3.135 Exclusive-OR circuit using NAND gates form.

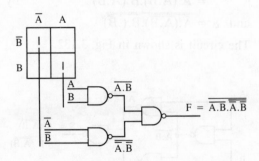

Points to remember

- Graphical analysis shows how an amplifier amplifies and demonstrates problems of cut-off and saturation
- D.C. analysis determines the d.c. operating or Q point for devices in circuits
- Equivalent circuit analysis predicts the gain and impedance levels of circuits including devices
- A.C. analysis predicts the variation of gain and impedance levels with signal frequency
- Frequency response can be shown by Bode plots or polar plots
- Transient analysis predicts the behaviour of a circuit when a switched signal is applied
- Logic gates may be assembled to produce a required function and the result expressed algebraically using Boolean algebra
- Circuit minimisation is possible using Boolean algebra rules and/or Karnaugh mapping techniques
- Any circuit may be modified to a form using AND-OR-INVERT (A-O-I) or inverting logic only

EXERCISES 3

3.1 A device has a VI characteristic which may be approximated to a straight line passing through the VI points 0, -5 mA; $+3$ V, 0. If it is connected in series with a d.c. supply E and a resistor R, determine the Q point for the following E, R values: (a) $+6$ V, 500 Ω; (b) $+6$ V, 2kΩ; (c) -3 V, 1.2 kΩ.

(25 mA, 4.42 V; 1.2 mA, 3.65 V; -3.5 mA, 0.9 V)

3.2 A device has the VI characteristic given by the points on the following table.

V (V)	-4	-2	0	$+2$	$+4$	$+6$	$+8$
I (mA)	-0.3	-0.25	0	$+1$	$+3$	$+4.6$	$+5.1$

If the device is connected in series with a supply E of 10 V and a resistor R, what value of R is required to give a Q point of (a) 4 V, 3 mA; (b) 1.25 V, 0.25 mA. (2 kΩ; 17.2 kΩ)

3.3 The device described in exercise 3.2 is connected in series with a d.c. supply of 4 V, a resistor of 2 kΩ and an a.c. supply with a peak value of (a) 3 V or (b) 6 V. Determine in each case, the extreme values of the device voltage. (3.1, 0.6; 4, -1.6)

3.4 The device described in exercise 3.2 is connected in series with a d.c. supply of 8 V, two resistors of 3 kΩ and 1 kΩ respectively and an a.c. signal of peak value 2 V. In parallel with the 3 kΩ resistor is a capacitor with negligible reactance at the signal frequency. Determine the extreme values of device voltage. (3.5, 1.3)

3.5 A three-terminal device has the I_B, V_{AB} characteristics given in the table below. The voltage V_{GB} applied between the control terminal G and the common terminal B selects the characteristics as shown. The characteristics can be approximated to straight lines.

V_{AB} I_A for $V_{GB} =$	−0.2	−0.1	0	+0.1	+0.2	+0.3
0	0	0	0	0	0	0
1	0.02	0.08	0.2	0.4	0.6	0.8
7	0.1	0.26	0.5	0.8	1.1	1.4

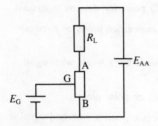

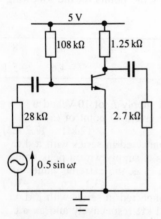

Fig. 3.136 Circuit for Ex. 3.5.

The device is connected in the circuit shown in Fig. 3.136
(a) Determine the operating or Q point for the following components and conditions.

E_{AA}	6 V	E_G	0	R_L	10 kΩ
	2 V		+0.3 V		1 kΩ
	4 V		+0.1 V		500 Ω

(b) If E_{AA} is 8 V, determine the necessary R_L and E_G for the following Q points.

I_A	0.32 mA	V_{AB}	3.0 V
	0.06 mA		4.0 V
	1.02 mA		6.0 V
	0.18 mA		0.5 V

(3.1, 0.3; 0.78, 0.61; 3.4, 0.58; 14, 0; 67, −0.2; 42, 0.1; 1.96, 0.2)

3.6 The circuit described in exercise 3.5 is modified by the addition of an a.c. signal source e_g, in series with E_G, If $e_g = 0.25 \sin \omega t$, $E_{AA} = 10$ V, determine the voltage gain v_{ab}/e_g for the following valueof E_G and R_L.

E_G	0.1 V	R_L	10 kΩ
	0 V		10 kΩ
	0 V		14.8 kΩ
	0.1 V		5 kΩ

(14.7; 15.3; 18.3 10.4)

3.7 The circuit in exercise 3.6 is modified by the addition of a capacitor of negligible reactance in series with a resistor R_C between A and earth. Determine the voltage gain for the following condition. E_{AA} 8 V, R_L 16.7 kΩ, E_G 0 V, R_C 10 kΩ. For what amplitude of e_g would the output become excessively distorted?   (9.5, greater than 0.2 V peak)

3.8 The BJT shown in Fig. 3.137 has the characteristics given in exercise 3.7. Determine: (a) the Q point; (b) the peak to peak voltage output, the voltage gain v_{ce}/v_{be}.   (2 mA, 2.5 V;   1.4 V, 28)

Fig. 3.137 BJT circuit for Ex. 3.8.

3.9 A two-terminal device has an *IV* characteristic which may be approximated to three straight lines between the points 0, 0; 12.5 mA, 2.5 V; 2.5 mA, 3.5 V; 20 mA, 5.5 V. It is connected in series with a variable supply E and a 240 Ω resistor. Sketch a graph showing how the device current will vary when E is increased steadily from 0 to 6 V and then back to 0. (0, 0; 12.5, 5.5; 8, 6; 2.5, 4.2; 9.2, 4.2 0, 0.)

3.10 In the circuit shown in Fig. 3.138, the voltage applied between A and B may be either +6 V or −6 V. In each case, calculate the voltage V_x and the currents in the 2 kΩ resistor and from the supply.

(5.3, 2.3, 3.96; −3, 0, −2.97)

Fig. 3.138 Diode resistor circuit for Ex. 3.10.

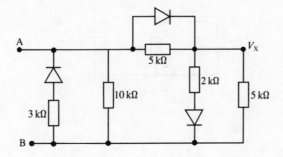

3.11 A square-wave voltage of ±5 V is applied to the input of the circuit shown in Fig. 3.139. Calculate the voltage levels for the output square wave V_o. (+2.23, −3.37)

Fig. 3.139 Diode resistor circuit for Ex. 3.11.

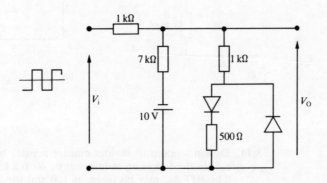

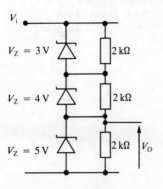

Fig. 3.140 Zener diode circuit for Ex. 3.12.

3.12 A varying positive voltage is applied to the circuit shown in Fig. 3.140. Calculate the input voltage levels at which the three zener diodes turn on and hence the corresponding levels of V_o.

(9, 10.5, 12; 3, 3.5, 5)

3.13   For each level of the BJT circuits shown in Fig. 3.141, calculate the Q point. In all cases, the H_{FE} may be taken as 150.

(2.4, 1.7;   10.4, 2.5;   0.074, 4.1;   0.2, 2.7;   2.76, 3.2.)

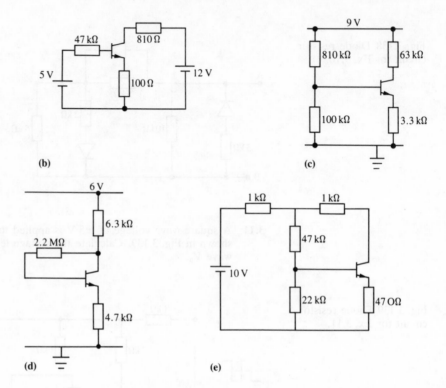

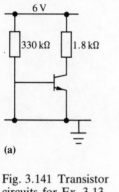

Fig. 3.141 Transistor circuits for Ex. 3.13.

3.14   Design a potential divider emitter resistor bias circuit for a BJT to satisfy the following requirements, R_L 6.8 kΩ, I_C 0.5 mA, V_{CE} 2 V. The BJT h_{FE} may be taken as 120 and the available d.c. supply is 6 V.   (R_E 2.7 kΩ, R_1 100 kΩ, R_2 56 kΩ)

3.15   In the circuit shown in Fig. 3.142, V_i is, at different times +3 V, +10 V, or +15 V. In each case, determine the Q point of the transistor.   (0, 6;   0.53, 2.3;   0.82, 0.1)

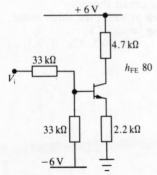

Fig. 3.142 Circuit for Ex. 3.15.

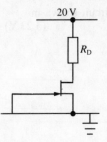

Fig. 3.143 JFET circuit for
Ex. 3.16.

3.16 The JFET in the circuit in Fig. 3.143 has V_p −3 V and I_{DSS} 12 mA. If
R_D is (a) 750 kΩ or (b) 1.9 kΩ, calculate the Q points.

(12, 11; 9.8, 1.38)

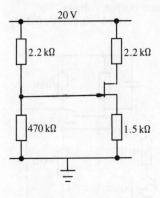

Fig. 3.144 JFET amplifier
circuit for Ex. 3.17.

3.17 Calculate the Q point for the JFET circuit in Fig. 3.144. Take V_p as
−3 V and I_{DSS} as 12 mA. (3.3, 7.8)

3.18 Determine the d.c. V_o indicated in Fig. 3.145, (a) with $V_{in} = 0$ or
(b) with $V_{in} = +0.1$ V. (−1.92, −3.2.)

Fig. 3.145 Emitter coupled
amplifier circuit for
Ex. 3.18.

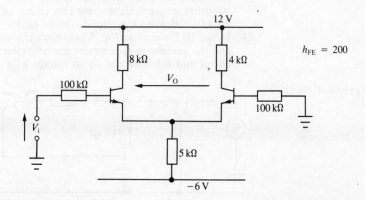

3.19   Calculate the voltage at the emitter of Tr$_2$ in the circuit shown in
Fig. 3.146.   (3.23 V)

Fig. 3.146 Direct coupled
transistor circuit for
Ex. 3.19.

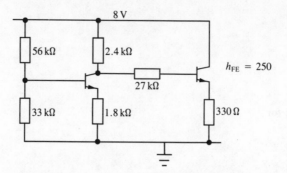

3.20   The BJT in each of the circuits shown in Fig. 3.147 has h_{ie} 3.5 kΩ, h_{oe}
2 × 10^{-5} S and h_{fe} 25. Calculate the indicated output voltage.
(0.68 V, 0.177 V, 0.3 V)

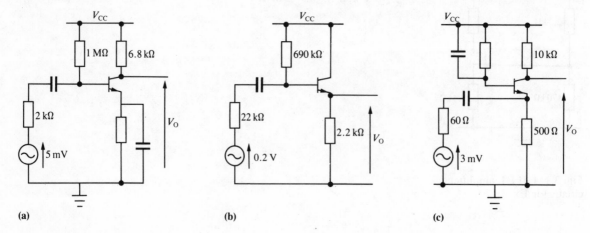

Fig. 3.147 Transistor
circuits for Ex. 3.20.

3.21   The transistors in the circuit in Fig. 3.146 have h_{fe} 250 and h_{ie}
respectively of 4 kΩ and 600 Ω; h_{oe} is negligible. Draw a complete
equivalent circuit and hence calculate the circuit voltage gain.   (−0.97)

3.22   A three-stage amplifier employs FETs in the common source
configuration. Each transistor has g_m 6 mA/V and r_o 150 kΩ; each
stage has an equivalent shunt bias circuit of 42 kΩ and a drain load of
2.7 kΩ. Calculate the circuit voltage gain.   (−3630)

3.23   In the BJT circuit in Fig. 3.148, both transistors have h_{fe} 120, h_{ie} 2 kΩ
and the remaining parameters are negligible. Draw an equivalent
circuit and calculate the signal voltage gain.   (173)

Fig. 3.148 Transistor circuit
for Ex. 3.23.

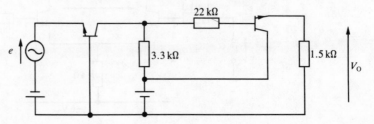

3.24 A 'black box' amplifier has an open-circuit voltage gain of -500 and input and output impedances respectively of 50 kΩ and 5 kΩ. An external load of 5 kΩ is connected between the output and earth and a 100 kΩ resistor is connected between output and input. Draw an equivalent circuit for this arrangement and calculate the output voltage and input current when a 10 mV signal is applied to the input.

(1.42 V, 14.5 μA)

3.25 A parallel combination of a 5 kΩ resistor and a 0.05 μF capacitor is connected in series with (a) a 1 kΩ resistor or (b) a 0.1 μF capacitor. A 1 V signal at a frequency of 10^4 r/s is connected across the combination. By means of phasor diagrams, calculate the voltage across the 0.05 μF capacitor.

(0.77 ∠−23°, 0.66 ∠8°)

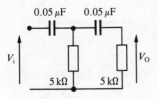

Fig. 3.149 *RC* circuit for Ex. 3.25.

3.26 In the circuit in Fig. 3.149, the frequency of the voltage V_{in} is either (a) 10^4 r/s or (b) 10^3 r/s. In each case, by phasor diagram construction, determine the ratio V_o/V_i.

(0.157 ∠49°, 0.011 ∠147°)

3.27 If the voltage applied to the circuit in Fig. 3.150 is $2 \sin 5000t$, draw a phasor diagram representing the circuit currents and voltages and hence determine the voltage V_o.

(6.64 ∠−86.5°)

Fig. 3.150 *RLC* circuit for Ex. 3.27.

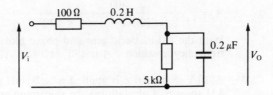

3.28 Repeat exercise 3.25 using j notation algebra.

3.29 Repeat exercise 3.26 using j notation algebra.

3.30 Repeat exercise 3.27 using j notation algebra.

3.31 Determine for the circuit shown in Fig. 3.151 the current flowing in the 1 kΩ resistor.

(1.63 ∠−27° mA)

Fig. 3.151 Circuit for Ex. 3.31.

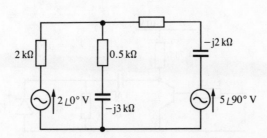

3.32 For each of the circuits shown in Fig. 3.152, determine the ratio V_o/V_i and arrange the result in the frequency response form.

$$\left(\frac{0.95}{(1 + j\omega/21\,000)}, \frac{0.166(1 + j\omega/100)}{(1 + j\omega/600)}\right)$$

Fig. 3.152 Circuits for Ex. 3.32.

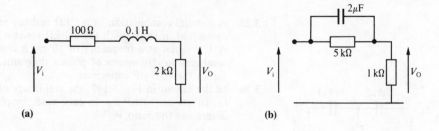

(a)　　　　　　　　　　　　　　　　　　(b)

3.33 The frequency responses for two systems are given by the following expressions:

(a) $$\frac{500}{\left(1 - \dfrac{j200}{\omega}\right)^2\left(1 + \dfrac{j\omega}{10^4}\right)\left(1 + \dfrac{j\omega}{5 \times 10^4}\right)}$$

(b) $$\frac{250(1 + j\omega/50)}{j\omega(1 + j\omega/800)(1 + j\omega/3000)}$$

Plot a graph for each showing the Bode straight line approximation for the gain response and hence find the unity gain frequencies.

$$(8.9, 5 \times 10^5; \ 3160)$$

3.34 An amplifier has a frequency response given by:

$$A = \frac{400}{(1 + j\omega/10^3)(1 + j\omega/4 \times 10^4)^2}$$

With the aid of Bode gain and phase plots, determine (a) the frequency at which $\angle A$ is $180°$ and (b) $|A|$ at this frequency.

$$(4 \times 10^4, 10)$$

3.35 A 10 V, 8 ms pulse is applied to a 20 mH inductor in series with a 4 Ω resistor. Calculate (a) the maximum voltage across the resistor and (b) the times at which v_r is 2 V. $\qquad$ (7.98;　1.1 ms, 14.9 ms)

3.36 A circuit consisting of a 0.02 μF capacitor in series with a 75 kΩ resistor has the following voltage signal applied to it; starting from zero, the voltage is switched to $+4$ V for 1 ms, then to -8 V for 2 ms and finally back to zero. Determine the voltage waveform across the resistor and find the maximum positive and negative values of this voltage. $\qquad$ $(-9.9, +5.38)$

3.37 The voltage applied to the input of the circuit shown in Fig. 3.153 is initially zero and then a single 4 V, 10 μs pulse is applied. Determine the output waveform, stating the maximum positive value and the time by which it has returned to approximately zero. $\qquad$ (4.5, 30 μs)

Fig. 3.153 Transistor switching circuit for Ex. 3.37.

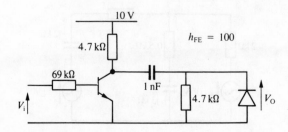

3.38 Reduce the block diagram shown in Fig. 3.154 to determine the voltage ratio V_o/V_i.

(68.1)

Fig. 3.154 Block diagram for Ex. 3.38.

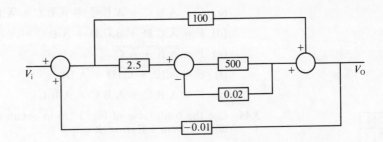

3.39 Use Boolean algebra rules to simplify the expression:

$$F = A.B + A.B.\overline{C} + \overline{B}.C + A.\overline{B}.C + A.B.C$$

Check the truth table of the above expression with that of the simplified circuit to confirm the accuracy of the simplified circuit.

3.40 From the truth table of the original expression for exercise 3.39 determine an expression for all combinations of output variables for which $F = 0$. Simplify this expression and again confirm the truth table of the simplified circuit against the original.

3.41 An industrial process uses logic circuitry to check that temperature, level and pressure of a fluid are at the required values. A buzzer is to sound if all three variables are outside their limits; a red lamp is to light if any two variables are outside their limits; and an amber lamp is to light if any one variable is outside its limit. Design a logic circuit to fulfil the requirements stated. Let T = temperature at correct value, $\overline{T}$ = temperature above a set value, L = correct fluid level, $\overline{L}$ = above a set level, P = correct pressure, $\overline{P}$ = above the set pressure level, B = buzzer ON, R = red lamp ON, A = amber lamp on.

3.42 A logic function is defined by

$$F = \overline{A}.\overline{B}.\overline{C}. + \overline{A}.B.\overline{C} + \overline{A}.\overline{B}.C$$

Simplify the expression and produce the simplified circuit using NAND gates only.

3.43 Derive the truth table for the original expression in exercise 3.42 and confirm that the truth table for the simplified circuit using NAND gates is the same.

3.44 A logic function is given by:

$$F = A.\overline{B}.C + \overline{A}.\overline{B}.C + A.B.\overline{C} + A.\overline{B}.\overline{C} + A.B.C$$

Simplify the expression and produce a simplified circuit using NOR gates only.

3.45 Derive the truth table for the original expression in exercise 3.44 and confirm that the truth table for the simplified circuit using NOR gates is the same.

3.46 A logic function is given by:

$$F = \overline{A}.\overline{B}.C + A.\overline{B}.C + A.\overline{B}.\overline{C} + \overline{A}.B.\overline{C}$$

Derive the truth table and from the table produce the A-O-I logic circuit:
(a) using sum of products logic circuit.
(b) using product of sums logic circuit.
Confirm your circuits in each case by comparing their truth table with the original.

3.47   Use Karnaugh maps to simplify expression in the following cases:

(a) $F = A.B.C + \overline{A}.B + A.B.\overline{C}$

(b) $F = A.B.\overline{C} + \overline{A}.B.C + \overline{A}.B.\overline{C}$

(c) $F = A.B.\overline{C} + A.\overline{B}.\overline{C}. + \overline{A}.\overline{B}.C + \overline{A}.B.C$

(d) $F = \overline{A}.C.D + B.\overline{C}.D + \overline{A}.B.\overline{C} + A.B.C + A.\overline{C}.D + \overline{A}.B.D$

(e) $F = \overline{\overline{A}.\overline{B} + \overline{A}.C}$

(f) $F = A.(\overline{\overline{B} + \overline{C}.(D + \overline{A}.B)})$

(g) $F = A.B.\overline{C} + A.B.C + A.\overline{B}.C$

3.48   Use the truth table of Fig. 3.155 to obtain a NAND-only circuit directly from a Karnaugh map.

A	B	C	F
0	0	0	1
0	0	1	1
0	1	0	0
0	1	1	1
1	0	0	1
1	0	1	1
1	1	0	0
1	1	1	1

Fig. 3.155 Truth table for logic function of exercise 3.48

3.49   Use the truth table of Fig. 3.156 to obtain a NOR-only circuit directly from a Karnaugh map.

A	B	C	F
0	0	0	0
0	0	1	1
0	1	0	0
0	1	1	1
1	0	0	1
1	0	1	0
1	1	0	0
1	1	1	0

Fig. 3.156 Truth table for logic function of exercise 3.49

3.50   Use the truth table of Fig. 3.157 to obtain an A-O-I circuit directly from a Karnaugh map.

A	B	C	F
0	0	0	0
0	0	1	0
0	1	0	1
0	1	1	1
1	0	0	1
1	0	1	1
1	1	0	0
1	1	1	0

Fig. 3.157 Truth table for logic function of exercise 3.50

4 Digital Circuits

The principal learning objectives of this chapter are to:

A **digital circuit** is one in which the signals are represented by discrete values. Most digital circuits are binary systems in that the signals have only *two* discrete values. Examples of digital signals represented by truth tables in which the signal was assumed to be either a logic level 1 or level 0 have already been discussed in Chapter 3. Such signals are usually represented by voltage levels; for example, +5 V could represent logic level 1 and 0 V the logic level 0. These voltage levels are examples of the use of what is known as **positive logic**, where the 1 level is at a higher value than the 0 level. Negative logic circuits exist but unless otherwise stated all logic will be assumed positive. The basis of digital circuitry is the use of semiconductor active devices such as diodes, bipolar junction transistors and FETs as switches. The use of an active device as a switch is essentially the same as any other type of switch in that when the switch is ON, or made, the output follows the input and when the switch is OFF, or broken, the output is isolated from the input. For an ideal switch there should be zero resistance when ON and an infinite resistance when OFF. Under these conditions there is no volt drop across the ON switch when it passes current and no leakage current through it when OFF. An ideal switch characteristic is represented by Fig. 4.1. Additionally, the time taken for the switch to change from one state to the other should be zero.

Figure 4.1 could be taken as the characteristic of an ideal diode where if the applied voltage $V > 0$ (forward voltage) a forward current flows having a value dependent on the applied and external load. If the applied voltage $V < 0$ then the reverse current is zero and the voltage across the device is the applied voltage V. No diode is ideal but approximations using piecewise linear circuits could be made to allow for the forward volt drop across the diode and the resistance of the diode when ON. Such a characteristic is shown in Fig. 4.2.

Typically, for a silicon device the value for $V_{ON} \cong 0.6$ V and the ON resistance, given by the slope of the forward characteristic, is about 5 Ω.

Digital circuits incorporating switches are often referred to as gates, using the analogy that gates may be opened or closed according to a set of input conditions that can influence the output. Gate circuits built exclusively from diodes are no longer in common use but a brief discussion on their application is appropriate since diodes form the basis of bipolar and field-effect devices anyway, and combinations of transistors and diodes are still used in modern circuitry.

Although digital circuits are in the main constructed using integrated circuits, and although there is no real necessity to be aware of the contents of the digital elements, some consideration of the construction of the gates may be beneficial. No deep analysis of circuit behaviour will be found in the following sections but the points raised will assist in an understanding of limitations of certain gates and of the parameters such as switching speed, fan-out, power consumption, etc., which may be found in manufacturers' literature regarding the devices being marketed.

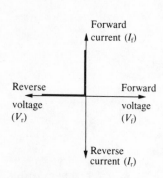

Fig. 4.1 Ideal diode characteristic.

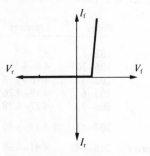

Fig. 4.2 Approximate piecewise linear representation of a diode characteristic.

Diode logic

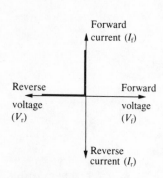

Fig. 4.3 Two-input diode-resistance OR gate.

Using diodes and resistance a basic OR gate can be constructed as shown in Fig. 4.3. Assuming that the input voltage that can be supplied to A or B is 0 V (logic 0) or +5 V (logic 1) it follows that if either A or B or both are at +5 V the associated diode conducts and the output is at logic 1. This either ignores the forward volt drop across the conducting diode or allows that the logic 1 level can be less than +5 V by the diode volt drop, i.e. V_{out} could be 4.4 V and still represent a logic 1. It can easily be checked that the circuit of Fig. 4.3 represents an OR function since the output will only be at 0 V (logic 0) when both diodes are off and this only occurs when A and B are both at 0 V (logic 0).

A basic AND gate is shown in Fig. 4.4. In this arrangement,

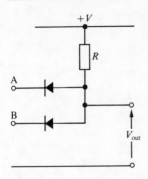

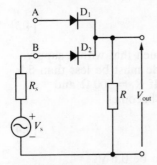

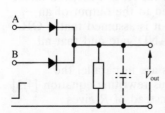

Fig. 4.4 Two-input diode-resistance AND gate.

Fig. 4.5 Two-input diode-resistance OR gate showing the effect of source resistance.

Fig. 4.6 Two-input diode-resistance OR gate showing the effect of stray capacitance.

when either or both inputs are at logic 0 (0 V) the associated diode conducts and the output is at logic 0. This either ignores the forward voltage drop across the diode or allows that the logic 0 level can be as high as +0.6 V. Only when both inputs A and B are at logic 1 (+5 V) will the diodes be off and the output rise to logic 1 (+5 V) which is the requirement for an AND gate.

It has been assumed for the gates of Figs. 4.3 and 4.4 that the voltage applied to the inputs is from a source of zero source resistance. This of course is not necessarily true and taking source resistance into account, as suggested in Fig. 4.5, could affect voltage levels in the circuit.

In this circuit it is assumed for simplicity that A = 0 V and that the forward resistance of diode D_2 is small compared to R_s and R so can be neglected. Since a volt drop across R_s is produced by diode forward current it follows that the voltage at B may be less than the +5 volts assumed for a logic 1 input. In fact, using the Kirchhoff loop formed by the diode D_2, R_s, R and V_s the output voltage can be calculated as

$$V_{OUT} = \frac{(V_s - V_D).R}{R + R_s} \qquad [4.1]$$

where V_D = forward volt drop across D_2.

Assuming a typical value for R_s of 1 kΩ and that $R = 5$ kΩ then equation [4.1] can be solved, using the previously given voltage levels, to give an output voltage V_{OUT} where

$$V_{OUT} = \frac{(5 - 0.6) \times 5 \times 10^3}{(5 + 1) \times 10^3} = \frac{4.4 \times 5}{6} \text{ V} = 3.67 \text{ V} \qquad [4.2]$$

This assumes $V_D = 0.6$ V.

This would suggest that a logic 1 could be even lower than assumed earlier and raises the question as to what is an acceptable value to which V_{OUT} could fall while still being recognised as a logic level 1. Obviously a choice of resistance values could make a difference here, since if R is chosen to be very much greater than R_s then the effect of R_s is reduced. However, a large value of R may itself introduce a problem. It has been assumed that a change in voltage level at the input to the gate would instantly produce a change at the output. However, stray capacitance is ever-present in such circuits and effectively shunts R as shown in Fig. 4.6.

Thus any transition at the output will be exponential in form with a time constant of CR seconds. Since little can be done about the value of C it is desirable to minimise R to enable rise time of the output to be as rapid as possible. The delay in V_{OUT} reaching its threshold voltage compared to the change in input voltage level is known as propagation delay. A suitable value for R to achieve suitable output voltage level and reduce propagation delay is a compromise and one which has not been satisfactorily resolved, being one of the reasons for the decline in popularity of the diode register logic.

As suggested in Chapter 3, a possible combinational logic circuit is the sum of products arrangement where a group of AND gates is followed by an OR gate. Using diode resistance logic, a circuit arrangement is shown in Fig. 4.7.

Considering the circumstances where all AND diodes in Fig. 4.7

Fig. 4.7 Diode-resistance
AND gate driving a diode-
resistance OR gate.

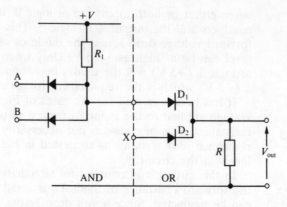

have inputs at logic 1 (i.e. +5 V) the value of V_{OUT} is given by

$$V_{OUT} = \frac{(V - V_{D_1}).R}{R + R_1} \qquad [4.3]$$

The current flow through R_1, D_1, and R is such that with a supply voltage of $V = +5$ V the voltage at D_1 anode must be less than 5 V so that all AND diodes are reverse biased. If $R_1 = 500\ \Omega$ and $R = 5\ k\Omega$ as before then the output voltage becomes

$$V_{OUT} = \frac{(5 - 0.6) \times 5}{0.5 + 5}\ V = 4.0\ V \qquad [4.4]$$

where $V_{D_1} =$ volt drop across D_1, assumed to be 0.6 V.

Equation [4.4] gives a better value for logic 1 than achieved from equation 4.2 for the OR gate because a negligible value for R_s has been assumed. Obviously, in the circuit of Fig. 4.7 if either A or B fell to a logic 0 then the input to diode D_1 is at logic 0 (0.6 V) and with D_1 reverse biased the output is at 0 V unless input X is at logic 1 so that D_2 conducts to give a logic 1 output.

Having established that a possible value for R_1 is 1 kΩ, the effect of more than one OR gate being connected to the output of an AND gate can be considered. In Fig. 4.8, it is assumed that n OR gates are connected to the output of the AND gate and that all other OR gate inputs are at logic level 0.

Since in Fig. 4.8 all OR gate resistances are in parallel the effective parallel resistance is $(R/n)\ \Omega$. Thus rewriting equation [4.3] with the effective parallel resistance R/n instead of R gives

$$V_{OUT} = \frac{(V - V_D)R/n}{R_1 + R/n} \qquad [4.5]$$

$$V_{OUT} = \frac{(5 - 0.6)5/n}{0.5 + 5/n} \qquad [4.6]$$

If we assume that 3 V is the minimum value that can be accepted as a logic 1 level then equation [4.6] could be rewritten with $V_{OUT} = 3$ V to give a value for n; i.e.

$$3\left(0.5 + \frac{5}{n}\right) = (4.4)\frac{5}{n}$$

giving $n = 4\frac{2}{3}$

Taking n to the nearest *lowest* whole number gives a figure of four OR gates that could safely be fed from the AND gate without

Fig. 4.8 An illustration of the loading effect on AND gate when used to drive more than one OR gate. All gates use diode-resistance logic.

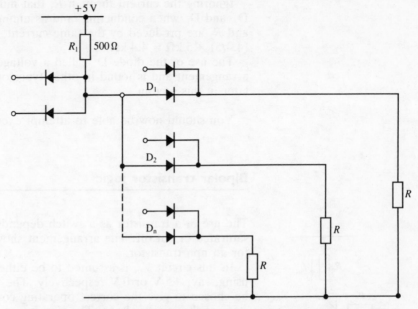

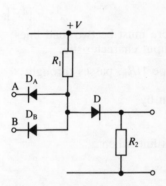

Fig. 4.9 Modification to the original two-input AND gate to allow correct logic 1 and 0 levels to be obtained at the gate output.

taking the output of the AND gate below its minimum allowed voltage level for logic 1 state. The value of n is the fan-out for the AND gate.

It might appear from equation [4.6] that making R_1 very much less than 500 kΩ would greatly improve the fan-out. R_1 however, must also act as a current limiter for any of the AND diodes when the input is logic 0 and a minimum value of R_1 must be present to ensure the diode current does not exceed its maximum value.

So far the circuit arrangements for the AND or OR gate have produced logic 1 levels of between 3.67 and 4.0 V and 0.6 V for logic 0. A simple modification to the original circuits allow these levels to be 5.0 V and 0 V respectively. Consider the AND gate of Fig. 4.9.

In this circuit an extra diode D and resistance R_2 have been added. Under the condition that either or both inputs A and B are logic 0 the associated diode D_A or D_B conducts and the anode of D is at 0.6 V. Thus the volt drop across D, which is forward biased, allows a 0 V output for the logic 0 state.

When both inputs A and B are at logic 1 (+5 V), the value of $+V$ and resistances R_1 and R_2 must be such that the anode of D is at +5.6 V, i.e. the diodes D_A and D_B still conduct but the value of voltage at D anode being at +5.6 V gives $V_{OUT} = +5.0$ V, allowing for a 0.6 volt drop across D. If we assume a supply voltage of +10 V for V and $R_2 = 5$ kΩ then a suitable value for R_1 can be calculated.

Since

$$V = V_{OUT} + V_D + V_{R_1}$$

and

$$10 = 5 + 0.6 + V_{R_1}$$

Therefore

$$V_{R_1} = 4.4 \text{ V}$$

Ignoring the current through R_1 that must flow through the diodes D_A and D_B when conducting, and assuming the volt drops across R_1 and R_2 are produced by the same current, then the value of R_1 is $(4.4/5) \times 5 \text{ k}\Omega = 4.4 \text{ k}\Omega$.

The use of the diode D, called a voltage offset diode, is a useful arrangement and is found in other types of gates, as will be seen later in this section.

You should now be able to attempt exercises 4.1 to 4.5

Bipolar transistor logic

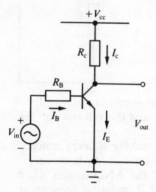

Fig. 4.10 Basic transistor switch using an npn device.

The use of a transistor as a switch depends on a device being saturated or cut off. The arrangement shown in Fig. 4.10 is typical for an npn transistor.

In this circuit V_{IN} is assumed to be either a logic 1 or logic 0 using, say, +5 V or 0 V respectively. The values for R_B and R_C must be chosen to give the correct operating conditions regardless of the state of the input voltage. The circuit may be analysed for base–emitter and collector–emitter conditions and the load lines produced superimposed on the respective device characteristics. Considering the base–emitter circuit and using Kirchhoff's voltage law gives

$$V_{IN} - V_{BE} = I_B R_B \qquad [4.7]$$

As V_{IN} can take two possible values there must be two load lines that can be superimposed on the device input characteristic:

(1) When $V_{IN} = 0$ V, the load line, of slope $1/R_B$, passes through the origin of $V_{BE} = 0$ V, $i_B = 0$ A
(2) When $V_{IN} = 5$ V, the load line is given by

$$5 - V_{BE} = i_B R_B$$

and two points can be defined for this line when

(a) $V_{BE} = 0$, $i_B = 5/R_B$ A
(b) $i_B = 0$, $V_{BE} = 5$ V

Assuming a device where $R_B = 50 \text{ k}\Omega$, a suitable input characteristic with load line for the +5 V input voltage condition is shown in Fig. 4.11(a).

From the intersection of the base load line with the device input characteristic it can be seen that the voltage at the base–emitter junction is 0.6 V and the base current 88 μA when $V_{IN} = +5$ V.

The two values of V_{BE}, and i_B, obtained above for $V_{IN} = 0$ V and +5 V, can be used on the device output characteristic to determine the collector current and output voltage V_{CE} for both values of V_{IN}. Figure 4.11(b) should make this clear. The load line in Fig. 4.11(b) has been drawn using the Kirchhoff voltage law around the collector–emitter loop, i.e.

$$V_{CC} - V_{CE} = i_C R_C$$

giving two suitable points on the output load line when:

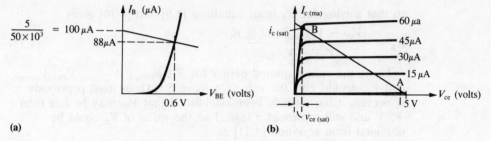

(a)

(b)

Fig. 4.11 Input and output characteristics and load lines for the basic transistor switch.

(a) $V_{CE} = 0$, $i_C = V_{CC}/R_C$ A
(b) $i_C = 0$, $V_{CE} = V_{CC}$ V

Assuming the value of R_C of 1 kΩ then the output load line can be drawn to intersect the V_{CE} axis at +5 V and the i_C axis at 5 mA (V_{CC}/R_C).

The points A and B define the two possible output conditions when using the transistor as a switch. At point A the base current $i_B = 0$ characteristic is used, giving an output voltage of ≅ +5 V and a small collector current due to minority charge carrier leakage current. At point B the base current is 88 μA and from the output characteristic it can be seen that this value of i_B is enough to allow intersection of the load line with the characteristic curve below the 'knee'. This gives an output voltage $V_{CE(sat)}$ of about 0.3 V and a collector current $i_{C(sat)}$ of about 4.7 mA.

It should be noted from Fig. 4.11(b) that the forward current gain h_{FE} of the device is given by 4 mA/40 μA = 100 for a given value of V_{CE} in the active region.

Because the device is operated with a base curent of 88 μA with an output current of 4.7 mA the device is said to be saturated, and the saturated value of forward current gain is given by

$$h_{FE(sat)} = \frac{4.7 \text{ mA}}{88 \ \mu\text{A}} \cong 53$$

It follows that to cause the transistor to saturate the base current must be at least (1/53) of the maximum collector current.

The arrangement described above is an **inverter** gate since when $V_{IN} = 0$ V (logic 0), the transistor is off and $V_{CE} = +5$ V (logic 1), and when $V_{IN} = +5$ V, the transistor is on with $V_{CE} = V_{CE(sat)} = 0.3$ V (≅ logic 0). This basic form of inverter is known as a **resistor-transistor logic** (**RTL**) NOT gate.

Consider again the circuit of Fig. 4.10 under saturation conditions, i.e. $V_{IN} = +5$ V. Then

$$i_{C(sat)} = \frac{V_{CC} - V_{CE(sat)}}{R_C} \tag{4.8}$$

and

$$i_{B(sat)} = \frac{i_{C(sat)}}{h_{FE(sat)}} = \frac{V_{CC} - V_{CE(sat)}}{h_{FE(sat)} R_C} \tag{4.9}$$

but

$$i_{B(sat)} = \frac{V_{IN} - V_{BE(sat)}}{R_B} \tag{4.10}$$

so that solving for R_B from equations [4.9] and [4.10] gives

$$R_B = \frac{(V_{IN} - V_{BE(sat)}) \cdot h_{FE(sat)}R_C}{V_{CC} - V_{CE(sat)}} \qquad [4.11]$$

Using the values quoted earlier for R_C, $h_{FE(sat)}$, $V_{CE(sat)}$, V_{IN}, $V_{BE(sat)}$ and V_{CC} would give the value for R_B of 50 kΩ, as used previously. However, it has already been established that V_{IN} may be *less* than +5 V and still represent a logic 1 so the value of R_B could be obtained from equation [4.11] as

$$R_B = \left(\frac{(3 - 0.6) \times 53 \times 1}{5 - 0.3}\right) k\Omega \cong 27 \text{ k}\Omega$$

This value for R_B would thus ensure that saturation of the transistor results when $V_{IN} = +3$ V; i.e. from equation [4.10]

$$i_{B(sat)} = \frac{(+3 - 0.6)}{27} \text{ mA} = 88 \text{ } \mu A$$

while if V_{IN} should be +5 V then also from equation [4.10]

$$i_{B(sat)} = \frac{(+5 - 0.6)}{27} \text{ mA} = 163 \text{ } \mu A$$

which is well above the required saturation value but still produces an output current of $i_{C(sat)} \cong 4.7$ mA.

The design of the inverter circuit has assumed zero loading effect on the gate. Since the inverter in practice is normally loaded by another circuit then, assuming the effective load of the circuit is purely resistive, the effect can be shown by Fig. 4.12.

The total current flowing in R_C is now no longer i_C only but must contain a component which flows in R_L, the external load; i.e.

$$i_T = i_C + i_L \qquad [4.12]$$

and since

$$V_{CE} = i_L R_L \qquad [4.13]$$

then applying Kirchhoff's voltage law to the series R_C device circuit gives

$$V_{CC} - V_{CE} = i_T R_C \qquad [4.14]$$

or $V_{CC} - V_{CE} = (i_C + i_L)R_C \qquad [4.15]$

(substituting for i_T in equation [4.12])

and since

$$i_L = \frac{V_{CE}}{R_L}$$

$$V_{CC} - V_{CE} = i_C R_C + V_{CE}\frac{R_C}{R_L}$$

or

$$V_{CC} - V_{CE}\left(1 + \frac{R_C}{R_L}\right) = i_C R_C$$

so that

$$i_C = \frac{V_{CC}}{R_C} - V_{CE}\left(\frac{R_L + R_C}{R_L R_C}\right) \qquad [4.16]$$

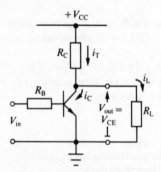

Fig. 4.12 Loading effect on the basic transistor inverter.

The term $(R_L + R_C)/R_L R_C$ in equation [4.16] is the sum of $1/R_L$ and $1/R_C$. In the event of R_L being infinitely large (R_L is open circuit so that the gate is not loaded) then equation reduces to

$$i_C = \frac{V_{CC} - V_{CE}}{R_C} \qquad\qquad [4.17]$$

which is similar to that of equation [4.8], which considered saturation conditions only.

Equation [4.16] is the load line equation for the loaded circuit, and once again two points can be established to define the line:

(a) $V_{CE} = 0$, $i_C = V_{CC}/R_C$

(b) $i_C = 0$, $V_{CE} = V_{CC}\left(\dfrac{R_L}{R_L + R_C}\right)$

The device output characteristics and the 'loaded' load line are shown in Fig. 4.13. The 'unloaded' load line is also shown for comparison purposes.

Fig. 4.13 Effect of 'loading' a transistor inverter in terms of load line on the output characteristics of the device.

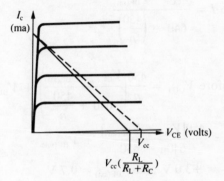

The effect of the load resistor in the device ON state is to reduce the value of base current needed to saturate the device and the unsaturated value of V_{CE} is also reduced. Both these effects are desirable. However, in the OFF state the value of output voltage has been reduced from the unloaded value of V_{CC} to the loaded value of $V_{CC}[R_L/(R_L + R_C)]$. The extent to which the logic 1 level is reduced in the loaded condition depends on the relative values of R_C and R_L with ideally $R_L \gg R_C$. However, R_C should be as large as possible to reduce device power consumption so that once again choice of values is a compromise.

Resistor-transistor logic The basic inverter has already been discussed (p. 00). Consider the circuit of Fig. 4.14. If any input is at logic 1 the associated transistor conducts and the output falls to logic 0. Only if *all* inputs are at logic 0 will the output go to logic 1. The circuit is therefore a NOR gate. This arrangement is a standard for this type of logic and typical values for V_{CC}, R_C and R_B are +3 V, 640 Ω and 450 Ω respectively.

Again, there is a loading problem if the gate is loaded with following RTL (resistor-transistor logic) gates. Figure 4.15 suggests

Fig. 4.14 Basic three-input NOR gate resistor–transistor logic.

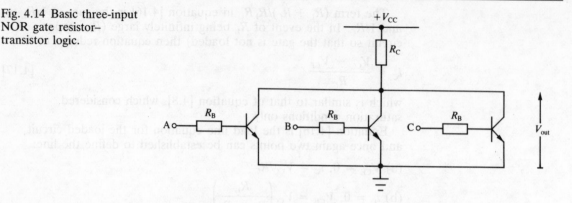

an equivalent loading effect of n succeeding gates when the output of the driving gate is at logic 1 or 3.0 V. From Fig. 4.15

$$V_{OUT} = \frac{450}{n}I + V_{BE(sat)}$$ [4.18]

where $I = \dfrac{V_{CC} - V_{BE(sat)}}{640 + \left(\dfrac{450}{n}\right)}$

Therefore $V_{OUT} = \dfrac{450}{n}\left(\dfrac{V_{CC} - V_{BE(sat)}}{640 + \dfrac{450}{n}}\right) + V_{BE(sat)}$

$$V_{OUT} = \frac{450[V_{CC} - V_{BE(sat)}]}{640n + 450} + V_{BE(sat)}$$ [4.19]

If $V_{CC} = +3.0$ V and $V_{BE(sat)} = 0.7$ V

$$V_{OUT} = \frac{450 \times 2.3}{640\,n \times 450} + 0.7$$

If $n = 0$, as for an unloaded circuit, then
$V_{OUT} = 2.3 + 0.7 = 3.0$ V
if $n = 1$, $V_{OUT} = 1.65$ V
$n = 2$, $V_{OUT} = 1.3$ V, etc

Fig. 4.15 Loading effect on an RTL gate.

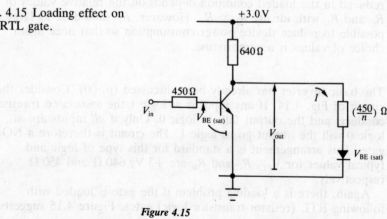

Figure 4.15

As V_{OUT} decreases, the minimum voltage that will drive a succeeding gate into saturation (0.7 V) is approached, leaving only a small voltage difference between this value and the actual output voltage. For example, if $n = 3$, $V_{OUT} = 1.14$ V so that a voltage spike of only 0.44 V (1.14 − 0.7) would cause the circuit to operate incorrectly.

Assuming the gate of Fig. 4.15 has a high voltage at the input specified as 1.14 V (to allow for loading effect) then the base current can be calculated from

$$I_B = \frac{1.14 - 0.7}{0.45} \cong 1 \text{ mA}$$

where $V_{BE(sat)}$ is assumed = 0.7 V, and for a low output of $V_{CE(sat)} = 0.3$ V the collector current can be calculated from

$$I_C = \frac{3.0 - 0.3}{0.64} \cong 4.22 \text{ mA}$$

so that the required transistor forward current gain is

$$h_{FE}(\text{min}) \geqq \frac{I_C}{I_B} \cong 4.22$$

In fact this value for h_{FE} has a built-in safety factor since a typical RTL NOR gate of this type would have a specified fan-out of 5 (i.e. $n = 5$).

Diode transistor logic

A basic DTL (diode transistor logic) NAND gate is shown in Fig. 4.16. This is simply a DRL AND gate followed by an inverter. A major problem in the use of DRL circuits is overcome by the use of the inverter stage and the gain it provides. For a typical NAND gate of Fig. 4.16, values of V_{CC}, R_1, R_2 and R_C are +5 V, 2 kΩ, 4 kΩ and 5 kΩ respectively.

Fig. 4.16 Basic two-input DTL NAND gate.

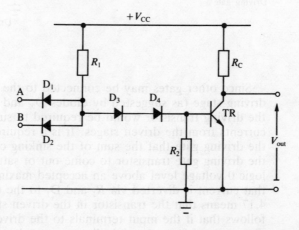

The use of diodes D_3 and D_4 is necessary to prevent the possibility of the 0.7 V, that would otherwise be at the transistor base if either A or B were at logic 0 (0 V), from turning the transistor on. Diodes D_3 and D_4 are effectively in series with the base–emitter diode of the transistor so that to turn the transistor on requires $3 \times 0.7 \text{ V} = 2.1 \text{ V}$ at the anode of D_3. Under the condition that the anode of D_3 is at 0.7 V (for either A or B at logic 0) each of the three diodes must share the voltage so that the $(0.7 \text{ V})/3 \cong 0.23 \text{ V}$ across the base–emitter junction of the transistor is insufficient to turn it on. Under this condition the output voltage is approximately +5 V (logic 1) so that either, or both, inputs A and B at 0 V (logic 0) gives the logic 1 output which is the characteristic of a NAND gate. The resistance R_2 is chosen to be large relative to the resistance of the base–emitter diode when conducting (i.e. R_2 can be neglected when TR is on) and small relative to the input resistance of the non-conducting base–emitter diode (i.e. R_2 shunts input of TR when TR is off). Thus R_2 ensures that the transistor is off even if the value at A or B input is not exactly 0 V.

The DTL gate of Fig. 4.16 is a **current sinking** gate in that if either input is low (produced by the output voltage of a saturated transistor in a driving stage) current from $+V_{CC}$ is diverted via either D_1 or D_2, depending which is at logic 0, to earth via the driving stage transistor. This is illustrated in Fig. 4.17.

Fig. 4.17 The effect of current 'sinking' using DTL when a driving gate at logic 0 is connected to a driven stage.

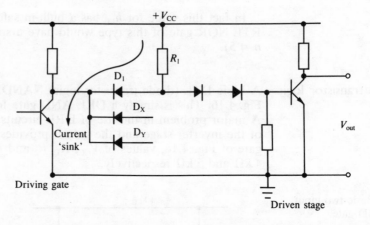

Since other gates may be connected to the logic 0 output of the driving stage (as suggested by diodes D_X and D_Y in Fig. 4.17) then the driving transistor would be required to sink the sum of all currents from the driven stages. It is a requirement of the fan-out of the driving gate that the sum of the sinking current must not allow the driving gate transistor to come out of saturation and raise the logic 0 voltage level above an accepted maximum level. The fact that current is diverted via R_1 and D_1 to the driving stage in Fig. 4.17 means that the transistor in the driven stage is kept off. It follows that if the input terminals to the driven stage are disconnected from the sink then current through R_1 will divert to the driven stage transistor, turning it on. Thus the inputs to the DTL NAND gate need not be high on all inputs due to voltage levels produced by the driving gate; simply leaving the input terminals unconnected will also achieve the same result. Figure 4.18 shows a transfer characteristic for a DTL gate.

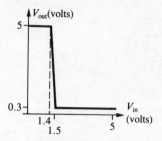

Fig. 4.18 Transfer characteristic for a DTL gate.

Since 2.1 V is required at the anode of D_3 in the DTL gate of Fig. 4.16 for the transistor to conduct, it follows that the voltage at A or B must be 0.7 V less than this value or 1.4 V. Saturation is achieved at 0.1 V higher than this, at 1.5 V, while all voltages higher than 1.5 V will drive the transistor further into saturation. The voltages are identified in Fig. 4.18.

Noise margin

The effect of what is known as **noise immunity**, or **noise margin**, can be gauged from Fig. 4.18. Suppose, for example, V_{in} to the gate is at 0.3 V when the gate is being driven from a saturated driving transistor. From the transfer characteristic the output voltage is +5 V and it would require a change in voltage at the input from 0.3 V to 1.4 V to effect a change in the output voltage state. Thus a noise input spike of 1.1 V would be necessary before the gate switched state; such a voltage is a measure of the noise margin for the low input state. However, if loading of the driving transistor caused the logic 0 level to rise to, say, 0.5 V then the noise margin is reduced to 0.9 V.

Similarly, for a high input condition of, say, 5 V the output voltage is at 0.3 V from the transfer characteristic. It would require a negative voltage noise spike of 3.5 V (+5 V − 1.5 V) to cause the gate to change state. Thus this is the noise margin for the high input state. Again, if loading of the driving stage causes logic 1 level to fall to, say, a minimum value of +3 V the noise margin is reduced to only 1.5 V.

Fan-out

When considering the number of gates that can be driven, of similar type to the DTL driving gate, account need only be taken of the effect when the driving gate output is low. If the transistor of the driving gate is off then V_{OUT} is +5 V and the input diodes of the driven gates are reverse biased; thus the driven gates do not deliver current to the driving gate when the latter's output is high (logic 1).

Figure 4.19 shows the current path through a driving gate when V_{OUT} is low. Only one driven gate is shown in Fig. 4.19. For n such gates the current into the driving gate is nI_1 and total collector current of the driving transistor is

$$I_c = I + nI_1 \tag{4.20}$$

where I is the current required to saturate the driving transistor, and

$$I = \frac{V_{CC} - V_{CE(sat)}}{R_C} \tag{4.21}$$

$$I = \left(\frac{5 - 0.3}{5}\right) \text{mA} = 0.94 \text{ mA}$$

Fig. 4.19 Circuit used to calculate the fan-out for a driving DTL gate.

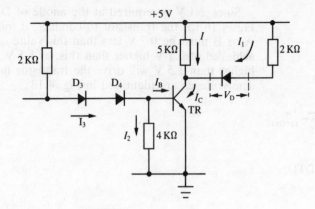

The current I_1 is found from

$$I_1 = \left(\frac{V_{CC} - V_{CE(sat)} - V_D}{2}\right) \text{mA} \qquad [4.22]$$

$$I_1 = \left(\frac{5 - 0.3 - 0.7}{2}\right) \text{mA} = 2 \text{ mA}$$

From the input to the transistor

$$I_3 = I_2 + I_B \qquad [4.23]$$

and assuming saturation with $V_{BE(sat)} = 0.7$ V

$$I_2 = \left(\frac{V_{BE(sat)}}{R_2}\right) \text{mA}$$

$$I_2 = \frac{0.7}{4} = 0.175 \text{ mA}$$

Volt drop across diodes D_3 and D_4 in series is 1.4 V so that the anode of D_3 is at 2.1 V and

$$I_3 = \left(\frac{V_{CC} - 2.1}{2}\right) \text{mA}$$

$$I_3 = \left(\frac{5 - 2.1}{2}\right) \text{mA} = 1.45 \text{ mA}$$

Substituting the values for I_2 and I_3 in equation [4.23] gives

$$I_B = I_3 - I_2$$

$$I_B = (1.45 - 0.175) \text{ mA} = 1.275 \text{ mA}$$

Assuming a transistor with an h_{FE} of 30 the value of I_c is given by $h_{FE}I_B$, so that

$$I_c = (30 \times 1.275) \text{ mA} = 38.25 \text{ mA}$$

Substituting this value for I_c, the value of I from equation [4.21] and the value for I_1 obtained in equation [4.22] into equation [4.20] gives

$$38.25 = 0.94 + 2n$$

Therefore $n = 18.65$

so that a fan-out of 18 may be obtained.

Under the conditions determined above, with a collector current of 38.25 mA there are several possible variations in parameter values that could prevent transistor saturation and hence allow the low output voltage to rise above the minimum value specified. Variations in V_{CC}, h_{FE}, R_C or R_1 come into this category. To allow for this effect a lower value of I_c may be used to ensure transistor saturation even with parameter variations. If a transistor collector current of 25 mA maximum was specified instead of the 38.25 mA calculated above the fan-out would be reduced. The new fan-out would be

$$n = \frac{25 - 0.94}{2} \simeq 12$$

Transistor–transistor logic

Using modern integrated circuit techniques, where it is possible to fabricate multi-emitter transistors, the logic group known as TTL (transistor–transistor logic) has evolved. This has had an enormous impact on logic circuit design and, with extensive use and the consequent mass production, the price has been very competitive. A basic TTL gate is the NAND gate and a typical circuit is shown in Fig. 4.20.

Fig. 4.20 Basic two-input TTL NAND gate.

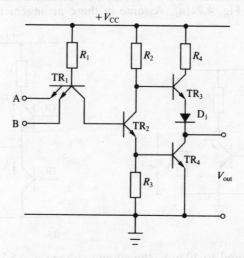

The input circuit of Fig. 4.20 is similar to that of the DTL NAND gate of Fig. 4.16, with TR_1 replacing the input diodes D_1 and D_2 of the DTL gate. TR_2 acts as a phase-splitter giving an input to TR_4 which is the inverse of that applied to the base of TR_3.

When A or B or both are at logic 0 the current flowing in R_1 is diverted via the associated emitter connection to that input. The base–emitter current that flows in TR_1 causes it to saturate and its collector is also at logic 0 level. This low voltage applied to TR_2 base means that TR_2 is off and hence its emitter voltage is low and its collector voltage is high. Thus TR_3 is turned on while TR_4 is off and the output voltage is at logic level 1.

When both A and B are at logic 1 the current R_1 is diverted through the collector of TR_1 to the base of TR_2 and saturates it. Thus TR_2 emitter goes high and its collector goes low causing in turn TR_3 to go off and TR_4 to go on. Thus the output falls to the logic 0 level.

Diode D_1 is used because, under the conditions when TR_4 and TR_2 are both on and saturated, the voltage at the output terminal is 0.3 V while at the base of TR_4 (and emitter of TR_2) the voltage is 0.7 V. Since TR_2 is saturated with $V_{CE(sat)}$ of 0.3 V, the voltage of TR_2 collector is 1.0 V. Thus the voltage at TR_3 base is 1.0 V, giving a potential difference between TR_3 base and the output of 0.7 V which is insufficient to turn on both TR_4 *and* D_1. Without D_1, however, the potential difference would be sufficient to turn TR_4 on.

The output circuit of TR_3, TR_4 and diode D_1 is referred to as a 'totem-pole' output because of the arrangement whereby TR_3 acts as an active load for TR_4. The resistive load of a transisitor, such as R_C in Fig. 4.16 of the DTL NAND gate is often referred to as a passive pull-up resistor and the transistor will not carry current when the output is high. Using a transistor as an effective load for the output transistor gives active pull-up, and for the TTL NAND gate of Fig. 4.20 when the output is high (TR_4 off) TR_3 is saturated and will pass current. The use of active pull-up enables faster circuit switching, requires less power and enables the stage to drive heavier loads.

Consider the passive pull-up arrangement feeding a load as shown in Fig. 4.21(a). Assume in these arrangements that I_C is to be

Fig. 4.21 Comparison of
TTL output stage load
resistors: (a) passive and
(b) active, pull-up resistors.

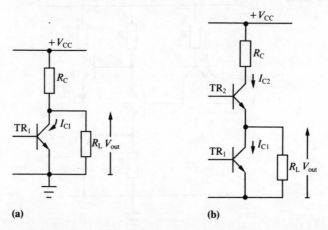

(a)   (b)

limited to 10 mA, the supply voltage is +5 V and that the minimum value for V_{OUT} in the high state if to be +4 V. In the circuit of Fig. 4.21(a) when TR_1 is on then

$$I_{C(max)} = \frac{+V_{CC}}{R_C}$$

(assuming $V_{CE(sat)} = 0$ V and TR_1 does not have to sink any load current). Therefore

$$R_C = \left(\frac{5}{10}\right) k\Omega = 500 \ \Omega \ \text{(minimum)}$$

When TR_1 is off then

$$V_{OUT} = V_{CC}\left(\frac{R_L}{R_C + R_L}\right) \qquad\qquad [4.24]$$

or

$$4 = 5\left(\frac{R_L}{0.5 + R_L}\right)$$

from which $R_L = 2 \text{ k}\Omega$ (minimum).

By contrast, considering the circuit of Fig. 4.21(b) when TR_1 is off then TR_2 is on, so that with a maximum collector current of 10 mA:

$$I_{C_2} = \frac{V_{CC}}{R_C + R_L}$$

(which assumes $V_{CE(sat)}$ for TR_2 is 0 V).

$$10 = \frac{5}{R_C + R_L}$$

or $R_C + R_L = 500 \, \Omega$ (minimum).

Also, since $V_{OUT} = 4$ V for high state and from equation [4.24]

$$V_{OUT} = V_{CC}\left(\frac{R_L}{R_L + R_C}\right)$$

then $4 = 5\left(\dfrac{R_L}{0.5}\right)$

and $R_L = 400 \text{ k}\Omega$ (minimum)

so that $R_C = 100 \, \Omega$ (minimum).

This shows that active pull-up allows the load to be reduced by a factor of 5 compared to the passive pull-up arrangement. This means the fan-out can be increased using an active load. Also the active circuit can sink 10 mA when delivering the logic 0 output voltage; in this case TR_1 is on and TR_2 is off.

TTL gates are available in a variety of types, the three most popular being standard, low-power and high-speed. The gates are basically the same for each of the ranges, the difference being due to the values of resistors R_1 R_2 R_3 and R_4 of the NAND circuit of Fig. 4.20. Table 4.1 shows typical values.

Table 4.1 Comparison of resistance values for low-power (L), standard (STD) and high-speed (H) TTL packages

	R_1 (kΩ)	R_2 (kΩ)	R_3 (kΩ)	R_4 (kΩ)
Low power	40	20	12	500 Ω
Standard	4	1.6	1.0	130 Ω
High speed	2.8	760 Ω	470 Ω	58 Ω

For a fast switching speed it is essential that the stray capacitance associated with the transistor output stage be rapidly charged or discharged. The time constant associated with the charge or discharge process depends on the value of R_4. Because of the effect of the totem-pole output either TR_3 or TR_4 is on at any time and hence R_4 can be made small, giving the gate a low output resistance in both high and low output states.

When the TTL output is high and undergoing a transition to low, as TR_2 turns on rapidly and turns TR_4 on, there is a period of time when both TR_3 *and* TR_4 are conducting. The delay in turning off TR_3 is due to the hole storage phenomenon in the base of a saturated transistor and the time it takes to sweep out the stored charge; TR_3 and TR_4 conducting simultaneously results in a current spike from the supply, the value of which is limited by R_4. The effect of the current spike can produce significant changes in the power supply voltage level and affect the amount of gate dissipation. Additionally, the spike may be looked upon as a noise signal.

The values of R_2 and R_3 will determine the possible value of the current spike while R_2 and R_4 determine the dissipation. Reducing R_2 and R_4 will increase current, and hence dissipation, but will increase speed. Choice of values is a compromise and the values for the standard range of gates reflect this.

Switching speed may be increased by the use of a diode clamping technique which prevents the output transistor from being driven too hard into saturation. This will obviously reduce the time taken to sweep out the stored charge in the base region. The circuit of Fig. 4.22(a) shows the position of a diode clamp between base and collector of the transistor.

Fig. 4.22 Schottky diode: (a) circuit arrangement to prevent transistor saturation; (b) constructional arrangement to produce the Schottky diode; (c) device symbol for Schottky transistor.

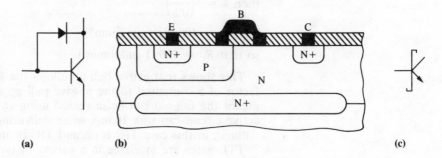

(a)   (b)   (c)

The diode used in Fig. 4.22 is a Schottky diode, and the Schottky transistor is easily fabricated by simply extending the metal base contact so that it also connects with the collector region as shown in Fig. 4.22(b). The anode of the Schottky diode is the metal while its cathode is the collector region. Since no minority carriers can exist in the metal the diode is a majority carrier device with no stored charge; it also has a lower forward volt drop (about 0.4 V) than a silicon pn junction. When the transistor is driven towards saturation current will be diverted via the Schottky diode rather than the transistor collector thus reducing stored charge and speeding the switching speed. A range of Schottky TTL circuits where the diodes are fabricated with the transisitors is available. The Schottky transistor symbol is shown in Fig. 4.22(c).

A typical transfer characteristic for a TTL gate is shown in Fig. 4.23. All the time the input voltage is below 0.7 V the logic level at the output will be about 3.6 V, corresponding to a logic level 1. When point **a** is reached TR_1 begins to supply current to the phase splitter TR_2. Between **a** and **b** TR_2 operates at a fixed gain (R_2/R_3) because TR_4 has not yet begun to conduct. At point **b** the voltage at the base of TR_4 is large enough to turn TR_4 on and the gain of the phase splitter increases. TR_4 goes into saturation, TR_3

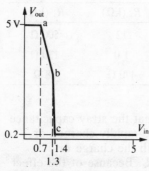

Fig. 4.23 Typical transfer characteristic of a TTL gate.

switches off and the output voltage falls to the saturation value of about 0.3 V.

For TTL circuits with a supply voltage of +5 V the logic 1 level is between 2.4 V minimum and 3.6 V while the logic 0 is between 0 V and 0.8 V maximum. A standard TTL input would require about 40 μA to keep it at logic 1 level and a current of about 1.6 mA for logic 0. The output can deliver 40 μA and 16 mA in these two states so a standard figure for the fan-out is 10.

TTL is a popular logic family with many circuits available in different type categories such as standard, high-speed, low-power, etc. The circuits have tended to standardise around the Texas Instruments 74 series although some manufacturers produce their own numbering schemes for their TTL circuits. Figure 4.24 shows the arrangement for two TTL circuits, the SN7400 which is a quad two-input NAND gate and the SN7402 which a quad two-input NOR gate. (The term 'quad' indicates 4 devices per manufacturer's package.)

Fig. 4.24 Pin-out diagrams for typical TTL circuits: (a) the 7400 quad two-input NAND gate; (b) the 7402 quad two-input NOR gate.

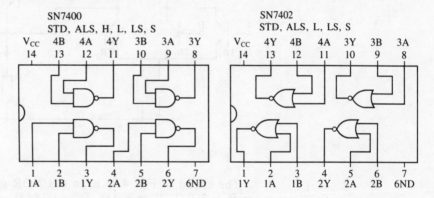

As mentioned in Fig. 4.24, the variations are quoted for a particular circuit, where applicable, to show that the particular circuit is available in the different type categories, i.e.

STD. STANDARD range
LS   Low power Schottky which combines the low-power range
      with Schottky characteristics
H   High-speed range
L   Low-power range
S   Schottky range
ALS   Advanced with combination of LS characteristics

In some circuits the range type is included in the number if that circuit is only available in that type category; for example, the SN74H101 is a J-K flip-flop only available in the high-speed category.

Emitter coupled logic

Emitter coupled logic (ECL), sometimes called current-mode logic (CML) or emitter-emitter coupled logic (E²CL), is very high speed switching logic because the transistors are not allowed to saturate. Switching speeds as fast as 1 ns are possible with this logic although the power dissipation level is high and it is sensitive to temperature variations. Fan-out is good, being about 30.

The basis of the circuit is an emitter coupled amplifier, as shown in Fig. 4.25, consisting of TR_1 and TR_2 and a reference amplifier TR_3. The common emitter resistor is large in comparison with the transistor load resistors so that the current flowing in R_E is essentially constant. Since either TR_1 (and/or TR_2) or TR_3 will always be on, the current through R_E will flow through TR_1 (and/or TR_2) or TR_3, i.e. the current will switch between transistors according to their state of conduction.

Fig. 4.25 Basic two-input ECL OR/NOR gate.

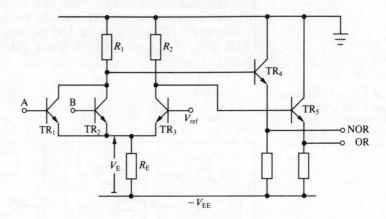

The circuit of Fig. 4.25 is an OR/NOR gate operating with values of $R_1 = 260\ \Omega$, $R_2 = 300\ \Omega$, $R_E = 1.18\ k\Omega$ and $-V_{EE} = -5.2$ V. The value for V_{ref} is -1.18 volts. Logic levels are $0 = -1.58$ V, and $1 = -0.75$ V.

When inputs A and B are low TR_1 and TR_2 will be off and TR_3 on and in the active region. Assuming V_{be} for TR_3 in the active region is 0.75 V then

$$V_E = V_{ref} - V_{BE} \qquad [4.25]$$

Therefore $V_E = -1.18 - 0.75 = -1.93$ V

and current I_E is given by

$$I_E = \frac{V_E - V_{EE}}{R_e} \qquad [4.26]$$

Therefore $I_E = \left[\dfrac{-1.93 - (-5.2)}{1.18}\right]$ mA $= 2.77$ mA

All of this current must pass through R_2 since TR_1 and TR_2 are off. Hence, the voltage at TR_3 collector is

$$V_{C_3} = -(0.3) \times (2.77)\ V = -0.83\ V$$

while V_{C_1} (or V_{C_2}) is at 0 volts.

Since base–emitter voltage across TR_4 and TR_5 is also 0.75 V then the voltage at the output is TR_5 is $(-0.83 - 0.75)$ V $= -1.58$ V while the voltage at the output of TR_4 is $(0 - 0.75)$ V $= -0.75$ V.

If now one or both inputs A or B go high to -0.75 V then the associated transistor conducts and

$$V_E = V_{IN} - V_{BE}$$

Therefore,

$$V_E = -0.75 - 0.75 = -1.5 \text{ V}$$

and from equation [4.26]

$$I_E = \left[\frac{-1.5 - (-5.2)}{1.18} \right] \text{mA} = 3.14 \text{ mA}$$

All of this current flows through R_1 and none through R_2 since TR_3 is cut off. Thus

$$V_{C_1} \text{ (or } V_{C_2}) = -(0.26) \times (3.14) \text{ V} = -0.82 \text{ V}$$

and via TR_4 the output voltage is $(-0.82) - (0.75)$ V $= -1.57$ V while the voltage at TR_5 emitter is $(0 - 0.75)$ V $= -0.75$ V.

The output from TR_4 emitter is thus $\overline{A + B}$, representing a NOR function, while the output from TR_5 emitter is the complement $(A + B)$, giving an OR function.

As may be expected, since ECL is a non-saturating logic the propagation delay is very small, typically 2 ns when the gate is lightly loaded. Loading increases the delay but not by a very marked degree. Because the device is always taking current from the supply the power dissipation per gate is quite high, typically 50 mW per gate. The use of emitter followers in the output stage of the gate assists in the reduction in gate switching speed by virtue of the low output impedance of the emitter follower, which means that charging and discharging times of output capacitance is reduced. The fan-out of the gate is large, typically up to 30 gates, being limited by the need of the gate to act as a current source when the output is at logic 1. Current sourcing means that the drawing gate must supply current to the driven gates. Under these conditions with a large number of gates connected to the driving gate output, the effective emitter resistance of the driving gate is reduced and the logic 1 output level will fall. The total stray capacitance also increases with the number of driven gates, so reducing the discharge time. (The charging of the stray capacitance is via the low output impedance of the emitter follower when the output is rising, but when the output is falling the capacitance must discharge through the driving gate emitter resistor which is typically 1.5 kΩ. It is this which can cause delays in the discharge time when the fan-out is large.) The ECL gate is unusual in that the transistor collectors are taken to ground and the emitters to a negative voltage $(-V_{EE})$. Amongst the reasons for arranging the power supply levels in this way, rather than the 'conventional' way of $+V_{CC}$ for the collectors and ground for the emitters, is protection for the output stages in the event of the output being shorted to earth. In a conventional circuit, shorting the output would place all of the supply voltage across the output emitter follower transistor which would probably result in its destruction. With the ECL arrangement, however, a short across the output merely puts the full supply voltage across the output 1.5 kΩ resistor.

You should now be able to attempt exercises 4.6 to 4.15.

Wired logic

It is sometimes convenient to connect the outputs of certain gates together directly. Figure 4.26 shows the outputs of two gates whose outputs have been connected together. In the circuit of Fig. 4.26 if both TR_1 and TR_2 are off then V_o will be high (approx. $+V_{CC}$) while if either TR_1 or TR_2 is saturated then V_o will be low (approx. 0.3 V). This sort of arrangement is called **wired logic** or **collector logic** since the collectors are wired together. The truth table of Fig. 4.26(b) shows that the arrangement gives a wired-AND arrangement; the effect of an AND gate can thus be simply produced without the need to provide an actual gate.

Consider the current in each of the output transistors of Fig. 4.26(a) for a saturated transistor:

$$I_{o_1} = I_{o_2} = \frac{V_{CC} - V_{CE(sat)}}{R_C}$$

where $R_C = R_{C_1} = R_{C_2}$.

The arrangement shows that R_{C_1} and R_{C_2} are effectively in parallel which means that more power is dissipated when only one transistor is on. Connecting gates together in this fashion can thus be wasteful. However, some gates are available *without* the collector load so that wired logic can be used by connecting the open-collector terminals together and connecting them, via a suitable resistor, to the power supply. Only gates with passive pull-up resistors should then be used with wired logic. If active pull-up transistors are used then, when one of the individual outputs is low and the other high, a continuous short circuit is applied to the power supply via the wired logic. Such a large current can be tolerated on a transient basis but to be applied continuously would cause destruction of the circuit. Figure 4.27 shows such a condition using the totem-pole output stages of TTL gates as the active pull-up circuits.

Consider the logic functions produced by wired logic. Assume that the output stages of Fig. 4.26(a) are of NAND gates with inputs to each gate being A, B and C, D respectively. Thus, if the gates are not wired

$$V_{o_1} = \overline{A.B}$$

$$V_{o_2} = \overline{C.D}$$

and since $V_o = V_{o_1}.V_{o_2}$ [4.27]

then $V_o = \overline{A.B}.\overline{C.D}$ [4.28]

or using De Morgan's theorem

$$V_o = \overline{A.B + C.D}$$ [4.29]

The circuit arrangement is shown in Fig. 4.28.

The symbolic representation of equations [4.28] or [4.29] is shown in Fig. 4.28. The outputs of the NAND gates are shown connected *through* an AND gate to indicate that wired logic and not an actual AND gate is used.

For a NOR gate arrangement using the same inputs

$$V_{o_1} = \overline{A + B}$$

$$V_{o_2} = \overline{C + D}$$

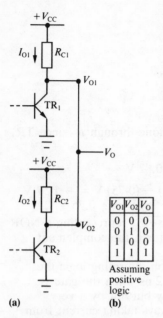

V_{O1}	V_{O2}	V_O
0	0	0
0	1	0
1	0	0
1	1	1

Assuming positive logic

(a)   **(b)**

Fig. 4.26 Wired logic: (a) circuit arrangement; (b) truth table.

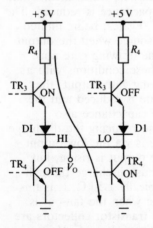

Fig. 4.27 Arrangement showing the potentially dangerous current flow that can arise using TTL totem-pole output stages in wired logic.

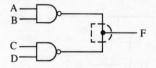

Fig. 4.28 Circuit symbol for wired-AND logic.

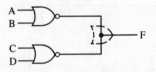

Fig. 4.29 Circuit symbol for wired-OR logic.

and since $V_o = V_{o_1}.V_{o_2}$ from equation [4.27]

$$V_o = \overline{A + B}.\overline{C + D}$$

and by De Morgan's theorem

$$V_o = \overline{A}.\overline{B}.\overline{C}.\overline{D}$$

$$V_o = \overline{A + B + C + D} \qquad [4.30]$$

As equation [4.30] shows, the resulting output is the NOR function of all input signals. This means the wired-OR arrangement can be used to increase the fan-in of a NOR network. The circuit arrangement of equation [4.30] is shown in Fig. 4.29.

You should now be able to attempt exercises 4.16 to 4.18.

Unipolar (MOSFET) logic

MOSFET (or MOS) devices are metal oxide semiconductor devices which can be constructed with n- or p-channel fabrication.

P-channel devices

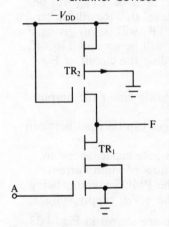

Fig. 4.30 A simple p-channel MOS inverter.

Since a p-channel transistor turns on when a negative voltage is applied to the gate then it is usual in p-channel devices to use negative logic. In the case of the simple inverter shown in Fig. 4.30, with the input at 0 V (logic 0) the device is off and the output is high ($\simeq -V_{DD}$ or logic 1). When the input goes to logic 1, the device conducts and the output is approximately 0 V, or logic 0.

In Fig. 4.30 the devices used are the enhancement type which have no physical conducting channel between source and drain. A channel is induced by the application of a suitable negative gate voltage. Transistor TR_2 is connected as an active pull-up load. TR_2 is saturated by connecting the gate to the $-V_{DD}$ supply rail. The insulated gate gives a very high input impedance for the device, indicating a possibly large fan-out. However, each device has parasitic capacitance and since the total capacitance is the sum of the driven gates there are long time constants associated with changes in output levels and the unipolar devices are very slow. Typical propagation delay for a p-type MOS logic gate is 100 ns.

Possible circuits for p-channel MOS NAND and NOR gates are shown in Figs. 4.31(a) and (b). It can be seen from Fig. 4.31(a) that both A and B must be at logic 1 before the output can fall to logic 0, so that the circuit provides the NAND function. From Fig. 4.31(b), if either A or B is at logic 1 the output will fall to logic 0. Only if A = B = 0, will the output go high, which is the requirement of a NOR gate.

CMOS circuits

CMOS is complementary symmetry MOS based on a combination of p-channel and n-channel devices. A CMOS inverter circuit is shown in Fig. 4.32. In this arrangement the active load is p-channel while the inverter device is n-channel, or vice versa. Note that with the use of a substrate lead the substrate should, for p-channel devices, always be connected to the most *positive* rail of the supply while for

Fig. 4.31 PMOS circuits for: (a) two-input NAND gate; (b) two-input NOR gate.

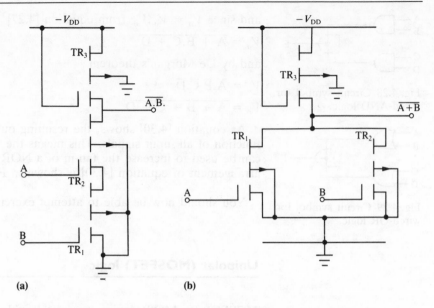

(a)   (b)

Fig. 4.32 Simple inverter using complementary MOS circuitry.

and n-channel device the substrate should be connected to the most *negative* rail of the supply. This assumes, as in this case, that the substrate terminal is not being used as an input terminal.

For a CMOS circuit a voltage at or above about 70% of $+V_{DD}$ will be logic 1 while a voltage below about 30% of $+V_{DD}$ will be logic 0. Thus if A in Fig. 4.32 is at logic level 0, TR_2 will be on (gate negative with respect to $+V_{DD}$) and TR_1 will be off giving F at logic 1. Conversely, if A is at logic 1 TR_1 will be on and TR_2 off, giving F at logic 0. The main points regarding the circuit of Fig. 4.32 can be summarised as follows:

- The current drawn from the supply during either the output HIGH or LOW state is minimal.
- Varying values of power supply voltages can be used between 3 V and 18 V.
- Except during the periods when input gate capacitances are being charged/discharged there is no flow of input current.
- The noise margin is better than for the PMOS inverter being typically 45% of the value used for the $+V_{DD}$ supply voltage.

Circuits for CMOS NAND and NOR gate are shown in Fig. 4.33 (a) and (b).

For the circuit of Fig. 4.33(a), the output voltage is only at logic 0 if both inputs are at logic 1. Under these input conditions TR_1 and TR_2 are on while TR_3 and TR_4 are off. Any other combination of inputs will isolate the output from the earth line and connect it via TR_3 or TR_4 to supply, giving a logic 1. Thus the circuit functions as a NAND gate.

For the circuit of Fig. 4.33(b), the output is only at logic 1, if both inputs are at logic 0, then TR_3 and TR_4 are on and TR_1 and TR_2 are off. Any other combination of inputs will isolate the output from the supply rail and connect it via TR_1 or TR_2 to earth giving a logic 0. Thus the circuit functions as a NOR gate.

CMOS logic appears similar to TTL circuits with active pull-up loads since no resistance loads are used. Thus the circuits are

Fig. 4.33 CMOS circuits
for: (a) two-input NAND
gate; (b) two-input NOR
gate.

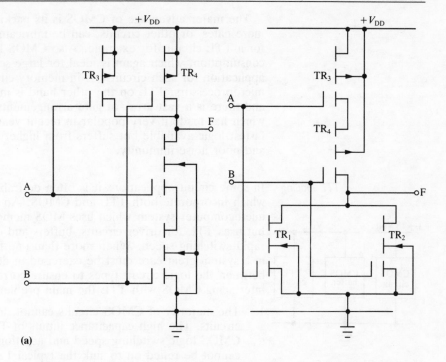

(a)

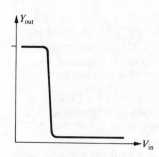

Fig. 4.34 Transfer
characteristic for a CMOS
gate.

capable of producing a relatively low impedance drive when in
either logic state, enabling it to run at higher speeds compared to,
say, PMOS circuits. Fan-out is also very large because of low output
impedance and high input impedance. However, the circuit during
its switching cycle passes through a stage where both p- and n-
channel transistors are on, giving rise to current spikes and
consequent noise (this is similar to the problem found with the
totem-pole output in TTL circuits). The transfer characteristic for
CMOS is shown in Fig. 4.34. No voltage levels have been shown
because the circuits can operate over a range of supply voltages, as
previously mentioned. The curve has a sharp 'knee' giving good
noise immunity.

Many logic functions are now available as integrated circuits in
the CMOS 4000 series. Figure 4.35 shows the arrangement for two
such circuits, 4001 quad two-input NOR gate and 4011 quad two-
input NAND gate.

Fig. 4.35 Pin-out diagrams
for typical CMOS circuits:
(a) the 4001 quad two-input
NOR gate; (b) the 4011
quad two-input NAND
gate.

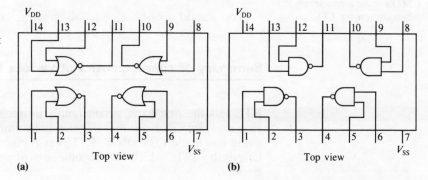

(a) (b)

The major advantage of CMOS is its packing density, i.e. many more gates, or other circuits, can be fabricated on a single chip than for a TTL circuit, for example. Also CMOS has a low power consumption, which again is ideal for large scale circuits. A major application for such circuitry is in memory circuits and microprocessors. TTL on the other hand is much faster than CMOS and there is a vast array of gate arrangements available in this logic which has made it very popular in recent years. ECL is in fact the fastest logic available but suffers from higher power consumption and poor noise immunity.

TTL loading   In some circuit applications it is often desirable to design systems which incorporate both TTL and CMOS. An example of this is a microcomputer system which uses MOS memory and microprocessor but uses TTL for driver circuits, buffers and decoders because of its rapid switching speed. When more than one family of logic is used in a system great care must be exercised in designing interfaces between the logic circuit types to ensure correct operation. When interfacing CMOS with TTL the main problems are:

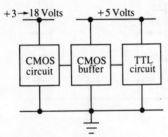

Fig. 4.36 The use of a CMOS buffer interface when driving TTL with CMOS.

1.   The majority of CMOS circuits cannot drive standard TTL circuits. The high capacitance inputs of TTL will affect the CMOS logic switching speed and additionally the CMOS circuit cannot be relied on to sink the typical 1 mA of currents from a low TTL input without coming above the 0.5 V maximum specified for a TTL input.

2.   TTL requires 5 V whereas CMOS can operate from 3 to 18 V. If both types of logic are operated at 5 V the TTL high output of about 3.6 V cannot be used to drive a CMOS gate requiring an input of 5 V. Some sort of interface circuit is required to allow for the differing voltage levels under this condition.

The first problem above can be solved by the use of CMOS buffers such as the 4049 and 4050, which are hex inverting and hex non-inverting buffer circuits respectively, designed to drive two TTL inputs. Figure 4.36 shows a possible arrangement.

The second problem can be solved by using a pull-up resistor of approximate value 1 kΩ to bring the TTL output up to +5 V from its normal value of +3.6 V. Figure 4.37 shows the arrangement.

When using a CMOS power supply voltage other than +5 V an interface circuit such as an inverter is included, to act as a voltage translator to change the 3.6 V of the TTL high output to the required value of the CMOS circuit (between 3 and 18 V).

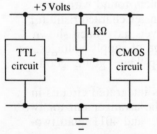

Fig. 4.37 The use of a pull-up resistor to obtain correct CMOS logic input levels when driven by TTL.

You should now be able to attempt exercises 4.19 to 4.26.

Summary of characteristics of various logic families

RTL was the first logic circuit family in integrated circuit form. However, its disadvantages of poor noise immunity and low logic swing has made it obsolete. RTL was replaced initially by DTL and latterly by TTL, which has become almost universal for bipolar type circuits.

The relative merits of TTL and CMOS have already been mentioned in the previous section. Reference to Table 4.2 gives pertinent details regarding the more usually specified parameters for each family.

Table 4.2 Comparison of various logic family parameters

	RTL	DTL	TTL	ECL	PMOS	CMOS†
Propagation delay (nanosecs)	12	25	10	2	100	40
Power dissipation (mW/gate)	20	10	10	50	5 mW (logic 0) 0 (logic 1)	5 nW
Noise margin (volts)	0.4	1.0	1.2	0.4	2.0	2.25
Power supply (volts)	5	5	5	−5.2	−20	5
Threshold voltage (volts)	0.7	1.4	1.4	−2.15	−5	2.25
Logic 0 voltage (volts)	0.3	0.3	0.3	−1.55	−3	0
Logic 1 voltage (volts)	2.5	4.0	3.6	−0.75	−11	4.9
Fan-out	5	8	10	30	5	25

† Supply voltage for CMOS is from 3V to 18V. Value of 5 V chosen for comparison with TTL.

It should be borne in mind that these parameters are typical and should not be taken as hard and fast values. Values for logic 1 and 0, for example, depend on factors such as loading and have certain minimum and maximum values respectively which should not be exceeded for successful operation.

You should now be able to attempt exercises 4.27 and 4.28.

Flip-flops and latches

A flip-flop may be defined as a 1-bit memory element (or **latch**) having two outputs which take up complementary states, i.e. 0 and 1, when a signal is applied to the input. The output condition would then be retained until another input signal combination causes the output state to change.

Some of the most widely used flip-flops such as the SR (set-reset), JK, D and T-type will be discussed in this section. The JK is the most versatile of the various types of flip-flop since it can be used as a basic element to generate all of the other types.

The SR flip-flop The **SR flip-flop** may be constructed from NOR gates, as shown in Fig. 4.38(a). The outputs Q and $\overline{Q}$ must be complementary and since each output is cross-connected back to form an input then these two inputs must always be complementary.

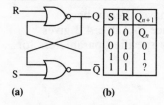

Fig. 4.38 The SR flip-flop:
(a) circuit diagram using
NOR gates; (b) truth table.

S	R	Q_{n+1}
0	0	Q_n
0	1	0
1	0	1
1	1	?

Consider the case when $Q = 1$, $\overline{Q} = 0$ when the inputs $S = R = 0$. Since the gates are NOR gates then if either of the two inputs to a gate is logic 1 the output must be logic 0. If both inputs are logic 0 the output is logic 1. Thus for $S = R = 0$. the output values for Q and $\overline{Q}$ remain the same, since with $Q = 1$ and $S = 0$ as the two inputs to one of the NOR gates the output $(\overline{Q})$ must be 0. Similarly with $\overline{Q} = 0$, $R = 0$ as the two inputs to the other NOR gate the output (Q) must be 1. Thus the outputs remain the same after the input conditions have been realised (Q_{n+1}) as they were prior to those input conditions being applied (Q_n). The outputs are said to be **inhibited**. The same conditions apply if $Q = 0$, $\overline{Q} = 1$ and $S = R = 0$ was applied.

If R stays at 0 but S goes to 1 when $Q = 0$, $\overline{Q} = 1$, then Q changes to 1 since the S input signal at logic 1 must change $\overline{Q}$ to 0 and this feedback signal, together with $R = 0$, produces $Q_{n+1} = 1$. It can be shown that regardless of the state of Q prior to the input conditions $S = 1$, $R = 0$ being applied, the value $Q_{n+1} = 1$ always applies *after* the $S = 1$, $R = 0$ condition is established at the input. Under these conditions the Q output is said to be 'set'.

If now S goes to 0 but R goes to 1, then it follows that whatever the initial values of Q and $\overline{Q}$, the value of Q_{n+1} is always 0. This is known as the reset (or clear) condition.

If S and R go to logic 1 simultaneously then both outputs try to fall to logic 0 and it is impossible to predict which output will in fact take up the 0 condition and which will rise to the 1 state. This condition is thus indeterminate. The truth table for the SR flip-flop is shown in Fig. 4.38(b).

The SR flip-flop may also be produced using NAND gates, as shown in Fig. 4.39(a). Inverters are required for the NAND version as shown. The conditions for a NAND gate output to fall to logic level 0 is that both inputs must be logic level 1, all other input conditions giving a logic 1 output. Thus if $S = R = 0$ with $Q = 0$, $\overline{Q} = 1$, both inputs to the top NAND gate are 1 and thus the output (Q) is 0. Also the inputs to the bottom NAND gate must be complementary, since $Q = 0$, and hence the output $(\overline{Q})$ is 1. Thus $S = R = 0$ gives the inhibit condition just as for the NOR gate circuits. Inspection will show that the truth table of Fig. 4.38(b) also applies to the NAND circuit of Fig. 4.39(a). The inverters may also be NAND gates since a single input NAND gate performs the invert function. The SR flip-flop symbol is shown in Fig. 4.39(b).

The SR flip-flop may be controlled by a control signal or clock pulse so that the S and R signals can be clocked into the flip-flop at a predetermined time. Figure 4.40(a) shows a possible arrangement. For the NAND circuit of Fig. 4.39(a) no extra gates are necessary to produce the clocked version since the inverters in that circuit can be used. Figure 4.40(b) shows a clocked SR flip-flop symbol.

Fig. 4.39 The SR flip-flop:
(a) circuit diagram using
NAND gates; (b) device
symbol.

Fig. 4.40 Clocked SR flip-
flop: (a) circuit diagram;
(b) device symbol.

Output waveforms for an SR flip-flop are shown in Fig. 4.41(a) for an unclocked circuit and Fig. 4.41(b) for the clocked circuit. The advantage gained by using the clock input is that the logic levels at the S and R input only take effect when the clock input goes high.

Fig. 4.41 Typical timing diagrams for the SR flip-flop: (a) unclocked circuit; (b) clocked circuit.

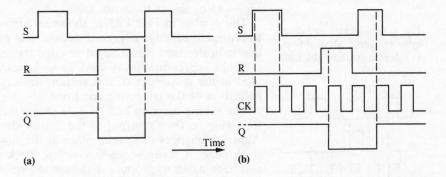

The combination of S and R inputs will determine the output state in both types of circuit but the clock in the second type of circuit will determine *when* the changes occur. In both types of circuits indeterminacy will occur if S = R = 1 and this input combination should be avoided.

Master-slave flip-flop

The **master-slave flip-flop** is a more complicated circuit than the basic SR flip-flop but it can operate at higher speeds. The circuit is basically two SR flip-flops in series with the clock pulse to the second flip-flop (the slave) inverted by comparison with the pulse applied to the first flip-flop (the master). This arrangement means that the outputs from the master flip-flop can be changed but the output from the slave remains unchanged. Figure 4.42(a) shows the circuit arrangement while Fig. 4.42(b) shows the clock pulse

Fig. 4.42 Master-slave SR flip-flop: (a) circuit diagram using NAND gates; (b) clock waveform.

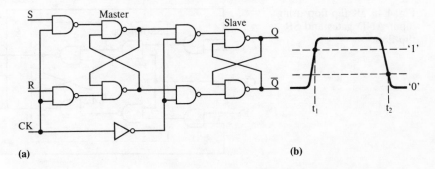

waveform. When the clock pulse rises to logic 1 level the information at the master input is gated to the output of the master but is not applied to the slave. When the clock pulse falls to logic 0 level the master output data is gated to the slave output while the master is inhibited. Data is gated to the output of the flip-flop on the trailing edge of the clock pulse. This is known as **negative-edge triggering** and is a useful application which will be examined in more detail in later circuits.

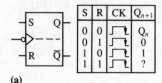

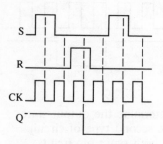

S	R	CK	Q_{n+1}
0	0	⊓	Q_n
0	1	⊓	0
1	0	⊓	1
1	1	⊓	?

(a)

Fig. 4.43 Clocked negative-edge triggered SR flip-flop: (a) device symbol; (b) truth table.

Fig. 4.44 Timing waveforms for clocked negative-edge triggered SR flip-flop.

Since the inputs to the flop-flop do not need to remain static during the period of the clock pulse, as in the original SR flip-flop, but can be changed ready for the next cycle while data is transferred from the master to slave flip-flop, the operating speed is higher. Figure 4.43(a) shows the symbol for the master-slave circuit while Fig. 4.43(b) shows the truth table.

The symbol in Fig. 4.43(a) shows an arrowhead at the clock input to represent an edge-triggered device. The circle on the input CK line indicates that it is a negative-edge triggered device. This symbolic representation is used by some manufacturers and will be used in the remainder of this section since it provides a good visual indication of the triggering condition.

The timing diagram of Fig. 4.44 shows the effect of negative-edge triggering on the Q output of the master-slave SR flip-flop. Although changes in Q take place at the time of the negative-going clock edge, it really needs a complete clock pulse going low to high and back again to produce a change. Some manufacturers show this by a small sketch of a complete clock pulse in the clock column of the truth table. This has been done in Fig. 4.43(b).

The JK flip-flop

The major disadvantage of the SR flip-flop is the indeterminate condition assumed by the outputs when both S and R inputs are taken to logic 1 simultaneously. The **JK circuit** is a modification of the SR circuit which removes the indeterminacy. The difference in the JK circuit compared to the SR verson is the feedback paths from the Q and $\overline{Q}$ outputs and the provision of AND gates at the input. Figure 4.45 should make the arrangement clear.

Fig. 4.45 JK flip-flop using input AND gates and SR flip-flop.

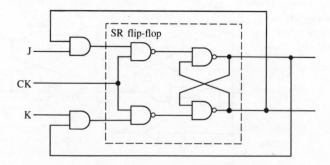

Suppose for Fig. 4.45 the output condition is that $Q = 1$, $\overline{Q} = 0$. Then if $J = K = 1$ the inputs to the SR flip-flop are $S = 0$, $R = 1$ and the clock pulse will result in the output levels changing to $Q = 0$, $\overline{Q} = 1$. These new output levels fed back to the input AND gates means that the next time a clock pulse arrives it is the other input to the SR flip-flop that is at logic 1, i.e. $S = 1$, $R = 0$, and the output changes back to $Q = 0$, $\overline{Q} = 1$ on the next clock pulse. This process will repeat since only one of the input AND gates is ever

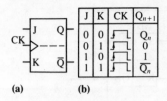

(a)   **(b)**

Fig. 4.46 JK flip-flop: (a) device symbol; (b) truth table.

enabled at a given time, by virtue of $J = K = 1$ and the state of the feedback output signal, and this alternates with successive clock pulses. This change in output state following a clock pulse when $J = K = 1$ is known as **toggling**.

The circuit symbol and truth table are shown in Fig. 4.46(a) and (b) respectively. With this type of circuit arrangement, because the change in output state is fed back directly to the input of the circuit, then if the clock pulse remains at level 1, the outputs will continue to change. This oscillation can be prevented if the clock pulse remains high long enough to initiate transition in the output state but falls to a low level before the transition is complete. One way of achieving this is to use edge-triggering circuits.

Edge-triggering circuit

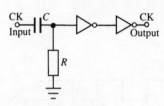

Fig. 4.47 Circuit arrangement for the generation of narrow clock pulses.

In those applications where it is necessary to generate narrow clock pulses (as suggested earlier) a simple arrangement using a differentiating *CR* network and TTL inverters may be used. Figure 4.47 shows a possible circuit arrangement.

It is assumed that the clock input to the circuit of Fig. 4.47 is a train of wide pulses and, assuming TTL voltage levels, the amplitude varies from approximately 0 V to a high of 3.6 V. The *CR* circuit differentiates these pulses to produce positive and negative spikes. The output is a narrow clock pulse coincident with the positive spike as shown in Fig. 4.48.

As a general rule the width of the output clock pulse is about 70% of the time constant of the differentiating circuit so that for $C = 470$ pf and $R = 330 \, \Omega$ the output clock pulse width is approximately 100 ns.

A negative-edge trigger could easily be produced from the circuit of Fig. 4.47 by simply placing another inverter between the clock input and the differentiating circuit.

Fig. 4.48 Waveforms for the circuit of Fig. 4.47 showing the reduced width output pulses.

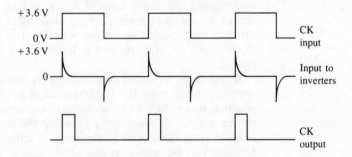

Master-slave JK flip-flop

Another solution to the problem of clock pulse widths producing output oscillations is to use the master-slave arrangement with the JK circuit. This gives a buffer between input and output which prevents output oscillations. Figure 4.49 shows the arrangement.

The circuit symbol is the same as that shown in Fig. 4.46(a) except that the circuit of Fig. 4.49 has more than one set of inputs, which is possible with this type of circuit. The truth table for this circuit is also the same as that shown in Fig. 4.46(b) although now the change of stage at the output is produced by the negative-going edge of the clock pulse. It should be noted that if the clock pulse stays at logic 1 (or 0) then regardless of the state of the JK inputs

Fig. 4.49 Master-slave JK
flip-flop using SR flip-flop
elements.

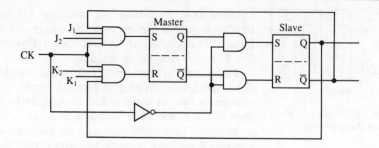

the output will be inhibited. Transition at the output will only occur on the negative edge of the trigger clock pulse in this circuit. Possible timing diagrams for master-slave JK flip-flop are shown in Fig. 4.50(a) and (b).

Fig. 4.50 Timing diagrams
for the master-slave JK flip-
flop: (a) for various values
of J and K; (b) in toggle
mode when J = K = 1.

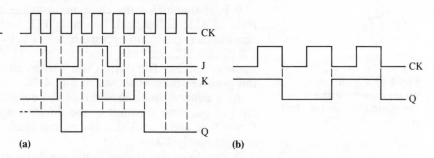

Figure 4.50(b) shows the effect on the output waveform of maintaining the input state at J = K = 1 to produce a toggle mode circuit where the circuit changes state after every clock pulse. The effect is to produce an output waveform at *half* the frequency of the clock. Such an arrangement is known as a divide-by-two ($\div 2$) circuit.

It has been assumed that transitions at the output occur simultaneously with the triggering edge of the clock pulse. In practice this is not so since there is a propagation delay within the device which, although only delaying the output by a few nanoseconds compared to the time of triggering, could limit clocking frequencies, especially in cascaded circuits. Also there are specified times during which inputs should be held stable for correct circuit operation; these times include:

(a) pulse width (t_w), which is the minimum time during which the clock level, either high or low, should be held stable; times may vary according to the clock level;
(b) set-up time (t_{su}), which is the time during which logic levels must be held stable at input terminals before the clock changes;
(c) hold time (t_h) which is the time for which logic levels should be held stable after the clock changes.

Typical values for a TTL JK circuit (74110) and a CMOS circuit (4027B) also operating at +5 V supply level are shown in Table 4.3.

Table 4.3 Timing comparison between a typical TTL circuit and a CMOS circuit (all times in nanoseconds)

Device	t_w (high)	t_w (low)	t_{su}	t_h
74110	25	25	20	5
4027B	80	95	95	15

Preset and clear

Many of the commercial flip-flops have two additional inputs which enable the outputs of the flip-flop to be set independently of the state of the clock and the normal inputs. An input of this kind which sets the Q output of the flip-flop to 1 independently of the clock is called a preset input, while the input which sets Q to 0 independently of the clock is called a clear input.

Consider the all-NAND clocked RS flip-flop preset and clear facilities as shown in Fig. 4.51(a). In this arrangement the inputs preset and clear would be active *low*, i.e. a negative-going pulse must be applied to affect the output state. This is particularly useful in a TTL arrangement when, in the absence of a physical connection, the inputs assume a logic 1 level and this would allow the flip-flop to function as a conventional SR flip-flop. Note that the condition when preset = clear = 0 should not be allowed to occur.

The circuit symbol is shown in Fig. 4.51(b). The circle on the preset and clear inputs indicates that those inputs are active low and this is why in Fig. 4.51(a) the inputs are shown as $\overline{\text{preset}}$ and $\overline{\text{clear}}$; the bar indicates the input is active low.

Fig. 4.51 Clocked SR flip-flop with preset and clear inputs: (a) circuit using NAND gates; (b) device symbol.

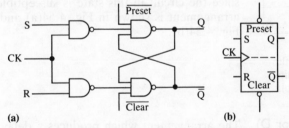

(a)   (b)

The flip-flops, which are simply an extension of the SR type anyway, also use the preset and clear inputs as Fig. 4.52 shows. This

Fig. 4.52 SN74H106 Dual JK flip-flop package using TTL.

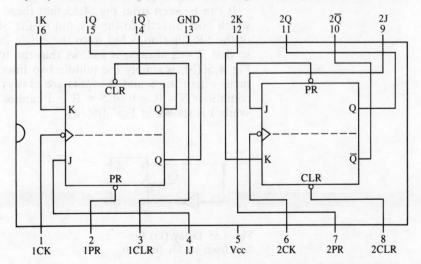

is a typical TTL circuit involving two JK flip-flops, with preset and clear inputs, on a single chip.

The trigger (or T) flip-flop

The **trigger (T) flip-flop** is a one-input flip-flop where a logic 1 on the input causes the output to toggle or 'trigger' after the receipt of a clock pulse. A T flip-flop is easily constructed from an SR flip-flop as shown in Fig. 4.53(a). To eliminate the problem of output oscillation should the input remain at 1 while the clock stays high, the SR flip-flop should be a master-slave type. The circuit truth table is shown in Fig. 4.53(b).

Fig. 4.53 Trigger (T) flip-flop: (a) circuit layout using an SR flip-flop; (b) circuit truth table.

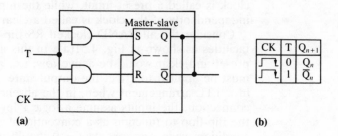

(a)

(b)

A JK flip-flop can be converted to a T-type merely by connecting the J and K inputs to a logic level 1. This condition could also be achieved by leaving the inputs unconnected since they will then assume a logic level 1; this latter condition is not always satisfactory since the circuit in this state is susceptible to noise. A circuit arrangement is shown in Fig. 4.54(a) and a symbol for the T-type in Fig. 4.54(b).

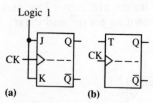

(a)　　　　(b)

Fig. 4.54 Trigger (T) flip-flop: (a) using a JK flip-flop; (b) device symbol.

The delay (or D) flip-flop

The arrangement which produces a delay of one clock period is a **delay (D) flip-flop**. A circuit using the SR flip-flop as a basis is shown in Fig. 4.55.

It can be seen from Fig. 4.55 that there is only one input line which goes directly to the S input of the SR flip-flop. The R input of the SR flip-flop is fed from the single input line via an inverter so that $R = \bar{S}$ always. It follows that the truth table, shown in Fig. 4.56(b), is simply the middle two lines of the SR flip-flop truth table where the S and R inputs are always complementary, i.e. the conditions $S = R = 0$ or $S = R = 1$ cannot exist. A typical device symbol is shown in Fig. 4.56(a).

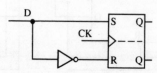

Fig. 4.55 Delay (D) flip-flop from an SR flip-flop.

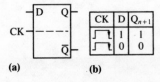

(a)　　　　(b)

Fig. 4.56 Delay (D) flip-flop: (a) device symbol; (b) truth table.

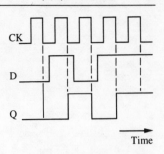

Fig. 4.58 Delay (D) flip-flop timing diagram.

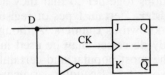

Fig. 4.57 Delay (D) flip-flop using a JK flip-flop.

Again, the D-type can be constructed from a JK flip-flop as shown in Fig. 4.57. A possible timing diagram for a D-type is shown in Fig. 4.58. The delay of one clock period can be clearly seen from these waveforms. Because the input to the D-type flip-flop may contain serial data (i.e. data in serial binary digit or bit form), the device is often referred to as a data flip-flop.

Common integrated circuit D-type flip-flops in both TTL and CMOS together with their truth tables are shown in Fig. 4.59 and 4.60 respectively. The TTL device is the SN7474 in which the set and clear inputs are referred to as set direct (SD) and clear direct (CD) respectively. The CMOS device is the 4013 which has the set (S) and reset (R) terminals labelled accordingly.

Fig. 4.59 SN7474 D-type TTL flip-flop: (a) circuit symbol; (b) truth table.

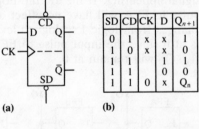

(a)

SD	CD	CK	D	Q_{n+1}
0	1	x	x	1
1	0	x	x	0
1	1	↑	1	1
1	1	↑	0	0
1	1	0	x	Q_n

(b)

Fig. 4.60 CMOS 4013 D-type flip-flop: (a) circuit symbol; (b) truth table.

(a)

S	R	CK	D	Q_{n+1}
1	0	x	x	1
0	1	x	x	0
0	0	↑	1	1
0	0	↑	0	0
0	0	0	x	Q_n

(b)

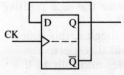

Fig. 4.61 D-type flip-flop used as a trigger (T) flip-flop.

The T-type may be constructed from a D-type by simply connecting the $\overline{Q}$ output back to the D input, as shown in Fig. 4.61.

You should now be able to attempt exercises 4.29 to 4.40.

Binary counters

Grouping flip-flops together so that they act as a data store produces a **register**. Certain types of register can be used to count pulses and are known as **counters**. Although, in general, any of the flip-flops already described may be used in a counter the JK flip-flops are the most popular and certainly the most flexible. There are basically two types of counter, asynchronous and synchronous.

Asynchronous counters have the pulses to be counted applied at one end of the counter and the binary output can record a count of 0 or 1. When the count exceeds 1 (i.e. binary 2) then the stage must reset to 0 and a carry of 1 is transferred to the next stage. Similar conditions apply for the next stage in that it too can only record an output of 0 or 1 and to exceed this count will mean resetting the output to 0 and transferring the carry of 1 to the next stage and so on. Thus a carry bit is seen to **ripple-through** the counter until the count is complete. For this reason asynchronous counters are often referred to as ripple-through counters.

Synchronous counters have the clock pulses applied to each stage of the counter simultaneously so that all output changes occur simultaneously. This is useful in those cases where the propagation delay associated with the ripple-through counter may be a problem.

Asynchronous counters

Consider the three-stage circuit of Fig. 4.62 using JK flip-flops connected in the toggle mode. Assume that all outputs are initially reset to zero by an input going low in the clear line prior to the input signal appearing. If the JK flip-flops are master-slave then the input pulse to FFA will have no effect on the Q_A output until the pulse changes from logic 1 to logic 0 level. However, the output Q_B remains at 0 since the input pulse to FFB has gone from 0 to 1. Similarly, Q_C will remain at 0.

Fig. 4.62 Asynchronous three-stage binary counter using JK flip-flops in the toggle mode.

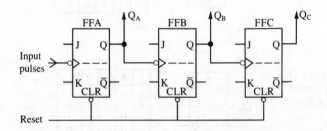

When the second input pulse has arrived and gone from logic 1 to logic 0, then FFA output again changes state and returns to 0. However, this is a negative transition and as such will cause the output of FFB to toggle so that Q_B goes to 1. Since this is a positive transition Q_C will be unaffected and remains at 0.

At the end of the third input pulse, Q_A again toggles, this time to logic 1 but Q_B and Q_C retain their values of 1 and 0 respectively.

At the end of the fourth input pulse, Q_A toggles and falls to 0. The transition from 1 to 0 at the input to FFB causes Q_B to fall to 0 and this transition in turn causes FFC to toggle so that Q_C goes to 1.

The procedure repeats until, as the truth table and waveforms of Fig. 4.63 show, the counter resets itself and begins to repeat the count should the input signals still be present.

Fig. 4.63 Asynchronous three-stage binary counter: (a) truth table; (b) timing waveforms.

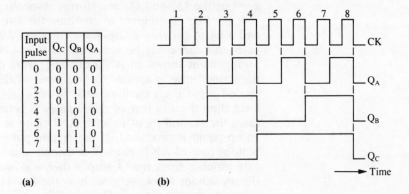

Input pulse	Q_C	Q_B	Q_A
0	0	0	0
1	0	0	1
2	0	1	0
3	0	1	1
4	1	0	0
5	1	0	1
6	1	1	0
7	1	1	1

(a)   (b)

Inspection of the truth table of Fig. 4.63 shows that the values of Q_A Q_B and Q_C in binary form correspond to the decimal value of the input pulses counted. Such a representation is known as a pure binary count and since the counter counts 'up' from 0 to its maximum value, in this case of 7, the counter is known as an **up-counter**.

Since the flip-flops used have outputs that are complementary it follows that if the counter were initially preset (i.e. all Q outputs set to 1) the $\overline{Q}$ must be 0 and if these outputs were used to trigger the following stage then a **down-counter** could be produced. Figure 4.64 shows the circuit while Fig. 4.65 shows the truth table and the waveforms.

Fig. 4.64 Asynchronous three-stage binary counter connected as a *down-counter.*

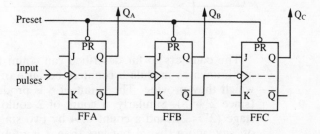

Fig. 4.65 Asynchronous three-stage binary down-counter: (a) truth table; (b) timing waveforms.

Input pulse	Q_C	Q_B	Q_A
0	1	1	1
1	1	1	0
2	1	0	1
3	1	0	0
4	0	1	1
5	0	1	0
6	0	0	1
7	0	0	0

(a)   (b)

In the waveforms of Fig. 4.65(b) it must be remembered that FFB and FFC are triggered by the negative transition of the $\overline{Q}$ output, which of course is the complement of the Q output, so that the changes in FFB and FFC outputs appear to occur on the leading edge of the Q_A and Q_B waveforms respectively.

It may be equirement to combine the two circuits of Figs. 4.62 and 4.64 to provide an **up-down counter**. With the aid of external gates this can easily be achieved, as Fig. 4.66 shows. The arrangement shows an AND-OR triggering circuit where the triggering pulse is applied to one input of the AND gate while the second input is an enable pulse. If the enable pulse is at logic 1 level then the top row of AND gates is activated and since, in this case, the second input to each AND gate is from the Q output then an up-count is produced. If the enable pulse is at logic 0 then the bottom row of AND gates is activated and because the other AND gate input is from the $\overline{Q}$ output then a down-counter is produced. Binary up-down counters, such as the SN74191 which is a four-element counter, are readily available on a single chip. Because the triggering circuits between FFA and FFB and between FFB and FFC are sum of products circuits the circuit may be redrawn using NAND gates only.

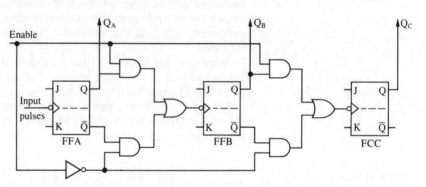

Fig. 4.66 Asynchronous three-stage (modulo 8) binary *up-down* counter.

The counters so far described can count from 0 to 7 inclusive and are known as **modulo 8 counters** because of the total count of 8 which they provide. The count of 8 is produced from three stages (since $2^3 = 8$). Similarly, a count of 2 could be produced by one stage ($2^1 = 2$), and a count of 4 by two stages ($2^2 = 4$). If a count of, say, more than 4 but less than 8 is required then three stages will still be required but the circuit will be modified with external gating to modify the count. There are two basic methods available for modifying the count:

1. *Reset.* In this case an external gate is used to sense the final value of the modified count and the output of the gate is used to reset all flip-flops to zero.
2. *Feedback.* In this arrangement the circuit is modified by extra gates which can be used to inhibit or alter the state of certain of the flip-flops in order to reset at the required count.

Consider the requirement to produce a modulo 5 counter. Then three flip-flops are still required but modified to reset after a count of 4. The circuit of Fig. 4.67 shows the reset method. The gate is a

Fig. 4.67 Asynchronous modulo 5 counter using reset.

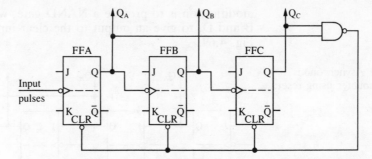

NAND gate that will only produce a logic 0 output, which is the required signal to the clear input, when a count of $5(Q_A = 1$, $Q_B = 0$, $Q_C = 1)$ is realised. In this arrangement the count of 5 is temporarily at the output terminals before the NAND gate output can cause the outputs to fall to zero. If this is a problem then this arrangement should not be used. The feedback method is shown in Fig. 4.68.

Fig. 4.68 Asynchronous modulo 5 counter using feedback: (a) circuit diagram; (b) truth table.

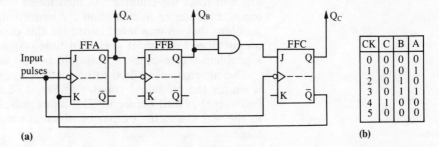

CK	C	B	A
0	0	0	0
1	0	0	1
2	0	1	0
3	0	1	1
4	1	0	0
5	0	0	0

(a)　　　　　　　　　　　　　　　　　　　　　　(b)

From the truth table of Fig. 4.68(b) it can be seen that the output of FFC will change to 1 after the condition $Q_A.Q_B = 1$. Normally the value of Q_C would remain at 1 after this change but, since it is required to reset the output to 0 after the receipt of input pulse 5, then FFC must be modified. This can be done by connecting the J input of FFC from an AND gate output which gives the function of $Q_A.Q_B$ and by clocking FFC directly from the input. At the end of input pulse 3 the output of the AND gate goes to 1 and this is transferred to Q_C at the end of the fourth input pulse, since with K unconnected the original Q_C value of 0 will change to 1 as the flip-flop is then in the toggle mode. The output of the AND gate falls to 0 after the fourth input pulse so that the flip-flop is again in the reset mode $(J = 0, K = 1)$ so that at the end of the fifth input pulse the Q_C value returns to 0.

Also, it can be seen that there is feedback from $\bar{Q}_C$ to the J and K input of FFA. Since $\bar{Q}_C = 1$ until the end of input pulse 4 then FFA toggles with successive input pulses as required. $\bar{Q}_C$ goes to 0 at the end of input pulse 4 so that for the fifth input pulse $J_A = K_A = 0$ and FFA is in the inhibit mode. Thus the end of the fifth input pulse will leave Q_A unchanged at 0.

Modulo 10 (or decade) counter

Since three flip-flops give a count of 8 and four flip-flops give a count of 16 $(= 2^4)$ then for a modulo 10 counter four flip-flops are necessary with external gating to modify the count. The simplest modification is to provide a NAND gate, with inputs from flip-flops B and D, to give an output to the clear inputs. A circuit is shown in Fig. 4.69.

Fig. 4.69 Asynchronous decade counter using reset.

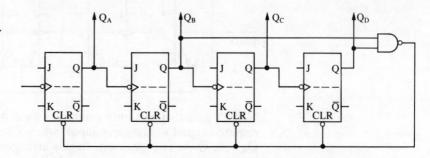

Since it is only on the count of 10 ($Q_A = 0$, $Q_B = 1$, $Q_C = 0$, $Q_D = 1$) that B = D = 1 then it is only on the count of 10 that a clear pulse is generated, and the output of the NAND gate going low will reset the counter. As mentioned earlier, when modifying a modulo 8 counter to a modulo 5 counter using reset, there is a problem that an unwanted count (in this case, 10) exists on the output lines for a short period of time. Again, if this is likely to be a problem the reset method should not be used.

The alternative method using feedback is shown in Fig. 4.70. This is simply the modulo 5 counter of Fig. 4.68 with a toggle flip-flop ($\div 2$ stage) placed before it. The input pulses to be counted are fed to the $\div 2$ stage, the output of which is used to clock the modulo 5 stage.

Fig. 4.70 Asynchronous decade counter using feedback.

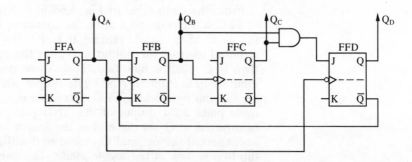

The truth table and waveforms for the counter of Fig. 4.70 are shown in Fig. 4.71. As can be seen from Fig. 4.71 the 'weighted' binary values of the outputs Q_A, Q_B, Q_C and Q_D gives a value equivalent to the decimal value of the number of input pulses counted. Since the count is from 0 to 9 inclusive in this case, then the unique representation of the four bits to an equivalent decimal number can be utilised to give a visual indication of the decimal count. The use of the four bits in this way gives a binary coded decimal (BCD) count.

A commercially available decade counter is the SN7490. This circuit is internally divided into two sections giving a $\div 2$ and a $\div 5$ representation. The circuit arrangement is shown in Fig. 4.72(a) and the truth table for the reset inputs in Fig. 4.72(b).

Fig. 4.71 Asynchronous decade counter: (a) truth table; (b) timing waveforms.

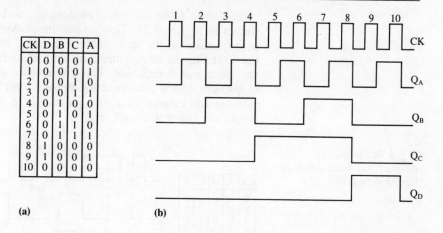

CK	D	B	C	A
0	0	0	0	0
1	0	0	0	1
2	0	0	1	0
3	0	0	1	1
4	0	1	0	0
5	0	1	0	1
6	0	1	1	0
7	0	1	1	1
8	1	0	0	0
9	1	0	0	1
10	0	0	0	0

(a)　　　　　　　　　　(b)

Fig. 4.72 SN7490 TTL decade counter: (a) package; (b) truth (function) table.

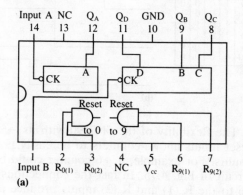

Input A	NC	Q_A	Q_D	GND	Q_B	Q_C
14	13	12	11	10	9	8

Reset inputs				Outputs			
$R_{0(1)}$	$R_{0(2)}$	$R_{9(1)}$	$R_{9(2)}$	D	C	B	A
1	1	0	X	0	0	0	0
1	1	X	0	0	0	0	0
X	X	1	1	1	0	0	1
X	0	X	0	←count→			
0	X	0	X	←count→			
0	X	X	0	←count→			
X	0	0	X	←count→			

1	2	3	4	5	6	7
Input B	$R_{0(1)}$	$R_{0(2)}$	NC	V_{cc}	$R_{9(1)}$	$R_{9(2)}$

(a)　　　　　　　　　　(b)

The 7490 can easily be arranged to produce the circuit of Fig. 4.70 by connecting the input pulses to be counted to pin 14 (input A) and making an external link between pin 12 (A output) and pin 1 (B input). Pins 12, 8, 9 and 11 provide outputs Q_A, Q_B, Q_C and Q_D respectively.

Alternatively, the modulo 5 could be placed first in the counting sequence to be followed by the ÷2 circuit. This could be achieved by connecting the input pulses to pin 1 (B input) and making an external connection between pin 11 (Q_D) and pin 14 (A input). A simplified version of this arrangement is shown in Fig. 4.73.

Fig. 4.73 Using the 7490 as a bi-quinary counter.

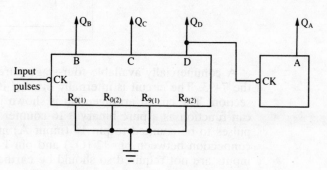

The truth table for this arrangement, and the waveforms, are shown in Figure 4.74. This arrangement does *not* give a BCD count, as inspection of the truth table will show; the 'weighted' values of the four bits do have a decimal equivalent from 0 to 4 inclusive but not from 5 to 9 inclusive. This circuit configuration is known as a **bi-quinary** and is preferred for those circuit applications where a symmetrical output waveform pattern is required, such as a frequency counter circuit, etc.

Fig. 4.74 Bi-quinary counter: (a) truth table; (b) timing waveforms.

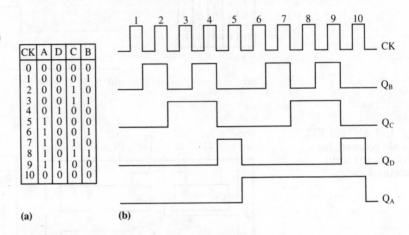

CK	A	D	C	B
0	0	0	0	0
1	0	0	0	1
2	0	0	1	0
3	0	0	1	1
4	0	1	0	0
5	1	0	0	0
6	1	0	0	1
7	1	0	1	0
8	1	0	1	1
9	1	1	0	0
10	0	0	0	0

(a)   (b)

The flexibility of the SN7490 with its ÷5 and ÷2 sections and reset inputs allows counters to be easily produced with modified counts. For example, a ÷6 counter could be produced using the circuit of Fig. 4.75. In fact the connections from Q_B and Q_C which go to the $R_o(1)$ and $R_o(2)$ inputs produce a reset counter without the need for an external gate since the gate is internal to the 7490 chip. In this case $Q_B = Q_C = 1$ uniquely on the count of 6 so, giving a reset after the binary equivalent of decimal 6 has appeared on the output lines for a short period of time.

Fig. 4.75 Modulo 6 counter using the 7490.

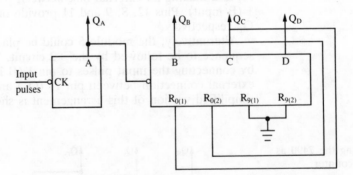

A commercially available four-stage pure binary counter (÷16) is the 7493. The circuit is internally divided into a ÷2 and a ÷8 section. The circuit arrangement is shown in Fig. 4.76. This circuit can function as a pure binary ÷16 counter by taking the input pulses to be counted to pin 14 (input A) and making an external connection between pin 12 (Q_A) and pin 1 (B input). The reset inputs are not required so should be earthed.

Fig. 4.76 SN7493 pure
binary modulo 16 counter
package.

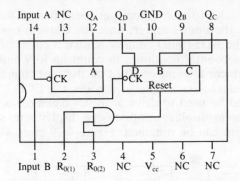

A modified count can easily be produced using the reset method
by using the reset inputs. For example, the circuit of Fig. 4.77 will
produce a ÷11 counter. Since $R_o(1)$ and $R_o(2)$ must both be logic 1
to reset the outputs to 0, the arrangement shown in Fig. 4.77 will
give a ÷11 count. Here Q_D is connected directly to $R_o(2)$, and the
output of an AND gate, whose inputs are Q_A and Q_B, is connected
to $R_o(1)$. The count of 11 is the only value between 0 and 11 where
$Q_D = 1$ *and* $Q_A.Q_B = 1$ so that a reset is produced after the binary
equivalent of decimal 11 has appeared on the output lines for a
short period.

Fig. 4.77 Using the 7493 to
provide a modulo 11
counter.

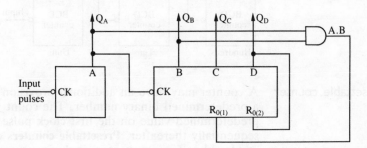

Cascading counters   Used as a BCD counter the 7490, or equivalent decade counter,
could be cascaded to give larger decimal counts. Suppose it is
required to build a counter to count from 000 to 999 before
resetting. Three stages of the 7490 would be required, with the Q_D
output of that counter representing units being used to clock the
next counter, which represents tens. The Q_D output of the tens
counter would likewise be used to trigger the 7490 to be used to
count hundreds. The arrangement is shown in Fig. 4.78.

Fig. 4.78 Cascading the
7490 decade counter to give
larger decimal counts.

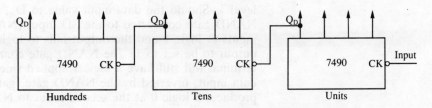

Each of the counters can count from 0 to 9 and, in the case of, say, the units counter, once the count is complete a carry is passed to the next (tens) counter so that a count of ten is recorded. The units counter continues to count up to 9 and when resetting produces a count of 20 (2 on the tens counter and 0 on the units counter). Via a suitable decoder the BCD output of each counter could be used to drive a display element such as an LED seven-segment display, gas-discharge display, etc. so that a direct visual output can be obtained. Figure 4.79 gives a possible circuit.

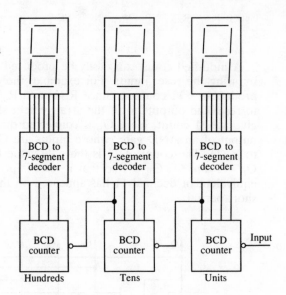

Fig. 4.79 Possible visual display arrangement via decoder/driver circuits from a cascaded decade counter circuit.

Presettable counter

A counter may have an additional provision for entering arbitrarily a predetermined binary number. The count will commence at the predetermined value on the first clock pulse and will count sequentially thereafter. **Presettable counters** can also be usefully employed as temporary storage devices since after the loading of the predetermined number the clock may be disabled for a required period of time, thereby preventing changes in the entered value. Asynchronous preset allows entry of a predetermined number without regard to the state of the clock. A possible arrangement is shown in Fig. 4.80. In this circuit the data strobe input controls the input of data into the counter. In the absence of a data strobe input the outputs of the NAND gates connected to the JK flip-flop SD and RD2 inputs will be high so that those inputs are unaffected and the counter will behave as a 'normal' BCD counter. Also, in the absence of the data strobe input the values of data input lines D_A, D_B, D_C and D_D are immaterial. When the data strobe input goes low then NAND gates 1 to 8 have one of their two inputs at logic level 1. Should the data input value to D_A, say, be logic 1 then the NAND gate connected to the SD input (NAND gate 1) has both inputs at logic 1 and the output falls to logic 0 which causes the Q_A output to be set to 1. The NAND gate connected to the RD2 terminal will still have a logic 1 applied since the logic 1 from the data input, inverted by the NAND gate (gate 9 in this case) produces a logic 0 at the second input to NAND gate 2. If the data

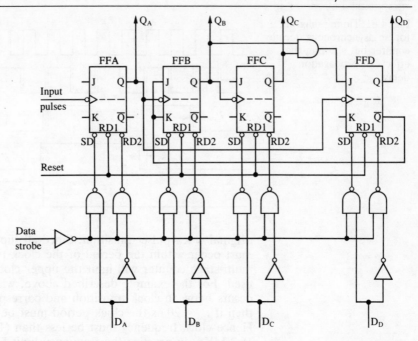

Fig. 4.80 Asynchronous presettable decade counter.

input value to D_B, say, is logic 0 via the inverter (NAND gate 10), NAND gate 4 has two logic 1 inputs and hence the output falls to 0, causing the Q_B output to be reset to 0. NAND gate 3 is effectively disabled since one of its inputs is logic 0. Thus via the NAND gates 1 to 8 and 9 to 12, the Q outputs can be forced to the values of the data input lines. In the asynchronous entry system the preset information may be entered and the count restarted in approximately the delay time of the binary elements.

Propagation delay

For all the asynchronous counters described so far it has been assumed that changes in state of any of the flip-flop elements occur at the same instant of time that the input pulses change from high to low state (negative-edge triggering assumed). This in fact is not the case since each flip-flop relies on changes in level of the preceding flip-flop to cause its output to change. Each flip-flop has a propagation delay which, although small in itself, may cause problems in multi-stage counters. The effect may be explained by examining the decade counter output waveforms of Fig. 4.71(b) which have been redrawn in Fig. 4.81 allowing for the propagation delay.

From Fig. 4.81 it can be seen that on the trailing edge of the clock pulse 4, Q_A changes from high to low but after a propagation delay of t_{pd} ns. The change in Q_A in turn causes Q_B to change from high to low with another propagation delay of t_{pd} ns with respect to its clock input (Q_A) giving a total delay of $2t_{pd}$ with respect to the original clock pulse. Similarly, Q_C will change t_{pd} ns after the change in the input clock pulse. Q_D in this case does not suffer even more delay because Q_D is triggered from the output of Q_A, and hence only suffers a delay of $2t_{pd}$ ns. Obviously, in a pure binary counter employing four stages Q_D would be delayed $4t_{pd}$ with respect to the

Fig. 4.81 Timing diagram for an asynchronous decade counter but showing the effects of propagation delay.

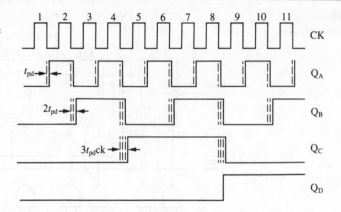

original clock pulse transition. Since all flip-flop output transitions must occur within the period of the clock pulse the total delay in a multi-stage counter will limit the upper clock frequency that can be used. For the example described above, where a total delay of $3t_{pd}$ occurs between clock transition and corresponding Q_C transition, then if $t_{pd} = 20$ ns the clock period must be greater than 60 ns. Hence clock frequency must be less than $(1/60 \times 10^{-9})$ Hz or 16.7 MKz. In practice the frequency limit for a four-bit counter in standard TTL is 16 MHz.

A side-effect of the propagation delay is the production of invalid output states. In the process of changing from, say, 3 to 4 in the decade counter, whose waveforms are shown in Fig. 4.81, there are instants between the negative transition of the third clock pulse and the establishment of the count of 4 at the Q outputs when values other than 0011 (3) and 0100 (4) exist. Inspection of the waveforms show that for a period of time t_{pd} ns after the negative transition of the fourth clock pulse the value at the Q outputs is 0011 (3); then Q_A does fall to 0 but for a further period of t_{pd} nanoseconds before Q_C rises to 1 during which there is a count of 0000 (0). Thus there are periods between the wanted state of 3 and 4 when invalid states 2 and 0 exist. This may well be an insignificant problem in those applications where the counter drives a visual display because slowly changing display elements would not be affected by invalid states lasting for only a few nanoseconds. However, in applications where the invalid state could produce problems (frequency synthesis circuits for example), then there is a need to provide a counter which does not suffer from invalid states. Such a counter could be a synchronous counter.

Synchronous counters A synchronous counter is an arrangement of flip-flops and logic gates in which all flip-flops are clocked simultaneously. The advantage is that a synchronous counter does not suffer false states during its output sequence. On the other hand, there is a requirement for extra gates between flip-flops to produce the correct sequence.

Since the flip-flops in a synchronous counter are controlled by the same clock pulse the cumulative propagation delay of the ripple counter is eliminated. The maximum frequency in this case is limited only by the delay of one flip-flop plus any extra produced by the gate circuits.

The design of synchronous counters for the larger count values is more difficult than for the asynchronous counter but can be simplified by the use of the Karnaugh mapping technique. Since the design aspect is beyond the scope of this book such techniques will not be discussed and the reader is referred to specialist texts where such design criteria may be found.

Synchronous pure binary counter A possible circuit of a modulo 16 counter is shown in Fig. 4.82(a) and the truth table in Fig. 4.82(b).

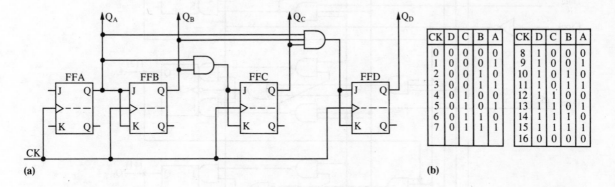

CK	D	C	B	A
0	0	0	0	0
1	0	0	0	1
2	0	0	1	0
3	0	0	1	1
4	0	1	0	0
5	0	1	0	1
6	0	1	1	0
7	0	1	1	1

CK	D	C	B	A
8	1	0	0	0
9	1	0	0	1
10	1	0	1	0
11	1	0	1	1
12	1	1	0	0
13	1	1	0	1
14	1	1	1	0
15	1	1	1	1
16	0	0	0	0

(a) (b)

Fig. 4.82 Synchronous modulo 16 counter: (a) circuit arrangement using JK flip-flops; (b) truth table.

From the truth table of Fig 4.82(b) it can be seen that Q_A must change state after every clock pulse and this can be achieved using the JK flip-flop in the toggle mode. The output Q_B can be seen to change following a transition of Q_A from 1 to 0 hence FFB can be driven by connecting J_B and K_B to the Q_A output.

Also from the truth table, it can be seen that Q_C changes following the condition that $Q_A = Q_B = 1$ and this can be achieved by driving FFC from Q_A and Q_B through an AND gate and the output of which feeds J_C *and* K_C. Finally the condition that Q_D changes state is fulfilled when $Q_A = Q_B = Q_C = 1$. Thus a three-input AND gate with those inputs and the output feeding both J_D and K_D will achieve the required result.

The arrangement of the AND gates in Fig. 4.82(a) used to provide inputs for FFC and FFD gives what is known as **parallel carry**. The carry in this case of course relates to the AND gate output going to a logic 1 state only when the preceding flip-flops are full, i.e. the logic circuits pass on a carry to the next stage. The parallel carry arrangement is shown again in Fig. 4.83(a).

Since the inputs Q_A and Q_B are common to both AND gates in Fig. 4.83(a) it follows that the circuit of Fig. 4.83(b) could be used. This circuit would seem to offer an advantage in that the AND gate which feeds FFD is now only a two-input gate whereas in the

Fig. 4.83 Logic circuit arrangement to give a carry to the synchronous modulo 16 counter: (a) parallel carry; (b) ripple carry.

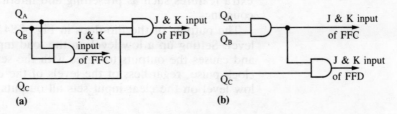

(a) (b)

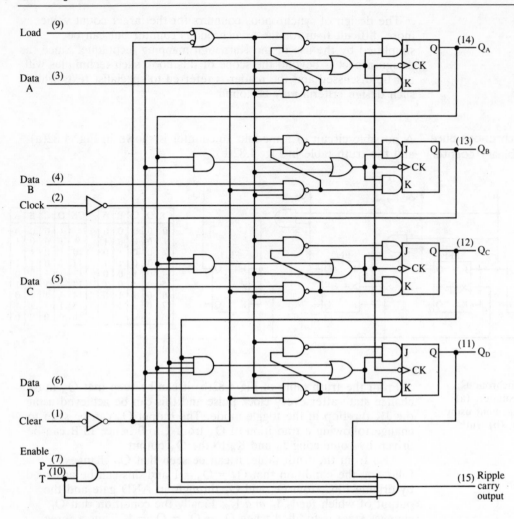

Fig. 4.84 SN74163 TTL
synchronous modulo 16
counter logic circuit
diagram.

parallel carry circuit it requires a three-input AND gate to produce
the same output. Should the modulo count be increased, say to 32,
then an extra AND gate will be required to fed FFE, and using the
parallel carry circuit the extra AND gate will require four inputs.
However, the circuit of Fig. 4.83(b), known as a **ripple carry** circuit
because the carry from one gate will ripple through succeeding
gates, is slower because of the extra delay introduced by using two
gates in tandem to generate the output to feed FFD; the problem
would be worse in higher count circuits.

A practical four-bit synchronous counter, the 74163, is shown in
Fig. 4.84. This circuit is basically the same as that of Fig. 4.82(a),
the extra logic gate circuitry being necessary to allow for the many
extra features such as presetting and internal look-ahead for fast
counting.

The outputs of the flip-flops in Fig. 4.84 may be preset to either
level. Setting up a low level on the load input disables the counter
and causes the outputs to agree with the set-up data, after the next
clock pulse, regardless of the levels of the enable inputs. Similarly, a
low level on the clear input sets all outputs low after the next clock

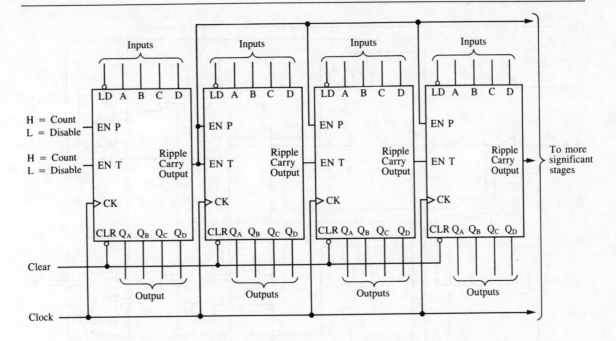

Fig. 4.85 Connections for a
four-stage synchronous
counter showing a method
of providing the look-ahead
carry facility.

pulse regardless of the level of the enable inputs. Parallel carry is
used in the 74163 for faster response and the device has a typical
maximum clock frequency of 32 MHz. The clock input is buffered
and triggers all flip-flops on the positive-going edge of the clock
input waveform.

Look-ahead carry uses an enable line which can be used, in
cascaded counter arrangements, to warn higher order counters that
the preceding counter is full. The enable input thus prepares the
higher order counter to change on the next clock pulse. In the
74163 both count enable inputs P and T must be high to count and
input T is used to enable the ripple carry output. A typical
arrangement of a four-stage synchronous counter is shown in
Fig. 4.85; this clearly shows the method of connecting the ripple
carry output and enable inputs to provide the look-ahead carry
facility.

Synchronous up-down counters

Up-down counters are available in synchronous circuits using
techniques similar to that described for the asynchronous up-down
counter of Fig. 4.66. The TTL circuit SN74LS169A is an example of
a synchronous up-down binary counter. This circuit also has
provision for look-ahead carry and is defined as programmable in
that it can be preset to either level. The direction of the count is
determined by the level of the up-down input; when the input is
high, the counter is an up-counter and when the input is low, the
counter is a down-counter. Input $\overline{T}$ is fed forward to enable the
carry output. The carry output thus enabled will provide a low-level
output pulse with a duration approximately equal to the high
portion of Q_A output when counting up and approximately equal to
the low portion of the Q_A output when counting down. The low-
level overflow carry pulse can be used to enable successive cascaded

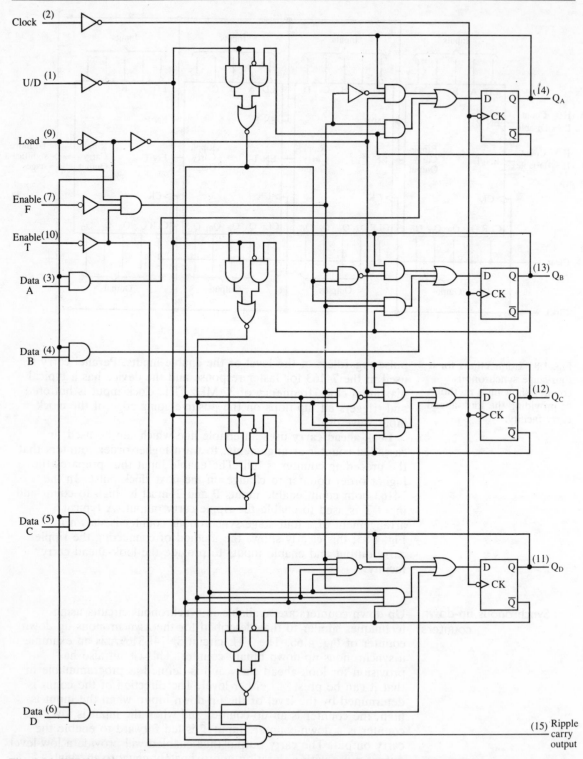

Fig. 4.86 SN74LS169A
synchronous up-down
binary counter logic
diagram.

circuits. The clock is buffered and triggers the flip-flops on the positive-going edge of the waveform. The circuit diagram is shown in Fig. 4.86.

Synchronous decade counter

A possible circuit arrangement is shown in Fig. 4.87. This is a modification of the binary counter circuit of Fig. 4.82(a). Assuming the counter is initially reset then the feedback signal from $\overline{Q_D}$ to the AND gate feeding FFB enables that gate so that the inputs to the J and K terminals of FFB are the same as the circuit of Fig. 4.82(a) until $\overline{Q_D}$ changes from high to low at the end of the eighth pulse. At this time the inputs to FFB are J = 0, K = 0. (Since J_B is fed from the AND gate which is now disabled and K_B is fed directly from Q_A.) At the end of the ninth clock pulse Q_A toggles to 1 and $Q_B = Q_C = 0$ while $Q_D = 1$. The J inputs to FFB and both inputs to FFC are now inhibited so will remain at 0. However, the inputs to FFD are J = 0, K = 1 (since J_D is fed from an AND gate which is disabled and K_D is fed directly from Q_A) so that at the end of the tenth clock pulse Q_D resets to 0. Since Q_A toggles to 0 at the same time the counter has reset at the end of the tenth pulse.

Fig. 4.87 Synchronous decade counter using JK flip-flops.

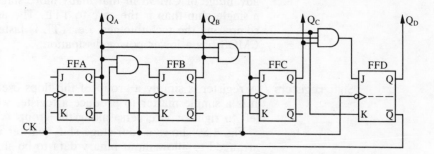

Decade counters, such as the 74160, are available with the basic circuit of Fig. 4.87 but with the added sophistication of preset capability and internal look-ahead carry, giving a circuit very similar to Fig. 4.84.

Other values of count are possible by modification of the circuit of Fig. 4.84 for counts less than 16 or by cascaded stages for very much larger counts. A simple three stage circuit modified to give a count of 6 is shown in Fig. 4.88(a) with the truth table in Fig. 4.88(b). Inspection of the truth table shows that FFA must toggle with successive clock pulses hence it is simply connected in the T-mode with J = K = 1. FFB output must change following the condition that $Q_A = 1$ except after the fifth clock pulse when Q_B

Fig. 4.88 Synchronous modulo 6 counter: (a) logic circuit arrangement; (b) truth table.

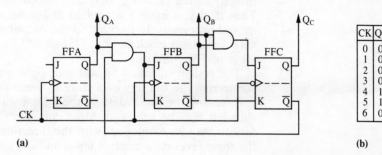

CK	Q_C	Q_B	Q_A
0	0	0	0
1	0	0	1
2	0	1	0
3	0	1	1
4	1	0	0
5	1	0	1
6	0	0	0

(a)

(b)

should be inhibited. This can be achieved by connecting the input to J_B and K_B from Q_A via a two-input AND gate, the other input to which is from $\overline{Q_C}$. Since $\overline{Q_C} = 1$ until the end of the fourth clock pulse, the input to J_B and K_B will follow Q_A until the end of the fourth clock pulse when $\overline{Q_C}$ becomes 0, inhibiting FFB from further change. The inputs to FFC will give $J_C = K_C = 1$ at the end of the third clock pulse when $Q_A = Q_B = 1$ so that at the end of the fourth clock pulse FFC toggles and Q_C goes to 1. Now $J_C = K_C = 0$, since $Q_A = Q_B = 0$, so that FFC is inhibited and remains at the $Q_C = 1$, at the end of clock pulse 5. Then $J_C = 0$, $K_C = 1$ since $Q_A = 1$, $Q_B = 0$ so that at the end of the sixth clock pulse Q_C resets to 0. Since Q_B was earlier inhibited at 0 and Q_A toggles to 0 at the end of the sixth clock pulse the counter is reset.

All of the counter circuits described in basic form could be built using TTL or CMOS flip-flops. Regarding the commercially available counters in the 74 series, these are TTL but similar circuits are available in CMOS. The 4029 for example is a synchronous up-down counter operating up to 5 MHz with a 5 V supply. The 4040 is a 12-stage binary counter with all 12 outputs available in a 16-pin integrated circuit (IC). This circuit illustrates the major advantage of CMOS in that many more stages can be fabricated on a single chip than is the case to TTL. The usual points apply when comparing the technologies, i.e. TTL is faster than CMOS while CMOS has a lower power dissipation.

Shift registers   A register is simply a group of flip-flops used to store binary data. It is a simple matter to produce a register which shifts the data from left to right or vice versa along the group.

Fig. 4.89 shows a four-stage shift register using JK flip-flops arranged to allow input binary data to be shifted to the right serially with successive clock pulses. The flip-flops are connected as D-types so that after the clock-pulse the datum at each flip-flop output is the same as that datum at its input prior to the clock pulse. The truth table in Fig. 4.89(b) shows the effect on the data bits for given binary values at the register input for several clock pulses.

Assuming that all flip-flop outputs are initially 0 then data presented at the input may be shifted right by the clock pulses. The data on the input line is connected directly to the J input of FFA but to the K input via an inverter. Thus the J and K inputs of the first flip-flop are complementary. This means that whatever the input on the J input to this stage is gated to the Q output at the end of the clock pulse. The same effect must occur at flip-flops B, C and D since the J and K inputs of these flip-flops are connected directly to the Q and $\overline{Q}$ outputs respectively of the preceding stage. Thus if, say, a logic 1 is applied to J input of flip-flop A it will be moved progressively through to the output of flip-flops B, C and D with successive clock pulses. This should be clear from the truth table of Fig. 4.89(b).

The system of Fig. 4.89 will shift right only. By appropriately connecting the outputs of a stage to the input of the preceding stage and feeding the input data to the J input of the final flip-flop, a shift left may be achieved. More commonly a shift-left or shift-right facility may be employed, with the direction of shift determined by the logic level on a control input and the connection between stages

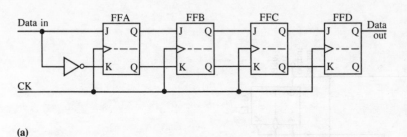

The table in the figure:

CK pulse	Data in	Output levels			
		A	B	C	D
0	X	0	0	0	0
1	1	1	0	0	0
2	1	1	1	0	0
3	0	0	1	1	0
4	0	0	0	1	1
5	1	1	0	0	1
6	1	1	1	0	0
7	0	0	1	1	0

(a)

(b)

Fig. 4.89 Four-stage shift register: (a) using JK flip-flops connected as D-types; (b) truth table showing the effect of the clock in shifting right one of the input bits to the data-in terminal.

made via logic gating circuits. A simple arrangement is shown in Fig. 4.90.

The control input for shift right, assuming it is logic level 1, allows the top row of the AND gates to be enabled and, since the control input to the bottom row of AND gates is inverted before connection, these gates are disabled. Thus the gating circuit allows shift-right data input via the OR gate to be applied to the flip-flop J terminal and, after clocking, at its output Q terminal, starting at FFA.

Assuming a logic 0 on the control input then the top row of AND gates is disabled and the bottom row enabled. Thus a shift-left data input may be applied, via the OR gate, to the input of the final flip-flop (FFD) and after clocking at its output Q terminal. This output is now connected via the logic gate network to the J input of FFC ready for the next clock pulse. Again the flip-flops shown are JK and it is assumed that data transfer occurs on the negative edge of the clock pulse.

Shift registers have many practical applications including that of a temporary storage element or buffer stage. This is especially useful in those computer applications where the processor, with its high-speed operation, may have to deal with slower peripherals. In the case of output data transfer from the processor to a peripheral the buffer register may store the output data, allowing the processor to proceed with other tasks, while waiting for the peripheral to accept the data. Since most peripherals deal with serial data (i.e. data clocked out one bit at a time) while the processor deals with parallel data (i.e. *all* bits in the data word shifted at the same instant) there is a need for a shift register that can handle parallel data. For data transfer from processor to peripheral the register would be parallel loaded and read out serially. This type of register is a parallel-in, serial-out device, frequently reduced to PISO. For data transfer from peripheral to processor the information is fed in serially and read out in parallel. Thus the register is a SIPO or serial-in, parallel-out device. Finally, shift registers are available in parallel-in, parallel-out (PIPO) forms.

A practical four-bit shift left/right register, the 74L99 is shown in Fig. 4.91. Although the flip-flops shown are SR types, all except the first stage operate in D-type mode, so that that only the middle two lines of the SR truth table applies. This circuit allows shift-right and shift-left processing using two clocks together with a mode control input. Shift right is achieved when clock 1 goes low with mode control low. Serial data for the shift right mode is entered at the J-$\overline{\text{K}}$ inputs. The first stage can operate as a J-$\overline{\text{K}}$, a D-type or T-type

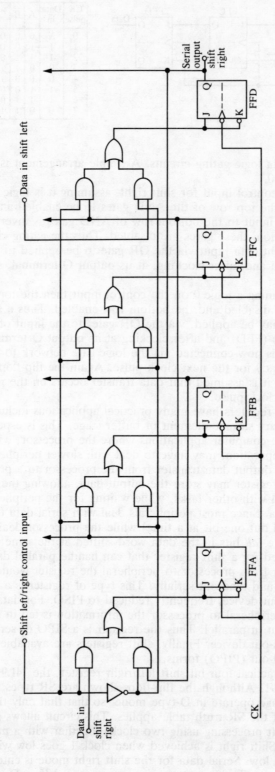

Fig. 4.90 Shift left/right
register logic circuit
diagram.

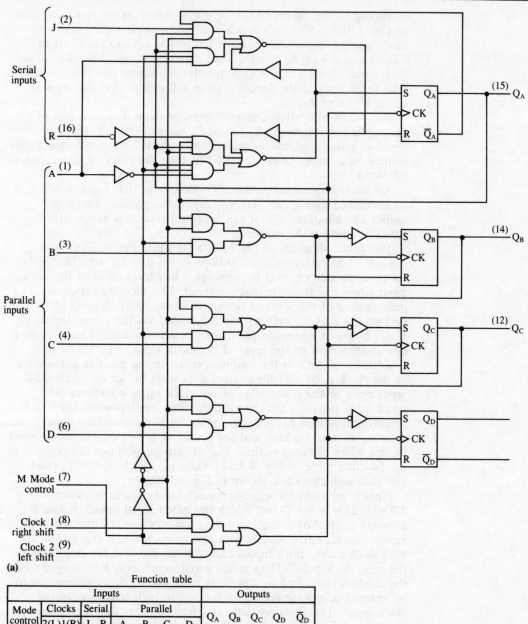

(a)

Function table

Mode control	Clocks 2(L) 1(R)		Serial J R		Parallel A B C D				Outputs Q_A Q_B Q_C Q_D $\bar{Q}_D$				
H	H	X	X	X	X	X	X	X	Q_{AO}	Q_{BO}	Q_{CO}	Q_{DO}	$\bar{Q}_{DO}$
H	↓	X	X	X	a	b	c	d	a	b	c	d	$\bar{d}$
H	↓	X	X	X	$Q_B\dagger$	$0_C\dagger$	$Q_D\dagger$	d	Q_{Bn}	Q_{Cn}	Q_{Dn}	d	$\bar{d}$
L	L	H	X	X	X	X	X	X	Q_{AO}	Q_{BO}	Q_{CO}	Q_{DO}	$\bar{Q}_{DO}$
L	X	L	L	H	X	X	X	X	Q_{AO}	Q_{AO}	Q_{Bn}	Q_{Cn}	$\bar{Q}_{Cn}$
L	X	↓	L	L	X	X	X	X	L	Q_{An}	Q_{Bn}	Q_{Cn}	$\bar{Q}_{Cn}$
L	X	↓	H	H	X	X	X	X	H	Q_{An}	Q_{Bn}	Q_{Cn}	$\bar{Q}_{Cn}$
L	X	↓	H	L	X	X	X	X	$\bar{Q}_{An}$	Q_{An}	Q_{Bn}	Q_{Cn}	$\bar{Q}_{Cn}$
↑	L	L	X	X	X	X	X	X	Q_{AO}	Q_{BO}	Q_{CO}	Q_{DO}	$\bar{Q}_{DO}$
↓	L	L	X	X	X	X	X	X	Q_{AO}	Q_{BO}	Q_{CO}	Q_{DO}	$\bar{Q}_{DO}$
↓	L	H	X	X	X	X	X	X	Q_{AO}	Q_{BO}	Q_{CO}	Q_{DO}	$\bar{Q}_{DO}$
↑	H	L	X	X	X	X	X	X	Q_{AO}	Q_{BO}	Q_{CO}	Q_{DO}	$\bar{Q}_{DO}$
↑	H	H	X	X	X	X	X	X	Q_{AO}	Q_{BO}	Q_{CO}	Q_{DO}	$\bar{Q}_{DO}$

†Shifting left requires external connection of Q_B to B, and Q_D to C. Serial data is entered at input D.

(b)

Fig. 4.91 SN74L99 four-bit shift left/right register: (a) logic circuit diagram; (b) truth (function) table.

according to the values of the inputs as shown in the function table of Fig. 4.91(b). Shift left is achieved when clock 2 goes low with mode control high. Serial data is fed in this case to the D input. Changes at the mode control input should normally be made while both clock inputs are low although the conditions described in the final three lines of the function table will ensure that the register contents are protected.

The 74L99 also allows parallel input of data. The four bits of data are applied to inputs A, B, C and D and clocked into the associated flip-flop with a high-to-low transition of clock 2 input with the mode control input high. During parallel loading the entry of serial data is inhibited.

An interesting circuit is the SN7496 five-bit shift register which is so connected that it can perform any of the possible four input-output combinations, i.e. it can be configured as a SISO, SIPO, PISO or PIPO shift register.

The circuit diagram of the 7496 is shown in Fig. 4.92(a) and consists of SR master-slave flip-flops which may be simultaneously set to a low output level by applying a low-level input to the clear input while the preset is inactive (low). This clearing action is independent of the level of the clock input. After clearing all outputs the register may be parallel loaded via the preset inputs A, B, C, D and E while the preset enable input is high. This process is also independent of the level of the clock input.

Transfer of data to the outputs occurs on the positive-going edge of the clock pulse and the correct data must be set up at each SR input prior to the rising edge of the clock input waveform. The serial input provides the data for the first flip-flop while the following flip-flops provide the inputs for the succeeding stages. The clear input must be high and the preset or preset enable inputs must be low when clocking occurs. The various possibilities are covered in the function table of Fig. 4.92(b) while typical clear, shift, preset and shift sequences are shown in Fig. 4.92(c).

Finally, an eight-bit serial-in, parallel-out register is shown in Fig. 4.93(a). This is the 74164, which has gated serial inputs A and B allowing control over the input data, since a low at either or both inputs inhibits entry of new data and resets the first flip-flop at the next clock pulse. Both inputs high will set the first flip-flop high at the next clock pulse. Data at the serial inputs may be changed while the clock is high, or low, but must meet the set-up requirements to be entered. Clocking occurs on the low-to-high transition of the clock input. The function table and typical clear, shift and clear sequences are shown in Fig. 4.93(b) and (c) respectively.

An advantage to be gained in using CMOS shift registers is the density of the circuitry for a single chip. The CD4031BM/CD4031BC 64-stage shift register is a good example of this. The block diagram is shown in Fig. 4.94. The device has two data inputs 'data in' and 'recirculate in', and a 'mode control' input. Data at the data input (when mode control is low) or data at the recirculate input (when mode control is high), which meets the set-up and hold time requirements, is entered into the first stage of the register and is shifted one stage at each positive transition of the 'clock'. Data output is available in both true and complementary form from the 46th stage. Both the data out (Q) and data out ($\overline{Q}$) outputs are buffered. The clock input is fully buffered but additionally a

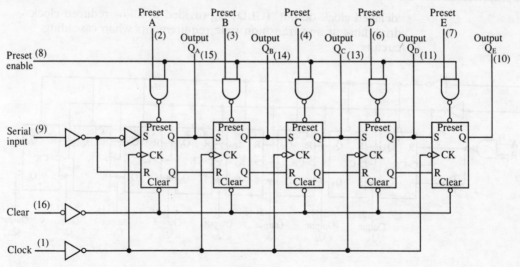

(a)

Function table

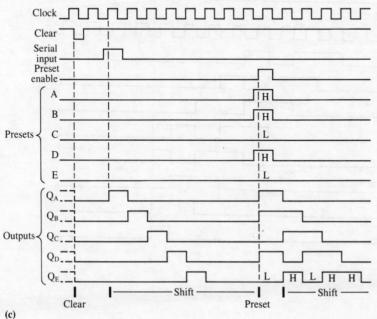

Inputs										Outputs				
Clear	Preset enable	Preset					Clock	Serial	Q_A	Q_B	Q_C	Q_D	Q_E	
		A	B	C	D	E								
L	L	X	X	X	X	X	X	X	L	L	L	L	L
L	X	L	L	L	L	L	X	X	L	L	L	L	L
H	H	H	H	H	H	H	X	X	H	H	H	H	H
H	H	L	L	L	L	L	L	X	Q_{AO}	Q_{BO}	Q_{CO}	Q_{DO}	Q_{EO}
H	H	H	L	H	L	H	L	X	H	Q_{BO}	H	Q_{DO}	H
H	L	X	X	X	X	X	L	X	Q_{AO}	Q_{BO}	Q_{CO}	Q_{DO}	Q_{EO}
H	L	X	X	X	X	X	↑	H	H	Q_{An}	Q_{Bn}	Q_{Cn}	Q_{Dn}
H	L	X	X	X	X	X	↑	L	L	Q_{An}	Q_{Bn}	Q_{Cn}	Q_{Dn}

(b)

H = high level (steady state), L = low level (steady state)
X = irrelevant (any input, including transition)
↑ = transition from low to high level
Q_{AO}, Q_{BO}, etc = level of Q_A, Q_B, etc, respectively before
the indicated steady-state input conditions were established.
Q_{AN}, Q_{Bn}, etc = the level of Q_A, Q_B, etc, respectively
before the most-recent ↑ transition of the clock.

(c)

Fig.4.92 SN7496 five-bit
shift register: (a) logic
circuit diagram; (b) truth
(function) table; (c) shift
sequences.

'delayed clock output' (CLD) is provided to allow reduced clock drive fan-out and transition time requirements when cascading circuits.

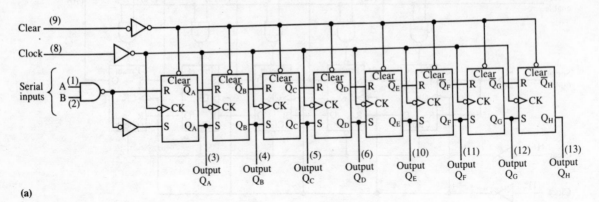

(a)

Function table

Inputs				Outputs		
Clear	Clocks	A	B	Q_A	Q_B . . .	Q_H
L	X	X	X	L	L	L
H	L	X	X	Q_{AO}	Q_{BO}	Q_{HO}
H	↑	H	H	H	Q_{An}	Q_{Gn}
H	↑	L	X	L	Q_{An}	Q_{Gn}
H	↑	X	L	L	Q_{An}	Q_{Gn}

(b)

H = high level (steady state), L = low level (steady state)
X = irrelevant (any input, including transitions)
↑ = transition from low to high level
Q_{AO}, Q_{BO}, Q_{HO} = the level of Q_A, Q_B, or Q_H, respectively, before the steady-state input conditions were established
Q_{An}, Q_{Gn} = the level of Q_A or Q_G before the most recent ↑ transition clock; indicates a one-bit shift

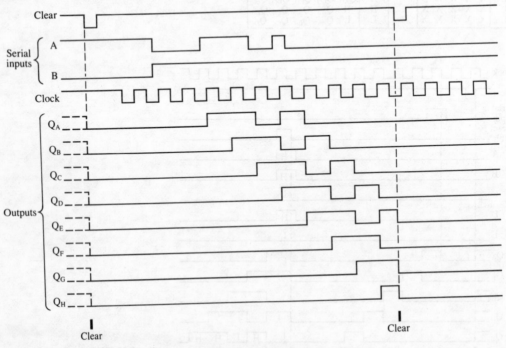

(c)

Fig. 4.93 SN74164 eight-bit serial-in, parallel-out register: (a) logic circuit arrangement; (b) truth (function) table; (c) clear and shift and clear sequences.

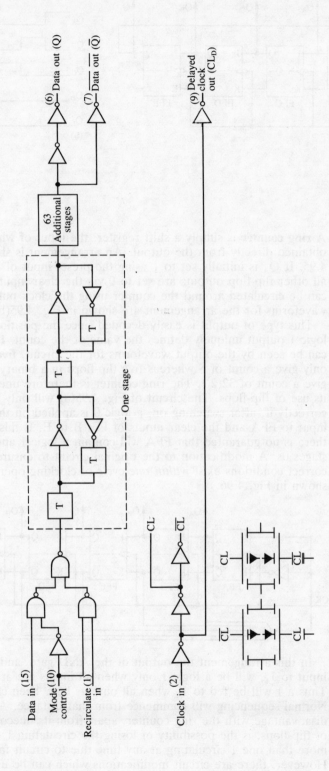

Fig. 4.94 CMOS
CD4031BM/CD4031BC 64-
stage shift register logic
circuit arrangement.

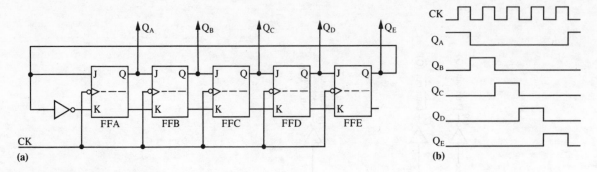

Fig. 4.95 Modulo 5 ring counter: (a) logic circuit arrangement; (b) timing waveforms.

Ring counters

A **ring counter** is simply a shift register, the input of which is obtained directly from the output. An arrangement is shown in Fig. 4.95. If Q_A is initially set to 1 using the preset input of FFA while all other flip-flop outputs are set to 0 via the clear inputs then the 1 can be circulated around the counter using the clock pulse. The waveforms for the arrangement are shown in Fig. 4.95(b).

This type of output is easily decoded since the position of the logic 1 output uniquely defines the value of the count. However, as can be seen by the output waveforms for the circuit, five flip-flops only give a count of 5 whereas five flip-flops in a binary counter can give a count of $32(2^5)$. The ring counter is therefore uneconomic in its use of flip-flops. The circuit of Fig. 4.95(a) will only operate correctly if, after switching on, a logic 0 is applied on the preset input to FFA and the clear inputs of FFs B to E. If this is not done there is no guarantee that FFA will contain a logic 1 and the other stages 0s. A modification to the original circuit to ensure that the correct conditions exist *within one cycle* of clocking operations is shown in Fig. 4.96.

Fig. 4.96 Modified modulo 5 ring counter to provide a 'self-starting' circuit. No inputs are required to the flip-flop preset and clear input terminals.

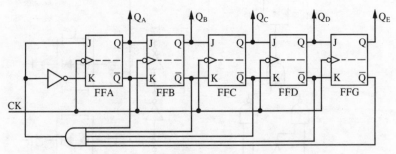

In this arrangement the output of the AND gate, and hence the input to J_A, will be a logic 1 only when all inputs are at logic 1. Thus a 1 will be fed to J_A when all outputs have been clocked to 0. Normal sequencing will commence from that instance. A basic disadvantage with the ring counter, apart from its uneconomic use of flip-flops, is the possibility of losing the circulating 1 or of having more than one 1 circulating at any time due to circuit faults. However, there are circuit modifications which can be used to detect either of these faults.

Circuits for additional logic 1 states

Since the function

$$F = A.B + A.C + B.C \qquad [4.31]$$

will only record a logic 1 output if any AND-ed combination of two variables has 1s at the input then, for normal operation, the output $F = 0$.

However, equation [4.31] is only for three stages; to cater completely for four stages would require six two-input AND gates while five stages would require ten two-input AND gates and so on.

If instantaneous indication is not essential then a simplified equation can be used. For example, the equation

$$F = (A + B).C \qquad [4.32]$$

will give an indication if either A or B *and* C is 1 but not if A and B is 1. However, the indication in this latter case would occur one clock pulse later. The function circuit for the detection of equation [4.32] is shown in Fig. 4.97(a) and, since universal inverting logic may be required, it is also shown converted to NAND gate form in Fig. 4.97(b).

Fig. 4.97 Ring counter logic circuitry to detect the presence of extra logic 1 output states: (a) basic circuit; (b) using only NAND gates.

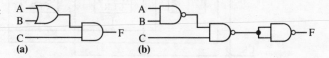

In Fig. 4.97(b) it has been possible to eliminate two NAND gates that would otherwise have been necessary by taking the connections from FFA and FFB from the inverting outputs. Similarly, for the four-stage circuit the indication logic function could be reduced to

$$F = (A + B).(C + D)$$

giving at worst one clock pulse delay, while for the five-stage circuit:

$$F = (A + B).(C + D).E$$

would be suitable.

Circuits for additional logic 0 states

For a three-input variable system a logic 1 will be obtained at the output if *all* the inputs are 0 using a NOR gate since

$$F = \overline{A + B + C} \qquad [4.33]$$

The arrangement is shown in Fig. 4.98(b). By De Morgan's theorem equation [4.33] can be modified to

$$F = \overline{A}.\overline{B}.\overline{C}$$

$$\text{or } F = \overline{\overline{\overline{A}.\overline{B}.\overline{C}}} \qquad [4.34]$$

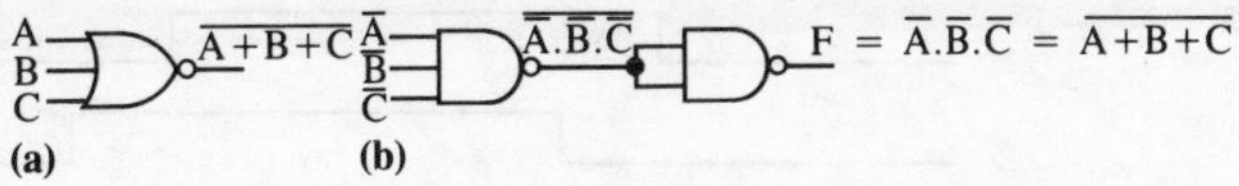

Fig. 4.98 Ring counter logic circuitry to detect the presence of extra logic 0 output states: (a) basic circuit; (b) using only NAND gates.

The circuits of Figs 4.95(a) and 4.96 use one feedback line from the final Q output to the first-stage J input and connect to the first-stage K input via an inverter. Since the $\overline{Q}$ output of the final stage is the complement of the signal fed back to the J input of the

Equation [4.34] is the basis of the NAND gate indication circuit of Fig. 4.98(b) which takes as its input the $\overline{Q}$ outputs of all three stages. Extension of the concept to cover four or more stages gives a result similar to equations [4.33] and [4.34].

first stage, the inverter can be dispensed with and a second feedback line connected as shown in Fig. 4.99. This circuit has been reduced to three stages for simplicity but the principle is the same regardless of the number of stages used. Preset and clear inputs have also been shown in this arrangement, which therefore is not self-starting.

Fig. 4.99 Modulo 3 ring counter without an inverter at the input stage.

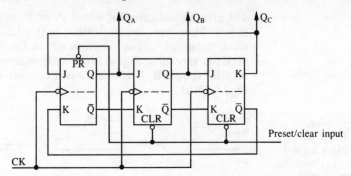

The twisted ring or Johnson counter

As the name suggests, the feedback which completes the ring in a **twisted ring counter** is twisted in that the feedback signal to the J input of the first stage is taken from the $\overline{Q}$ output of the final stage instead of the Q output as before. Figure 4.95(a) has been redrawn in Fig. 4.100(a) showing this modification. Assuming that for the

Fig. 4.100 Modulo 10 twisted ring, or Johnson, counter: (a) logic circuit arrangement; (b) timing waveforms.

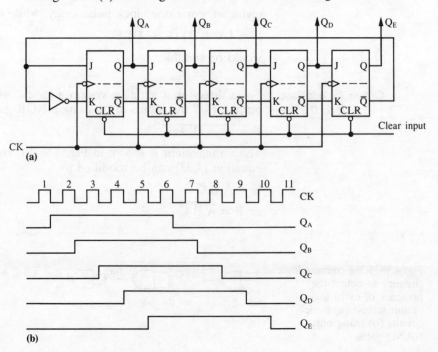

twisted ring counter all outputs are initially cleared then the outputs are at 0, but the input to the J terminal of the first stage must be 1 since it is taken from the $\overline{Q}$ output of the last stage. Thus the first clock pulse will place a 1 at the Q output of the first stage. However, the $\overline{Q}$ output of the final stage will remain at 1 until the clock has passed the initial 1 to the output of the final stage. This means that after the first clock pulse the input to FFA continues to be logic 1, and with successive clock pulses the Q_A output remains at logic 1 until the value of $\overline{Q}_E$ changes. Thus the twisted ring counter, if initially cleared, will sequentially fill with logic 1s and then begin to remove them starting from the first stage until reset occurs with all stages cleared. The waveforms for the circuit of Fig. 4.100(a) are shown in Fig. 4.100(b) and should make the sequence clear.

As Fig. 4.100(b) shows, there are 10 unique states using a five-element twisted ring counter, so that the circuit of Fig. 4.100(a) could be used as a decade counter. The arrangement is obviously more economical in the use of flip-flops than the original ring counter arrangements.

The technique of using *both* output terminals of the final stage to feed back to the input terminals of the first stage, and hence dispensing with the input inverter, can be applied here and a three-stage circuit using this technique is shown in Fig. 4.101(a) while the truth table is shown in Fig. 4.101(b).

Fig. 4.101 Modulo 6 twisted ring counter: (a) logic circuit arrangement; (b) truth table.

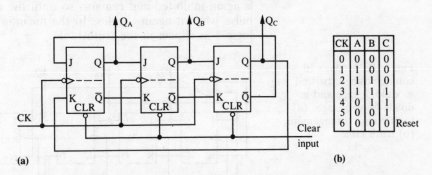

CK	A	B	C	
0	0	0	0	
1	1	0	0	
2	1	1	0	
3	1	1	1	
4	0	1	1	
5	0	0	1	
6	0	0	0	Reset

(a)

(b)

The circuit using three stages gives a six-state sequence. This can easily be modified to a five-state sequence by lifting the feedback connection from the Q_C output and replacing it at the Q_B output. All other connections remain the same. The circuit arrangement is shown in Fig. 4.102(a). Since the change at the first stage output is initiated by its input levels at the J and K terminals, moving the feedback connection from Q_C to Q_B output means that Q_A goes to logic 1 at the end of the second clock pulse instead of at the end of the third pulse as in the original circuit. Thus the input conditions for FFA at the end of the second clock pulse are now $J_A = 1$, $K_A = 1$ and hence at the end of the third clock pulse FFA toggles to give a logic 0 output. $Q_B = Q_C = 1$ at this time. Successive clock pulses will shift this 0 sequentially to the right until the counter is cleared. The truth table of Fig. 4.102(b) should make this clear.

From the circuit of Fig. 4.102(a) it would be possible to follow the circuit with a ÷2 JK stage to give a decade counter. This would not give a BCD count but could still be decoded to give the correct unique count value. Additionally only four flip-flops are required,

Fig. 4.102 Modulo 5 twisted ring counter: (a) logic circuit arrangement; (b) truth table.

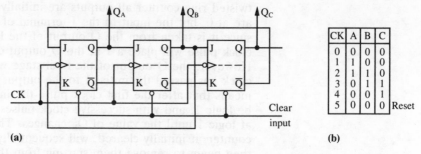

CK	A	B	C
0	0	0	0
1	1	0	0
2	1	1	0
3	0	1	1
4	0	0	1
5	0	0	0	Reset

(a)   (b)

which is the same as for the previous decade counters and hence the arrangement is not uneconomic in the use of flip-flops. As a bonus, only one logic gate is required to achieve the count as is shown in Fig. 4.103(a). The truth table is shown in Fig. 4.103(b). The output from the NOR gate is 0 if either, or both, inputs are 1 and rises to 1 when both inputs are 0. From the truth table of Fig. 4.103(b) it can be seen that since the inputs to the NOR gate are Q_B and $\overline{Q}_C$ the output of the NOR gate is logic 0 until the end of the fourth clock pulse. Thus, assuming the counter is initially reset, Q_D will remain at 0 until the end of the fourth clock pulse since the flip-flop is inhibited. At the end of the fifth clock pulse the J_D and K_D inputs are both 1 so Q_D toggles to 1. However, the output of the NOR gate falls to 0 again at the end of the fifth clock pulse so that FFD is again inhibited and remains so until the end of the tenth clock pulse when it again toggles. In the meantime its value for Q_D is logic 1 as shown in the truth table.

Fig. 4.103 Modulo 10 counter using a twisted ring modulo 5 stage and a divide-by-two stage: (a) logic circuit arrangement; (b) truth table.

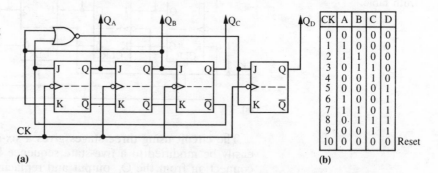

CK	A	B	C	D
0	0	0	0	0
1	1	0	0	0
2	1	1	0	0
3	0	1	1	0
4	0	0	1	0
5	0	0	0	1
6	1	0	0	1
7	1	1	0	1
8	0	1	1	1
9	0	0	1	1
10	0	0	0	0	Reset

(a)   (b)

You should now be able to attempt exercises 4.41 to 4.69.

Other circuit functions

Extending the concept of digital analysis of Chapter 3, where it was seen that the basic function of combinational logic circuitry is to produce an output for certain input conditions, various circuit functions can be produced, including multiplexers, decoders/ demultiplexers, driver circuits and arithmetic operators. It is the purpose of this section to examine briefly such circuit functions with reference wherever possible to available MSI/LSI circuitry. It is not the intention to produce a section on the logic design behind such circuitry and if you require such an approach then there are many textbooks which can be consulted. It is hoped rather to provide a guide to the types of circuit available and the reasons why they are provided.

Multiplexers and encoders

Multiplexers

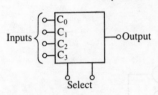

Fig. 4.104 Multiplexer block diagram.

An arrangement where several inputs can be connected but only one input, under the influence of a select input, will have access to the output is called a **multiplexer**. In this respect the multiplexer is a combinational logic switch which is controlled by a logic signal. A block diagram representation of a 4-to-1 multiplexer is shown in Fig. 4.104.

Two select inputs are required in this case because there are four inputs; three select lines would be needed for eight inputs etc., since it is the combination of select input variables which enables any one of the inputs to connect to the output. Figure 4.105 is a possible logic circuit arrangement for a 4-to-1 multiplexer.

The logic equation which provides the switching function is given from Fig. 4.105 as

$$Y = C_0.\overline{A}.\overline{B}. + C_1.\overline{A}.B + C_2.A.\overline{B} + C_3.A.B \qquad [4.35]$$

It follows from equation [4.35] that if $A = B = 0$ then

$$Y = C_0$$

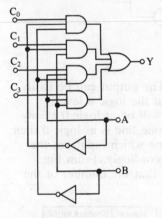

Fig. 4.105 Logic circuit arrangement of a 4-to-1 multiplexer.

since the AND functions $C_1.\overline{A}.B$, $C_2.A.\overline{B}$ and $C_3.A.B$ will all give outputs of 0 and the output follows the logic input C_0. Similarly, if $A = 0$, $B = 1$ then

$$Y = C_1 \text{ etc.}$$

Figure 4.106 shows the logic circuit of 74153 dual 4-to-1 multiplexer. The duplicated circuit is essentially similar to that of Fig. 4.105 except for the provision of a strobe terminal that allows for an enable input which, if held high, keeps the output low regardless of the state of the other inputs. When the strobe is low then normal multiplexing can occur; the strobe is useful for providing cascading, i.e. N lines to n lines rather than N lines to 1 line as in the basic circuit.

Other multiplexer formats are available, such as 16-to-1 (74150), 8-to-1 (74251 and others), quadruple 2-to-1 (74157 and others), and more. Multiplexer circuits have many applications in digital circuits including computers.

Encoders

The circuit of an **encoder** has, like the multiplexer, many input lines but the output is a coded pattern which identifies each of the inputs. Consider a simple requirement where three inputs are present at the encoder and two outputs are used to indicate which of the inputs is at logic 0. The block diagram is shown in Fig. 4.107(a).

It is assumed in this simple arrangement that only one input can be at logic level 0 at any instant and that the enable input is low.

Fig. 4.106 SN74153 dual 4-
to-1 multiplexer logic circuit
diagram.

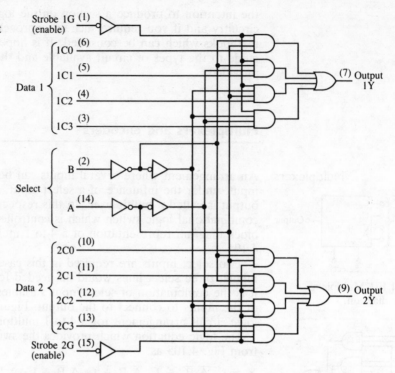

The truth table is shown in Fig. 4.107(b). The output can be easily decoded to determine which input lines held the logic 0 level.

The assumption that only one input line will be at logic 0 at any one time is not always valid. If more than one line is at logic 0 then a **priority encoder** must be used to determine which input has the highest priority and set the output levels accordingly. From Fig. 4.107(b) the results of the truth table show that the number of the

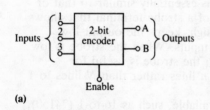

Inputs {
1
2
3
} → 2-bit encoder
→ A
→ B
} Outputs

Enable

(a)

Inputs			Enable	Outputs	
1	2	3	E	A	B
X	X	X	1	0	0
0	1	1	0	0	1
1	0	1	0	1	0
1	1	0	0	1	1

(b)

Fig. 4.107 Two-bit encoder:
(a) block diagram; (b) truth
(function) table.

Inputs			Enable	Outputs	
1	2	3	E	A	B
X	X	X	1	0	0
0	0	1	0	1	0
0	1	0	0	0	1
0	1	1	0	0	1
1	0	0	0	1	1
1	0	1	0	1	1
1	1	0	0	1	1

Fig. 4.108 Truth table for a
priority encoder. The value
of the binary weighted
decimal number on the
output lines gives an
indication of the *highest*
value input line at logic 0.

input line which is at 0 is indicated as a binary weighted decimal number on the two output lines. This could be extended to a situation where more than one input line is at logic 0 by giving the

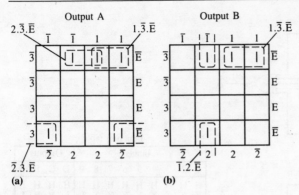

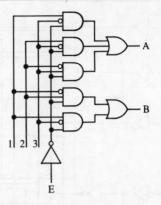

Fig. 4.109 Karnaugh maps to produce the priority encoder output line logic circuit requirements: (a) for output A; (b) for output B. Fig. 4.116 Truth (function) table for the Gray code.

Fig. 4.110 Logic function to produce the required priority encoder output line requirements.

output line a weighted value equivalent to a logic 0 on the highest value input line. The truth table of Fig. 4.108 should make this clear.

The logic circuit required to produce these output line requirements from the given input line levels can be deduced using Karnaugh mapping techniques. For output A the map required is shown in Fig. 4.109(a) while that for output B is shown in Fig. 4.109(b). The inputs considered for each map are 1, 2, 3 and enable, which is active low.

The enable (E) variable has been shown on the maps in Fig. 4.109 even though the only value considered is active low (i.e. $\bar{E}$). The maps could be simplified to three-variable maps and then simply add $\bar{E}$ as a requisite input to each resulting logic circuit. From the maps it can be seen that

$$A = 1.\bar{3}.\bar{E} + 2.\bar{3}.\bar{E} + \bar{2}.3.\bar{E} \qquad [4.36]$$

$$B = \bar{1}.2.\bar{E} + 1.\bar{3}.\bar{E} \qquad [4.37]$$

The required logic function from equations [4.36] and [4.37] can be built as shown in Fig. 4.110.

The 74147 is a 10 line (decimal) to 4 line (BCD) priority encoder, which will encode the highest order data line. In fact only nine data lines are encoded since the implied decimal zero condition requires no input conditions; zero is encoded with all nine data lines at logic 1 level. The function logic is shown in Fig. 4.111(a) and the function table in Fig. 4.111(b).

The application for the 74147 specifically includes keyboard encoding, but in general priority encoders are used wherever there is a need to connect a number of data lines to a smaller number for encoding purposes; specifically decimal to BCD has been mentioned but binary to octal is also possible.

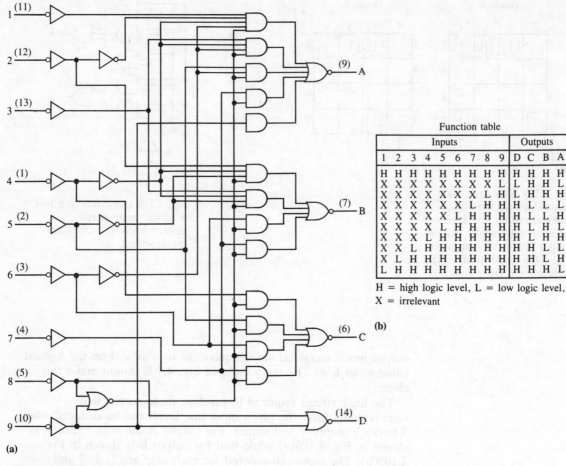

Function table

Inputs									Outputs			
1	2	3	4	5	6	7	8	9	D	C	B	A
H	H	H	H	H	H	H	H	H	H	H	H	H
X	X	X	X	X	X	X	X	L	L	H	H	L
X	X	X	X	X	X	X	L	H	L	H	H	H
X	X	X	X	X	X	L	H	H	H	L	L	L
X	X	X	X	X	L	H	H	H	H	L	L	H
X	X	X	X	L	H	H	H	H	H	L	H	L
X	X	X	L	H	H	H	H	H	H	L	H	H
X	X	L	H	H	H	H	H	H	H	H	L	L
X	L	H	H	H	H	H	H	H	H	H	L	H
L	H	H	H	H	H	H	H	H	H	H	H	L

H = high logic level, L = low logic level,
X = irrelevant

(b)

(a)

Fig. 4.111 SN74147 10-to-4
line priority encoder: (a)
functional logic circuit; (b)
function (truth) table.

Demultiplexers and decoders

In essence the circuits of demultiplexers and decoders are the opposite of those discussed earlier in this section. For example, a **demultiplexer** can take one input and transfer the data on that input line to the correct one of several output lines. Where multiplexers and demultiplexers are used together, as for example at each end of a single data line, then the select function must be synchronised at both the multiplexer and demultiplexer to ensure correct routing of the data.

The **decoder** can be the reverse of the encoder circuits. The 10-line to 4-line encoder discussed in the previous section has its counterpart in the 4-line to 10-line, BCD to decimal, decoders which are commercially available. In addition, there are a variety of decoder circuits designed for specific purposes. For example,

Binary			Octal
C	B	A	
0	0	0	00
0	0	1	01
0	1	0	02
0	1	1	03
1	0	0	04
1	0	1	05
1	1	0	06
1	1	1	07

(a)

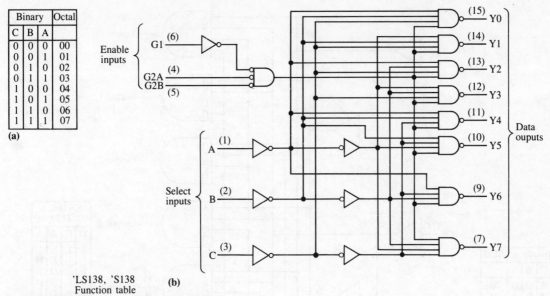

(b)

'LS138, 'S138
Function table

Inputs					Outputs							
Enable		Select										
G1	G2*	C	B	A	Y0	Y1	Y2	Y3	Y4	Y5	Y6	Y7
X	H	X	X	X	H	H	H	H	H	H	H	H
L	X	X	X	X	H	H	H	H	H	H	H	H
H	L	L	L	L	L	H	H	H	H	H	H	H
H	L	L	L	H	H	L	H	H	H	H	H	H
H	L	L	H	L	H	H	L	H	H	H	H	H
H	L	L	H	H	H	H	H	L	H	H	H	H
H	L	H	L	L	H	H	H	H	L	H	H	H
H	L	H	L	H	H	H	H	H	H	L	H	H
H	L	H	H	L	H	H	H	H	H	H	L	H
H	L	H	H	H	H	H	H	H	H	H	H	L

*G2 = G2A + G2B
H = high level, L = low level, X = irrelevant
(c)

Fig. 4.112 Binary to octal decoder: (a) truth table; (b) functional logic diagram; (c) function table. (b) and (c) use the SN74LS138 TTL circuit.

decoders can be used to indicate the number of input variables that are 'true' or 'false', i.e. if a three-input decoder is used with one output, that output can be made true if, say, an odd number (1 or 3) of the input variables is true. Decoders can also indicate majority or minority variable states. For example, a three-input circuit could be used as a 'majority' decoder to give a true output if any two of its three input variables is true. Another variation on the decoder theme has already been mentioned, namely the code converter, where it is required to convert from, say, decimal to BCD, binary to octal, etc. Some of these decoder circuits will be discussed in this section.

Decoders A **binary to octal decoder** would use the truth table of Fig. 4.112(a) from which any of the octal values 0 to 7 inclusive are represented by a three-variable binary weighted signal. Thus output line 0 will be true if $A = B = C = 0$. If the output line is active low then a NAND gate can be used with inputs $\overline{A}$, $\overline{B}$ and $\overline{C}$. Similarly, output line 1 will be true if $A = 1$, $B = 0$, $C = 0$ and a NAND gate with inputs A, $\overline{B}$ and $\overline{C}$ will be uniquely low if this combination is

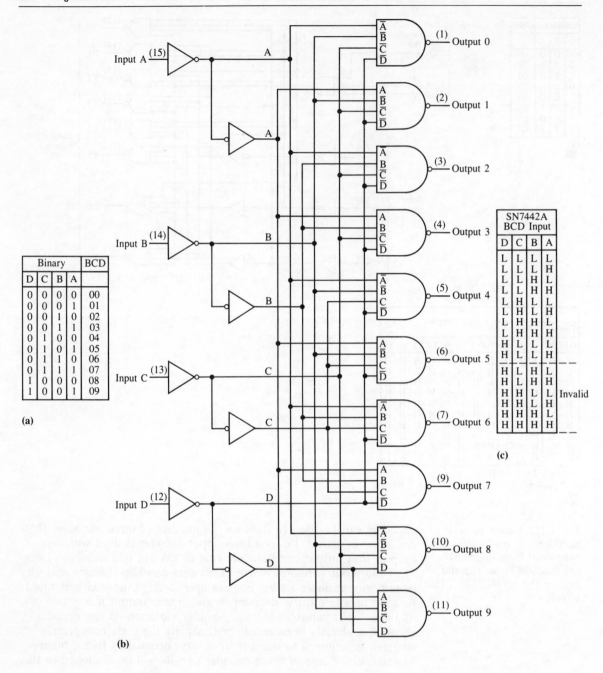

Binary				BCD
D	C	B	A	
0	0	0	0	00
0	0	0	1	01
0	0	1	0	02
0	0	1	1	03
0	1	0	0	04
0	1	0	1	05
0	1	1	0	06
0	1	1	1	07
1	0	0	0	08
1	0	0	1	09

(a)

(b)

SN7442A BCD Input			
D	C	B	A
L	L	L	L
L	L	L	H
L	L	H	L
L	L	H	H
L	H	L	L
L	H	L	H
L	H	H	L
L	H	H	H
H	L	L	L
H	L	L	H
H	L	H	L
H	L	H	H
H	H	L	L
H	H	L	H
H	H	H	L
H	H	H	H

Invalid

(c)

Fig. 4.113 BCD-to-decimal decoder: (a) truth table; (b) functional logic diagram; (c) function table. (b) and (c) use the SN7442A TTL circuit.

present at its input. Similar reasoning applies for all of the eight output circuits. In a practical arrangement, such as the 74138 3-to-8 line decoder, enable inputs are included to simplify cascading and/or data reception. The logic diagram for the 74138 is shown in Fig. 4.112(b) together with the function table in Fig. 4.112(c) which shows the required logic state that must be present on the gate enable inputs to give the required output.

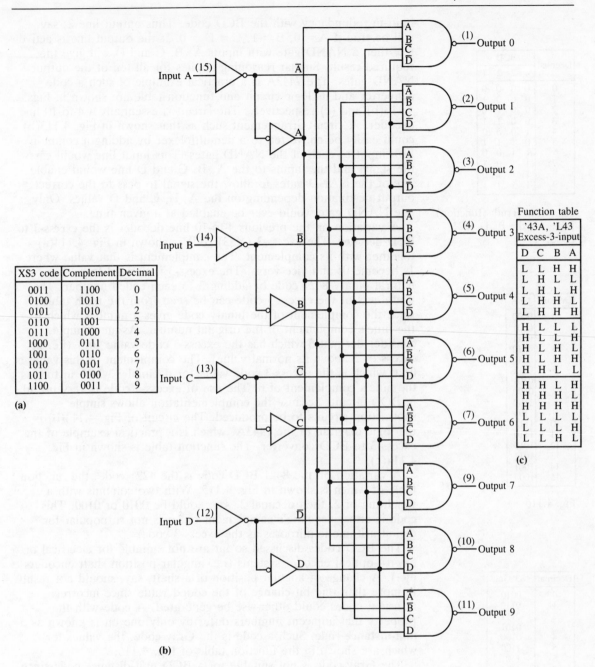

(a)

XS3 code	Complement	Decimal
0011	1100	0
0100	1011	1
0101	1010	2
0110	1001	3
0111	1000	4
1000	0111	5
1001	0110	6
1010	0101	7
1011	0100	8
1100	0011	9

Function table

'43A, 'L43 Excess-3-input			
D	C	B	A
L	L	H	H
L	H	L	L
L	H	L	H
L	H	H	L
L	H	H	H
H	L	L	L
H	L	L	H
H	L	H	L
H	L	H	H
H	H	L	L
H	H	L	H
H	H	H	L
H	H	H	H
L	L	L	L
L	L	L	H
L	L	H	L

(c)

(b)

Fig. 4.114 XS3 to BCD decoder: (a) truth table; (b) functional logic diagram; (c) function table. (b) and (c) use the SN7443A TTL circuit.

A BCD-to-decimal decoder would use the truth table of Fig. 4.113(a) from which any of the decimal values 0 to 9 inclusive can be represented by a four-variable binary weighted signal. The weighted signal in this case uses the natural or 8421 code, so called because this gives the value of the 'weighting' to each of the four digits. Since four digits can give 2^4 or 16 combinations, there is

Decimal	BCD			
	4	2	2	1
0	0	0	0	0
1	0	0	0	1
2	0	0	1	0
3	0	0	1	1
4	1	0	0	0
5	0	1	1	1
6	1	1	0	0
7	1	1	0	1
8	1	1	1	0
9	1	1	1	1

Fig. 4.115 Truth (function) table for the BCD 4221 code.

Decimal	Gray
0	0000
1	0001
2	0011
3	0010*
4	0110
5	0111
6	0101
7	0100
8	1100
9	1101
10	1111
11	1110
12	1010*
13	1011
14	1001
15	1000

Fig. 4.116

Decimal	XS3 Gray
0	0010
1	0110
2	0111
3	0101
4	0100
5	1100
6	1101
7	1111
8	1110
9	1010
10	1011
11	1001
12	1000
13	0000
14	0001
15	0011

Fig. 4.117 Truth (function) table for the XS3 Gray code.

built-in redundancy with the BCD code. Thus output line 2, say, will be true if $A = 0$, $B = 1$, $C = D = 0$. If the output line is active low then a NAND gate with inputs $\overline{A}$, B, $\overline{C}$ and $\overline{D}$ will give the required result. Similar reasoning applies for all ten of the output NAND gates. The 7442A is a practical example of such a code converter and its logic circuit and function table are shown in Figs. 4.113(b) and (c) respectively. The circuit is essentially a 4-to-10 line decoder. A circuit arrangement such as that shown in Fig. 4.113(b) could easily be converted to a demultiplexer by adding a common fifth input to each of the NAND gates. This input line would carry the signal and the inputs to the A, B, C and D line would enable *one* of the NAND gates to allow the signal to pass to the correct output destination depending on the A, B, C and D values. Only one NAND gate would ever be enabled at a given time.

A variation on the previous 4-to-10 line decoder is the **excess-3 to BCD decoder**. The excess-3 (XS3) code is shown in Fig. 4.114(a) together with its complement. The complement is that value where 1s become 0s and vice versa. The excess-3 is derived from the natural binary 8421 code by adding 3 to each coded number. The usefulness of the excess-3 code can be seen from the truth table since the complement of the binary code gives a number which is the nine's complement of the original number. As an example, consider decimal 6 which has the excess-3 code value 1001 (i.e. in excess of 3 above its normal value). The complement of this value is 0110 which is the excess-3 code value for decimal 3 which in turn is the nine's complement of 6. The use of excess-3 code is very useful in BCD circuits because the complementation allows simple subtraction circuits to be produced. The circuit of Fig. 4.114(b) is the logic diagram of the 7443A, which is a practical example of the excess-3 to BCD converter. The function table is shown in Fig. 4.114(c).

A variation on the 8421 BCD code is the 4221 code, the function table of which is shown in Fig. 4.115. With two columns with a weighting of 2, then decimal 2, say, could be 0010 or 0100. This code is also gives nine's complement but it is not so popular for digital arithmetic purposes as the excess-3 code.

The BCD codes discussed so far are not suitable for electrical or electro-optical encoder systems (i.e. angular position shaft encoders etc.). A change of angular position of a shaft, say, should not result in more than one bit change of the coded value since incorrect transient codes could otherwise be generated. A code with the property that adjacent numbers differ by only one bit is known as a **unit-distance code**. Such a code is the Gray code, the values for which are shown in the function table of Fig. 4.116.

The Gray code is not suitable for a BCD unit-distance code since it is not possible to change from decimal 9 to decimal 0 with only one bit change. However, a code which does have the property of only one bit change for the transition from 9 to 0 is the excess-3 Gray code shown in Fig. 4.117. This arrangement will still have the unit-distance property since the BCD excess-3 Gray code is simply those values between the asterisks of Fig. 4.116. The 7444 provides an excess-3 Gray to decimal converter; the logic circuit is shown in Fig. 4.118(a) with the function table in Fig. 4.118(b).

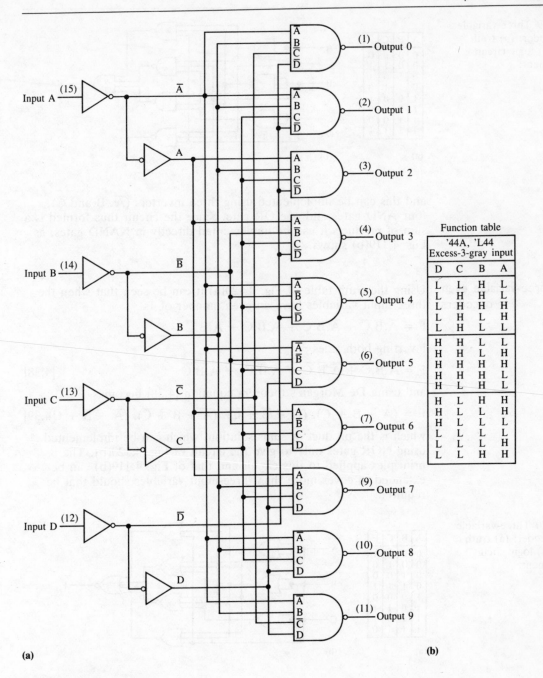

Function table

| '44A, 'L44 Excess-3-gray input | | | |
D	C	B	A
L	L	H	L
L	H	H	L
L	H	H	H
L	H	L	H
L	H	L	L
H	H	L	L
H	H	L	H
H	H	H	H
H	H	H	L
H	L	H	L
H	L	H	H
H	L	L	H
H	L	L	L
L	L	L	L
L	L	L	H
L	L	H	H

(a)

(b)

Fig. 4.118 SN7444 XS3
Gray to decimal converter:
(a) functional logic
diagram; (b) function table.

Three-variable odd decoders

For a three-variable input the truth table of Fig. 4.119(a) shows that the output is required to be 1 when the three input variables have an odd number of 1s i.e.

$$F = \overline{A}.\overline{B}.C + \overline{A}.B.\overline{C} + A.\overline{B}.\overline{C} + A.B.C$$

Fig. 4.119 Three-variable odd decoder: (a) truth table; (b) logic circuit arrangement.

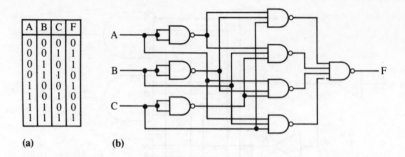

A	B	C	F
0	0	0	0
0	0	1	1
0	1	0	1
0	1	1	0
1	0	0	1
1	0	1	0
1	1	0	0
1	1	1	1

(a)   (b)

and this can be implemented using three inverters $(\overline{A}, \overline{B}$ and $\overline{C})$, four AND gates and one OR gate. Since the circuit thus formed is a sum of products it can be implemented directly in NAND gates, as Fig. 4.119(b) shows.

Three-variable even decoders

Using the truth table of Fig. 4.120(a) it can be seen that when the three input variables have an even number of 1s

$$\overline{F} = \overline{A}.\overline{B}.\overline{C} + \overline{A}.B.C + A.\overline{B}.C + A.B.\overline{C}$$

inverting both sides

$$F = \overline{\overline{A}.\overline{B}.\overline{C} + \overline{A}.B.C + A.\overline{B}.C. + A.B.\overline{C}} \qquad [4.38]$$

and using De Morgan's theorem equation [4.38] becomes

$$F = (A + B + C).(A + \overline{B} + \overline{C}).(\overline{A} + B + \overline{C}).(\overline{A} + \overline{B} + C) \qquad [4.39]$$

which is the product of sums solution, which can be implemented using NOR gates only to give the circuit of Fig. 4.120(b). The principles applied to this circuit and that of Fig. 4.119(b) can be extended to cover more than three-input variables should that be required.

Fig. 4.120 Three-variable even decoder: (a) truth table; (b) logic circuit arrangement.

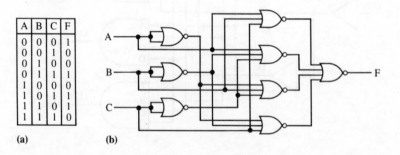

A	B	C	F
0	0	0	1
0	0	1	0
0	1	0	0
0	1	1	1
1	0	0	0
1	0	1	1
1	1	0	1
1	1	1	0

(a)   (b)

Majority decoders

Using the truth table of Fig. 4.121(a) the condition for two or more of the input variables to be 1 gives the equation

$$F = \overline{A}.B.C + A.\overline{B}.C + A.B.\overline{C} + A.B.C \qquad [4.40]$$

Using Boolean algebra rules or mapping techniques, equation [4.40] reduces to

$$F = A.B + B.C + C.A$$

and this can be implemented, using NAND gates, in the circuit of

Fig. 4.121 Majority decoder: (a) truth table; (b) NAND gate circuit implementation; (c) A-O-I circuit implementation; (d) Karnaugh map of the logic function.

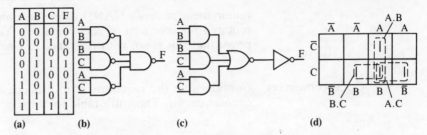

A	B	C	F
0	0	0	0
0	0	1	0
0	1	0	0
0	1	1	1
1	0	0	0
1	0	1	1
1	1	0	1
1	1	1	1

(a)   (b)   (c)   (d)

Fig. 4.121(b). It can also be implemented directly from the Karnaugh map as an A-O-I circuit and this is also shown in Fig. 4.121(c). To assist in checking that both circuits can be derived from the map, the Karnaugh map is shown in Fig. 4.121(d).

Minority decoders In this arrangement the output is at logic 1 level if any one of the inputs does not agree with the other two inputs. The truth table is shown in Fig. 4.122(a).

Using the logic 0 outputs the equation can be written as

$$\overline{F} = \overline{A}.\overline{B}.\overline{C}. + A.B.C \qquad [4.41]$$

inverting both sides

$$F = \overline{\overline{A}.\overline{B}.\overline{C} + A.B.C} \qquad [4.42]$$

and using De Morgan's theorem

$$F = (A + B + C). (\overline{A.B.C}) \qquad [4.43]$$

$$F = A.(\overline{A.B.C}) + B.(\overline{A.B.C}) + C.(\overline{A.B.C}) \qquad [4.44]$$

inverting both sides

$$\overline{F} = \overline{A.(\overline{A.B.C}) + B.(\overline{A.B.C}) + C.(\overline{A.B.C})} \qquad [4.45]$$

and using De Morgan's theorem

$$\overline{F} = \overline{A.(\overline{A.B.C})}.\overline{B.(\overline{A.B.C})}.\overline{C.(\overline{A.B.C})} \qquad [4.46]$$

from which

$$F = \overline{\overline{A.(\overline{A.B.C})}.\overline{B.(\overline{A.B.C})}.\overline{C.(\overline{A.B.C})}} \qquad [4.47]$$

Equation [4.47] can be implemented using NAND gates as shown in Fig. 4.122(b).

Fig. 4.122 Minority decoder: (a) truth table; (b) NAND gate circuit implementation.

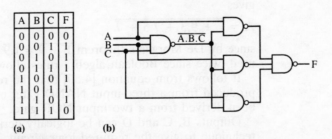

A	B	C	F
0	0	0	0
0	0	1	1
0	1	0	1
0	1	1	1
1	0	0	1
1	0	1	1
1	1	0	1
1	1	1	0

(a)   (b)

The equation of [4.47] produces a minimised decoding circuit since converting the circuit directly to NAND gates from the Karnaugh map of the truth table of Fig. 4.122(a) would give a requirement of *seven* NAND gates (assuming a NAND gate is required for inversion of each variable). You should be able to prove that this is so, with reference to Chapter 3 where necessary.

Code converters

Consider first the requirement to produce a decimal-to-8421 BCD counter circuit. The truth table for each code is shown in Fig. 4.123.

Fig. 4.123 Truth table for decimal-to-BCD conversion.

Decimal inputs										BCD outputs			
9	8	7	6	5	4	3	2	1	0	D	C	B	A
0	0	0	0	0	0	0	0	0	0	0	0	0	0
0	0	0	0	0	0	0	0	1	0	0	0	0	1
0	0	0	0	0	0	0	1	0	0	0	0	1	0
0	0	0	0	0	0	1	0	0	0	0	0	1	1
0	0	0	0	0	1	0	0	0	0	0	1	0	0
0	0	0	0	1	0	0	0	0	0	0	1	0	1
0	0	0	1	0	0	0	0	0	0	0	1	1	0
0	0	1	0	0	0	0	0	0	0	0	1	1	1
0	1	0	0	0	0	0	0	0	0	1	0	0	0
1	0	0	0	0	0	0	0	0	0	1	0	0	1

Inspection of the truth tables gives the condition that output A is 1 when the input for decimal lines, 1, 3, 5, 7 and 9 are logic 1. For output B to be logic 1 the inputs 2, 3, 6 and 7 must be 1. For output C to be logic 1 the inputs 4, 5, 6 and 7 must be 1 and finally, for output D to be logic 1 the inputs 8 and 9 must be 1.

Using two-input NOR gates, if both inputs are logic 0 the output is 1 and for NAND gates if any input is logic 1 the output is 0. Thus a NOR-NAND combination is suitable in this case for providing the correct values for outputs A, B, C and D for any of the possible decimal input values. A possible circuit is shown in Fig. 4.124.

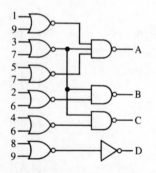

Fig. 4.124 Decimal-to-BCD code converter logic arrangement.

The output for A in Fig. 4.124 is given by

$$A = 1 + 3 + 5 + 7 + 9 \qquad [4.48]$$

which can be produced from a NAND gate with inputs $\bar{1}$, $\bar{3}$, $\bar{5}$, $\bar{7}$, and $\bar{9}$ since

$$A = \overline{\bar{1}.\bar{3}.\bar{5}.\bar{7}.\bar{9}} \qquad [4.49]$$

which be De Morgan's theorem simplifies to equation [4.48].

Modifying equation [4.49], again using De Morgan's theorem, gives

$$A = \overline{\overline{1 + 9}.\overline{3} + \overline{7.5} + \overline{7}} \qquad [4.50]$$

since by De Morgan's theorem $\overline{1 + 9} = \bar{1}.\bar{9}$ etc. Function 7 has been used twice since Boolean algebra rules allow this: i.e. A.A = A, etc.

It follows from equation [4.50] that the required output A can be produced from a three-input NAND gate, each input having itself been derived from a two input NOR gate.

Outputs B, C and D can be logically derived using the same technique to give the required logic circuit. There is no NAND gate associated with output D. In fact and OR gate would have sufficed here but, since inverting logic has been used throughout, a NOR gate followed by an inverter has been utilised.

Example 4.1   *Convert the four-bit code to Gray code.*

The truth tables of each code are shown in Fig. 4.125. From the truth tables it can be seen that each of the Gray code values can be 1 for a combination of binary values B_0, B_1, B_2 and B_3. For example, $G_0 = 1$ when $\overline{B}_3.\overline{B}_2.\overline{B}_1.B_0 = 1$ and $\overline{B}_3.\overline{B}_2.B_1.\overline{B}_0 = 1$ etc. The Karnaugh map can be used for all possible combinations to allow simplification of logic requirements.

Fig. 4.125 Truth table for the conversion of a four-bit binary code to Gray code.

Decimal	Binary				Gray			
	B_3	B_2	B_1	B_0	G_3	G_2	G_1	G_0
0	0	0	0	0	0	0	0	0
1	0	0	0	1	0	0	0	1
2	0	0	1	0	0	0	1	1
3	0	0	1	1	0	0	1	0
4	0	1	0	0	0	1	1	0
5	0	1	0	1	0	1	1	1
6	0	1	1	0	0	1	0	1
7	0	1	1	1	0	1	0	0
8	1	0	0	0	1	1	0	0
9	1	0	0	1	1	1	0	1
10	1	0	1	0	1	1	1	1
11	1	0	1	1	1	1	1	0
12	1	1	0	0	1	0	1	0
13	1	1	0	1	1	0	1	1
14	1	1	1	0	1	0	0	1
15	1	1	1	1	1	0	0	0

The maps for G_0, G_1, G_2 and G_3 are shown in Fig. 4.126.

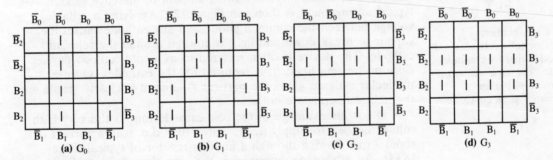

(a) G_0 (b) G_1 (c) G_2 (d) G_3

Fig. 4.126 Karnaugh maps used to determine the minimised logic circuit requirement for the code conversion of four-bit binary to Gray code: (a) Gray code bit G_0; (b) Gray code bit G_1; (c) Gray code bit G_2; (d) Gray code bit G_3.

From the Karnaugh maps it can be seen that

$$G_0 = \overline{B}_0.B_1 + B_0.\overline{B}_1 = B_0 \oplus B_1 \qquad [4.51]$$

$$G_1 = B_1.\overline{B}_2 + \overline{B}_1.B_2 = B_1 \oplus B_2 \qquad [4.52]$$

$$G_3 = B_2.\overline{B}_3 + B_3.\overline{B}_2 = B_2 \oplus B_3 \qquad [4.53]$$

$$G_4 = B_3. \qquad [4.54]$$

Thus the equations of [4.51] to [4.54] can be converted to the logic circuit of Fig. 4.127.

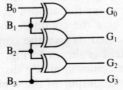

Fig. 4.127 Logic circuit to implement four-bit binary to Gray code conversion.

Decoder/drivers   The output of a decoder may be required to feed some kind of numeric display element which can visually indicate the output state of the decoder. Whatever type of indicator device is used it is important that the amount of current required to flow through the indicator lamp can be safely handled by the decoder output, whether the device acts as a current sink or current source. To make sure that device current can be safely handled a **driver stage** could be used between the decoder output and the indicating element. Each driver may well be a transistor which can be switched by the decoder outputs but will additionally supply a large current to the display element. TTL driver circuits are available in IC form and can easily be connected in circuit providing there is a suitable matching between the driver, in terms of voltage and current, and the indicator lamp. Alternatively, some decoders have driver outputs specifically built in to the single IC package. The type of numeric display devices that can be used can be broadly split into two categories: indicator tubes and seven-segment displays.

The numeric indicator tube, often called the **Nixie tube**, is a gas discharge device. The Nixie tube was developed by the Burroughs Corporation and although the name is a registered trademark it is often used as a generic term to cover this type of device. The tube consists of ten metal pieces shaped into the form of the required decimal digits with each piece being used as a separate cathode. All cathodes share a common anode. The assembly is enclosed within a glass tube in which there is a small amount of mercury vapour. A typical arrangement is shown in Fig. 4.128. Applying a suitable voltage between the common anode and any cathode causes the vapour to break down and light is emitted around the relevant cathode. The device is often referred to as a cold-cathode device since no heating filament is required for the cathodes. Once a particular cathode glows the particular numeral is clearly visible and the unlit numerals are not seen.

A major disadvantage of the cold-cathode tube is the very high voltage needed to supply it, typically 170 V d.c. with a current of about 7 mA flowing through a limiting resistor of typical value 15 kΩ. An advantage, however, is the very clear digit produced. The 74141 is a BCD-to-decimal decoder/driver which is specifically designed for feeding a Nixie tube type display. The outputs are buffered so as to withstand the high voltages needed to operate a Nixie tube. This can be clearly seen on the logic diagram of Fig. 4.129. The outputs are protected by zener diodes which can hold the output voltage to a safe level.

The seven-segment display is the most popular arrangement and can be obtained using **light-emitting diode** (LED) elements, gas-discharge or **liquid crystal display** (LCD) elements. The diagram of Fig. 4.130 shows the arrangement of the seven segments usually labelled a to g inclusive and the decimal point, DP, when provided.

Each segment is an individual LED and has an anode and cathode lead. In some cases the anodes of all seven segments are connected together to give the common anode type display, or the cathodes may be connected to give the common cathode type display. The arrangement is shown in Fig. 4.131.

In TTL circuitry the common anode display is usually preferred. For this arrangement the output of the logic circuit, or driver, must go low to provide segment illumination. For a particular number to

Fig. 4.128 Numeric indicator tube cathodes. All cathodes share a common anode. The arrangement is enclosed within an evacuated glass envelope.

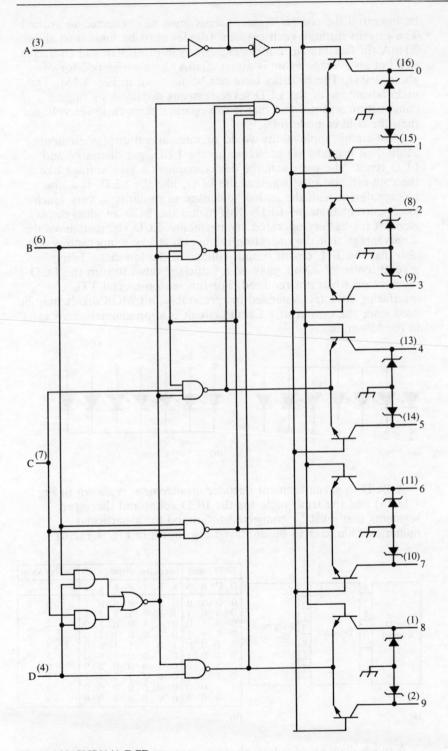

Fig. 4.129 SN74141 BCD-
to-decimal decoder/driver
logic circuit for driving a
numeric indicator tube
display.

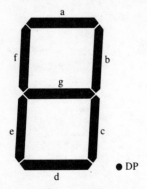

Fig. 4.130 Seven-segment display layout arrangement.

be indicated the correct segment leads must be connected to ground. The current through each segment (diode) must be limited to about 20 mA (in fact the current depends on type of display and character size but an average value is about 20 mA) by a series resistor of about 250 Ω. The resistors have not been shown in Fig. 4.131. The main advantage of the LED seven-segment display is its rugged construction and the fact that it can operate at much lower voltages than the cold-cathode tube.

Most circuit applications would require several display elements. Multi-digit displays are available for the LED, gas discharge and LCD types. The gas discharge device requires a high voltage like the cold-cathode tube whereas the LCD, like the LED, is a low voltage device with the added advantage of requiring a very much lower current than the LED. This makes the LCD an ideal display element for battery operated arrangements. LCD elements have the disadvantage that the operation is sluggish at low temperatures and also that the TTL circuit output voltage in the low state, being several tenths of a volt, may be a sufficient value to turn the LCD segment on when not required. For this reason special TTL interfacing circuitry is needed or, preferably, a CMOS driver may be used since the output of a CMOS circuit is approximately zero volts in the low state.

Fig. 4.131 Seven-segment display connections: (a) common anode; (b) common cathode.

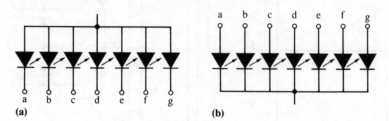

(a)

(b)

A BCD to seven-segment decoder arrangement is shown in Fig. 4.132(a) and the truth table for the BCD count and the seven segments that must be connected to ground for a particular numerical character to be displayed are shown in Fig. 4.132(b).

Fig. 4.132 BCD to 7-segment decoder; (a) block diagram; (b) truth table.

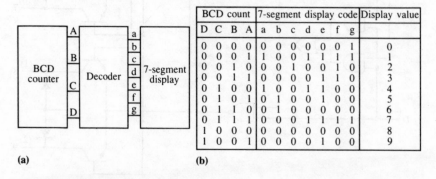

BCD count				7-segment display code							Display value
D	C	B	A	a	b	c	d	e	f	g	
0	0	0	0	0	0	0	0	0	0	1	0
0	0	0	1	1	0	0	1	1	1	1	1
0	0	1	0	0	0	1	0	0	1	0	2
0	0	1	1	0	0	0	0	1	1	0	3
0	1	0	0	1	0	0	1	1	0	0	4
0	1	0	1	0	1	0	0	1	0	0	5
0	1	1	0	0	1	0	0	0	0	0	6
0	1	1	1	0	0	0	1	1	1	1	7
1	0	0	0	0	0	0	0	0	0	0	8
1	0	0	1	0	0	0	0	1	0	0	9

(a)

(b)

A block diagram of a typical seven-segment decoder/driver is shown in Fig. 4.133. In addition to the BCD input terminals and the seven-segment output terminals a to g there are shown three terminals marked RB1, RB0 and LT. When connecting the IC in circuit with a seven-segment display element, if the RB1 terminal

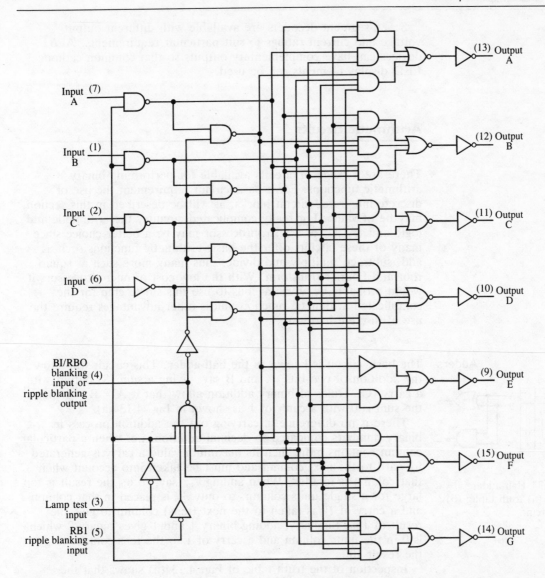

Fig. 4.133 SN7447 BCD to 7-segment decoder driver functional logic diagram.

(ripple blanking input) is connected to logic 0 the decoder will blank the display of numeral 0, i.e. it will only indicate 1 to 9 inclusive. Hence, for a BCD input of 0000 the element display stays blank. If the RB0 (ripple blank output) is used in multi-digit circuits with several seven-segment decoders interconnected with separate display elements, then if RB1 is at logic 0 and the BCD input is 0000, output RB0 switches to 0. Otherwise RB0 logic level is 1.

LT (lamp test) can be used to test the display segments. If LT is at logic 0 all segments will light as decoder outputs all go to 0.

An example of a TTL decoder/driver for BCD to seven-segment display is the 7447. The outputs of this circuit are buffered to a voltage rating of 15 V and a current rating of 20 mA so that the circuit can directly drive an LED display. The logic circuit arrangement of the 7447 is shown in Fig. 4.133. When the 7447 is used with LED displays, current limiting resistors of typical value 330 Ω must be used in each of the output lines.

Seven-segment decoders are available with different output voltage and current ratings to suit particular requirements. Also devices can have complementary outputs so that common cathode LED display elements can be used.

Arithmetic circuits

There are a range of circuits available for performing binary arithmetic functions. For a basic circuit requirement the use of discrete integrated circuit blocks, as will be described in this section, may be of value. For really complicated circuitry the arithmetic and logic unit (ALU) of a microprocessor may be a better choice since many of these circuits offer the basic arithmetic functions such as add, subtract, multiply and divide, plus many more such as square root and Fourier transform. With the low cost of microprocessors it would probably be advantageous to use this single chip for the complicated arithmetic function unless other advantages require the use of discrete hardware.

Adders The basic circuit is known as the **half-adder**. This circuit will allow the addition of two bits, A and B say, giving a sum (S) output with a carry (C). Rules of binary addition allow that if $A = B = 1$ then the sum is 0 with a carry of 1 as shown in Fig. 4.134(a).

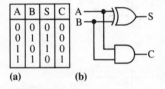

A	B	S	C
0	0	0	0
0	1	1	0
1	0	1	0
1	1	0	1

(a)   (b)

Fig. 4.134 Half-adder circuit: (a) truth table; (b) logic circuit.

There is no difference in carrying out the addition process in binary numbers compared to decimal addition, i.e. when a particular column contains more then its maximum value a carry is generated forward to the next column and must be taken into account when that column is totalled. When adding 7_{10}, say, to 6_{10} the result is too large for a single units column, so only 3_{10} is placed in that column and a carry of 1_{10} is taken to the next (tens) column to give the result of 13_{10}. Similarly, adding binary 1 and 1 gives binary 2 which is 0 in the units column and a carry of 1 to the tens column to give the result 10_2.

Inspection of the truth table of Fig. 4.134(a) shows that the function S is satisfied by the expression

$$S = A.\overline{B} + \overline{A}.B = A \oplus B$$

so that S is satisfied using an Exclusive-OR circuit. Also

$$C = A.B$$

so that C can be catered for with a single two-input AND gate. The basic circuit can thus be constructed as shown in Fig. 4.134(b). A half-adder on its own is of little practical use since it needs to cater for a third input which could be the carry from another addition process. A block diagram of a full adder, together with its truth table, is shown in Fig. 4.135.

From the truth table it can be seen that for the sum

$$S = \overline{A}.\overline{B}.C_{in} + \overline{A}.B.\overline{C}_{in} + A.\overline{B}.\overline{C}_{in} + A.B.C_{in}$$

$$\text{or } S = (\overline{A}.B + A.\overline{B}).\overline{C}_{in} + (A.B. + \overline{A}.\overline{B}).C_{in}$$

$$\text{or } S = (A \oplus B).\overline{C}_{in} + (\overline{A \oplus B}).C_{in}$$

Fig. 4.135 Full-adder
circuit: (a) block diagram;
(b) truth table.

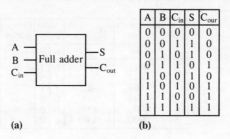

A	B	C_{in}	S	C_{our}
0	0	0	0	0
0	0	1	1	0
0	1	0	1	0
0	1	1	0	1
1	0	0	1	0
1	0	1	0	1
1	1	0	0	1
1	1	1	1	1

(a)   **(b)**

giving $S = A \oplus B \oplus C_{\cdot in}$ [4.55]

Also from the truth table it can be seen that for the carry out

$$C_{out} = \overline{A}.B.C_{in} + A.\overline{B}.C_{in} + A.B.\overline{C}_{in} + A.B.C_{in}$$

or $C_{out} = (A \oplus B).C_{in} + A.B$ [4.56]

The complete circuit is shown in Fig. 4.136.

Fig. 4.136 Full-adder logic
circuit arrangement.

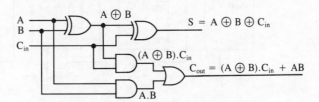

The concept of adders has so far only considered the addition of two single bits, A and B, with a possible carry-in bit. If, however, it is required to add n bits, where n is the number of bits in a given binary number, then either n full-adders must be employed with each full-adder taking care of a particular one of the n bits, together with the carry-in, or one full-adder can be used and the bits fed in sequentially. The former technique gives a parallel adder while the latter technique gives a serial, or sequential, adder. The principle of the serial adder is simpler and will be dealt with first.

Consider the circuit of Fig. 4.137(a). The inputs A and B are the least significant bits in two shift registers (not shown) at a particular moment and they are to added to the carry of the previous addition process to yield a sum and a new carry. The sum is passed as the most significant bit to a shift register while the carry is delayed in the D-type flip-flop by one clock period before being applied as the new carry-in at the full-adder input.

The input registers are now shifted right by one bit, placing new values at the A and B inputs of the full-adder. These inputs are combined with the previous addition carry and the sum and carry-out are processed as described previously. Note that $n + 1$ sum bits are generated by two n-bit inputs. Refer to Fig. 4.137(b) which shows in truth table form the effect of adding two four-bit numbers using five clock pulses. The numbers to be added are binary 0110 (decimal 6) and binary 0101 (decimal 5) giving a result of binary 1011 (decimal 11) as expected.

The serial adder is simple, requiring only one full-adder, but slow because bits are added one at a time. A four-bit parallel adder is shown in Fig. 4.138 and is faster because each pair of the four-bits is added simultaneously. The problem with this arrangement is that

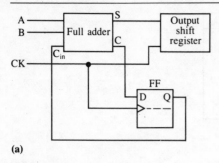

(a)

CK input	Input shift register	Input shift register	Adder input		D-type output	S	C	Output shift register contents
	A	B	A	B				
0	0110	0101	0	1	0	1	0	00000
1	0011	0010	1	0	0	1	0	10000
2	0001	0001	1	1	0	0	1	11000
3	0000	0000	0	0	1	1	0	01100
4	0000	0000	0	0	0	0	0	10110
5	0000	0000	0	0	0	0	0	01011

(b)

Fig. 4.137 Serial adder: (a)
logic circuit arrangement;
(b) truth (function) table.

Fig. 4.138 Four-bit parallel
adder.

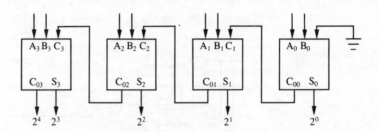

when a carry is generated it must appear at the input of the adder
serving the next significant bit and, in worst-case conditions, a carry
may have to ripple through the full length of the adder. For this
reason this circuit may be referred to as a ripple-through adder.

Example 4.2 *Using the circuit of Fig. 4.138 work out the effect of adding two four-bit
numbers given by A = 1101 (decimal 13) and B = 1011 (decimal 11). What
would be the total delay for such an addition?*

The addition process is such that for the two numbers quoted a carry
must ripple through *all four* adders. This can be seen by summing A and B

	binary	decimal
A	1 1 0 1	1 3
+B	1 0 1 1	+ 1 1
S	1 1 0 0 0	2 4
C	1 1 1 1	

It follows that after the first addition process the output from the least
significant adder is correct (since $C_0 = 0$ always). However, the output from
the next significant adder (S_1) is incorrect since it could not take into
account the carry-in (C_1) from the summing of A_0 and B_0. A second
addition process is then necessary to allow for the carry-in (C_1) from the
first stage, but the output from the *next* stage is still incorrect etc. The total
delay is thus proportional to the total number of stages, four in this case.

The delay time of the ripple-through adder may be reduced using
the concept of look-ahead carry. From equation [4.56] the output
carry from any full-adder stage was established as

$$C_{out} = (A \oplus B).C_{in} + A.B$$

Fig. 4.139 Four-bit parallel adder with 'look-ahead carry' facility.

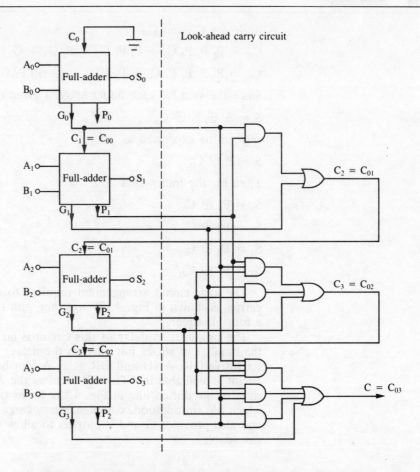

The term $A \oplus B$ is referred to as the propagation (P) term while the term $A.B$ is referred to as the generate (G) term so that equation [4.56] can be rewritten as:

$$C_{out} = P.C_{in} + G \qquad [4.57]$$

Referring back to the diagram in Fig. 4.138 of the four-stage parallel adder, the propagation and (carry) generate terms are

$$G_0 = A_0.B_0, \ P_0 = A_0 \oplus B_0 \qquad [4.58(a)]$$

$$G_1 = A_1.B_1, \ P_1 = A_1 \oplus B_1 \qquad [4.58(b)]$$

$$G_2 = A_2.B_2, \ P_2 = A_2 \oplus B_2 \qquad [4.58(c)]$$

$$G_3 = A_3.B_3, \ P_3 = A_3 \oplus B_3 \qquad [4.58(d)]$$

While the carry-outs are given by

$$C_{00} = P_0.C_0 + G_0 \qquad [4.59(a)]$$

$$C_{01} = P_1.C_1 + G_1 \qquad [4.59(b)]$$

$$C_{02} = P_2.C_2 + G_2 \qquad [4.59(c)]$$

$$C_{03} = P_3.C_3 + G_3 \qquad [4.59(d)]$$

Substituting for C_1 ($=C_{00}$) in equation [4.59(b)] gives

$$C_{01} = P_1.P_0.C_0 + P_1.G_0 + G_1 \qquad [4.60]$$

while for C_{02} and C_{03}

$$C_{02} = P_2.P_1.P_0.C_0 + P_2.P_1.G_0 + P_2.G_1 + G_2 \qquad [4.61]$$

$$C_{03} = P_3.P_2.P_1.P_0.C_0 + P_3.P_2.P_1.G_0 + P_3.P_2.G_1 + P_3.G_2 + G_3 \quad [4.62]$$

Since the sum for each adder stage is given in equation [4.55] as

$$S = A \oplus B \oplus C_{in}$$

this can be expressed as

$$S = P \oplus C_{in} \qquad [4.63]$$

Then for the four stages

$$S_0 = P_0 \oplus C_0 \qquad [4.64(a)]$$

$$S_1 = P_1 \oplus C_{00} \qquad [4.64(b)]$$

$$S_2 = P_2 \oplus C_{01} \qquad [4.64(c)]$$

$$S_3 = P_3 \oplus C_{02} \qquad [4.64(d)]$$

The logic circuit arrangement required to generate the look-ahead carries is shown in Fig. 4.139 together with the full-adder circuits for a four-bit adder.

The propagation delay of this circuit is no longer dependent on the number of stages but only on the delays produced by an Exclusive-OR, AND and OR gate. A four-bit look-ahead carry circuit is available in TTL MSI form as the 74182. The circuit to the right of the dotted line in Fig. 4.139 forms the basis of the 74182 which has, in addition, complementary carry outputs to those shown and also provides G and P outputs to allow the circuit to be cascadable.

Subtractors   The function of a subtractor circuit is basically similar to that of an adder circuit with a carry-in replaced by a borrow-in and a borrow-out replacing the carry-out. The resultant sum is replaced by a difference.

However, it would be unusual to find a pure subtractor circuit in practice. It is far more practical to provide a circuit which combines the process of addition and subtraction. For this reason an adder circuit can be used to perform a subtraction process; the two's complement of the number to be subtracted (the subtrahend) is added to the other number (the minuend). The two's complement can be obtained by inverting each bit and then adding 1 to the least significant bit (see App. 2). A circuit which performs the function of adder/subtractor is shown in Fig. 4.140 which utilises a four-bit full-adder circuit similar to that of Fig. 4.138 and available in TTL MSI form as the SN7483 or 74283.

In this arrangement the logic level on the control line determines the function performed by the circuit; a logic 0 on the control line activates gate G_1 G_3 G_5 and G_7, which allow the B inputs to be fed directly to the adder circuit so that the outputs are equal to the sum of A and B. A logic 1 on the control line allows gates G_2 G_4 G_6 and G_8 to function so that the complement of the B inputs is fed to the adder. Since C_1 is simultaneously equal to logic 1 the result is the same as adding the two's complement of number B to A.

Fig. 4.140 Four-bit full adder/subtractor circuit.

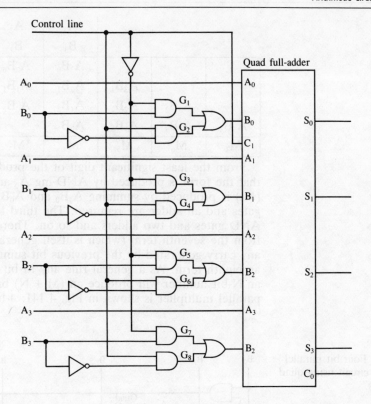

Multipliers

The rules of Boolean mulitplication follow those of decimal numbers in that the multiplicand is multiplied by successive digits of the multiplier and the product results from a series of partial products. In binary multiplication the process is simplified since each digit can be 1 or 0 so that each partial product is either equal to the multiplicand or is zero.

When multiplying 4 by 3 in binary, summing the partial products gives a product that is the binary equivalent of 12

$$4_{10} = 0\ 1\ 0\ 0$$

$$3_{10} = 0\ 0\ 1\ 1$$

so that $0\ 1\ 0\ 0 \times 0\ 0\ 1\ 1$ becomes

$$
\begin{array}{r}
0\ 1\ 0\ 0 \\
\times\ 0\ 0\ 1\ 1 \\
\hline
0\ 1\ 0\ 0 \\
0\ 1\ 0\ 0\ 0 \\
0\ 0\ 0\ 0\ 0\ 0 \\
0\ 0\ 0\ 0\ 0\ 0 \\
\hline
0\ 0\ 0\ 1\ 1\ 0\ 0 = 12_{10}
\end{array}
$$

In general terms, to multiply a number $A_3\ A_2\ A_1\ A_0$ by a second number $B_3\ B_2\ B_1\ B_0$ can produce the following product

			A_3	A_2	A_1	A_0
			B_3	B_2	B_1	B_0
			A_3B_0	A_2B_0	A_1B_0	A_0B_0
		A_3B_1	A_2B_1	A_1B_1	A_0B_1	
	A_3B_2	A_2B_2	A_1B_2	A_0B_2		
A_3B_3	A_2B_3	A_1B_3	A_0B_3			
M_7	M_6	M_5	M_4	M_3	M_2	M_1

From the least significant digit of the product (M_1) it can be seen that the form is produced by AND-ing A_0 and B_0. The second term (M_2) is produced by summing A_1B_0 and A_0B_1 so that two AND gates and an adder are required. The third term requires three AND gates and two adders and so on. There may by a carry-out from the seventh term (which is itself generated by adding A_3B_3 to any carry generated by the previous bit summing action) resulting in an eighth term. As a general rule any M-bit number multiplied by an N-bit number will produce an $(M + N)$ bit product. A four-bit parallel multiplier is shown in Fig. 4.141; 4-bit × 4-bit parallel binary

Fig. 4.141 Four-bit parallel multiplier circuit using quad full-adders.

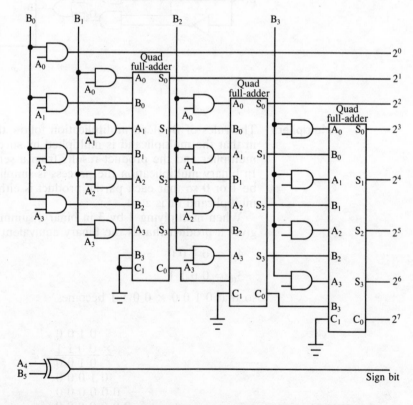

multipliers are available in MSI TTL logic as the 74284/5. The connections between these chips to produce a 4 × 4 multiplier are shown in Fig. 4.142.

Fig. 4.142 Cascading four-bit parallel binary multipliers to produce larger binary outputs.

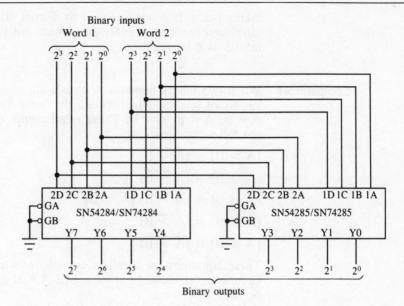

Dividers The rules of Boolean division follow those of the procedure used to perform a 'long division' of decimal numbers. Because the binary digit can only be 1 or 0, however, the division process is simpler although it is longer than that of decimal division. As a general rule when an m-bit dividend is divided by an n-bit divisor, the quotient of q has $m - n + 1$ digits provided that the dividend is an integer multiple of the divisor or q is reduced to an integer.

When dividing 12_{10} by 4_{10} using binary digits, it can be shown that the quotient is the binary equivalent of decimal 3

$$12_{10} = 1\ 1\ 0\ 0$$

$$4_{10} = \quad\ 1\ 0\ 0$$

so that 1100 ÷ 100 becomes

```
              1 1
1 0 0 ) 1 1 0 0 (
        1 0 0
1 0 0 ) 0 1 0 0
        1 0 0
        0 0 0 0
```

Solution is 11 remainder 0 i.e. decimal 3, remainder 0.

A combinational logic circuit can be produced to perform such subtraction in a manner similar to that of the multiplication circuit, since the binary division is a process of successive subtraction and four-bit subtractor circuits could be utilised. Initially the divisor is subtracted from the *n* most significant bits of the dividend and then the divisor is subtracted from the remainder augmented by the next least significant bit of the dividend. The resulting difference is carried as a remainder as long as it is not negative. If the difference is negative then the subtraction is not performed and the next least significant bit is added to the remainder prior to another subtraction taking place. It is not intended to discuss divider circuits further and interested readers are referred to specialist texts should further information be required.

Comparators When two binary numbers A and B are compared there can be a variety of relationships between the numbers, i.e. $A > B$, $A \geqslant B$, $A = B$, $A \leqslant B$, $A < B$. These relationships can be shown to obey the following equations

$$(A > B) = \overline{(A \leqslant B)} \qquad\qquad [4.65(a)]$$

$$(A \geqslant B) = \overline{(A < B)} \qquad\qquad [4.65(b)]$$

$$(A = B) = \overline{(A > B)}.\overline{(A < B)} \qquad\qquad [4.65(c)]$$

$$(A \leqslant B) = \overline{(A > B)} \qquad\qquad [4.65(d)]$$

$$(A < B) = \overline{(A \geqslant B)} \qquad\qquad [4.65(e)]$$

These equations are Boolean expressions which, given the circumstances, must be true; e.g. if $A \leqslant B$ is true then $A > B$ must be false and so on.

If A and B are one-bit numbers then the magnitude relationships can be translated to logic statements as the two-function Karnaugh maps of Fig. 4.143 show. Thus for $A > B$ to be true it is necessary for $A = 1$, $B = 0$ or $A.\overline{B} = 1$. Similarly, for $A < B$ to be true then B must equal 1 and $A = 0$ so that $\overline{A}B = 1$. It follows from equation [4.65(c)] that, for $A = B$ to be true, $\overline{(A > B)}.\overline{(A < B)}$ must be true, and this can be seen from the Karnaugh map of Fig. 4.143(c). This relationship is interesting since it produces the Exclusive-NOR function $\overline{A \oplus B}$.

A one-bit circuit which produces the magnitude conditions $A > B$, $A = B$, $A < B$ as outputs is shown in Fig. 4.144(a); the symbol for the circuit is shown in Fig. 4.144(b).

The single-bit comparator can be expanded to enable serial comparison of n-bit numbers by using the circuit of Fig. 4.145. The two numbers requiring comparison are loaded into shift registers (not shown) and the three D-type flip-flops connected to the output lines are cleared. The D-type flip-flops are used as status flags. The numbers to be compared are clocked out of the shift register a bit at a time and fed in to the one bit comparator starting with the most significant bit. An output on the $A > B$ or $A < B$ will set its flip-flop which in turn inhibits the clock. For $A = B$ the clock is allowed to continue. If at the end of an n-bit comparison the $A > B$ and $A < B$ flip-flops have not been set then the two numbers are equal.

The serial, or sequential, comparator is simple but slow. If fast comparison is required then a parallel comparator may be used. As in the serial circuit the principle of comparison for a parallel comparator is to compare the most significant bit of each number and to proceed to the next most significant bit if the required condition is not initially found. For example, if comparing two four-bit numbers $A_3 A_2 A_1 A_0$ and $B_3 B_2 B_1 B_0$ then initially A_3 is compared with B_3. Suppose it is required to check if $A > B$ and $A_3 = 1$, $B_3 = 0$; it follows then that $A > B$ and the comparison is completed. If, however, $A_3 = 0$, $B_3 = 1$ then $A \not> B$ and again comparison is complete. If $A_3 = B_3 = 1$ or $A_3 = B_3 = 0$ then A_2

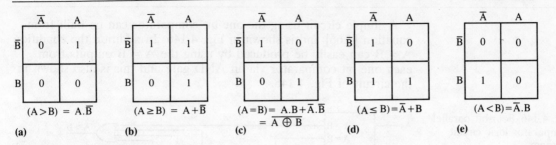

Fig. 4.143 Karnaugh maps for comparator functions: (a) A > B; (b) A ≥ B; (c) A = B; (d) A ≤ B; (e) A < B.

Fig. 4.144 One-bit comparator: (a) logic circuit; (b) block diagram.

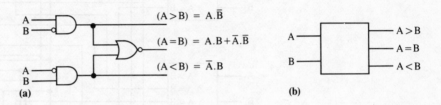

Fig. 4.145 Serial *n*-bit comparator.

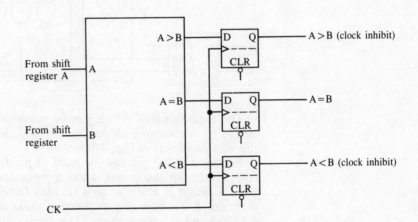

must be compared with B_2 and so on till the comparison is completed. Using the above process as a guide it is possible to write the logic equations for A > B as:

$$A > B = (A_3 > B_3) + (A_3 = B_3).(A_2 > B_2)$$
$$+ (A_3 = B_3).(A_2 = B_2).(A_1 > B_1)$$
$$+ (A_3 = B_3).(A_2 = B_2)(A_1 = B_1).(A_0 > B_0)$$
[4.66]

From equation [4.65] and the maps of Fig. 4.143 the logic functions may be rewritten as

$$A > B = A_3.\overline{B}_3 + \overline{(A_3 \oplus B_3)}.(A_2.\overline{B}_2)$$
$$+ \overline{(A_3 \oplus B_3)}.\overline{(A_2 \oplus B_2)}.(A_1.\overline{B}_1)$$
$$+ \overline{(A_3 \oplus B_3)}.\overline{(A_2 \oplus B_2)}.\overline{(A_1 \oplus B_1)}.(A_0.\overline{B}_0) \quad [4.67]$$

A simple circuit using the one-bit comparator can be built from equation [4.66] and is shown in Fig. 4.146. In addition the condition $A = B$ can easily be produced by using the $A = B$ outputs from each one-bit comparator via an AND gate and this is also shown in the circuit of Fig. 4.146.

Fig. 4.146 Four-bit parallel comparator logic circuit diagram.

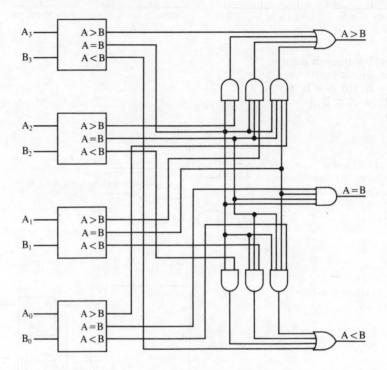

The logic function $A < B$ can be produced from equation [4.66] simply by transposing all As and Bs, and the logic circuit to produce this is also shown in Fig. 4.146.

Equation [4.76] can also be used to produce a four-bit magnitude comparator but the circuit is too complicated to reproduce here. Such a circuit is available in TTL MSI form as the 7485. A block diagram of this chip and its function table are shown in Figs 4.147(a) and (b) respectively. This circuit allows comparison of straight binary and natural (8421) BCD codes and is fully expandable to any number of bits without external gates being required. Words of greater length may be compared by connecting comparators in cascade with the $A > B$, $A = B$, $A < B$ outputs of a less significant bit stage being connected directly to the $A > B$, $A = B$ and $A < B$ inputs of the stage that processes more significant bits.

Binary rate multipliers

Consider the circuit of Fig. 4.148(a). There are three T-type flip-flops with three monostable circuits triggered on the rising edge of the input waveform. The T-type flip-flops will toggle on the negative edge of the input clock waveforms as shown in Fig. 4.148(b) while the outputs from the monostables are triggered from the leading edge of the input waveforms.

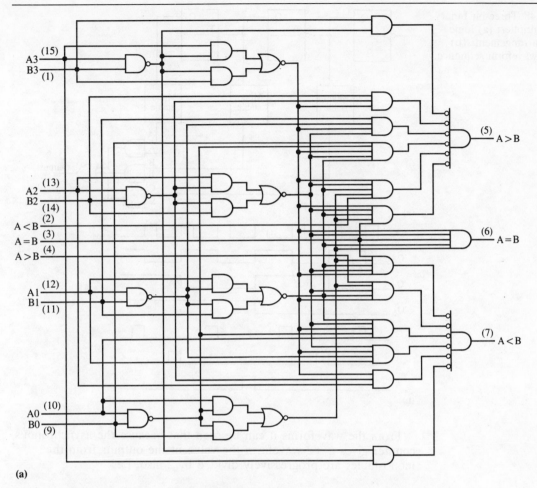

(a)

Function tables

Comparing inputs				Cascading inputs			Outputs		
A3.B3	A2.B2	A1.B1	A0.B0	A > B	A < B	A = B	A > B	A = B	A > B
A3 > B3	X	X	X	X	X	X	H	L	L
A3 < B3	X	X	X	X	X	X	L	H	L
A3 = B3	A2 > B2	X	X	X	X	X	H	L	L
A3 = B3	A2 < B2	X	X	X	X	X	L	H	L
A3 = B2	A2 = B2	A1 > B1	X	X	X	X	H	L	L
A3 = B3	A2 = B2	A1 < B2	X	X	X	X	L	H	L
A3 = B3	A2 = B2	A1 = B1	A0 > B0	X	X	X	H	L	L
A3 = B3	A2 = B2	A1 = B1	A0 < B0	X	X	X	L	H	L
A3 = B3	A2 = B2	A1 = B1	A0 = B0	H	L	L	H	L	L
A3 = B3	A2 = B2	A1 = B1	A0 = B0	L	H	L	L	H	L
A3 = B3	A2 = B2	A1 = B1	A0 = B0	L	L	H	L	L	H

'85, 'LS85, 'S85

A3 = B3	A2 = B2	A1 = B1	A0 = B0	X	X	H	L	L	H
A3 = B3	A2 = B2	A1 = B1	A0 = B0	H	H	L	L	L	L
A3 = B3	A2 = B2	A1 = B1	A0 = B0	L	L	L	H	H	L

'L85

A3 = B3	A2 = B2	A1 = B1	A0 = B0	L	H	H	L	H	H
A3 = B3	A2 = B2	A1 = B1	A0 = B0	H	L	H	H	L	H
A3 = B3	A2 = B2	A1 = B1	A0 = B0	H	H	H	H	H	H
A3 = B3	A2 = B2	A1 = B1	A0 = B0	H	H	L	H	H	L
A3 = B3	A2 = B2	A1 = B1	A0 = B0	L	L	L	L	L	L

H = high level, L = low level, X = irrelevant

(b)

Fig. 4.147 SN7485 four-bit comparator: (a) functional logic diagram; (b) function (truth) table.

Fig. 4.148 Three-bit binary rate multiplier: (a) logic circuit arrangement; (b) timing waveform sequence.

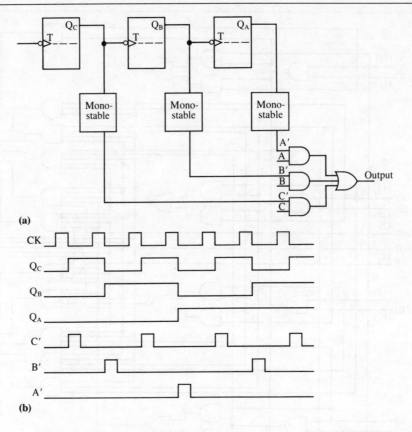

(a)

(b)

From the waveforms it can be seen that because the asynchronous counter gives a ÷2 waveform the value of the outputs from the monostables are progressively divided by 2 also, i.e.

$$C' = \frac{C}{2} \text{ (where C is the input clock rate)} \qquad [4.68(a)]$$

$$B' = \frac{C}{2^2} \qquad [4.68(b)]$$

$$A' = \frac{C}{2^3} \qquad [4.68(c)]$$

The inputs A, B and C are bits from a control word and when their value is 1 the associated gate is activated so that pulses produced by the clock are summed by the output OR gate. The circuit shown is slow and has all the problems associated with asynchronous counters. The 7497 is a synchronous six-bit binary rate multiplier using a six-stage parallel counter and associated logic circuitry in MSI TTL. The circuit diagram of this device is shown in Fig. 4.149(a) and a simplified block diagram in Fig. 4.149(b).

The binary rate input ('M') is applied to the six inputs A to F while the input frequency is applied to the clock. The device is enabled with clear, enable and strobe inputs low and under these conditions the output frequency at the Z output is given by

$$f_{OUT} = \frac{M.f_{in}}{64} \qquad [4.69]$$

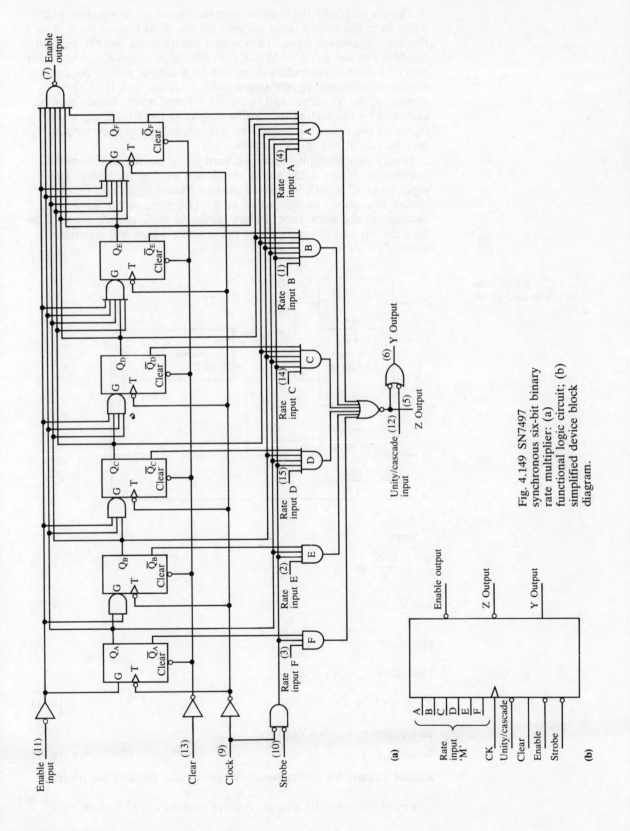

Fig. 4.149 SN7497 synchronous six-bit binary rate multiplier: (a) functional logic circuit; (b) simplified device block diagram.

The six stages of the counter give the factor 64 in equation [4.69] while M is the binary input number on the A to F lines, with A as the least significant input. Thus with a binary word 001000, say, on the M input the value of M is 8 (i.e. 001000 is decimal 8) and so for every 64 clock input pulses there will be 8 output pulses. As can be seen from the logic circuit diagram the Y output is a NAND-ed version of the Z output and the unity/cascade input. Stages may be cascaded by connecting the enable output to the strobe and enable inputs of the following stage. The sub-multiple frequency outputs may be taken fron the Y outputs.

Binary rate multipliers may be used for a variety of arithmetic operations. Figure 4.150 shows the device used as an adder. The input clock C is split into three distinct phases C_1 C_2 and C_3 to ensure that pulses cannot arrive at the OR gate, or the up-down counter, at the same time. Pulses produced at Z_3 and Z_4 are used to feed the up and down inputs of the counter which will stabilise when $Z_3 = Z_4$.

Fig. 4.150 Binary rate multipliers used as an adder.

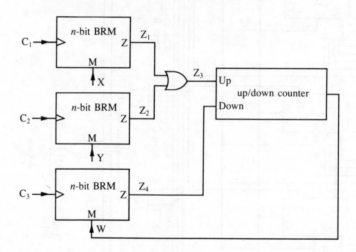

Since

$$Z_1 = \frac{C.X}{2^n}$$

and $Z_2 = \dfrac{C.Y}{2^n}$ [4.70]

and $Z_4 = \dfrac{C.W}{2^n}$

then since $Z_3 = Z_1 + Z_2$

$$Z_3 = \frac{C}{2^n}(X + Y)$$ [4.71]

and equating Z_3 with Z_4 gives

$$W = X + Y$$ [4.72]

Similar circuits for subtractors, multipliers and dividers are possible.

You should now be able to attempt exercises 4.70 to 4.98.

Memories

Semiconductor memories fall into two basic categories: **random access memory (RAM)**, perhaps better described as read/write memory, and **read only memory (ROM)**. For RAM there is a matrix array of flip-flop elements, each one forming a **memory cell**. The title random access was chosen since the time taken to gain access to any individual cell is the same regardless of its position in the array. For this type of memory it is possible to determine the contents of any cell by a read operation or to change the contents to a new value by means of a write operation. The read operation is non-destructive because reading will not alter the contents of the cell, which remains the same before and after the read operation is carried out. The type of memory is, however, volatile since removing the power supply to the cells will cause the contents to be destroyed; reconnecting the power supply causes each cell to take up a particular state, i.e. logic 0 or 1, but that state is unlikely for all cells to be the same as before the power was removed. ROM is also a matrix array of cells but the array is non-volatile so that the contents of cells, written by a manufacturer or by a customer will be the same and can be read to give the value of the contents. In the normal way the ROM cannot be written into since its function is to always provide a predetermined fixed value for its contents. However, some ROMs can have their contents altered although it is a lengthy and complex process. By a process known as **field programming** users can buy a ROM containing all 1s, or all 0s, in each cell and by an electrical process cause certain cells to change value to obtain the required contents. Such a memory is a **programmable read only memory** or **PROM**; a PROM cannot have its memory changed after it has been programmed but there is a special kind of PROM where this can be done. The **erasable PROM**, or **EPROM**, can have its contents wiped out by exposing the memory to ultraviolet light to penetrate to the actual chip below. Quartz is used because it allows the passage of UV light, unlike glass and other materials which can allow the visible light to pass but are opaque to UV. Another type of EPROM is the **electrically alterable PROM**, or **EAPROM**, which has the contents changed by electrical means rather than using UV light.

The means of locating a particular memory cell is by input lines known as **address lines** or, collectively, as the **address bus**. A sequence of logic levels on the address lines uniquely identifies, and addresses, one particular memory cell, i.e. if 16 address lines are available and all lines are at logic 0 a particular cell, identified by the value 0000 0000 0000 0000 on the address lines, is accessed. Similarly, the data contained in the accessed memory location may be read (or written into in the case of RAM) by means of **data lines** or, collectively, the **data bus**.

The methods of implementing the read/write operations will be discussed in the following sections.

Random access memory cells

The basic cell is the flip-flop and a typical arrangement utilising bipolar transistors is shown in Fig. 4.151. Suppose when power is switched on that transistor TR_1 is on and TR_2 is off. This state

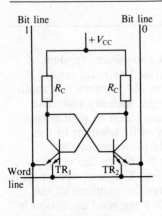

Fig. 4.151 Basic static memory cell using bipolar junction transistors.

would be maintained as long as the power supplies remain on and the cell is not written into; thus the cell retains information or has memory. Normally the word line is held at or near 0 V while the bit lines are at about 1.5 V. Thus the cell current flows to earth via the word lines and not the bit lines. The cell is not selected under this condition. In order to read the contents of a cell the word line is taken high to, say, 3 V so that the current from the conducting transistor is now switched from the word line to either bit line 0 or bit line 1. With a current sensing amplifier in each bit line it is an easy matter to determine the state of the sensed cell.

To perform a write operation the cell is selected by raising the word line high and raising one of the bit lines high, say to 3 V, while forcing the other bit line low, say to 0.7 V. This forces one transistor into cut-off and the other into saturation. Suppose initially TR_1 was on and TR_2 off and that the write operation is intended to change the state of the cell. It is thus necessary to raise bit line 1 and lower bit line 0 while simultaneously raising the word line.

A basic unipolar flip-flop is shown in Fig. 4.152(a). TR_1 and TR_2 are the memory transistors while the TR_3 and TR_4 are active loads. The bit lines are connected to the cell via transistors TR_5 and TR_6; these transistors are open-circuit unless, for NMOS devices, the word line is taken to a positive voltage. Once turned on, the state of the cell can be sensed via TR_5 and TR_6 on the bit lines. To write into the cell with TR_5 and TR_6 on it is necessary to force the bit lines to a 1 or 0 state as required. This type of cell is easily converted to the X-Y select type by simply having two transistors between each output of the memory device and the bit lines; one transistor in each branch being the X-select and the other the Y-select as Fig. 4.152(b) shows.

The cells described are known as **static cells**. Alternative cells which use the charge stored on a capacitor as an indication of the state of the cell are known as **dynamic cells**. Because stored charge can leak away in time these cells require constant **refreshing** to maintain the state of charge. Dynamic cells are available in both bipolar and unipolar techniques but unipolar is preferred due to their reduced cell size and much improved dissipation. For that reason bipolar dynamic memories will not be discussed further. Consider the simple three-transistor cell of Fig. 4.153. The capacitor is in fact the gate–source capactior of the memory transistor TR_1. To write into the cell transistor TR_2 is turned on so that the input data can charge or discharge C. To read from the cell the data-out line is precharged and TR_3 is turned on. If a logic 1 was stored on C then TR_1 will be on and the data-out line will be discharged. Sensing amplifiers can detect this discharge to indicate the state of the cell. The cell is refreshed by first reading out to determine the state and then reading in via the data-in line after amplification of the output. It has been determined that the refresh interval must be not less than 2 ms or the data contents may be lost.

A wide range of static RAM chips are available in both bipolar and unipolar technology. They range from small 64-bit memories of about 35 ns cycle time (the time taken to perform a complete read or write operation on the memory) up to 8 K capacity with cycle times of about 450 ns.

For dynamic unipolar RAM chips the power consumed is much less in its quiescent mode and only rises to large proportions when

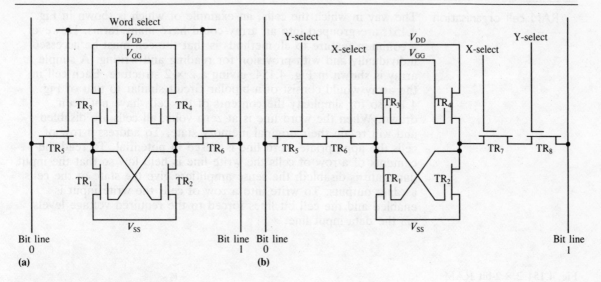

(a)

(b)

Fig. 4.152 Basic static memory cell using unipolar transistors: (a) with single word select line; (b) with X-select and Y-select lines.

Fig. 4.153 Three-transistor dynamic memory cell using unipolar technology.

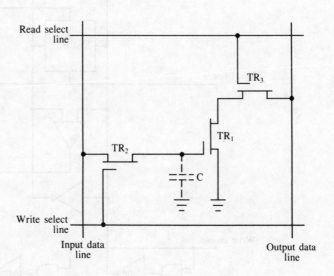

the memory is being accessed. This is a considerable advantage compared to bipolar RAMs which, because one transistor of the flip-flop will always be conducting, will take a large power drain from the supply *all* the time. Because of the combined effect of larger packing densities and lower quiescent power requirements the memory capacity in dynamic RAMs ranges from 4 K up to 64 K. Cycle times vary from 350 ns to 500 ns but typically quiescent powers are 20 mW, rising to 450 mW when operating.

RAM cell organisation

The way in which the cells, an example of which is shown in Fig. 4.151, are grouped into an array could have many forms. However, a common feature to all methods is that the cell must be accessed individually and with provision for reading and writing. A simple array is shown in Fig. 4.154, giving a 2 × 2 structure. Each cell in the array would consist of a bipolar circuit similar to that of Fig. 4.151 so for simplicity the contents of the cells have not been drawn. When the word line is at zero volts, all cells are disabled and will retain their original memory state. To address a row of cells the appropriate word line is raised in potential. To read the contents of a row of cells the write line is held low so that the input data gate is disabled; the sense amplifiers give the state of the cells as data outputs. To write into a row of cells the write input is enabled and the cell bit lines forced to the required voltage levels via the data input line.

Fig. 4.154 2 × 2-bit RAM cell organisation using static bipolar junction transistors.

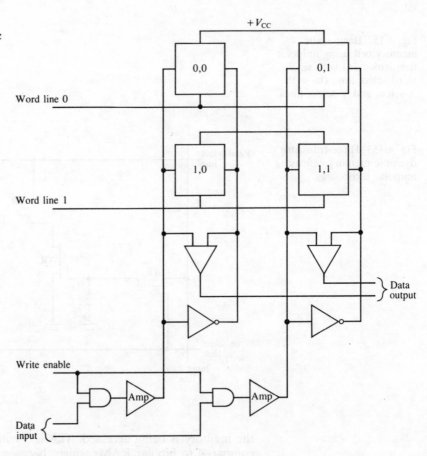

The arrangement of Fig. 4.154 means that the contents of a row are accessed at any one time and not just a single cell. To access a single cell would require a modified cell construction with two word lines, labelled X and Y, so that the flip-flop of Fig. 4.151 would have an extra emitter for both transistors as shown in Fig. 4.155.

Fig. 4.155 Static memory cell using bipolar junction transistors giving X- and Y-select lines.

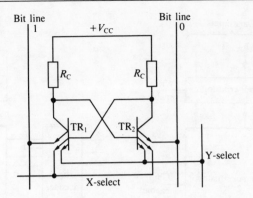

In a multi-cell array all horizontal cells in the matrix would be connected to one of the X-select lines (X_0, X_1, X_2 etc.) with all cells connected vertically by Y-select lines (Y_0, Y_1, Y_2 etc.). Thus raising any one X-select line (X_1 say) and any one Y-select line (Y_2 say) would uniquely select the one cell that is at the intersection of the X_1 and Y_2 lines on the matrix. To obtain a two-bit 'word' using this technique would require two such planes like the one described above with all connections identical for each plane but each cell having individual input and output data lines. More infomation regarding this technique will be given later in this section. For large arrays the number of word lines, or X- and Y-select lines, would be very large and for each line to be given an individual pin on the encapsulated memory chip would be impossible. Instead a decoder is used to reduce the number of inputs required; a block diagram of a complete RAM using word line and X- and Y-select inputs is shown in Figs 4.156(a) and (b).

It is usual practice to define a RAM chip in terms of the number of bits, regardless of the configurations of the bits into 'words'; thus, a 64 bit, 256 bit, 1024 bit (or 1 K), 4096 bit (4 K) and 8192 bits (8 K). The trend is to increase the bit capacity for single chips, thus reducing cost. TTL RAM circuits include the SN74189A 64-bit RAM and the SN74201 256-bit RAM. The 74189A is configured as 16 × 4 bits, using a circuit arrangment similar to that of Fig. 4.156(a), while the 74291 is configured as 256 × 1 bit using a circuit arrangement similar to that of Fig. 4.156(b). A simplified pin-out diagram of the SN74189A is shown in Fig. 4.157.

The address select inputs can be the address lines of a computer system which in this case gives four address line inputs and, via a decoder, 16 outputs only one of which can be activated for a particular combination of 0s and 1s on the four address lines. Each of the 16 rows of four cells in the 74189A is thus accessed every time one of the 16 output lines from the decoder goes high. To perform a read operation the memory enable input must be held low and the write enable high when selecting the required address. The data output is the *complement* of the contents of the four cells accessed. To perform a write operation the information at the data input lines will be stored in the four selected cells, specified by a particular address, if both memory (or chip) enable and write enable inputs are held low.

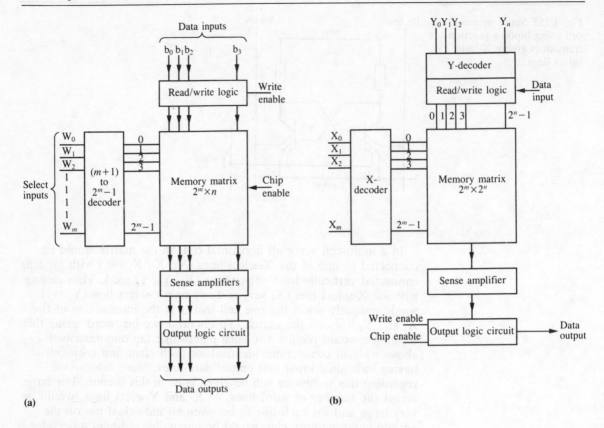

(a)   **(b)**

Fig. 4.156 Block diagram circuit arrangement of memory chips with decoding circuits: (a) using word line select; (b) using X- and Y-select.

Fig. 4.157 Block diagram arrangement of the SN74189A 64-bit RAM. This is arranged as 16×4-bits.

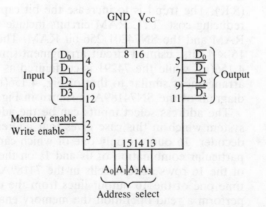

A diagram of the SN74201 is shown in Fig. 4.158. As can be seen, the array is organised as a 16×16 matrix requiring a 1-of-16 row decoder using four input address lines and a 1-of-16 column decoder using four input address lines (since $2^4 = 16$ and $16 \times 16 = 256$).

Fig. 4.158 SN74201
256 × 1-bit RAM chip
block diagram.

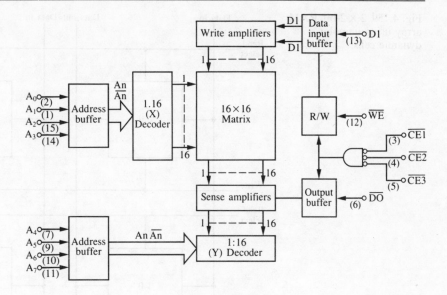

To perform a read operation the write enable input must be high and all three memory enable inputs low. When any of the memory enable inputs go high the output will be in a high impedance state. The data output will the complement of the original data input. To perform a write operation all memory enable inputs and the write enable must go low. With the write enable input low the data output is in a high impedance state; this is necessary when a number of outputs may be connected to a common data bus so that the high impedance output state will neither drive nor load the bus line.

The simplified circuit of Fig. 4.159 shows a 2 × 2 RAM array using unipolar dynamic cells. Each of the memory cells is identical to that of Fig. 4.153 and have not been redrawn here to keep the diagram simple. The circuit involving precharging and write enable has been drawn for one of the columns, the circuit for the other column being identical.

The data lines are precharged prior to a read operation. For a read operation the state of the data out line will therefore retain its charge, if the memory transistor is off, or be discharged if the memory transistor is on. This concept was discussed earlier when the dynamic cell was introduced. For a write operation the memory capacitor is taken to the state of the data-in line. Note that if the write operation places a logic 1 in the cell, i.e. capacitor charged, then during a read operation the data output line is discharged so that there is inversion between the data in and data out lines. For refresh, a read cycle is followed by a write cycle during which the write enable input will go to logic 1 turning on TR_1. The effect of TR_2 is to ensure inversion between the data in and data out lines for, if the data out line is at logic 1, TR_2 conducts and connects the data-in line to ground. Thus the cell capacitor is refreshed with its correct value. Because of the arrangements for refresh, whole rows of cells are refreshed simultaneously so that the refresh operation need not be a lengthy one.

Fig. 4.159 2 × 2-bit RAM array using unipolar dynamic cells.

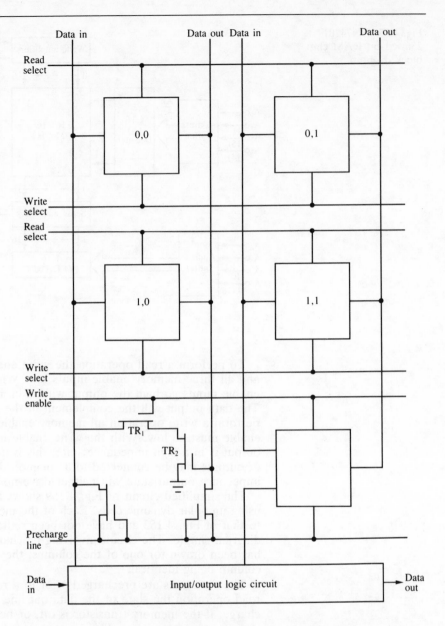

Considering again the 74189A 64-bit RAM, which can be arranged to give a 16 × 4 bit configuration, it is possible to use two such chips to give a 16 × 8 bit configuration as shown in Fig. 4.160. All inputs, such as address lines, memory enable and write enable, are common to both chips while the two groups of four data lines can be combined to give an eight-bit data arrangement which could be connected to the data bus of a computer system.

To increase the memory capacity of the 74189A to give, say, 128 bits arranged as 32 × 4 bits the circuit of Fig, 4.161 could be used. In this case the same address lines (A$_0$ to A$_3$ inclusive) are used for both chips and also the write enable lines are commoned. The difference in this case is that the memory enable line is used to select *which* of the two chips is actually accessed. By feeding one

Fig. 4.160 Expanding the four-bit output of the SN74189A to give an eight-bit word.

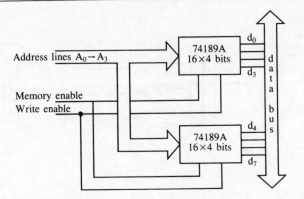

Fig. 4.161 Expanding the memory size using the SN74189A to give 32 × 4-bits.

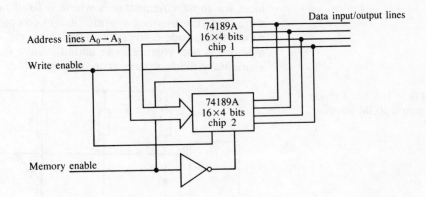

chip directly via the memory enable input and the other chip by the same input, via an inverter, then only one chip can ever be accessed at any one time. That is, if a logic 0 is on the memory enable line then chip (1) is accessed while if a logic 1 is on the memory enable line then chip (2) is accessed, since the inverter will produce a logic 0 at the memory enable input of the chip.

The memory enable input in this arrangement could be connected to one of the address lines, say A_4. The arrangement effectively gives a 32 × 4 bit array and five address lines would give access to any of 32 locations (since $2^5 = 32$). To increase the memory capacity further, to 64 × 4 bits say, requires some form of decoder so that only one of the individual chip memory enable inputs, 4 in this example, can go low at any given time. A 2-to-4 decoder could be used, fed by two address lines, A_4 and A_5 say, to provide this type of output. A possible decoder logic circuit and its truth table are shown in Figs 4.162(a) and (b) respectively.

Increasing the memory size has the effect of reducing access speed since the interconnections produce extra capacitance. The effect of extra logic gates in the decoding circuit also produces delays so that large memories tend to be slower than smaller ones. However, in terms of external decoder complexity it might be preferable, if large memory size is required, to go for a single chip since the decoder circuits required to access the larger array of such a chip are already in the chip.

Fig. 4.162 2-to-4 decoder
for memory enable: (a)
logic circuit; (b) truth table.

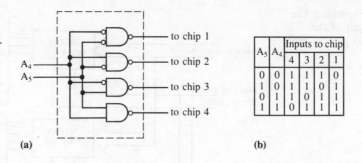

(a)

A_5	A_4	Inputs to chip			
		4	3	2	1
0	0	1	1	1	0
1	0	1	1	0	1
0	1	1	0	1	1
1	1	0	1	1	1

(b)

Read only memory circuits

This is also, like RAM, a matrix array of memory cells but unlike RAM these cells are capable of being read only. The arrangement is ideal for those circumstances where a fixed memory is required to store, say, the monitor routine of a microcomputer, frequently used subroutines, code-conversion translators, etc. The fact that the memory is non-volatile is an added bonus. Consider first the elementary diode matrix array of Fig. 4.163.

Fig. 4.163 4 × 4 diode
matrix ROM array.

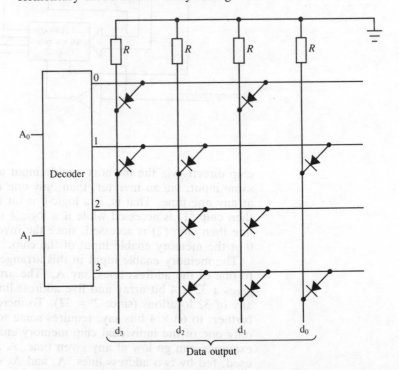

The word line in this case is determined by the decoder and only one line can go high at any instant. A particular word associated with each word line is produced by omitting certain diodes from the matrix. Where a word line goes high then the bit lines with diodes connected to the word line will also go high; where no diode is connected the data output line will always be low. Thus if word line 0 is raised to logic 1 the data word output will be 1010 while if word line 1 is raised high the data output word becomes 1101 etc. In practice, not diodes but transistor cells in both bipolar and unipolar technologies are used. High-volume application ROMs are made using photographic masks and etching/diffusion processes in much the same way as ordinary integrated circuits. The siting of a 'diode' therefore depends on whether the mask allows the transistor to be processed at the particular intersection of the word and bit lines. An arrangement using bipolar transistors is shown in Fig. 4.164.

Fig. 4.164 ROM array using bipolar junction transistors.

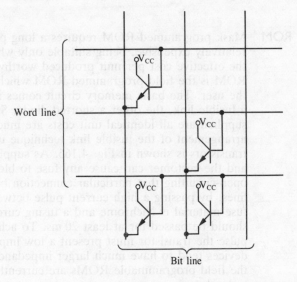

Fig. 4.165 SN74187 1024-bit ROM arranged as 256 × 4 bits.

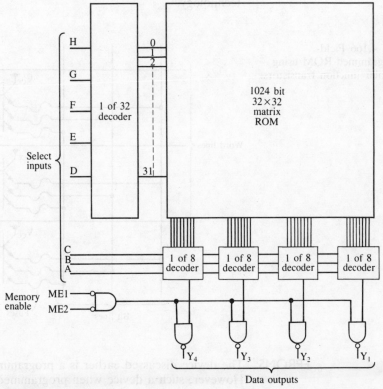

An example of a TTL/MSI ROM is the SN74187 1024-bit ROM which is arranged as a 256 × 4 bit element. The array is actually 32 × 32 bits but since the output gives four bits simultaneously the column decoders are grouped into four groups of eight. Thus only three address lines (A, B, and C) are needed to uniquely define each group of eight columns while five address lines (D to H) are needed to uniquely define each possible row. Memory enable inputs must go low to access the memory. The circuit diagram is shown in Fig. 4.165.

Field programmed ROM   Mask programmed ROM requires a long programming time and is relatively expensive, being suitable only where high volume makes the effective cost per unit produced worthwhile. An alternative ROM is the field programmed ROM which can be programmed by the user. The basic memory circuit comes in two versions, one using a fusible link, the other a shorted diode. Since the ROMs as supplied are all identical unit costs are much reduced. A possible arrangement of the fusible link technique using multi-emitter transistors is shown in Fig. 4.166. As supplied, all fuses are intact and the customer can cause any fuse to blow, thus effectively open-circuiting that particular connection between word and bit lines, by passing a high current pulse between those lines. Typical fuse material is Nichrome and a fusing current of about 20 mA should be passed for at least 20 ms. To achieve such a large current pulse the transistor must present a low impedance; since MOS devices tend to have much larger impedances than bipolar devices the field programmable ROMs are currently made using bipolar technology.

Fig. 4.166 Field-programmed ROM using bipolar junction transistors.

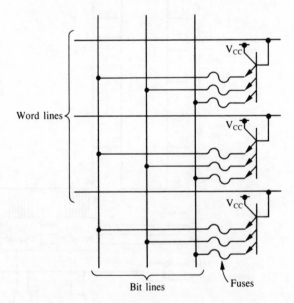

EPROMS   The device discussed earlier is a programmable ROM or PROM. However, such a device when programmed by the user cannot be reprogrammed unless the unlikely event occurs that it is only necessary to blow further fuses. Once a fuse has blown in the PROM it cannot be replaced. The kind of ROM that can be programmed by the user, and then where necessary reprogrammed,

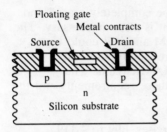

Fig. 4.167 Floating gate EPROM constructional details.

is known as an erasable PROM or EPROM. An EPROM may use p-channel enhancement MOS devices as the coupling element between the word and bit lines; the gates are surrounded by oxide as shown in Fig. 4.167 and in the normal state all FETs are non-conducting. By passing a large current between drain and source of the FET an avalanche breakdown is caused in the oxide and a negative charge accumulates in the gate because of trapped electrons. Removing the current leaves the electrons trapped because of the oxide surrounding the gate. The effect of the electrons however is to produce an induced channel between source and drain which allows the device to turn on. Individual cells can then be turned on by pulsing them while others may be left in the off state.

Since this cell has no gate terminal it is usual in this type of EPROM to place a second MOS device in series at each intersection of bit lines and word lines as drawn in Fig. 4.168. If the word line goes high the 'normal' FET conducts but unless the second cell has been pre-programmed to be on there will be no connection between the two lines.

Fig. 4.168 ROM memory array using the floating gate device of Fig. 4.167.

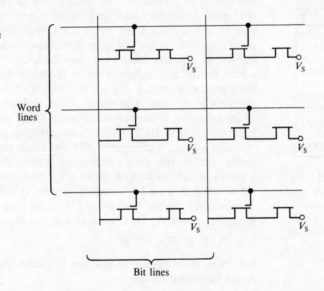

Word lines

V_S

V_S

V_S

V_S

V_S

V_S

Bit lines

To erase the contents of the ROM it is exposed to high-intensity ultraviolet light, via a quartz window in the chip encapsulation surface, which causes a breakdown mechanism to be set up in the oxide, allowing the trapped charge to be released. During normal read operations the voltage across the reprogrammable cell is much less than the value needed to program it so there is little danger of these voltages causing a build-up of a charge within the floating gate.

An alternative form a EPROM is the electrically alterable programmable ROM or EAPROM. This device can be programmed by applying large positive or negative voltages between an isolated gate and the substrate of suitably fabricated FETs which can turn on or off the device as required. The advantage of a EAPROM compared to the EPROM is that it can be selectively changed in given locations while the EPROM has to be completely erased and reprogrammed. A source of ultraviolet light is also unnecessary. The EAPROM is effectively a non-volatile RAM and if it could be produced with the required performance, reliability and cost parameters it could replace conventional RAM circuits. The technology at present is not able to meet these requirements and the EAPROM is only used in special cases.

You should now be able to attempt exercises 4.99 to 4.119.

Miscellaneous circuits

Parity generators/ checkers

Decimal value	Odd parity	Even parity
	8 4 2 1 P	8 4 2 1 P
0	00001	00000
1	00010	00011
2	00100	00101
3	00111	00110
4	01000	01001
5	01011	01010
6	01101	01100
7	01110	01111
8	10000	10001
9	10011	10010

Fig. 4.169 Odd and even parity bits used with a four-bit BCD word. The sum of the logic 1s must be odd, or even, to determine the value of the parity bit.

When a coded group of bits is transmitted from one point to another the possibility exists that a change in value of one bit in any symbol for any reason will generate another valid symbol that cannot be recognised as being in error. A BCD (8421) code has limited redundancy and if errors resulted in a number greater than the binary equivalent of decimal 10 to 15 then an error could be detected. However, adding a single bit to any code will make any two code symbols differ by at least two bits. The resulting error detecting code is known as a distance-2 code. The extra bit is known as a **parity bit** and it may be used to give **odd parity**, in which the total number of 1s in the code symbol is an odd sum, or **even parity**, in which the total number of 1s in the code symbol gives an even sum. The parity bit is *redundant* as far as transmission of information is concerned. Figure 4.169 shows the effect of adding a parity bit to a BCD (8421) code to give odd and even parity.

For a system which transmits the code group symbols in parallel mode, i.e. all bits in a symbol are transmitted simultaneously, then a series of Exclusive-OR gates may be used to generate the parity bit which can then be terminated with the data. From the work done earlier on the Exclusive-OR gate then for a four-input variable function where A, B, C and D are the inputs it is possible to write

$$F = A \oplus B \oplus C \oplus D \qquad [4.73]$$

but from the associative law for Boolean algebra equation [4.73] could be rewritten as

$$F = (A \oplus B) \oplus (C \oplus D) \qquad [4.74]$$

or as

$$F = [(A \oplus B) \oplus C] \oplus D \qquad [4.75]$$

The circuits to satisfy equations [4.74] and [4.75] can be built from two-input gates as shown in Figs 4.170(a) and (b) respectively.

The circuit of Fig. 4.170(a) is known as the tree-type circuit and is preferred to the circuit of Fig. 4.170(b), known as the unbalanced circuit, because there are fewer levels of gating with consequent reduction in propagation delay.

Examination of the output of either of these circuits will show that F is true if any one or any three of the input variables are true. Thus the output F will generate an even parity bit which can be transmitted to line together with the data word. If an odd parity is required then this parity bit can be generated from $\overline{F}$, i.e. the

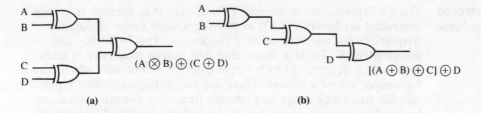

(a) $(A \otimes B) \oplus (C \oplus D)$

(b) $[(A \oplus B) \oplus C] \oplus D$

Fig. 4.170 Logic circuit for parity bit generation: (a) tree circuit; (b) unbalanced circuit.

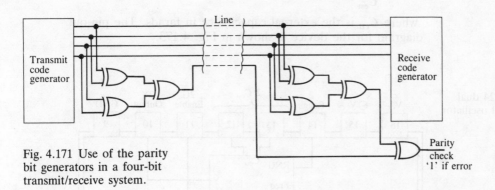

Fig. 4.171 Use of the parity bit generators in a four-bit transmit/receive system.

output of either circuit in Fig. 4.170 could be sent to line via an inverter. At the receiving end a similar Exclusive-OR circuit could generate a parity bit from the received data word and this receiver generated parity bit could then be checked against the *received* parity bit. If both bits are equal then the received code is assumed correct. The arrangement is shown in Fig. 4.171.

The SN74180 is a TTL nine-bit odd/even parity generator/checker. The circuit is shown in Fig. 4.172.

Fig. 4.172 SN74180 nine-bit odd/even parity generator/ checker functional logic circuit diagram.

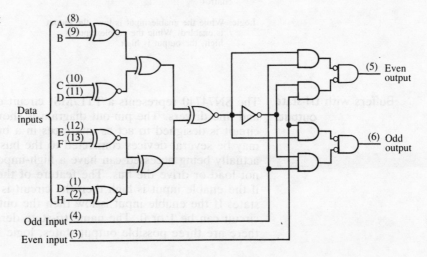

Voltage controlled oscillator

The SN74124 is one of several TTL devices that operate as **voltage controlled oscillators** (VCO) to provide a wide range of output frequencies for use with other TTL devices. The 74124 has two independent VCOs in a single chip, the output frequency of each VCO being determined by a single external component which may be a capacitor or a crystal. There are two voltage sensitive inputs, one for frequency range and one for frequency control, which can be used to vary the frequency of the output. For the S version of the device (SN74S124) the range of frequencies can extend from 0.12 Hz to 85 MHz. Used with an external capacitor the frequency of the output can be determined from the equation

$$f_o = \frac{5 \times 10^{-4}}{C_{ext}} \text{ (for the SN74S124)}$$

where C_{ext} is the external capacitance in farads. The pin-out diagram for the device is shown in Fig. 4.173.

Fig. 4.173 SN74124 dual voltage-controlled oscillator package.

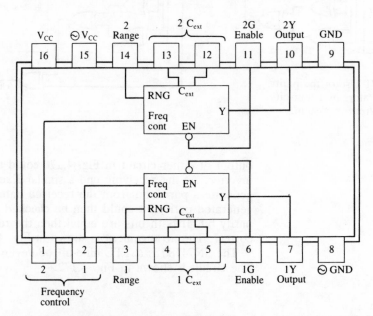

Logic: While the enable input is low, the output is enabled. While the enable input is high, the output is high.

Buffers with tri-state outputs

The SN74740 represents a TTL/MSI circuit offering eight buffers and line drivers. The pin-out diagram is shown in Fig. 4.174(a). This circuit is designed to act as interfaces in a bus system where there may be several devices connected to the bus and all those not actually being accessed can have a high-impedance state which does not load or drive the bus. The feature of the tri-state circuit is that if the enable input is high then the circuit is in the high-impedance state. If the enable input is low then the output of the tri-state circuit can be 1 or 0. The name tri-state derives from the fact that there are three possible output states, logic 1, logic 0 and high

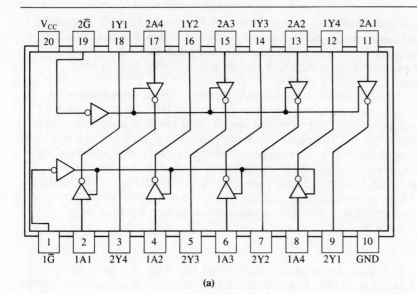

Enable input 1$\overline{G}$	1Y outputs
H	Z
H	Z
L	enabled (inverting)
L	enabled (inverting)

(b)

(a)

Fig. 4.174 SN74470 octal buffer/line driver: (a) package; (b) truth (function) table.

impedance. The outputs in the 74740 are inverted versions of the input as Fig. 4.174(b) shows for the 1Y outputs.

For TTL circuitry the tri-state condition allows that in the totem-pole output stage mentioned earlier a condition can exist where both the output transistor and its active load are OFF, leaving the output isolated from both the supply rails. In normal TTL of course a condition can exist where one or other of these output transistors is OFF but not both.

You should now be able to attempt exercises 4.120 to 4.125.

Timing circuits

In many applications the **timing of events** in digital systems is related to a common clock. In other situations, however, the independent time delay provided, say, by a monostable is required. This may be at an interface where pulses of varying amplitude and duration are converted to pulses of equal amplitude and duration. Other applications involving time delay occur in measurement and display systems. In each case the requirement is a controlled duration pulse initiated by a **trigger input** pulse as shown in Fig. 4.175.

Fig. 4.175 Use of a monostable to produce a time delay: (a) block diagram; (b) timing waveform.

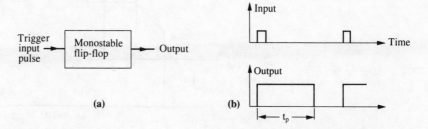

Within limits the trigger input can be varying in both amplitude and duration. In each case the output pulse has the controlled duration t_p and the amplitude can be chosen to fit the system requirement. In a different system t_p may range from a few microseconds to hundreds of seconds.

The function of a monostable flip-flop can be implemented in various ways including discrete component BJT circuits, logic gate circuits, operational amplifier circuits and purpose-built integrated circuit timers. The choice may depend upon a variety of factors such as availibility of power supplies and the required output levels. In this section the principle of operation is examined using the BJT circuit as a model. These principles are then applied to the analysis of logic gate variations and the operation of purpose-built timers. Consideration is given to operational amplifier flip-flops in Chapter 5.

Principle of operation

In each case the circuit, as with other flip-flops, has two states, one of which is stable. This means that in the absence of external signals the circuit is naturally held in the stable state. The timing is achieved by the change of charge on a capacitor through a resistor. This is initiated after the trigger input pulse has switched the circuit from the stable state to the quasi-stable state. The ensuing voltage change across the capacitor eventually trips the circuit back into the stable state.

The BJT monostable flip-flop

A circuit for a BJT one-shot with typical component values is shown in Fig. 4.176. For the analysis the stable state is first investigated and then the effect of the trigger pulse is considered. This descriptive analysis allows the essential waveforms to be drawn and, finally, the waveform timing can be analysed. **The stable state** can

Fig. 4.176 BJT monostable flip-flop.

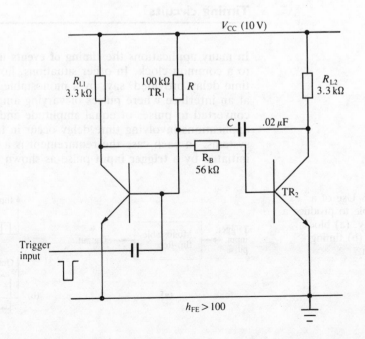

$h_{FE} > 100$

be identified by ignoring any capacitors and then considering the state of each transistor. In this case, TR_1 has a positive supply through R to V_{cc} and since the ratio R/R_L is less than h_{FE}, TR_1 will be on and saturated with V_{CE1} and V_{BE1} at the saturation levels of 0.1 V and 0.7 V respectively. This low collector voltage ensures that V_{BE2} is below the threshold of conduction (0.5 V) and TR_2 is therefore off, with V_{CE2} at the supply voltage value of 10 V. The capacitor C is then charged to $(10 - 0.7)$ V.

Switching occurs when the negative trigger pulse is applied, temporarily reducing V_{BE1}. The resulting reduction in I_{C1} raises V_{CE1} and hence V_{BE2}. If V_{BE2} passes the 0.5 V threshold, TR_2 starts to conduct and V_{CE2} falls causing a further reduction in V_{BE1}. This is a cumulative action (providing the loop gain is greater than unity) which is completed with TR_1 off and TR_2 on. The whole change in collector voltage (10 V to 0.1 V) is passed through C, since C cannot change its state of charge instantaneously, and V_{BE1} falls from 0.7 V to $0.7 - 9.9$, or it falls to -9.2 V.

The quasi-stable state is maintained so long as TR_1 is held off with a negative value of V_{BE1}. Initially one end of C is held at 0.1 V by V_{CE2} and the other is at the value of TR_1 base voltage which is -9.2 V. However, the base side of C can charge through R towards the supply voltage of $+10$ V. During the charging period V_{BE1} rises exponentially from -9.2 V towards $+10$ V and at a particular time will pass the 0.5 V threshold value. At this point TR_1 begins to conduct and initiates cumulative switching in the reverse direction. This returns the circuit to the stable state with the capacitance C virtually uncharged (0.5 V at one plate and 0.1 V at the other). The collector voltage V_{CE2} finally rises to V_{cc} as C is now charged through R_{L2}. The waveforms resulting from this action are shown in Fig. 4.177.

The important timing waveform is that for V_{BE1}; this is redrawn in Fig. 4.178, showing the *aiming* voltage and other relevant levels.

If the timing waveform is referred to $+10$ V aiming voltage level, we can write

$$v = V_1.e^{-t/CR} \text{ V}$$

but when this voltage equals V_2 the threshold value is reached, ending the quasi-stable state. Hence the equation

$$V_2 = V_1.e^{-t_p/CR} \text{ V}$$

may be used to find t_p. Inserting values

$$9.5 = (9.3 + 9.9).e^{-t_p/CR}$$

$$\text{or } \frac{19.2}{9.5} = e^{+t_p/CR}$$

$$\text{Giving } t_p = CR.\log_e\left(\frac{19.2}{9.5}\right)$$

$$= 0.704CR \text{ seconds}$$

In the circuit $CR = 2$ ms so that $t_p = 1.408$ ms.

Taking the more general case, the above result approximates to

$$V_{cc} = 2 V_{cc}.e^{-t_p/CR} \text{ V}$$

$$\text{or } t_p = CR.\log_e 2 = 0.69CR \text{ seconds}$$

Fig. 4.177 BJT monostable waveforms.

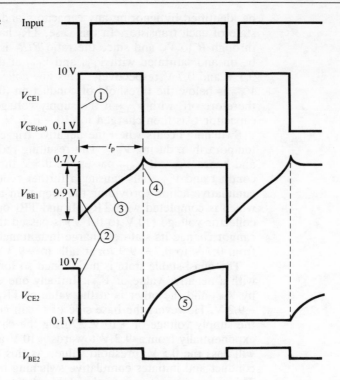

① Initial rise due to input trigger

② Fall due to cumulative switching action

③ Timing transient

④ V_{BE} threshold and turn-on transient

⑤ $R_{LS}C$ output transient

Fig. 4.178 Waveform at TR_1 base of the BJT monostable, showing the exponential rise in voltage from the instant of switching TR_1 off.

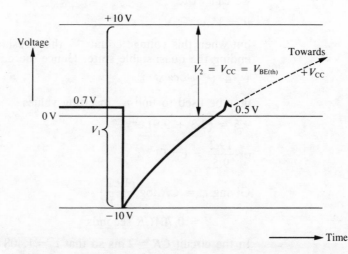

This approximation is valid for all circumstances except very low V_{cc} values. Finally, the collector transient for TR_2 can be taken as about three time constants of CR_L or about 0.2 ms. This is only likely to be important if another trigger pulse is applied within this time as this can cause a modification to the following output pulse duration.

Example 4.3 | *Design a BJT monostable to interface with a TTL gate and provide a delay of 0.1 ms. Assume the TTL gate has a maximum '0' input of 0.4 V; the current from the gate would be 0.1 mA under these conditions. The BJT will have normal voltage levels and an h_{FE} of 120 can be assumed.*

The circuit will be the same as that shown in Fig. 4.176 with the output to the TTL gate taken from TR_1 collector. V_{cc} will be +5 V to provide the required logic 1 level during the delay time t_p. To select a value for the load resistors we shall take a collector current of ten times the gate input current or 1 mA. Hence under maximum logic 0 conditions

$$R_L = \frac{5 - 0.4}{1} \text{ k}\Omega = 4.6 \text{ k}\Omega$$

The value assumed will be the nearest preferred value of 4.7 kΩ.

To ensure saturation for the on state, R must be less than $h_{FE}.R_L$ or 120×4.7 kΩ = 564 kΩ, but to allow for component and parameter tolerances a much smaller value is satisfactory. A value of 220 kΩ is thus chosen and the remaining components can be determined. R_B can also be 220 kΩ ensuring adequate I_{C2} for the quasi-stable on state of TR_2.

Finally for C, since

$$t_p = 10^{-4} = CR.\log_e 2$$

then

$$C = \frac{10^{-4} \times 10^9}{220 \times 10^3 \times 0.69} \text{ nF} = 0.66 \text{ nF}$$

A logic gate monostable

One form of logic gate monostable employs three NAND gates as shown in Fig. 4.179. Referring to earlier work, gates 1 and 2 can form a basic SR flip-flop. The starting point for analysis is to note that if R_i is of a high value, point E is effectively a logic 1. Now if the output is 0, capacitor C will be discharged through the diode, providing a 0 at point A. Point B will therefore be 1 but the 0 from the output ensures that D is also a 1, keeping the output from gate 1 at 0 as suggested.

Fig. 4.179 Logic gate monostable using NAND gates: (a) logic circuit; (b) input of a TTL NAND gate.

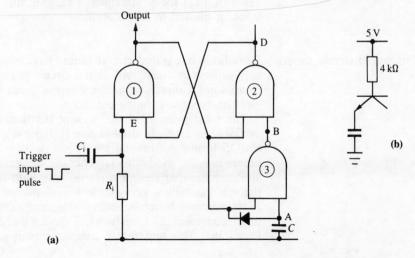

The negative input pulse switches the output to 1 and reverse biases the diode. Initially A remains at 0 and B at 1 and the SR flip-flop switches to make D a 0, holding the output at 1. Capacitor C now charges with the input sinking current from gate 3. When the voltage at A passes the threshold level for the gate (about 1 V), B falls to 0, resetting the output to 0 and returning the circuit to the stable state.

Analysis of the duration of t_p is not as precise as that for the last case as neither precise details of the internal construction of the gate nor the threshold levels will be known. A typical gate input circuit is shown in Fig. 4.179(b) and the basic waveforms in Fig. 4.180.

Fig. 4.180 Waveforms at point A and output of the logic gate monstable.

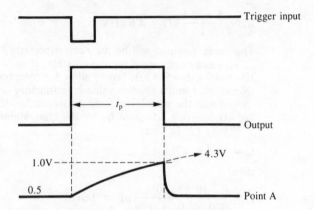

Point A starts from 0.5 V since that is the level to which C discharges through the diode. In the quasi-stable state C charges from 0.5 V towards +5 V V_{cc}, but allowing for the base–emitter volt drops the figure must be reduced to 4.3 V. Thus, referring once again to the final level, the threshold will be 3.3 V. Hence

$$3.8.e^{-t/CR} = 3.3 \text{ V}$$

or $t = CR.\log_e\left(\dfrac{3.8}{3.3}\right) = 0.14CR$ seconds

where R is the base resistor inside the TTL gate. Allowing for a value of 4 kΩ for a 'standard' TTL gate then $t = 560C$ seconds or $0.56C$ if quoted in microfarads.

Integrated circuit timers

Monolithic integrated circuit timers have a wide range of applications in linear and digital circuitry. In many cases these circuits are a direct economical replacement for mechanical and electromechanical timing devices.

The most popular of the present IC timers is the 555, which is available in an 8-pin **dual-in-line** (DIL) package in both bipolar and CMOS forms. Additionally a 14-pin package containing two 555 timers may be used; this is generally designated as the 556 timer although number designations do vary with manufacturer. The 555 timer is basically a very stable IC capable of being operated either as an accurate bistable, monostable or astable multivibrator. The timer comprises 23 transistors, 2 diodes and 16 resistors, as shown in Fig. 4.181. This apparently complex circuit can be reduced to the

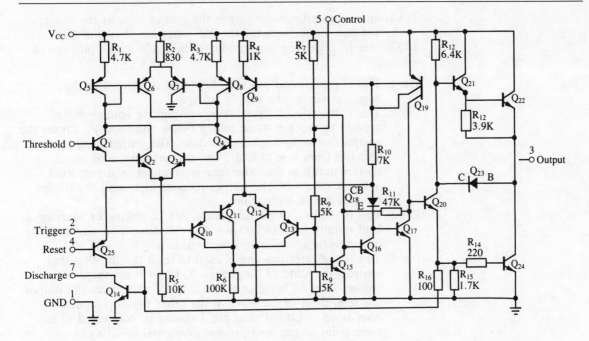

Fig. 4.181 555 timer schematic circuit diagram.

functional block diagram shown in Fig. 4.182 and consists of two comparators, a flip-flop, two control transistors and a high current output stage.

Fig. 4.182 555 timer block diagram.

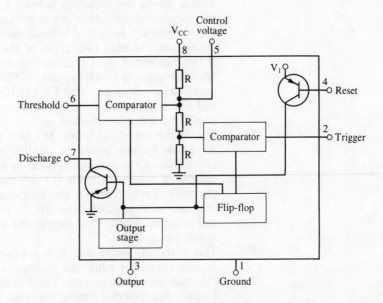

The comparators are actually operational amplifiers that compare input voltages to internal reference voltages which are generated by an internal voltage divider chain of three 5000 Ω resistors. The reference voltages provided are one-third of V_{cc} and two-thirds of V_{cc}. When the input voltage to either of the comparators is higher than the reference voltage for that comparator, the amplifier goes into saturation and produces an output signal to trigger the flip-flop.

The output of the flip-flop controls the output state of the timer.

The pin-out connections for the 555 timer are shown in Fig. 4.183. Details regarding connections to be made to the pins are as follows:

Pin 1. This is the ground pin and should be connected to the negative side of the supply voltage.

Pin 2. This is the trigger input. A negative-going voltage pulse applied to this pin when falling below one-third V_{cc} causes the comparator output to change state. The output level then switches from low to high. The trigger pulse must be of shorter duration than the time interval set by an external CR network otherwise the output remains high until the trigger input is driven high again.

Pin 3. This is the output pin and is capable of sinking or sourcing a load requiring up to 200 mA and can drive TTL circuits. The output voltage available is approximately $V_{cc} - 1.7$ V.

Pin 4. This is the reset pin and is used to reset the flip-flop that controls the state of output pin 3. Reset is activated with a voltage level of between 0 V and 0.4 V and forces the output low regardless of the state of the other flip-flop inputs. If reset is not required then pin 4 should be connected to the same point as pin 8 to prevent accidental resetting.

Pin 5. This is the control voltage input. A voltage applied to this pin allows device timing variations independently of the external timing network. Control voltage may be varied from between 45% to 90% of the V_{cc} value in monostable mode. In the astable mode the variation is from 1.7 V to the full value of the supply voltage. This pin is connected to the internal voltage divider so that a voltage measurement from here to ground should read two-thirds of the voltage applied to pin 8. If this pin is not used it should be bypassed to ground, typically using a 10 nF capacitor. This helps to maintain immunity from noise. The CMOS ICs for most applications will not require the control voltage pin to be decoupled and it should be left unconnected.

Pin 6. This is the threshold input. It resets the flip-flop and hence drives the output low if the applied voltage rises above two-thirds of the voltage applied to pin 8. Additionally a current of minimum value 0.1 μA must be supplied to this pin since this determines the maximum value of resistance that can be connected between the positive side of the supply and this pin. For a 15 V supply the maximum value of resistance is 20 MΩ.

Pin 7. This is the discharge pin. It is connected to the collector of an npn transistor while the emitter is grounded. Thus when the transistor is turned on, pin 7 is effectively grounded. Usually the external timing capacitor is connected between pin 7 and ground and is thus discharged when the transistor goes on.

Pin 8. This is the power supply pin and is connected to the positive side of the supply. The voltage applied may vary from 4.5 V to 16 V although devices which operate up to 18 V are available.

Fig. 4.183 555 timer pin-out diagram: (a) BJT device; (b) CMOS device.

Bipolar pin-out diagrams

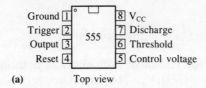

(a) Top view

C-MOS pin-out diagrams

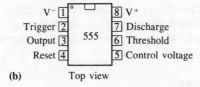

(b) Top view

An application of the 555 as a one-shot monostable is shown in Fig. 4.184. All components within the dotted lines are in the 555 timer itself. The switch S_1 is the grounded emitter npn transistor mentioned earlier. In the reset, or rest, condition the switch S_1 is closed and the voltage across the capacitor is grounded. The timing cycle is started by applying an external trigger pulse to set the flip-flop and open the switch S_1 across the capacitor. The capacitor voltage then rises exponentially towards V_{cc} with a time constant of CR seconds. When this voltage reaches an internally set threshold voltage V_{ref}, the voltage comparator changes state, resets the flip-flop, closes switch S_1 and ends the timing cycle. The output taken from the flip-flop corresponds to a timing pulse of duration t where

$$t = CR.\log_e\left(\frac{V_{cc}}{V_{cc} - V_{ref}}\right) \text{ seconds}$$

Since the internal threshold voltage V_{ref} is obtained from the supply voltage via a potential divider network, then

$$V_{ref} = V_{cc}\left(\frac{R_2}{R_1 + R_2}\right) \text{ V}$$

Fig. 4.184 One-shot monostable using the 555 timer.

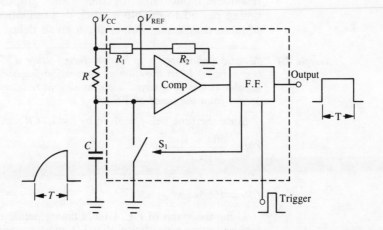

and the basic timing equation becomes independent of the supply voltage

$$t = CR.\log_e\left(1 + \frac{R_2}{R_1}\right) \text{ seconds}$$

The basic circuit of a monostable is shown in Fig. 4.185. In this case $V_{ref} = (2/3).V_{cc}$. The circuit triggers when a negative-going pulse applied to pin 2 reaches $\frac{1}{3}.V_{cc}$. The trigger pulse must be narrower than the desired output pulse. Once triggered the output remains set until the charge time has finished, even if triggered again during this period. The circuit can, however, be reset by applying a negative pulse to pin 4. The pulse duration t using the figures above gives a time of $1.1CR$ seconds if C is in microfarads and R is in megohms.

Fig. 4.185 Basic circuit of the one-shot monostable showing external components required.

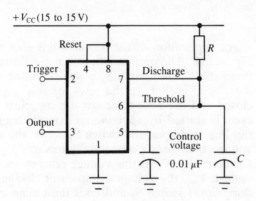

The basic monostable may be changed into a resettable circuit by connecting the trigger input to the reset. In this case a negative-going pulse applied to these inputs causes the timing capacitor to discharge, the output to go low and the cycle to recommence on the positive edge of the reset pulse. Thus the multivibrator can be stopped in the middle of a cycle and restarted. As for a conventional monostable the output pulse width is $1.1CR$ seconds duration. Shown in Fig. 4.186 are the circuit connections for the resettable monostable together with a graph of CR values against timing period to enable a quick determination to be made of suitable component values for a given delay.

Example 4.4   *A basic monostable is to be constructed using a 555 timer. If the trigger input is a pulse of 10 μs duration at a repetition frequency of 1 kHz, calculate suitable values of external capacitance and resistance to give a time delay output from the timer of 500 μs.*

Since the time delay is given by $t = 1.1CR$ seconds, then

$$CR = \frac{500 \times 10^{-6}}{1.1}$$

or

$$CR = 454.54 \ \mu\text{s}.$$

Using the graph of Fig. 4.186 a timing period of 500 μs could be achieved using a resistance of 10 kΩ and a capacitor of 0.045 μF. Selecting

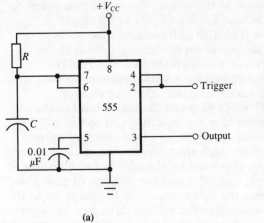

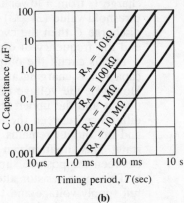

(a)

(b)

Timing period, T(sec)

Fig. 4.186 Resettable monostable using the 555 timer: (a) basic circuit; (b) graph of CR component values to give specific timing period.

a resistance value of 10 kΩ directly could give a value for C from

$$C = \frac{454.54}{10 \times 10^3} \mu F = 0.045 \ \mu F$$

A preferred value of 47 nF would be used in this case.

The 555 may be used as an astable multivibrator as the circuit of Fig. 4.187 shows. Once again the circuit within the dotted lines is the timer while R and C are the timing components. When the circuit is at rest the flip-flop is reset and the output level is high. The external capacitor C is initially uncharged but starts charging exponentially with a time constant of CR seconds. When the voltage across C reaches the upper threshold V_A, comparator 1 changes state and sets the flip-flop causing the output to go low. This in turn causes C to discharge exponentially with a time constant of CR seconds back to the lower threshold value V_B. When this threshold is reached comparator 2 triggers the flip-flop, causing it to reset and hence take the output high. Thus the cycle commences again and a

Fig. 4.187 Astable multivibrator using the 555 timer.

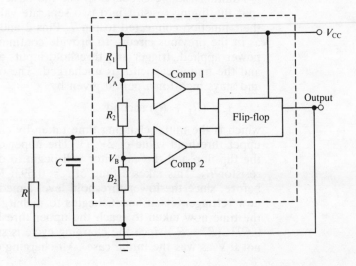

series of pulses are produced at the output. Since the time taken to charge C from a lower threshold value ($V_{cc}/3$) to the upper threshold value ($2V_{cc}/3$) is 0.693CR and a similar time is taken to discharge C then the cycle time is approximately 1.39CR seconds and the frequency of oscillation is approximately 0.72/CR Hz.

In this circuit the output should be a square wave since the capacitor is charging and discharging into the same resistor. However, the voltage difference between the upper level of the output and the upper threshold is not exactly equal to the voltage difference between the lower threshold and lower level of output, and the output remains in the high state longer than in the low level state. This situation can be corrected by the addition of a resistor R_2 between the positive supply rail (V_{cc}) and the junction of pins 2 and 6. This extra resistor allows the timing capacitor to charge up to the full supply voltage and thus makes the voltage difference between the upper level of output and upper threshold level equal to the voltage difference between the lower threshold and lower level of the output. The circuit of the modified astable is shown in Fig. 4.188.

Fig. 4.188 Modified astable multivibrator to give a squarer output waveform.

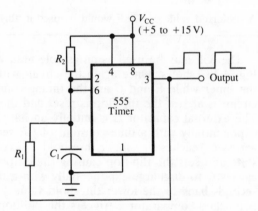

Another astable circuit is to use the basic monostable circuit but split the timing resistor into two separate values R_1 and R_2 with their junction connected to pin 7. Pins 2 and 6 are connected, just as in the previous circuit, to provide continuous triggering. With power applied, trigger and threshold inputs are both below $V_{cc}/3$ V and the timing capacitor is uncharged. The output voltage is high and stays high for a period given by

$$t_1 = 1.1C(R_1 + R_2) \text{ seconds}$$

which is the initial charging time taken by capacitor C to reach the upper threshold value of $2V_{cc}/3$. The upper comparator will trigger the flip-flop and the capacitor will begin to discharge through resistor R_2. This takes a time of $t_2 = 0.693CR_2$ seconds just as before, since the lower threshold level aimed at is $V_{cc}/3$. The timer now retriggers itself and C begins to recharge. Note, however, that the time now taken to reach the upper threshold level is $0.693C(R_1 + R_2)$ since the charging cycle is started from $V_{cc}/3$ and not 0 V as was the initial case. All charging cycle times will have

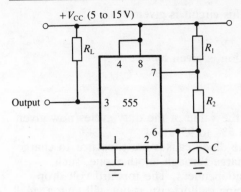

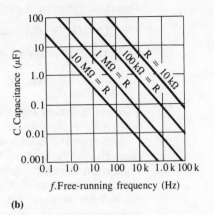

(a)

(b)

Fig. 4.189 Alternative astable multivibrator: (a) circuit; (b) graph of CR component values to give specific timing periods. Note that $R = R_1 + 2R_2$ in these graphs.

this value apart from the initial one. Thus the total time required to complete a charge and discharge cycle is

$$T = t_1 + t_2 = 0.693C(R_1 + 2R_2) \text{ seconds}$$

and the frequency of oscillation is $1/T$ so that

$$f = \frac{1.44}{C(R_1 + 2R_2)} \text{ Hz}$$

The circuit of this astable together with a graph of free-running frequency values for specified component values is shown in Fig. 4.189. The duty cycle, defined as the on time as a percentage of the total cycle time, is given in this case by the ratio

$$\frac{R_2}{R_1 + 2R_2}$$

As R_2 decreases the duty cycle approaches zero and as R_2 increases the duty cycle approaches 50%.

A circuit which extends the duty cycle range of the astable circuit is easily produced by the addition of a diode as Fig. 4.190 shows. The diode connected between pin 6 and pin 7 means that the timing capacitor can now only charge through resistor R_1 since during the charge cycle R_2 is short-circuited by the diode. During the discharge cycle only R_2 is involved, as is the case for the unmodified circuit.

Fig. 4.190 Modified astable multivibrator which gives an improved duty cycle.

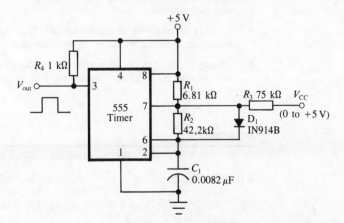

The total time for a cycle in this circuit is given by

$$T = 0.693(R_1 + R_2) \text{ seconds}$$

while the frequency of oscillation is given by

$$f = \frac{1.44}{C(R_1 + R_2)} \text{ Hz}$$

This modification can change the value of the duty cycle (now given by $R_1/(R_1 + R_2)$) to the range 5% to 95%.

Although the addition of the diode causes the capacitor to charge through one resistor and discharge through another one, such separation is not completely independent. The forward volt drop across D_1 is not in fact producing a short-cirucit but will give a small but infinite voltage. The effect of this can be eliminated by adding another diode, D_2, in series with R_2 in a direction opposite to that of D_1. All timing equations are unaffected by this change.

You should now be able to attempt exercises 4.126 to 4.135.

Points to remember

- Basic AND/OR gates may be produced using diodes and resistance logic (DRL)
- The basic RTL NOT gate may be used to produce NOR and NAND gates
- Emitter coupled logic (ECL) gives high speed of operation using non-saturating transistors
- Wired logic allows the use of open-collector gates
- MOS logic has great packing density
- Complementary symmetry (CMOS) circuits give very low power dissipation and good noise margin
- A flip-flop is a one-bit memory element with complementary outputs Q and $\overline{Q}$; many types can be constructed using NAND or NOR gates
- A counter is a group of flip-flops arranged to count the number of input clock pulses; it may be asynchronous of synchronous
- A shift register is a group of flip-flops used to store binary data; it can be used as a ring counter
- Memory may consist of read/write (RAM) or read only (ROM); it can be expanded in terms of bits or of memory size
- Timing circuits can produce a controlled duration pulse; the timing requirement may be met by a BJT monostable flip-flop or by IC timers

EXERCISES 4

4.1   For the circuit of Fig. 4.3 calculate the value of v_{out} if the inputs A and B are (i) both 0, (ii) both 1. Assume $+V$ is $+3$ V and R = 1 kΩ. Assume volt drop across a conducting diode is 0.6 V and logic levels are 0 = 0 V, 1 = +3 V.

(0 V, +2.4 V)

4.2 For the circuit of Fig. 4.4 calculate the value of v_{out} if the inputs A and B are (i) A = 0, B = 1, (ii) A = B = 1. Assume supply voltage is +3 V, $R = 1$ kΩ. Volt drop across a conducting diode may be taken as 0.6 V. Assume input logic levels are 0 = 0 V, 1 = +3 V.

(+0.6 V, +3 V)

4.3 For the circuit of Fig. 4.5 calculate the value of v_{out} if V_s can swing between 0 V and +3 V, $R = 1$ kΩ, $R_s = 500$ Ω, and the volt drop across the conducting diode is 0.6 V. Assume the input to diode A is at 0 V.

(+1.6 V)

4.4 For the circuit of Fig. 4.6, if R is 1 kΩ and the value of shunt capacitance is found to be 100 pF, what would be the effect on the output voltage if input A is suddenly switched from logic 0 (0 V) to logic 1 (+3 V)? Assume diode B has 0 V at its input. What would be the effect on v_{out} if the voltage at input A suddenly reverts to logic 0? Assume volt drop across a conducting diode is 0.6 V and resistance of the source is negligible. Sketch the output voltage waveforms you would expect for each input transition. What would be the propagation delay for such a gate?

4.5 Using the circuit of Fig. 4.8 with +V of 3 V and $R_1 = 500$ Ω find the fan-out of the driving gate is the value for R for each driven gate is 2 kΩ. Assume volt drop across a conducting diode is 0.6 V and that the minimum value of a logic 1 is 1.8 V. What would be the effect on fan-out of reducing the value of R_1? Is there any reason why the value of R_1 should be maintained at 500 Ω? ($n = 1$)

4.6 The circuit of Fig. 4.12 represents a 'loaded' RTL logic NOT gate. If V_{cc} is +5 V, $R_{\text{c}} = 1$ kΩ, $R_B = 50$ kΩ, find the value of R_L which would cause the logic 1 level of the output of the gate to fall from +5 V to +3 V. How many identical circuits can the NOT gate be assumed to be driving if the input resistance of a single driven stage is 12 kΩ? (1.5 kΩ, 8)

4.7 Using the circuit of Fig. 4.15 with $V_{\text{cc}} = +3$ V and $R_{\text{c}} = 1$ kΩ, $R_B = 1$ kΩ, find the fan-out of the stage if v_{out} can fall to +1.5 V and still be a logic 1. What is the minimum value h_{FE} for the stage if a logic 1 of 1.5 V appears at its input? Assume $V_{\text{ce(sat)}}$ for the transistor is 0.1 V and $V_{\text{be(sat)}} = 0.7$ V. ($n = 1$, 3.6)

4.8 A two-input RTL circuit is shown in Fig. 4.191. Sketch the output voltage as a function of the voltage input at A assuming input at B is zero. Assume $V_{\text{BE}} = 0.7$ V for the transistors, $V_{\text{ce(sat)}} = 0.2$ V and $h_{\text{FE}} = 10$.

Fig. 4.191 Two-input RTL circuit for Ex. 4.8.

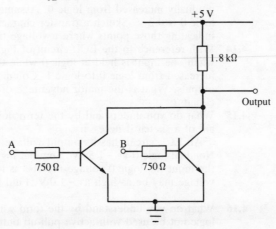

4.9 For the circuit of Fig. 4.16, V_{cc} is +5 V, $R_L = 2$ kΩ, $R_2 = 5$ kΩ, $R_1 = 4$ kΩ. The maximum collector current in the transistor is to be limited to 15 mA to provide a saturation voltage of +0.4 V maximum. Assuming A is at +5 V, B at 0 V, what is the value of the input current? What would be the fan-out for this gate? (0.975 mA, 13)

4.10 Draw the circuit diagram of a three-input TTL NAND gate and comment on the state of the output voltage levels for all eight of the input signal combinations. With reference to the resistances in your circuit explain the difference between standard and low power TTL circuits. What is a Schottky diode and what advantage is to be gained in the use of such a diode in TTL circuits?

4.11 With reference to the circuit of Fig. 4.20, it is required to sink 12 mA in the output transistor without V_{out} exceeding 0.4 V while the source current is to be 6 mA with V_{out} not being allowed to fall below 2.4 V. If the input current for such a stage could be -0.8 mA for V_{in} of 2.4 V and 100 μA for V_{in} of 2.4 V, find the fan-out for both output high and output low conditions.   (60, 15)

4.12 Discuss why active pull-up resistors are to be preferred to passive pull-up resistors for a TTL gate. Using the circuit of Fig. 4.192 find the value of external load that may be connected to the output terminals if I_c is to be limited to 12 mA, V_{cc} is +5 V, while $V_{ce(sat)}$ is 0.2 V. The minimum value for V_{out} in the high state is to be 2.4 V.
                                                                 (192 Ω)

Fig. 4.192 TTL output stage pull-up resistors: (a) passive; (b) active for Ex. 4.12.

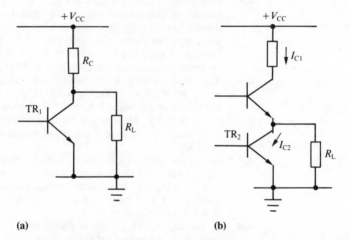

(a)   (b)

4.13 With reference to the TTL NAND gate of Fig. 4.20 describe the circuit action when one input is held at logic 1 while the second is gradually increased from logic 0. Assume a logic 1 is 4 V while a logic 0 is 0.2 V. Sketch a transfer characteristic for such a gate clearly indicating those points where a voltage transition occurs.

4.14 With reference to the ECL circuit of Fig. 4.25 describe circuit action when one input is held at logic 0 while the other is gradually increased from logic 0 to logic 1. Comment on the effect at *both* outputs. What is the major advantage of ECL compared to TTL circuits?

4.15 What do you understand by the term noise margin? Explain with the aid of a sketch if necessary.
     An ECL circuit has its output voltage levels specified as follows: Nominal logic 0 at -1.58 V, nominal logic 1 at -0.75 V, while for the input the logic 0 voltage may be as low as -1.3 V and the logic 1 voltage may be as high as -1.0 V. Find the noise margins.
                                                                 (0.55 V, 0.58 V)

4.16 What do you understand by the term wired logic? Why should wired logic not be used with active pull-up outputs?

4.17 If two gates such as those illustrated in Fig. 4.20 are connected as wired-AND gates what would be the output current if one gate output is high and the other low? Assume standard TTL with a single TTL gate load.   (22.3 mA)

4.18 An open-collector NAND gate is shown in Fig. 4.193. Two such gates are to be configured in wired-AND logic, to drive three standard TTL gates using a pull-up resistor. If V_{cc} is $+5$ V, maximum V_{out} for logic 0 is 0.4 V, minimum V_{out} for logic 1 is 2.4 V, find a suitable value for the pull-up resistor. NAND gate output current for logic 1 can be taken as 0.2 mA and for logic 0 as 16 mA. Driven gate input current for logic 0 can be taken as -1.6 mA and for logic 1 as 40 μA. $(170\ \Omega\text{–}5\ \text{k}\Omega)$

Fig. 4.193 An open collector NAND gate for Ex. 4.18.

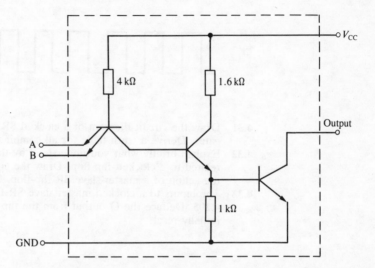

4.19 Using p-channel MOSFETs draw a circuit for a three-input NAND gate. Explain circuit action and derive the truth table.

4.20 Using p-channel MOSFETs draw a circuit for a three-input NOR gate. Explain circuit action and derive the truth table.

4.21 Using n-channel MOSFETs draw a circuit of a two-input NAND gate. Explain circuit action and derive the truth table.

4.22 Using n-channel MOSFETs draw a circuit of a two-input NOR gate. Explain circuit action and derive the truth table.

4.23 Using MOSFET devices design a two-input CMOS AND gate.

4.24 Discuss the difficulties encountered when interfacing TTL and CMOS gates. Suggest some practical examples of where such interfacing might occur.

4.25 A TTL gate is required to drive six CMOS gates. Comment on the suitability, or otherwise, of a direct connection. Would external components be required? If so, give reasons for their use.

4.26 A CMOS gate is required to drive more than one TTL gate. Comment on the practical difficulties and suggest remedies. Draw a sketch of your suggested arrangement.

4.27 Contrast the various logic family parameters and suggest applications for which each type might be most suitable.

4.28 ECL, TTL and CMOS are the most popular logic families. Suggest reasons for this and compare salient parameters of these types of logic families. Give applications for which each of these types may be most suited.

4.29 Draw the circuit diagram of a clocked SR flip-flop using NOR gates and explain circuit action for all combinations of input states.

4.30 The inputs to a clocked SR flip-flop constructed from NOR gates is as shown in Fig. 4.194. Deduce the Q output from the flip-flop assuming that Q is initially low. Assume positive-edge triggering.

Fig. 4.194 Clocked SR flip-
flop inputs for the Ex. 4.30.

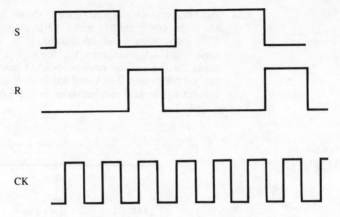

4.31 Draw the circuit diagram of a clocked SR flip-flop using NAND gates. Derive a truth table for all combinations of input states.

4.32 Explain briefly what you understand by the terms master and slave related to a clocked flip-flop. Draw the circuit diagram and explain the action of a master-slave SR flip-flop.

4.33 The inputs to a clocked master-slave SR flip-flop are as shown in Fig. 4.195. Deduce the Q output from the flip-flop assuming that Q is initially reset.

Fig. 4.195 Clocked SR
master-slave flip-flop inputs
for Ex. 4.33.

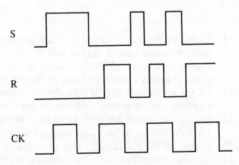

4.34 State the disadvantage inherent in the SR flip-flop and comment on how the JK flip-flop overcomes this disadvantage. Draw a circuit of a JK flip-flop to illustrate your answer.

4.35 Why, when using a JK flip-flop, should it be necessary to use edge triggering? Explain simply how edge triggering may be achieved.

4.36 Draw a circuit diagram and explain the action of a master-slave JK flip-flop. What is the advantage of the master-slave JK compared to the 'ordinary' JK circuit?

4.37 A master-slave JK flip-flop has the inputs shown in Fig. 4.196. Deduce the Q output assuming Q is initially reset.

4.38 What is the advantage in using flip-flops with preset and clear inputs? A master-slave JK flip-flop has the inputs as shown in Fig. 4.197. Deduce the value of the Q output with time if it is initially reset.

Fig. 4.196 JK master-slave
flip-flop inputs for example
4.37.

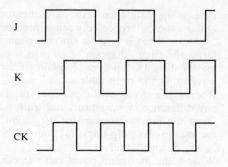

Fig. 4.197 JK master-slave
flip-flop inputs for Ex. 4.38.

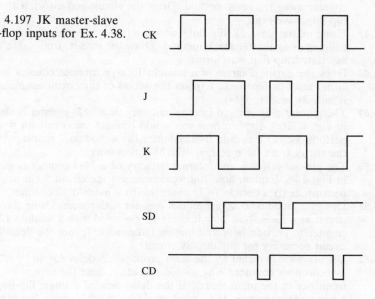

4.39 Using a JK flip-flop as a basis, produce (i) a T flip-flop, (ii) a D
flip-flop. Show all extra logic circuitry used and derive truth tables for
both types of flip-flop.

4.40 The waveforms of Fig. 4.198 are applied to the D-type flip-flop.
Deduce the Q output waveform assuming it is initially zero.

Fig. 4.198 D-type flip-flop
inputs for Ex. 4.40.

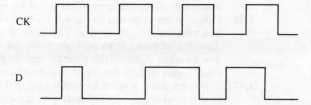

4.41 Discuss the difference between sequential and combinational logic. Discuss the difference between asynchronous and synchronous counters. What is a ripple-through counter?

4.42 Draw the circuit diagram of an asynchronous modulo 8 counter using master-slave JK flip-flops. Explain circuit action with the aid of waveforms and truth table.

4.43 Redraw the circuit of the previous question using T flip-flops. Is there any difference in waveforms and truth table?

4.44 Draw the timing diagrams for the following asynchronous circuits that use master-slave flip-flops: (a) four-bit binary up-counter (b) four-bit binary down-counter.

4.45 What is the maximum count value (modulus) for a counter using 3, 5, 7 and 9 flip-flops? Briefly discuss methods by which the value of the count of an *n*-stage counter may be reduced below its maximum value.   (8, 32, 128, 512)

4.46 Using master-slave JK flip-flops design an asynchronous modulo 6 counter using the reset method. Draw the circuit and associated flip-flop waveforms.

4.47 Using master-slave JK flip-flops design an asynchronous modulo 6 counter using the feedback method. Draw the circuit, truth table and associated flip-flop waveforms.

4.48 Draw the circuit diagram of a modulo 10 asynchronous counter using either reset or feedback. Explain the action of the circuit assuming it is initially reset to 0000.

4.49 The pin-out diagram and internal circuitry of a 7490 counter is shown in Fig. 4.72(a). Explain how you would connect the circuit for it to perform as: (a) a decade BCD counter (b) a modulo 7 counter. Use the circuit truth table of Fig, 4.72(b) if necessary.

4.50 The pin-out diagram and internal circuitry of a 7493 counter is shown in Fig. 4.76. Explain how you would connect the circuit for it to perform as (i) a decade BCD counter, (ii) a modulo 12 counter.

4.51 Design a modulo 60 asynchronous counter with preset. Using this circuit as a basis show how it could be cascaded with a modulo 12 counter to provide hour and minute indications. Ignore the decoding circuit necessary for any display circuit.

4.52 Discuss what is meant by the term propagation delay for an asynchronous counter. Why should this effect limit the upper frequency of the input signal? If the delay time of a single flip-flop in a five-stage counter is 18 ns what would be suitable upper frequency limit?   (13.88 MHz)

4.53 Draw a circuit diagram of a modulo 16 counter operating synchronously with JK flip-flops. Show, by using the truth table or circuit waveforms, the transitions that occur at each flip-flop output for the complete count. Assume the flip-flops are triggered on the positive edge of the clock waveform.

4.54 With respect to synchronous counters, what do you understand by (i) parallel carry (ii) serial, or ripple, carry. Illustrate your answer with suitable diagrams.

4.55 A four-stage synchronous counter of the type shown in Fig. 4.82(a) has a flip-flop propagation delay of 20 ns, the AND gates have a propagation delay of 5 ns and the clock period is 20 ns. Find the maximum clock frequency that could be used.   (22.22 MHz)

4.56 Design a synchronous modulo 8 up-down counter using JK flip-flops and additional logic circuits. The control signal to determine the direction of count is to be logic 1 for the up count and logic 0 for the down count. Assume the counter can be set to the correct initial count value before the input count commences.

4.57 Show, by using a suitable circuit and with reference to a truth table, how a modulo 16 synchronous counter can be converted to produce a modulo 10 counter. Assume JK flip-flops with negative edge triggering are to be used.

4.58 Discuss briefly the advantages/disadvantages of asynchronous and synchronous counters.

4.59 Draw the circuit of a three-stage shift register using D-type flip-flops. Show, using a truth table, how a logic 1 may be clocked through the register.

4.60 Discuss how, with the aid of a control input signal and suitable logic circuitry, a serial shift register may be connected to provide a shift left or shift right facility.

4.61 A four-bit serial shift register is to be used to temporarily store the data word 1011, where the least significant bit is on the right. Show, by reference to a circuit diagram and truth table, the transitions that occur at the flip-flop outputs with successive clock pulses. Use JK flip-flops.

4.62 Using a simplified block diagram show how a four-bit shift register may be configured as (i) serial-in, parallel-out, (ii) parallel-in, serial-out and (iii) serial-in, serial-out. Give possible applications where each of the above registers may be used in practice.

4.63 Using D-type flip-flops draw the circuit of a simple modulo 4 ring counter. Explain circuit action by reference to the output waveforms for a complete count sequence.

4.64 Design a self-starting four-stage ring counter using JK flip-flops. Draw a circuit diagram and explain circuit action with reference to output waveform transitions for a complete count sequence.

4.65 Design a four-bit Johnson counter using JK flip-flops. The starting count is to be 0000. Show the timing of the circuit for four input clock pulses.

4.66 Design a five-stage modulo 10 Johnson counter using JK flip-flops. Draw the circuit diagram and, with reference to output waveform transitions, explain circuit action.

4.67 Draw the circuit diagram of a modulo 6 Johnson counter and show, by reference to a truth table, how the count of six is obtained.

4.68 Modify the circuit of the previous example to produce a modulo 5 counter. Show also the modified truth table.

4.69 Using the modulo 5 Johnson counter and a $\div 2$ stage, produce the circuit diagram of a modulo 10 counter. Explain circuit action with reference to output waveform transitions for a complete count sequence.

4.70 Define what you understand by multiplexing. Design a logic system where a multiplexer may be used to select data from one of four input sources. Assume each input source will provide a four-bit word.

4.71 Using a four-bit multiplexer, design a circuit which has three input variables X, Y and Z and $F = \overline{X}.Y.\overline{Z} + \overline{X}.Y.Z + X.\overline{Y}.\overline{Z} + X.Y.Z$. Assume inputs X and Y can be used for data select inputs and Z and $\overline{Z}$ may appear as data inputs.

4.72 Draw the simple block diagram of a two-bit encoder using three inputs. If the enable input is low deduce a possible output combination for the encoder. What is the disadvantage of such a circuit and what could be done to overcome this disadvantage?

4.73 A logic network has three inputs and two outputs. The state of the outputs indicates which input has a logic 0 applied to it. If two or more of the inputs have logic 0, the state of the outputs should indicate the highest priority. Deduce a minimal logic circuit of the network to produce the required functions as described above.

4.74 Draw a logic circuit using NAND gates to produce a binary to octal decoder circuit.

4.75 Using the truth table of Fig. 4.113(a) for a BCD to decimal decoder, deduce a decoding circuit (utilising Karnaugh map if necessary) to implement the decode function. Remember that there are 'don't care' states associated with the four-bit binary code when used as BCD. These 'don't care' states could help minimise the decoding logic.

4.76 Design a logic circuit to provide a BCD to excess-3 code conversion using NAND or NOR gates.

4.77 Design a logic circuit to convert (i) a three-bit natural binary code into Gray code and (ii) vice versa.

4.78 Design a logic circuit to convert (i) Gray code to excess-3 code, (ii) excess-3 code to Gray code.

4.79 Why is the Gray code preferred in some instances to the BCD code? What is the reason for the use of excess-3 code compared to the Gray code?

4.80 Design a logic circuit to produce a logic 1 output when three input variables have an odd number of logic 1s. Use NOR gates only.

4.81 Design a logic circuit to produce a logic 1 output when three input variables have an even number of logic 1s. Use NOR gates only.

4.82 What do you understand by the terms minority decoder and majority decoder? Design a logic circuit, using three input variables, to give a logic 1 output if the conditions for (i) minority and (ii) majority are present at the input.

4.83 Design a decimal to BCD converter for active logic 0 levels on the input lines.

4.84 A Johnson code counter is to have its code converted to BCD before driving a seven-segment display. Deduce a suitable logic circuit, using minimisation techniques wherever possible, to achieve this objective.

4.85 A BCD counter is to be decoded before driving a seven-segment display. Deduce a suitable logic circuit to achieve the correct inputs to the display. Use minimisation techniques where possible to simplify the circuit.

4.86 What do you understand by the term 'half-adder'? Design a logic circuit using only NAND gates to perform the half-adder function.

4.87 Using half-adder circuits show how they may be interconnected to produce a full-adder circuit. Confirm circuit action by means of a truth table.

4.88 Design a four-bit serial adder using a single full-adder circuit. Use a shift register to store one of the addends and also the result. Use a D-type flip-flop to produce the required delay.

4.89 Design a four-bit parallel adder using registers which may be filled serially although the addition process is a parallel one. State the advantages of the parallel adder compared to the serial version.

4.90 State the differences in terms of wiring arrangement, speed and accuracy between a serial adder and a parallel adder.

4.91 What do you understand by the term 'look-ahead carry' associated with adder circuits? What is the advantage to be gained in the use of look-ahead carry circuitry?

4.92 Design a combinational logic circuit using Exclusive-OR, AND and OR gates to produce a full subtractor circuit.

4.93 Design a four-bit by four-bit multiplier circuit. Quad full-adders may be used as the basis of the circuit. Show all extra logic gates and interconnections.

4.94 Draw a logic circuit suitable for a one-bit comparator. Use Karnaugh maps to establish the conditions for the comparator outputs.

4.95 Draw a circuit diagram suitable for use as a four-bit comparator. Explain how the circuit would check $A > B$ if $A = 1101$ and $B = 1100$.

4.96 Using block diagrams for four-bit magnitude comparators show how three such blocks may be connected to provide a ten-bit comparator.

4.97 Draw a circuit diagram of a three-bit binary rate multiplier. Explain circuit action with reference to time-related waveforms wherever possible.

4.98 Use a block diagram for a four-bit binary rate multiplier and show, using an up-down counter, how it may be connected to function as an adder circuit.

4.99 Describe what you understand by 'memory' in an electronic sense. What is the purpose of memory?

4.100 Describe and compare read only memory and random access memory.

4.101 Describe briefly the difference between volatile and non-volatile memory.

4.102 Explain, with the aid of a sketch of a basic bipolar static memory cell, how the memory cell is accessed for both a read and write operation.

4.103 Explain the differences between static and dynamic memory. Mention advantages/disadvantages of each type.

4.104 Explain, with the aid of sketches where necessary, how a basic memory cell can be arranged with X and Y addressing inputs.

4.105 Draw a block diagram, and explain the action of, a circuit incorporating a memory matrix $2^m \times n$ using word select. Show all connections including address lines, data lines, read/write and chip select inputs.

4.106 Draw a block diagram, and explain the action of, a circuit incorporating a matrix of $2^m \times 2n$ using X and Y select. Show all connections including address lines, data lines, read/write and chip select inputs.

4.107 State what you would understand by the term 'decoder'. If a memory uses 16 word select lines, design a decoder with four address line inputs to access each line.

4.108 If an X and Y select memory has three X address line inputs and three Y address line inputs what is the memory capacity? (64)

4.109 If a RAM chip has 32 rows and 16 columns each with a single bit, find the number of address line inputs that are required to gain unique addressing of every location. (9)

4.110 Describe, with the aid of a sketch if necessary, the special requirements for reading and writing into a dynamic RAM cell. What is the advantage to be gained in using a dynamic cell as against a static one?

4.111 Using memory 'blocks' of 256×4 bits size, design a memory that gives 256×8 bit words. Show all address and data bus connections and control inputs.

4.112 Using memory 'blocks' of 128×1 bit size, design a memory that gives 128×8 bit words. Show all address and data bus connections and control inputs.

4.113 Using memory 'blocks' of 256×4 bit size, design a memory that gives $1 \text{ k} \times 4$ bit capacity. Show all address and data bus connections and control inputs.

4.114 What is the purpose of the chip select (memory enable) input? Design a decoder that will give the chip select input to eight identical chips using address lines A_{10} A_{11} and A_{12}.

4.115 Explain, with the aid of a sketch where necessary, what is meant by a mask programmed ROM.

4.116 Descibe briefly the differences between mask programmable and field programmable ROMs.

4.117 Describe the features of the following memory types:(i) RAM (ii) ROM (iii) PROM (iv) EPROM.

4.118 Describe briefly the differences between mask programmable and electrically programmable ROMs. State advantages of each type.

4.119 Sketch a typical electrically programmable ROM cell and state how it is programmed. How can such a cell be erased?

4.120 State what you understand by odd and even parity. Why should a parity bit be used in a data word?

4.121 The following 'words' are to be sent using odd parity. What is the value of the parity bit in each case? (i) _1010011 (ii) _1110011 (iii) _1000011 (iv) _0000101   (1; 0; 0; 1)

4.122 What is meant by a parity tree circuit? Will the use of a parity bit ensure error-free transmission of the data word?

4.123 A four-bit BCD word is to be transmitted using odd parity. Deduce the required truth table and draw the logic circuit which achieves this function.

4.124 Show how parity bit generators may be incorporated in a four-bit transmit/receive system. Draw the circuit and explain its action.

4.125 Define what you understand by a tri-state device. What advantages can be gained in practice by the use of the tri-state buffer circuit?

4.126 A bipolar transistor monostable flip-flop is constructed as shown in Fig. 4.199. In the stable state TR_1 is on and TR_2 off. The trigger pulse turns TR_2 on and TR_1 off. Describe circuit action after TR_2 is turned on. Deduce an equation for the time that TR_1 is turned off.

Fig. 4.199 A bipolar transistor monostable flip-flop for Ex. 4.126.

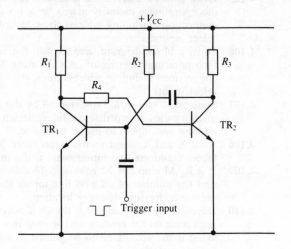

4.127 If for the circuit of Fig. 4.199, V_{cc} is +10 V, $R_1 = R_3 = 5.6\ k\Omega$, $R_2 = 100\ k\Omega$, $C = 0.01\ \mu F$, find the period for which TR_1 is turned off by the trigger pulse.   (0.69 ms)

4.128 Draw a logic gate monostable circuit using NAND gates and, assuming a 5 V supply rail is used, deduce the period of the timing pulse. State any assumptions made.

4.129 Refer to the block diagram of a 555 timer used as a one-shot multivibrator as shown in Fig. 4.184. Explain how the circuit operates. What is the importance of the 'internal' resistors R_1 and R_2?

4.130 For the one-shot multivibrator of Fig. 4.184 deduce the period of the timing pulse output if $C = 1.0\ \mu F$ and $R = 10\ k\Omega$.   (11 ms)

4.131 A basic monostable is to be constructed using a 555 timer. If the trigger input is a pulse of 5 μs duration at a repetition frequency of 750 Hz, calculate suitable values of external capacitance and resistance to give a time delay of 250 μs at the timer output.
                    (22 nF, 10 kΩ)

4.132 Figure 4.189 shows the 555 timer connected to operate as an astable multivibrator. What are the times for which the output is high and low? Base your calculations on the components R_1, R_2 and C.

4.133 For the circuit of Fig. 4.189 find the frequency of oscillation and the duty cycle if $R_1 = 1\ k\Omega$, $R_2 = 4.7\ k\Omega$ and $C = 0.01\ \mu F$.

(13.85 kHz, 0.452)

4.134 What is the function of the diode shown in Fig. 4.190?

4.135 For the circuit of Fig. 4.190 find the frequency of oscillation and the duty cycle if $R_1 = 1\ k\Omega$, $R_2 = 4.7\ k\Omega$ and $C = 0.01\ \mu F$.

(25.26 kHz, 0.175)

5 Analogue Circuits

The principal learning objectives of this chapter are to:

Analogue signals are those in which the information is contained in the amplitude of the voltages and currents. This may be in a d.c. form, a.c. form (i.e. sinusoidal with varying frequency and amplitude) or one of many other waveforms. The circuits used to process these signals will be ideally either linear, operating on signals of different amplitudes and d.c. levels in an identical manner, or non-linear in a controlled manner. These differences can be illustrated by the transfer characteristics and waveforms shown in Fig. 5.1.

The linear characteristic in Fig. 5.1(a) shows that a sinusoidal input results in a sinusoidal output regardless of amplitude or d.c. operating point. The characteristic in (b) is non-linear and its action depends on both amplitude and d.c. operating point. The sinusoidal input with zero bias is rectified (negative half-cycles removed) while that on the curved part of the characteristic is distorted, with positive and negative half-cycles amplified unequally.

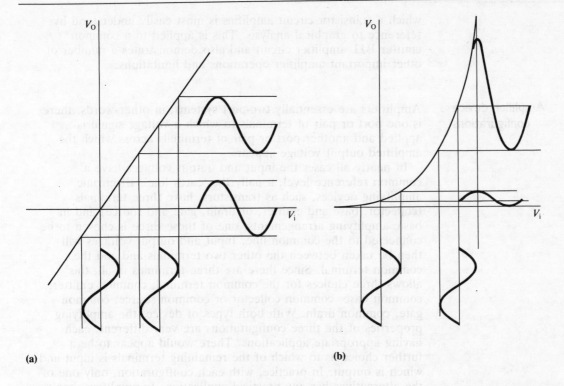

Fig. 5.1 Transfer
characteristic and signals:
(a) linear characteristic; (b)
non-linear characteristic.

Both linear and non-linear circuits are discussed in this chapter
but the most important class of analogue circuit, the amplifier,
should ideally be linear over a specific working range. This
chapter is split into five major sections: the first is concerned with
the analysis of basic amplifying circuits using graphical and
equivalent circuit methods for transistor amplifiers and introduces
the analysis of basic op-amp circuits; the following two sections
examine the effect of signal frequency on amplifier performance
and discuss large signal amplifiers; the principles of feedback are
considered in detail with applications to both discrete component
and op-amp circuits in the fourth section, and a wide range of
linear and non-linear op-amp circuits are described and analysed
in the final section.

SMALL-SIGNAL AMPLIFIERS

Transistor amplifiers

This section is concerned with the use of single transistors as the
active device in amplifying circuits. Such devices, both BJTs and
FETs, are used in a variety of arrangements or configurations
which must be defined before proceeding further. The way by

which a transistor circuit amplifies is most easily understood by reference to graphical analysis. This is applied to a common emitter BJT amplifier circuit and also demonstrates a number of other important amplifier operations and limitations.

Amplifier device configurations

Amplifiers are essentially **two-port systems**; in other words, there is one port or pair of terminals to which a voltage signal is applied and another port or pair of terminals across which the amplified output voltage appears.

In nearly all cases the input and output voltages have a common reference level, usually the 'earth line'. Electronic amplifying devices, such as transistors, have three terminals (collector, base and emitter, or drain, gate and source) and in basic amplifying arrangements, one of these three is chosen to be connected to the common line. Input and output voltages will then be taken between the other two terminals and and the common terminal. Since there are three terminals in all, this allows three choices for the common terminal, common emitter, common base, common collector or common source, common gate, common drain. With both types of device, the amplifying properties of the three configurations are very different, each having appropriate applications. There would appear to be a further choice as to which of the remaining terminals is input and which is output. In practice, with each configuration, only one of the alternatives has any practical application. In amplifier circuits, the output terminal will always have a load (usually resistance) connected from that terminal to the common line (either directly or through the d.c. supply rail). The two sets of three configurations are shown in Fig. 5.2. These circuits are simplified and the normal d.c. bias arrangements must be included to ensure a correct operating point for linear amplifying behaviour.

Although the common terminal specifies the configuration, it is easier to identify each by remembering the input and output terminals as follows:

	CE	CC	CB
	Common Emitter	Common Collector	Common Base
Input	base	base	emitter
Output	collector	emitter	collector

	CS	CD	CG
	Common Source	Common Drain	Common Gate
Input	gate	gate	source
Output	drain	source	drain

This identification can also be extended to modified circuits having additional resistance (or impedance) in the lead between the common terminals and earth. Some examples are shown in Fig. 5.3.

The first, circuit (a), has V_{in} applied to the gate and V_o from the drain; it is therefore in the common source (CS) configuration but with resistance R_S in the common lead. Circuit (b) has output from the collector and input to the emitter; it is therefore in the

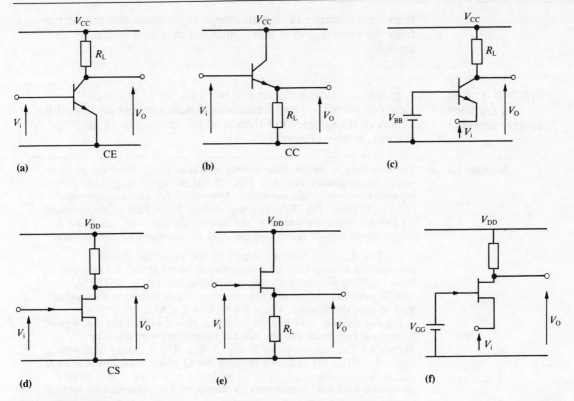

Fig. 5.2 Transistor
configurations: (a) common
emitter; (b) common
collector; (c) common base,
(d) common source; (e)
common drain; (f) common
gate.

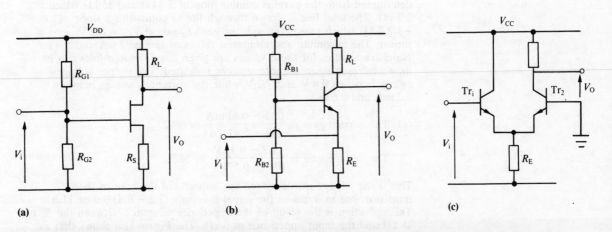

Fig. 5.3 Transistor
configurations modified by
additional resistance.

common base (CB) configuration but with R_B (the parallel
combination of R_{B1} and R_{B2}) in the common base lead. The two
transistors in the third circuit are operating in different
configurations: Tr_2 has the base earthed and is therefore in the
common base configuration with input to the emitter and output

from the collector. Tr_1 has the input to the base and the output from the emitter and is thus connected as a common collector amplifier.

Graphical analysis of a common emitter amplifier

A graphical analysis demonstrates a number of important amplifier properties and limitations. These are best illustrated by a numerical example. For details of the techniques of graphical analysis, refer to Chapter 3.

Example 5.1

The transistor in the amplifier circuit shown in Fig. 5.4 has the input and output characteristics shown in Fig. 5.5. At the signal frequency, the capacitive reactances are negligible. Determine: (a) The d.c. operating point or Q point. (b) The a.c. signal voltage at the base of the transistor. (c) The a.c. input impedance at the base of the transistor. (d) The a.c. output signal voltage at the collector. (e) The voltage and current gain.

(a) The d.c. load lines are drawn on the input and output characteristics using 10 V and, respectively, 470 kΩ and 3.3 kΩ. The input load line intersects the input characteristic at $V_{BE} = 0.65$ V, $I_B = 20$ μA. The output characteristic for 20 μA intersects the output load to give the Q point $V_{CE} = 3.4$ V, $I_C = 2$ mA.

(b) For the a.c. load line on the input characteristic, the a.c. signal, its internal resistance and the 470 kΩ resistor are simplified by Thévenin's theorem as shown in Fig. 5.4(b). The a.c. load line, with a slope of $-1/17.3$ kΩ is drawn through the Q point down to the zero I_B axis at approximately 1 V. The equivalent a.c. voltage of $0.3 \sin \omega t$ moves the load line horizontally as shown with the intersection moving up and down the input characteristic.

Reading from the characteristic, I_B moves between 5.8 μA and 32 μA and the corresponding V_{BE} change is from 0.57 V to 0.7 V, and a.c. signal of 0.13/2 V or $65 \sin \omega t$ mV.

(c) From (b) above, the signal current is $(32 - 5.8)/2$ or $13.1 \sin \omega t$ μA. The input impedance at the base is thus 65 mV/13.1 μA or 5 kΩ.

(d) The a.c. or signal load line on the output characteristic is determined from the parallel combination of 3.3 kΩ and 15 kΩ which is 2.7 kΩ. The load line is drawn through the Q point with a slope of $-1/2.7$ kΩ which passes through the zero I_C axis at $V_{CE} = 8.8$ V as shown. The maximum and minimum values of I_B are 32 μA and 5.8 μA. No characteristics for these values are given but estimated lines can be drawn in resulting in the intersections indicated. The output voltage swings between 0.4 V and 7.2 V while the current I_C swings between 3.2 mA and 0.6 mA.

(e) The current gain is $\dfrac{(3.2 - 0.6) \text{ mA}}{(32 - 5.8) \text{ μA}}$ or 99.2.

The voltage gain is $\dfrac{(7.2 - 0.4) \text{ V}}{(0.7 - 0.57) \text{ V}}$ or 52.3.

This is the voltage gain between the output and the base of the transistor, but in terms of the signal it is only $(7.2 - 0.4)/0.6$ or 11.3. This reduction is the result of the impedance mismatch between the R_s of 18 kΩ and the input impedance of 5 kΩ. The Figure also shows that as the signal voltage goes positive, I_B and I_C go to their maximum values and V_C goes to the minimum values. This shows a 180° phase shift or inversion between input and output. Thus strictly, the voltage gain, $A_V = -52.3$ or -11.3 depending upon where it is measured.

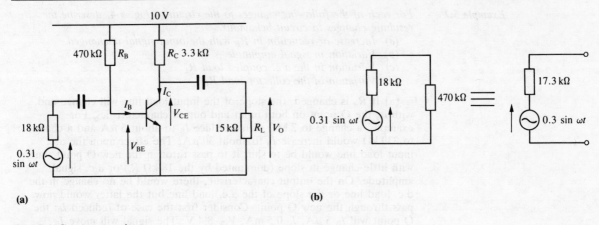

(a)

Fig. 5.4 Common emitter
amplifier circuit for
example 5.1.

(b)

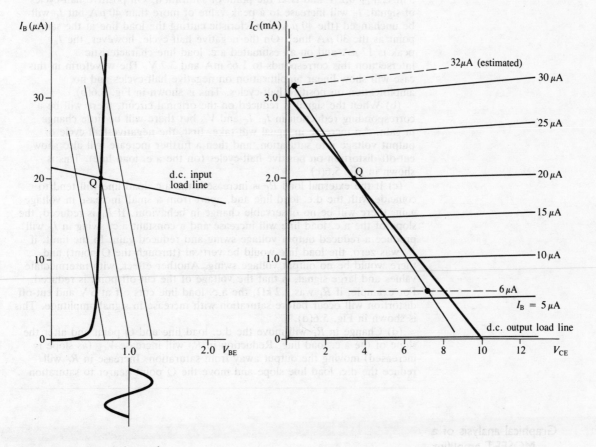

Fig. 5.5 Characteristics and
load lines for examples 5.1
and 5.2.

Example 5.2   *For each of the following changes to the circuit in Fig. 5.4, describe the resulting changes in circuit behaviour.*

   (a) Increase or reduction in R_B with the input signal unchanged.
   (b) Variation in signal amplitude.
   (c) Variation in the a.c. coupled load R_L.
   (d) Variation in the collector load R_C.

(a) If R_B is changed, the slope of the input load line will change and, with it, the Q point on both input and output characteristics. For example, a change to 2 MΩ would reduce I_B to about 5 μA and a change to 330 kΩ would increase I_B to about 30 μA. The effect upon the a.c. input load line would be to shift it to pass through the new Q point, but with little change in slope (dominated by the 18 kΩ R_s) or a.c. signal amplitude. On the output characteristic, there would be no change in the d.c. load line or the slope of the a.c. load line but the latter would now pass through the new Q point. Consider first the case of reduced I_B: the Q point will I_B 5 μA, I_C 0.5 mA, V_{CE} 8.4 V. The signal will move I_B between about 20 μA and zero with the negative half-cycle only about 5 μA and distorted with the device cut off at the signal peak. The equivalent I_C variation is between 2 mA and zero and the V_{CE} variation between 3.4 V and 9.7 V. The distorted waveform is shown in Fig. 5.6.

Consider a second case, with I_B increased to 30 μA; the Q point is I_C 3 mA, V_{CE} 0.2 V and is at the point of saturation. On positive half-cycles of signal, I_B will increase to a peak value of more than 40 μA but I_C will be unchanged (the 40 μA characteristic cutting the load line at the same point as the 30 μA line). On the negative half-cycle, however, the I_B peak is 17 μA and on an estimated a.c. load line characteristic intersection this corresponds to 1.65 mA and 3.7 V. The waveform in this case will show linear amplification on negative half-cycles and no amplification on positive half-cycles. This is shown in Fig. 5.6(b).

(b) When the signal is reduced on the original circuit, there will be a corresponding reduction in I_B, I_C and V_{ce} but there will be little change in gain. An increase in signal will take, first, the negative half-cycle of output voltage into saturation, and then a further increase will also show cut-off distortion on positive half-cycles (on the a.c. load line). This is shown in Fig. 5.6(c).

(c) If the external load R_L is increased, the a.c. load line will tend to coincide with the d.c. load line and, apart from a small increase in voltage gain, there will be no observable change in behaviour. If R_L is reduced, the slope of the a.c. load line will increase and a constant a.c. swing in I_C will produce a reduced output voltage swing and reduced gain. In the limit, if R_L was zero, the load line would be vertical (through the Q point) and there would be no output voltage swing. Another effect, with intermediate values and large signal, is that the voltage of the cut-off point is reduced. For example, if R_L was 2.2 kΩ, the a.c. load line cuts off at 6 V and cut-off distortion will occur before saturation with increase in signal amplitude. This is shown in Fig. 5.6(d).

(d) Change in R_C will move the d.c. load line and Q point and alter the slope of the a.c. load line. Reduction in R_C will increase V_{CE} (as slope is increased) moving the output away from saturation. Increase in R_C will reduce the d.c. load line slope and move the Q point nearer to saturation.

Graphical analysis of a MOSFET amplifier

Example 5.3   *An enhancement MOSFET has the characteristics shown in Fig. 5.7. It is to be used as a common source amplifier with a drain load R_D and a 20 V supply. Assuming the a.c. gate signal to have no d.c. content, select suitable*

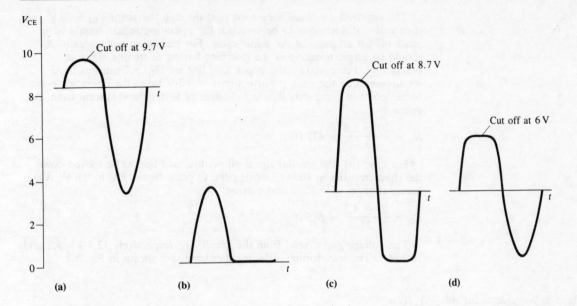

Fig. 5.6 Waveforms for the
solution of example 5.2.

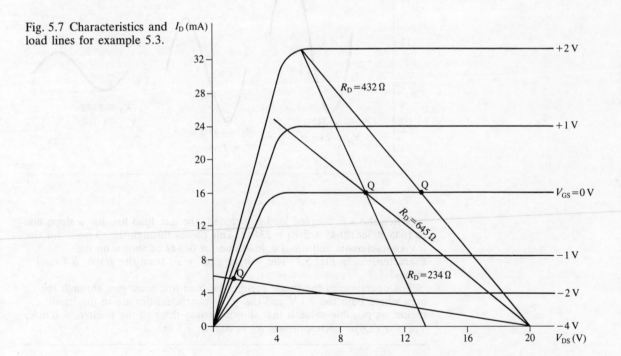

Fig. 5.7 Characteristics and
load lines for example 5.3.

*values of R_D for maximum gain if the peak value of the a.c. signal is (a) 2 V
and (b) 1 V. In each case determine the resulting voltage gain and output
waveform. Find also how these would be modified for case (b) if an
additional load of 368 Ω is capacitively coupled to the output. What is the
value of R_D if triode region operation is required when the gate signal is
between +1 V and −1 V?*

The required d.c. load lines must pass through the point $V_{DS} = 20$ V and, if excessive distortion is to be avoided, the operating region should be in pinch-off for all parts of the signal cycle. For maximum voltage gain, R_D should be large, resulting in the load line having as small a slope as practicable. For case (a), the upper load line on Fig. 5.7 satisfies these requirements as the peak positive signal of +2 V takes the operating point to the point of pinch-off. Reading changes of voltage and current from the graph

$$R_D = \frac{20 - 5.6}{33.3} = 432 \ \Omega$$

For case (b), the smaller signal allows the load line to be moved down the characteristics as shown, moving the Q point from 13 V to 9.6 V. Again taking changes in voltage and current

$$R_D = \frac{20 - 4.5}{24} = 645 \ \Omega$$

The voltage gains, read from the graph, are respectively $12.7/4 = 3.2$ and $10/2 = 5$. The waveforms, including d.c. level, are shown in Fig. 5.8.

Fig. 5.8 Waveforms for the solution of example 5.3.

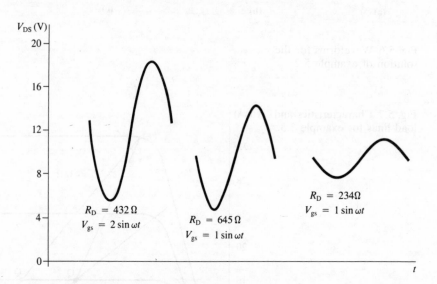

When the a.c. coupled load is included, the a.c. load line has a slope due to $(645 \times 368)/(645 + 368) = 234 \ \Omega$. This passes through the Q point for the 0 V characteristic and the d.c. load line for 645 Ω as shown on the characteristics in Fig. 5.7. The voltage gain, read from the graph, is found to be 1.7.

For operation in the triode region, the load line must pass through the area where both the +1 V and the −1 V characteristics are in the triode region. A possible value is that shown passing through the point $I_D = 6$ mA, $V_{DS} = 0$ V. The corresponding R_D is 20/6 or 3.3 kΩ.

Graphical analysis with a reactive load

In the previous examples in this chapter the series loads have been resistive. This situation will usually be true for practical circuits operating in their normal frequency range. At extremes of frequency, however, the loading will become complex or mainly reactive. Analysis of this situation by graphical methods is not easy, but examination of a simple case can illustrate the type of behaviour that can occur.

Consider a common emitter BJT circuit with a pure inductance in the collector lead. The circuit and characteristic are shown in Figs 5.9 and 5.10 respectively and numerical values are used to demonstrate the effects.

Fig. 5.9 Transistor with inductive load.

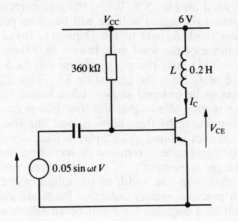

Fig. 5.10 Characteristics and load line for the circuit in Fig. 5.9.

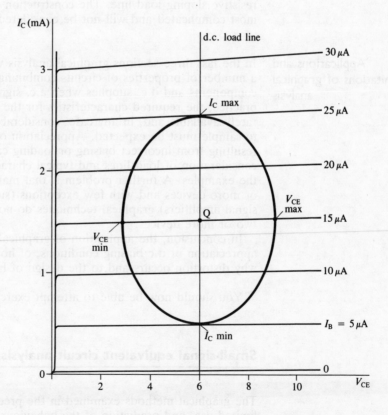

Since the d.c. resistance in series with the transistor is zero, the d.c. load line will be vertical from $V_{CE} = 6$ V. Calculating $I_B = (6 - 0.6)/360 = 15$ μA, the Q point is V_{CE} 6 V, I_C 1.5 mA. The input impedance can be taken as having the same value as that in example 5.1, i.e. 5 kΩ. The peak signal current is therefore 0.05/5 mA or 10 μA. The base current will swing between 5 μA and 25 μA and results in a sinusoidal collector current which is flowing in the collector load. The resulting a.c. voltage must be 90° out of phase with the current as the load is inductive. Thus, when the a.c. current is at its maximum and minimum values, the a.c. component of voltage will be zero; i.e. V_{CE} will be at 6 V. These two points are indicated on Fig. 5.9. When the a.c. current is zero (I_C at the d.c. Q point) the signal voltage will be at its peak positive and negative values (with respect to the Q point). These depend upon the reactance of the load ωL. If $\omega = 15\,000$ r/s ($f = 2.4$ kHz), $\omega L = 3$ kΩ, and the peak voltage will be 3 kΩ × 1 mA (the peak value of I_C from the graph) or 3 V. This allows two further load line points to be entered at $I_C = 1.5$ mA and $V_{CE} = 3$ V and 9 V. Examination shows that the four points can only be joined as a smooth curve by the elliptical load line shown. The a.c. Q point thus moves around the ellipse as indicated by the arrows. If the signal amplitude is reduced or increased the ellipse will be smaller or larger respectively. If the signal frequency is increased or reduced, only the width of the ellipse will change in the same way.

In practice, purely inductive loads are unlikely to occur and the result of a complex load will be an ellipse superimposed in a resistive sloping load line. The construction of such a load line is most complicated and will not be considered in this book.

Applications and limitations of graphical analysis

In the last three sections graphical analysis was used to demonstrate a number of properties of circuits combining transistors with passive components and d.c. supplies when a.c. signals are applied. In practice, the required characteristics for the device type in use are rarely available and, in any case, considerable variation from sample to sample must be expected. Appreciation of distorted waveforms resulting from incorrect biasing or loading can be much improved by consideration of load lines and typical characteristics, as shown in the examples. A further problem is that many circuits employ two or more devices and with few exceptions (such as push-pull large-signal amplifiers) graphical techniques do not allow the combining of two or more devices.

In conclusion, the application of graphical analysis is limited to an appreciation of d.c biasing conditions, of how amplifiers work, of why distortion occurs and to the design of large-signal amplifiers.

You should now be able to attempt exercises 5.1 and 5.2.

Small-signal equivalent circuit analysis of amplifiers

The graphical methods examined in the preceding sections have only limited use, and prediction of the behaviour of the many different amplifier circuits requires an alternative approach. Small-signal equivalent circuits provide a means by which amplifier circuits can be analysed using methods of electrical network analysis. Equivalent circuits for BJTs and FETs are discussed in Chapter 2 and the required techniques are demonstrated in Chapter 3. In this section the choice of equivalent circuit is discussed and the technique is applied to a wide range of commonly used basic transistor amplifier circuits.

Choice of equivalent circuit

Various equivalent circuits are possible but, before making a choice, two factors must be considered. These are the objectives of the analysis, and the information available concerning the devices and their equivalent circuit parameters.

A possible objective is the precise prediction of the behaviour of a particular device circuit. A more practical objective is the prediction of the general behaviour of the circuit configuration: Is the voltage gain several hundred; about 10; less than one and inverting etc? Is the input impedance high and the output impedance low or vice versa? Are these properties sensitive to a particular component value or to the d.c. bias conditions? A third objective might be to examine the circuit behaviour under limiting signal conditions, usually at high frequency.

The parameters of an equivalent circuit will vary widely between devices of the same type and for the particular device as bias conditions are changed. A precise study will only be possible if the parameters are measured for the particular device under the particular conditions of operation. This is only likely to be useful in a study of limiting conditions. In practice, manufacturers' published data for a particular device is limited to a range of h_{FE} for BJTs and I_{DSS} and V_P for FETs. In each case, the spread will be large (typically $\pm 50\%$). Additional information is available from the d.c. operating conditions, which can be calculated or estimated from the circuit components and d.c. supplies. This is particularly useful for the BJT as the essential parameter g_m is related directly to the d.c. bias collector current by $g_m = 40I_C$ A/V (or mA/V if I_C is quoted in mA) Since g_m is also given by h_{fe}/h_{ie}, the combination of a typical h_{fe} and the d.c. bias conditions gives a good approximation for h_{ie} (this is assuming that the d.c. h_{FE} and the small signal h_{fe} are the same). The second order h parameters h_{oe} and h_{re} can usually be ignored as any error by so doing is much less than errors introduced by tolerance on manufacturers' data and that on the passive circuit components, It can be useful however to be aware of the order of magnitude of these parameters as limiting conditions of performance can sometimes be understood by an appreciation of these factors.

Thus for the analysis of mid-band performance (low frequency performance for direct coupled amplifiers) we shall use the amplified equivalents shown in Fig. 2.23 for the BJT and Fig. 2.32 for FETs.

Later in this chapter (p.367–72), high frequency performance is studied using the equivalents shown in Figs 2.25, 2.26, and 2.32.

Single-stage BJT amplifier circuit analysis

The common emitter amplifier

The circuit for the **common emitter amplifier** is shown in Fig. 5.11(a) and a mid-band equivalent circuit is shown in Fig. 5.11(b). Since it is 'mid-band', the reactances of the capacitors are approximated to short circuit and any stray shunt capacitance is

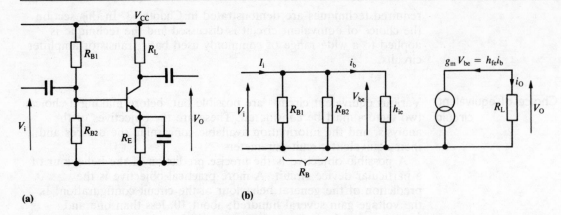

(a)   **(b)**

Fig. 5.11 The common emitter amplifier: (a) full circuit; (b) mid-band small-signal equivalent circuit.

taken as open circuit. The d.c. supply V_{CC} is regarded as short circuit to signals. Both the g_m and the h_{fe} form for the dependent current generator are indicated.

Following the standard procedure (p. 00):

$$V_{be} = V_i \text{ and } i_b = \frac{V_i}{h_{ie}}$$

$$V_o = -g_m V_i R_L \text{ or } \frac{h_{fe} V_i R_L}{h_{ie}}$$

therefore the voltage gain $A_v = \dfrac{V_o}{V_i} = -g_m R_L$ or $\dfrac{h_{fe}}{h_{ie}} R_L$ [5.1]

The first of these is the preferred form since g_m will be known from the inverting d.c. collector current ($g_m = 40 I_C$). This represents an inverting gain which with typical values will be between 20 and several hundred.

For the current gain A_i

$$I_o = -h_{fe} i_b \qquad\qquad\qquad [5.2]$$

but i_b is only a part of I_i, the remainder flowing in the combined R_B:

$$\left(R_B = \frac{R_{B1} \times R_{B2}}{R_{B1} + R_{B2}} \right)$$

Therefore, from equation [5.2]

$$I_o = I_i \times \frac{R_B}{R_b + h_{ie}} \times h_{fe}$$

and $A_i = \dfrac{I_o}{I_i} = \dfrac{-h_{fe} R_B}{R_B + h_{ie}}$

Alternatively, this may be written

$$A_i = \frac{-h_{fe}}{1 + h_{ie}/R_b} \qquad\qquad\qquad [5.3]$$

With practical values, R_B is usually much greater than h_{ie} and the error in taking $A_i = -h_{fe}$ will be less than that due to incorrect knowledge of h_{fe}. A further point is the whereabouts of the output current; in this case it is obvious, but in many circuits there will be an external load, often the input to another stage. In such cases, I_o will be the current flowing into that load towards earth.

The output impedance of any circuit is obtained by suppressing any external signal sources (replacing them by their internal resistances) and calculating the impedance 'looking in' at the output terminals. In this case, if the voltage source providing V_{in} is zero, then v_{be} and i_b will be zero as will the dependent generator. This leaves the output impedance simply as

$$Z_{out} = R_L \qquad\qquad\qquad [5.4]$$

Example 5.4 *Determine the output voltage and open circuit gain for the circuit shown in Fig. 5.12. The signal frequency is such that all capacitive reactances can be taken as zero. The h_{fe} for the BJT is quoted as 220.*

Fig. 5.12 Common emitter amplifier circuit for example 5.4.

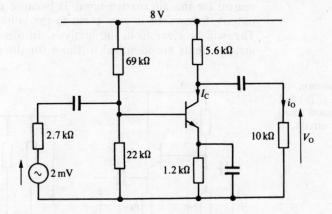

The first step is to find the d.c. collector current so that g_m and h_{ie} can be determined.

$$V' = \frac{8 \times 22}{22 + 69} = 1.93 \text{ V} \quad R' = \frac{69 \times 22}{69 + 22} = 16.7 \text{ k}\Omega.$$

$$I_C = \frac{1.93 - 0.7}{\dfrac{16.7 + 1.2}{220} + 1.2} = 0.96 \text{ mA}$$

(Note: $V_{CE} = 8 - 0.96(1.2 + 5.6) = 1.5$ V, so the BJT is operating in the linear region.) Now,
$g_m = 39I_C = 38.4$ mA/V

$$h_{ie} = \frac{h_{fe}}{g_m} = \frac{220}{38.4} = 5.7 \text{ k}\Omega$$

The effective load $R'_L = \dfrac{5.6 \times 10}{5.6 + 10} = 3.6 \text{ k}\Omega$

The voltage gain $A_v = -38.4 \times 3.6 = -138$.

But the input voltage V_{in} depends upon the total input impedance, which is the parallel combination of h_{ie} and the base bias resistors

$$\frac{1}{Z_{in}} = \frac{1}{5.7} + \frac{1}{69} + \frac{1}{22}$$

from which, $Z_{in} = 4.2$ kΩ.

$$V_i = 2 \text{ mV} \times \frac{4.2}{4.2 + 2.7} = 1.22 \text{ mV}$$

and $V_o = 1.22 \text{ mV} \times -138 = 0.17 \angle 180° \text{ V}$

For the overall current gain, the required ratio is

$$\frac{\text{Current into the 10 k}\Omega \text{ resistor}}{\text{Current from the generator}} = \frac{0.17/10}{0.002/6.9} = 58.7$$

This is appreciably less than h_{fe} as a result of the signal current lost in the bias resistors and into the 5.6 kΩ collector load.

The common collector amplifier

The circuit for a common collector amplifier or emitter follower is shown in Fig. 5.13(a) and the equivalent circuit in Fig. 5.13(b). The reason for the alternative name is because the voltage at the emitter output 'follows' or is very close to the voltage at the base input. This will be clear from the analysis. In this simple form the biasing arrangements are identical to those for the common emitter circuit.

Fig. 5.13 The common collector amplifier: (a) full circuit; (b) mid-band small-signal equivalent circuit.

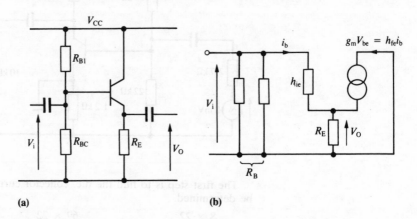

By inspection of the equivalent circuit, two equations can be written

$$V_i = i_b h_{ie} + V_o \qquad [5.5]$$

$$V_o = (i_b + h_{fe} i_b) R_E \qquad [5.6]$$

Substituting in equation [5.5], $V_i = i_b h_{ie} + i_b (1 + h_{fe}) R_E \qquad [5.7]$

The voltage gain is V_o/V_i and this is obtained by dividing equations [5.6] by [5.7] and cancelling i_b.

$$A_v = \frac{(1 + h_{fe}) R_E}{h_{ie} + (1 + h_{fe}) R_E}$$

For all practical purposes, $1 + h_{fe} = h_{fe}$ and dividing numerator and denominator by h_{fe} gives

$$A_v = \frac{R_E}{\dfrac{h_{ie}}{h_{fe}} + R_E}$$

But $h_{ie}/h_{fe} = 1/g_m$, therefore

$$A_v = \frac{R_E}{1/g_m + R_E} \tag{5.8}$$

This is the most convenient form for the result since R_E is a circuit component and g_m depends upon the d.c. conditions. This result represents a non-inverting voltage gain of less than one. With practical values, for most circuit arrangements, this figure will lie between 0.9 and 1.0 but lower values can occur, particularly in the two-transistor circuit known as an emitter coupled amplifier.

The input impedance at the base is obtained from equation [5.7] and is given by

$$Z_{in} = h_{ie} + (1 + h_{fe})R_E \tag{5.9}$$

This is a very useful result as it applies not only to this circuit but to any circuit 'seen' from the base with resistance or impedance in the emitter lead. This resistance (R_E in this case) is *amplified* by $(1 + h_{fe})$ and typical values will give a Z_{in} of 100 kΩ to 1 MΩ or more. High Z_{in} is often a desirable circuit property but, unfortunately in this case, the complete circuit input impedance includes the base bias resistors in parallel with that at the base.

The current gain A_i is obtained by inspection of the circuit and found to be

$$A_i = +(1 + h_{fe}) \tag{5.10}$$

Determination of the output impedance Z_{out} requires redrawing the equivalent circuit to include the suppressed signal source resistance R_S as shown (with some rearrangement) in Fig. 5.14. For the output impedance analysis, a voltage V is applied to the circuit and the resulting current I is calculated. R_E is not included in this calculation as this can be added in parallel to the final result. R_S and R_B can be combined to give an equivalent R'_S. Following the standard procedure, take V as the unknown and express i_b (for the dependent generator) in terms of this unknown.

Fig. 5.14 The common collector amplifier: equivalent circuit for the determination of Z_{out}.

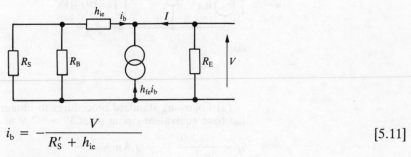

$$i_b = -\frac{V}{R'_S + h_{ie}} \tag{5.11}$$

Writing a nodal equation,

$$i_b + h_{fe}i_b + I = 0 \tag{5.12}$$

or $I = -(1 + h_{fe})i_b$

substituting in equation [5.11],

$$I = -(1 + h_{fe})\left(-\frac{V}{R'_S + h_{ie}}\right)$$

and $Z_{out} = \dfrac{V}{I} = \dfrac{R'_S + h_{ie}}{1 + h_{fe}} \tag{5.13}$

This result shows the opposite effect to that found in [5.9]. 'Looking in' at the emitter, impedance in series with the base has been divided by $(1 + h_{fe})$. The result tends to a very low output impedance.

Equation [5.13] may be rewritten with $(1 + h_{fe})$ approximated to h_{fe}

$$Z_{out} = \frac{R_S'}{h_{fe}} + \frac{h_{ie}}{h_{fe}} = \frac{R_S'}{h_{fe}} + \frac{1}{g_m} \qquad [5.14]$$

This result for a low value of R_S is simply $1/g_m$ which is determined by the d.c. operating conditions. As a typical value, a d.c. collector current of 1 mA results in a Z_{out} of 25 Ω.

In summary, the common collector amplifier has non-inverting gain of nearly one, a high input impedance and a low output impedance. This behaviour is particularly useful in 'matching' between systems with a high internal resistance and a load or other circuit with relatively low input impedance. This can be illustrated by means of an example.

Example 5.5   *The BJT in the emitter follower circuit in Fig. 5.15 has a typical h_{fe} of 300. Determine (a) the voltage gain and input impedance for this amplifier and (b) the Thévenin equivalent for the complete circuit at the output terminals.*

Fig. 5.15 Circuits for example 5.5: (a) full circuit; (b) equivalent output circuit.

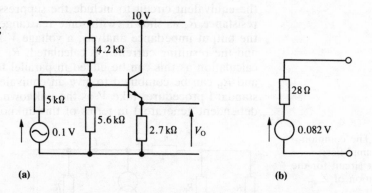

(a) Following standard procedures to obtain the d.c. collector current, the base equivalent circuit gives $V' = 5.7$ V and $R = 2.4$ kΩ. Therefore

$$I_C = \frac{5.0}{\dfrac{26.7}{300} + 2.7} = 1.8 \text{ mA}$$

Hence, $g_m = 72$ mA/V, $h_{ie} = \dfrac{300}{72} = 4.2$ kΩ

From equation [5.8],

$$A_v = \frac{2.7}{\dfrac{1}{72} + 2.7} = 0.994$$

From equation [5.9],

$$Z_{in} = 4.2 + 300 \times 2.7 = 814.3 \text{ kΩ}$$

But for the complete amplifier circuit, we must include the bias resistors (with a combined parallel value of 24 kΩ). Therefore

$$Z_{in} = \frac{24 \times 814.3}{838.3} = 23.3 \text{ k}\Omega$$

This demonstrates a limitation of the emitter follower stage if the requirement is a very high impedance; alternatives are discussed later in this section.

(b) For the Thévenin equivalent at the output, the open circuit output voltage and the output impedance are required. The first is obtained from the information in part (a) above. Allowing for the input impedance,

$$V_i = 0.1 \times \frac{23.3}{23.3 + 5} = 0.0823 \text{ V}$$

$$V_o = 0.994 \times 0.0823 = 0.082 \text{ V}$$

For Z_{out}, we use equation [5.14] and the equivalent source impedance R_s' is required. This is given by R_s in parallel with R_B

$$R_s' = \frac{5 \times 24}{29} = 4.13 \text{ k}\Omega$$

Then $Z_{out} = \dfrac{1}{70.2} + \dfrac{4.13}{300} = 0.028 \text{ k}\Omega$ or 28 Ω

Note: this is also simply $\dfrac{h_{fe} + R_s}{h_{fe}}$

The resulting equivalent circuit is shown in Fig. 5.15(b). This stresses the impedance conversion property of the common collector amplifier, as the original internal resistance of the signal source has been converted from 5 kΩ to 28 Ω with only the small drop in signal voltage level due to the fall in input impedance resulting from the shunt bias resistance.

The common base amplifier

A circuit for a common base amplifier is shown in Fig. 5.16(a). This arrangement uses the same d.c. bias circuit as that used for the common emitter amplifier and has the base 'decoupled' at the signal frequency by the capacitor C_B. The circuit is identified as **common base**, as the input and output are at the emitter and collector respectively. The small-signal equivalent is shown in Fig. 5.16(b).

Fig. 5.16 The common base amplifier: (a) full circuit; (b) mid-band small-signal equivalent circuit.

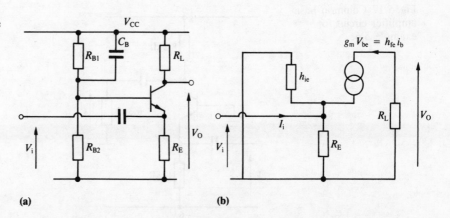

Following the usual equivalent circuit procedure, V_E can be selected as an unknown variable. The dependent generator variables are then

$$V_{be} = -V_E = V_i \qquad\qquad [5.15]$$

$$i_b = -\frac{V_i}{h_{ie}} \qquad\qquad [5.16]$$

Hence, $V_o = -g_m V_{be} R_L = g_m V_i R_L$

and the voltage gain $A_v = \dfrac{V_o}{V_i} = g_m R_L \qquad\qquad [5.17]$

This result has the same magnitude as that for the common emitter stage but it is non-inverting.

Writing a nodal equation at V_E,

$$I_i + i_b + h_{fe} i_b = \frac{V_i}{R_E}$$

Substituting from equation [5.16] and rearranging,

$$I_i = V_i \left(\frac{1}{R_E} + \frac{1 + h_{fe}}{h_{ie}} \right)$$

The term in brackets is the parallel admittance of R_E and $h_{ie}/(1 + h_{fe})$. R_E will be much greater than $h_{ie}/(1 + h_{fe})$ and so Z_{in} may be taken as

$$Z_{in} = \frac{h_{ie}}{h_{fe}} = \frac{1}{g_m} \qquad\qquad [5.18]$$

This is the same result as that for the Z_{out} of the common collector equation [5.14] when R'_S is zero. This is because, in both cases, the impedance is that seen 'looking in' at the emitter. In this case, the result is a very low input impedance. When V_s is suppressed to find Z_{out}, there is no i_b and therefore no $h_{fe} i_b$ generator. Thus, the output impedance is simply

$$Z_{out} = R_L$$

Example 5.6   *Determine the voltage and current gains and the input impedance of the BJT circuit shown in Fig. 5.17.*

Fig. 5.17 Common base amplifier circuit for example 5.6.

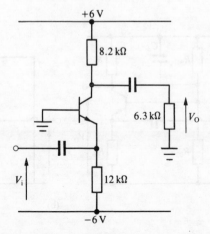

The circuit shown is common base with input and output at emitter and collector respectively. No information regarding the BJT is given but g_m, obtained from the d.c. conditions, will be sufficient in this case.

$$I_C = \frac{0 - 0.7 - (-6)}{12} = 0.44 \text{ mA}$$

Therefore

$$g_m = 39 \times 0.44 = 17.2 \text{ mA/V}.$$

From equation [5.18] $Z_{in} = \dfrac{1}{g_m} = 58 \ \Omega$ (in parallel with 12 kΩ)

From equation [5.17] $A_v = g_m R_L$

where in this circuit R_L is the parallel combination of 8.2 kΩ and 6.3 kΩ.

$$A_v = \frac{17.2 \times 8.2 \times 6.3}{8.2 + 6.3} = 61.3$$

For the current gain,

$$I_{in} = \frac{V_i}{0.058} \text{ mA},$$

$$I_{out} = \frac{61.3 \times V_i}{6.3} \text{ mA}$$

Therefore $A_i = \dfrac{61.3}{6.3} \times 0.058 = 0.56$

Single-stage FET circuit analysis

The common source amplifier

The common source amplifier is a basic amplifying FET circuit, very similar to the equivalent common emitter BJT circuit. The small-signal equivalent is even more simple (see Fig. 2.32) but the relationship between this and the d.c. conditions is more complex, as is the calculation of the d.c. conditions. Figure 5.18 shows a JFET common source amplifier and the small-signal equivalent.

Fig. 5.18 The common source amplifier: (a) full circuit; (b) mid-band small-signal equivalent circuit.

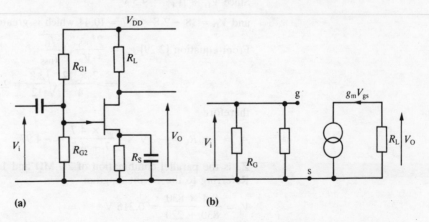

Since $V_{gs} = V_i$, $V_o = -g_m V_i R_L$

and $A_v = -g_m R_L$

Z_{in} is simply R_G, the parallel combination of R_{G1} and R_{G2}, which can be a very high resistance. Again, $Z_{out} = R_L$.

Example 5.7   *Calculate the output voltage for the JFET circuit in Fig. 5.19. The JFET is quoted as having I_{DSS} 12 mA and V_P −4 V. The circuit capacitances may be assumed to have zero reactance at the signal frequency.*

Fig. 5.19 JFET amplifier circuit for example 5.7.

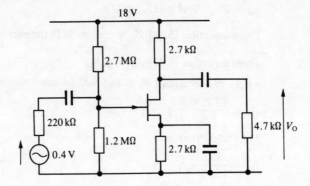

Determination of the d.c. conditions uses equation [2.25].

$$I_D = I_{DSS}\left(1 - \frac{V_{GS}}{V_P}\right)$$ [5.19]

and V_{GS} is given by $V_{GS} = V'_G - I_D R_S$

where $V'_G = \dfrac{18 \times 1.2}{1.2 + 2.7} = 5.5$ V

so $V_{GS} = 5.5 - 2.7 I_D$

Substituting into equation [5.19],

$$I_D = 12\left[1 - \frac{(5.5 - 2.7 I_D)}{-4}\right]^2$$

Rearranging and solving the quadratic gives I_D values fo 2.8 mA and 4.4 mA. Only the first result satisfies the conditions for pinch-off operation

$V_D > V_G + |V_P|$

Since $V_G + |V_P| = 9.5$ V

and $V_D = 18 - 2.8 \times 2.7 = 10.44$ which is greater than 9.5 V.

From equation [2.29] $g_m = \dfrac{2 I_{DSS}}{-V_P}\sqrt{\dfrac{I_D}{I_{DSS}}}$

$$= \frac{2 \times 12}{4}\sqrt{\frac{2.8}{12}} = 2.9 \text{ mA/V}$$

therefore

$$A_v = -g_m R_L = -2.9 \times \frac{2.7 \times 4.7}{2.7 + 4.7} = -4.97$$

Z_{in} is the parallel combination of 2.7 MΩ and 1.2 MΩ which is 830 kΩ. Referring to the circuit diagram

$$V_i = \frac{0.4 \times 830}{830 + 220} = 0.316 \text{ V}$$

$$V_o = 0.316 \times 4.97 = -1.57 \text{ V}$$

The common drain amplifier or source follower

As with the previous circuit, this arrangment can be compared with the BJT equivalent, the common collector amplifier, and it has very similar properties. A simple buffer circuit and its small-signal equivalent are shown in Fig. 5.20.

Fig. 5.20 The common drain amplifier: (a) full circuit; (b) mid-band small-signal equivalent circuit.

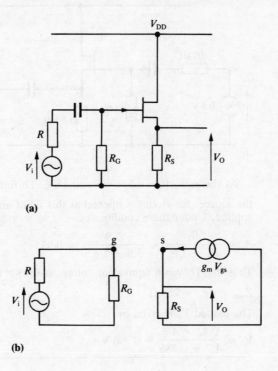

(a)

(b)

Referring to the equivalent circuit,

$$V_o = g_m V_{gs} R_S$$

$$V_{gs} = V_i - V_o = V_i - g_m V_{gs} R_S$$

or

$$V_i = V_{gs}(1 + g_m R_S)$$

Now, $A_v = \dfrac{V_o}{V_i} = \dfrac{g_m V_{gs} R_S}{V_{gs}(1 + g_m R_S)}$

$$= \frac{R_S}{R_S + 1/g_m} \qquad [5.20]$$

This result should be compared with the BJT equivalent equation [5.8].

Note that since R_G will normally be much larger than the signal source resistance, V_{in} is approximately the same as e. g_m is usually smaller than that for a BJT so the gain will not be quite as near unity. Z_{out} can be found in the same way as for the BJT circuit and is given by

$$Z_{out} = \frac{1}{g_m}$$

Example 5.8   *For the circuit shown in Fig. 5.21, calculate the equivalent generator seen at the source of the FET and hence find the output voltage V_o. The g_m for the FET can be taken as 3.5 mA/V.*

Fig. 5.21 Source follower circuit for example 5.8.

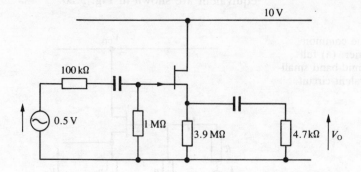

At the gate, $V_i = 0.5 \times \dfrac{1.1}{1} = 0.45$ V. To find the equivalent circuit at the source, the circuit is opened at this point and Thévenin's theorem is applied. Under these conditions

$$A_v = \frac{R_L}{R_L + 1/g_m} = \frac{3.9}{3.9 + 1/3.5} = 0.93$$

Thus the Thévenin equivalent voltage is $0.93 \times 0.45 = 0.42$ V.

$Z_{out} = i/g_m = 286 \ \Omega$

The loaded V_o is given by

$$V_o = \frac{0.42 \times 4.7}{4.7 + 0.286} = 0.396 \text{ V}$$

Modified single-stage amplifiers

The simple configurations analysed in the previous sections are often modified by the addition of resistance (or impedance) in the *common* lead. Such resistors are usually part of the d.c. biasing arrangements and are decoupled (effectively short-circuited) at mid-band signal frequencies by a shunt capacitor. It is useful, however, to investigate their effect at low frequencies when the decoupling is ineffective. Alternatively, such common lead components may be included as a part of a feedback network. Three cases will be considered, the common emitter and common base BJT circuits and the common source FET circuit.

Analysis of a common emitter stage with emitter resistance   A circuit and its small-signal equivalent are shown in Fig. 5.22. From the circuit

$$V_i = h_{ie}i_b + R_E(1 + h_{fe})i_b \tag{5.21}$$

hence $Z_{in} = \dfrac{V_i}{i_b} = h_{ie} + R_E(1 + h_{fe})$ $\tag{5.22}$

Fig. 5.22 A common
emitter amplifier with
emitter resistance: (a) full
circuit; (b) small-signal
equivalent circuit.

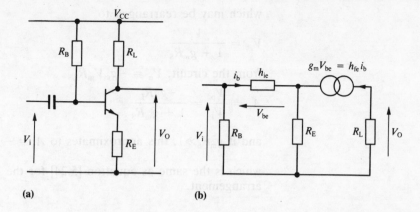

(a)   (b)

which is the same result as equation [5.9]. Again, from the figure

$$V_o = -h_{fe}i_bR_L$$

$$= \frac{-h_{fe}R_LV_i}{h_{ie} + (1 + h_{fe})R_E}$$

$$A_v = \frac{V_o}{V_i} = \frac{-h_{fe}R_L}{h_{ie} + (1 + h_{fe})R_E} \qquad [5.23]$$

But $h_{fe} \gg 1$ and $h_{fe}R_E$ is usually much greater than h_{ie}, so an
approximate result is

$$A_v = \frac{-h_{fe}R_L}{h_{fe}R_E} = \frac{-R_L}{R_E} \qquad [5.24]$$

This is a very convenient result and is quite satisfactory for typical
circuit values unless R_E is very small.

**Analysis of the common
source amplifier with
source resistance**

The circuit for a common source amplifier shown in Fig. 5.23
includes a source resistance R_S to stabilise the d.c. operating point
in the same way as R_E serves this function in a common emitter
amplifier. The equivalent circuit in (b) shows the two gate bias
resistors combined as R_G. From this circuit,

$$V_{gs} = V_i - g_mV_{gs}R_S$$

Fig. 5.23 A common source
amplifier with source
resistance: (a) full circuit;
(b) small-signal equivalent
circuit.

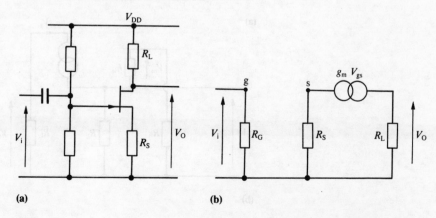

(a)   (b)

which may be rearranged to

$$V_{gs} = \frac{V_i}{1 + g_m R_S}$$

from the circuit, $V_o = -g_m V_{gs} R_L$

$$A_v = \frac{V_o}{V_i} = \frac{-g_m R_L}{1 + g_m R_S}$$

and if $g_m R_S \gg 1$, this approximates to $A_v = -\dfrac{R_L}{R_S}$

which is the same as equation [5.24] for the BJT equivalent arrangement.

Analysis of a common base amplifier with base resistance

The common base amplifier shown in Fig. 5.24(a) (input emitter, output collector) has the combined R_B between base and earth without a decoupling capacitor. Comparison between the equivalent circuit in Fig. 5.24(b) and that for the basic common base amplifier in Fig. 5.16(b) shows that the only difference is that R_B is added in series with h_{ie}. Thus equations [5.17] and [5.18] for voltage gain and input impedance can be modified accordingly. Both of these equations involve g_m, so if a modified g'_m is found, the analysis is complete. For the basic BJT,

$$g_m = \frac{h_{fe}}{h_{ie}}$$

Fig. 5.24 A common base amplifier with base resistance: (a) full circuit, (b) small-signal equivalent circuit.

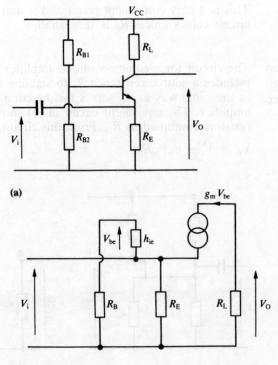

In this case, $g'_m = \dfrac{h_{fe}}{h_{ie} + R_B}$ [5.25]

The remaining results are then given by

$A_v = g'_m R_L$

$Z_{in} = \dfrac{1}{g'_m}$

Example 5.9

The transistor circuit shown in Fig. 5.25 is used as (a) a common emitter amplifier or (b) as a common base amplifier or (c) as a common collector amplifier by taking the inputs and outputs at the appropriate terminals. In each case calculate the voltage gain and input impedance. There are no decoupling capacitors and h_{fe} can be taken as 180.

In all cases, the g_m and h_{ie} will be required and these are obtained from the d.c. conditions.

$V' = \dfrac{20 \times 4.7}{37.7} = 2.5 \text{ V}, \; R' = \dfrac{33 \times 4.7}{37.7} = 4.1 \text{ k}\Omega$

Hence $I_C = \dfrac{2.5 - 0.7}{\dfrac{4.43}{180} + 0.33} = 5.1 \text{ mA}$

Therefore $g_m = 200$, and $h_{ie} = \dfrac{180}{200} = 900 \text{ }\Omega$

(a) Common emitter with R_E.

$A_v = \dfrac{-2.7}{0.33} = -8.2$ since $h_{ie} \ll h_{fe} R_E$

$Z_{in} = 0.9 + 180 \times 0.33 = 60.3 \text{ k}\Omega$

(b) Common base with R_B. R_B in this case is the parallel combination of 33 kΩ and 4.7 kΩ already found as R' of 4.1 kΩ. Find

$g'_m = \dfrac{180}{4.1 + 0.9} = 36 \text{ mA/V}$

Hence $A_v = 36 \times 2.7 = 97.2$

and $Z_{in} = \dfrac{1}{36} = 0.028 \text{ k}\Omega$ or 28 Ω

(Note that without R_B, these figures would have been 540 and 5 Ω.)

(c) Common collector with collector resistance. The presence of collector resistance has no effect upon normal emitter follower performance provided that it is not so large as to cause saturation. The transistor effectively isolates the collector circuit from the base and emitter circuits. Thus,

$A_v = \dfrac{0.33}{0.33 + 1/200} = 0.985$

and $Z_{in} = 0.9 + 180 \times 0.33 = 60.3 \text{ k}\Omega$

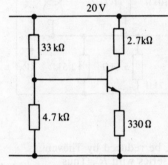

Fig. 5.25 Common base amplifier circuit for example 5.9.

20 V

33 kΩ

2.7 kΩ

4.7 kΩ

330 Ω

Additional single-transistor amplifier arrangements

Two other amplifier arrangements are sometimes used, with additional signal paths between collector and base or between emitter and base. Both are forms of feedback, as are circuits in Figs. 5.22 and 23. These could be analysed using feedback theory, but in this section the analysis is completed by equivalent circuit methods and the resulting properties are demonstrated by means of numerical examples.

Example 5.10   *The circuit shown in Fig. 5.26 is a bootstrapped emitter follower employing a BJT with an h_{fe} of 120. Assuming that capacitive reactances are negligible at signal frequencies, draw an equivalent circuit and determine the voltage gain, the input impedance and, assuming zero source impedance, the output impedance.*

Fig. 5.26 A bootstrapped emitter follower amplifier for example 5.10: (a) full circuit; (b) small-signal equivalent circuit.

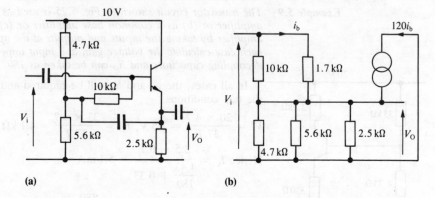

(a)   (b)

The base bias potential divider can, as usual, be reduced by Thévenin's theorem but the effective R_B includes 10 kΩ in series with R'. Thus

$$V' = \frac{10 \times 5.6}{4.7 + 5.6} = 5.4 \text{ V}, \qquad R_B = \frac{4.7 \times 5.6}{4.7 + 5.6} + 10 = 12.6 \text{ k}\Omega$$

and $I_C = \dfrac{5.4 - 0.7}{\dfrac{15.05}{120} + 2.5} \simeq 1.8 \text{ mA}$

therefore $g_m = 72 \text{ mA/V}$ and $h_{ie} = \dfrac{120}{72} = 1.7 \text{ k}\Omega$

The first equivalent circuit is drawn in Fig. 5.26(b). The parallel resistors can be combined and then the circuit is redrawn to give the form shown in Fig. 5.27(a). Writing nodal equations,

$$i_b = \frac{V_i - V_o}{1.7},$$

and $\dfrac{120(V_i - V_o)}{1.7} = V_o\left(\dfrac{1}{1.26} + \dfrac{1}{1.7} + \dfrac{1}{10}\right) - V_i\left(\dfrac{1}{10} + \dfrac{1}{1.7}\right)$

Collecting terms,

$$V_i\left(\frac{120}{1.7} + \frac{1}{10} + \frac{1}{1.7}\right) = V_o\left(\frac{120}{1.7} + \frac{1}{1.26} + \frac{1}{1.7} + \frac{1}{10}\right)$$

or $71.3V_i = 72.1V_o$

therefore voltage gain $A_v = \dfrac{V_o}{V_i} = 0.989$, which is typical for an emitter follower.

Now, $I_i = (V_i - V_o)\left(\dfrac{1}{1.7} + \dfrac{1}{10}\right)$

$= V_i(1 - 0.989)0.688$

therefore $Z_{in} = \dfrac{V_i}{I_i} = \dfrac{1}{0.688(1 - 0.989)} = 132 \text{ k}\Omega$

Fig. 5.27 Equivalent circuits for example 5.10.

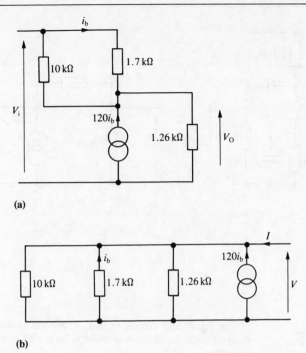

(a)

(b)

This result should be compared with that for the basic emitter follower in example 5.5, where the high input impedance for the transistor is masked by the shunt base bias resistors. The redrawn equivalent circuit for the output impedance calculation is shown in Fig. 5.27(b). Referring to this diagram,

$$i_b = -\frac{V}{1.7}$$

and $I + 120\left(-\dfrac{V}{1.7}\right) = V\left(\dfrac{1}{1.7} + \dfrac{1}{10} + \dfrac{1}{1.26}\right)$

or $I = V\left(\dfrac{121}{1.7} + \dfrac{1}{10} + \dfrac{1}{1.26}\right)$

$$Z_{out} = \frac{1}{\dfrac{121}{1.7} + \dfrac{1}{10} + \dfrac{1}{1.26}} = 0.0138 \text{ k}\Omega \text{ or } 13.8 \ \Omega$$

which is once again the approximate i/g_m for an emitter follower.

Example 5.11   *Determine the properties of the BJT amplifier shown in Fig. 5.28 given that the transistor h_{fe} is 250 and the capacitor reactances and negligible.*

Since the base bias resistor is connected through the collector load to the positive d.c. supply, the equation for determination of I_C is given by

$$12 - 0.7 = 3.3(I_C + I_B) + 100I_B + 1.2(I_C + I_B)$$

or $11.3 = 4.5I_C + \dfrac{104.5}{250}I_C$ and $I_C = 2.3$ mA

therefore $g_m = 90$ mA/V, $h_{ie} = \dfrac{250}{90} = 2.8$ kΩ

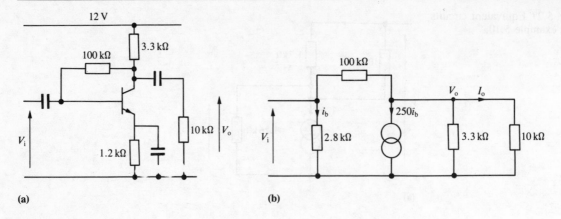

Fig. 5.28 Circuits for example 5.11: (a) full circuit; (b) small-signal equivalent circuit.

The equivalent circuit in Fig. 5.28(b) is in the correct form for nodal analysis with the unknown node voltages V_i and V_o. For the dependent generator,

$$i_b = \frac{V_i}{2.8}$$

for the V_i node,

$$I_i = V_i\left(\frac{1}{2.8} + \frac{1}{100}\right) - V_o\left(\frac{1}{100}\right)$$

and for the V_o node, $-250\dfrac{V_i}{2.8} = -V_i\left(\dfrac{1}{100}\right) + V_o\left(\dfrac{1}{100} + \dfrac{1}{3.3} + \dfrac{1}{10}\right)$

These can be simplified to

$$I_i = 0.37V_i - 0.01V_o$$

$$0 = 89.3V_i + 0.413V_o$$

Solving by determinants,

$$V_i = \frac{0.413V_i - 0}{0.37 \times 0.413 + 0.01 \times 89.3} = 0.39I_i$$

hence $Z_{in} = \dfrac{V_i}{I_i} = 0.39 \text{ k}\Omega$

$$V_o = \frac{-89.3V_i}{0.413}$$

and $A_v = -216$

$$I_o = \frac{-216V_i}{10} = -21.6V_i$$

therefore $A_i = \dfrac{-216V_i}{V_i/0.39} = -84.$

These results show that the circuit has normal voltage gain for a common emitter amplifier but both Z_{in} and A_i are much less than might be expected (typical CE amplifier has $A_i \simeq h_{fe}$ and $Z_{in} \simeq h_{ie}$).

Multi-device amplifier circuits

If the gain of a single-stage transistor amplifier is insufficient for a particular application, it is a simple matter to build a multi-stage amplifier to provide high gain. In practice, such amplifiers have a number of disadvantages when they are contructed as discrete component circuits and the high gain function is best provided by op-amp circuits. There are, however, a few configurations that are worth examination. The first of these is the Darlington connection of two (or more) transistors to provide an effective transistor with different parameters to a single device. The remaining circuits consist of groups of interconnected basic single stages.

Darlington connected transistors

The two transisitors shown in Fig. 5.29(a) are **Darlington connected**: the collectors are in parallel but the base of the second is fed from the emitter of the first.

Fig. 5.29 Darlington connected transistors: (a) connection circuit; (b) small-signal equivalent circuit.

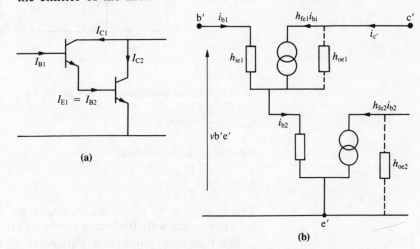

(a)

(b)

The equivalent circuit of the combination in Fig. 5.29(b) is arranged to have three external terminals b′, c′, e′ and the whole may be treated as a 'super' transistor (another name is the super α pair) with its own set of parameters. The h_{oe} components have been included as these will influence parameters which may not be negligible for the combined transistor. Initially, however, we can ignore these while determining h'_{ie} and h'_{fe} the essential h-parameters for the combination.

Writing equations from the circuit

$$i_{b2} = i_{b1} + h_{fe1}i_{b1} = i_{b1}(1 + h_{fe1})$$

$$v''_{be} = v_{be1} + v_{be2} = i_{b1}h_{ie1} + i_{b1}(1 + h_{fe1})h_{ie2}$$

Hence $h'_{ie} = \dfrac{v''_{be}}{i_{b1}} \simeq h_{ie1} + h_{fe1}h_{ie2}$ [5.26]

Also $i'_c = h_{fe1}i_{b1} + h_{fe2}i_{b1}(1 + h_{fe1})$ [5.27]

and $h'_{fe} = \dfrac{i'_c}{i_{b1}} = h_{fe1} + h_{fe2}(1 + h_{fe1}) \simeq h_{fe1}h_{fe2}$ [5.28]

From these results you can see that the current gain is very high and that the input impedance is also very high with h_{ie2} being 'amplified' by h_{fe1}.

The remaining h-parameters are determined with the input open circuit. This eliminates i_{b1} and therefore $h_{fe1}i_{b1}$. The remaining equivalent circuit is shown in Fig. 5.30.

Fig. 5.30 Equivalent circuit for reverse parameter calculation with Darlington connected transistors.

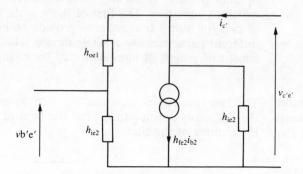

The reverse parameter h'_{re} is given by

$$h'_{re} = \frac{v''_{be}}{v''_{ce}} \text{ when } i_b = 0 \qquad [5.29]$$

Referring to the circuit,

$$h'_{re} = \frac{h_{ie2}}{h_{ie2} + 1/h_{oe1}} \qquad [5.30]$$

This result will be by no means negligible, having a value of perhaps 0.1 (compared with 10^{-5} for a single BJT).

The remaining admittance parameter h'_{oe} is given by

$$h'_{oe} = \frac{i'_c}{v'_{ce}} \text{ when } i'_b = 0.$$

From Fig. 5.30

$$i'_c = v''_{ce}h_{oe2} + h_{fe2}i_{b2} \qquad [5.31]$$

$$\text{but } i_{b2} = \frac{v''_{ce}}{h_{ie2} + 1/h_{oe1}} = \frac{v''_{ce}h_{oe1}}{1 + h_{oe1}h_{ie2}}$$

This result approximates to $v''_{ce}h_{oe1}$ which, when substituted into equation [5.31] results in

$$h'_{oe} = \frac{i'_c}{v'_{ce}} = h_{oe2} + h_{fe2}h_{eo1} \qquad [5.32]$$

This resulting amplification of h_{oe1} represents an output impedance of only a few hundred ohms which limits the usefulness of a Darlington pair for voltage amplification. In practice, Darlington circuits are widely used as current amplifiers in d.c. power supply circuits and the high input impedance property is used in the input stages of op-amp integrated circuits.

Example 5.12   *In the circuit shown in Fig. 5.31, the Darlington connected transistors both have an h_{fe} of 150 and the h_{oe} figures may be taken as 10^{-6} and 10^{-4} S for Tr_1 and Tr_2 respectively. Determine the d.c. conditions and hence the parameters of the combined transistor. With the aid of a small-signal equivalent circuit, determine the gain of the stage.*

Fig. 5.31 Darlington connected amplifier circuit for example 5.12.

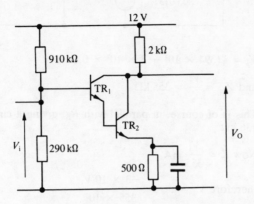

Determination of the d.c. conditions follows the normal pattern, allowing for two V_{BE} volt drops in series with the base circuit and that the ratio of I_{B1} to I_{E2} is h_{fe}^2.

$$V' = \frac{12 \times 290}{1200} = 2.9 \text{ V}, \quad R' = \frac{910 \times 290}{1200} = 220 \text{ k}\Omega.$$

Thus, $I_{C2} = \dfrac{2.9 - 1.4}{\dfrac{220.5}{22500} + 0.5} = 2.9 \text{ mA}.$

and $I_{C1} = \dfrac{2.9}{150} = 20 \ \mu A$

The corresponding values for g_m and h_{ie} are

$g_{m2} = 113.1 \text{ mA/V} \quad g_{m1} = 0.765 \text{ mA/V}$

$h_{ie2} = 1.32 \text{ k}\Omega \quad h_{ie1} = 196 \text{ k}\Omega$

Applying the results of the analysis of the Darlington pair

from equation [5.28], $h_{fe}' = 150^2 = 22\,500$

from equation [5.26] $h_{ie}' = 196 + 150 \times 1.32 = 394 \text{ k}\Omega$

from equation [5.32] $h_{oe}' = 10^{-4} + 150 \times 10^{-6} = 0.25 \text{ mS}$

from equation [5.30] $h_{re}' = \dfrac{132}{132 + 1000} = 0.0013$

The full *h*-parameter equivalent circuit can now be drawn as shown in Fig. 5.32. Referring to the equivalent circuit, the two variables are i_b' and v_{ce}'. From the output circuit,

$$v_{ce}' = -22\,500 i_b' \times \frac{1}{(0.25 + 0.5)10^{-3}} = -3 \times 10^7 i_b$$

In the input circuit,

$$V_i - 0.0013 v_{ce}' = 394 \times 10^3 i_b'$$

rearranging and substituting for v_{ce}',

Fig. 5.32 Small-signal equivalent circuit for example 5.12.

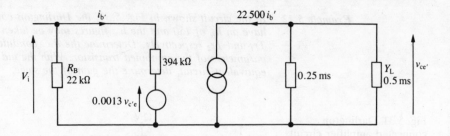

$$V_i = i_b'(394 \times 10^3 - 3 \times 10^7 \times 0.0013)$$

and $Z_{in} = \dfrac{V_i}{i_b'} = 355 \text{ k}\Omega$

This is of course, in parallel with R_B, giving a circuit input impedance of 136 kΩ.

Now $i_b' = \dfrac{V_i}{355}$ mA,

therefore $V_o = v_{ce}' = \dfrac{-3 \times 10^7 V_i}{355 \times 10^3}$

and $A_v = \dfrac{V_o}{V_i} = \dfrac{-3 \times 10^4}{355} = -84.5$

Thus the high effective h_{fe} has not produced a high voltage gain. However, the input impedance is much higher than that found in a typical common emitter stage, and the current gain is given by

$$A_i = -22\,500 \times \dfrac{0.5}{0.75} = -15\,000$$

A cascode amplifier

The common base BJT configuration, with very low input impedance, is rarely useful as a single-stage amplifier but its performance at high frequencies make it useful when it is combined with a common emitter stage in the arrangement known as a **cascode amplifier**. The circuit shown in Fig. 5.33 is a two-stage CE-CB amplifier. Analysis could be achieved using a single

Fig. 5.33 Cascode amplifier circuit for example 5.13.

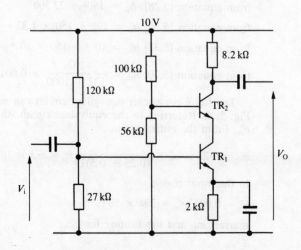

combined equivalent circuit but it is more convenient to use the results for the individual stages as follows. For the CB stage.

$$A_{v2} = +g_m R_L$$

$$Z_{in2} = 1/g_m$$

But this is the load R_{L1} for the CE stage, so the gain of the CE stage

$$A_{v1} = -g_m R_{L1} = \frac{-g_m}{g_m} = -1$$

and the overall voltage gain

$$A_v = A_{v1} \times A_{v2} = -g_m R_L \qquad [5.33]$$

The input impedance is simply h_{ie} for the CE stage. In practice, the results may be modified by undecoupled resistance in the base of the common base stage. This is now illustrated by a numerical example.

Example 5.13   *Determine the Q point for the two transistors and the overall voltage gain for the circuit shown in Fig. 5.33. The h_{fe} for both transistors can be taken as 200.*

The d.c. analysis for Tr_1 follows the usual procedure:

$$V' = \frac{10 \times 27}{147} = 1.84 \text{ V}, \quad R' = \frac{120 \times 27}{147} = 22 \text{ k}\Omega$$

$$I_{C1} = \frac{1.14}{\dfrac{24}{200} + 2} = 0.54 \text{ mA}$$

From the circuit, $I_{C2} = I_{C1} = 0.54 \text{ mA}$.

To find the Q points, V_{B2} can be found from the equivalent circuit at the base of Tr_2 together with I_{B2} from $I_{C/h_{fe}}$.

For Tr_2,

$$V' = \frac{10 \times 56}{156} = 3.6 \text{ V}, \quad R' = \frac{100 \times 56}{156} = 36 \text{ k}\Omega$$

The d.c. equation for this circuit if now given by

$$V_{B2} = 3.6 - I_{B2} \times 36 = 3.6 - \frac{0.54}{200} \times 36 = 3.5 \text{ V}$$

Hence $V_{E2} = V_{C1} = 3.5 - 0.7 = 2.8 \text{ V}$

then, $V_{CE1} = 2.8 - 0.54 \times 2 = 1.72 \text{ V}$

$$V_{CE2} = 10 - 0.54 \times 8.2 - 2.8 = 2.77 \text{ V}$$

Thus, both transistors are operating correctly in the linear mode with the same collector current.

$$g_{m1} = g_{m2} = 0.54 \times 39 = 21 \text{ mA/V}$$

$$h_{ie1} = h_{ie2} = \frac{200}{21} = 9.5 \text{ k}\Omega$$

Tr_2 is common base with an R_B of 36 kΩ in the base lead. The effective g_m for Tr_2 is now given by

$$g_m' = \frac{h_{fe}}{h_{ie} + R_B} = \frac{200}{36 + 9.5} = 4.4 \text{ mA/V}$$

The voltage gain $A_{v2} = 4.4 \times 8.2 = 36$.

The input impedance $Z_{in2} = 1/4.4 = 0.227$, which is also the load for Tr_1.

$A_{v1} = -21 \times 0.227 = -4.8$

Overall gain $= A_{v1} \times A_{v2} = -4.8 \times 36 = -173$.

Had the base of Tr_2 been decoupled, equation [5.33] would have applied, and this gives the same result.

An emitter coupled amplifier

Operational amplifiers, to be discussed in the next section, have differential inputs or, in other words, the function of the circuit is specified by

$V_o = A_d(V_1 - V_2)$ where A_d is the differential gain.

The reasons for this are discussed elsewhere (p.487–90) but the implementation of this function is based on the two-device circuit known (for BJTs) as an **emitter coupled** or **long-tailed pair amplifier**. In practice, the internal input stage of op-amps is always based on this arrangement.

A circuit is shown in Fig. 5.34. The differential output V_o is taken between the two collectors as $V_{o1} - V_{o2}$. Since this is a linear circuit, it can be analysed by use of the superposition principle (p.00), finding V_o separately due to e_1 and e_2 in turn and then adding the result. Alternatively, we can let e_2 be zero and find

$$A_d = \left(\frac{V_{o1} - V_{o2}}{e_1} \right) \qquad [5.34]$$

Fig. 5.34 An emitter coupled amplifier.

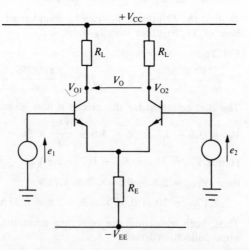

Under these circumstances, Tr_2 is a common base stage with input at the emitter. Tr_1 is both common collector with output at the emitter loaded by Z_{in2} and common emitter with V_{o1} at the collector. Thus for Tr_2,

$$V_{o2} = g_{m2}R_L V_E$$

and $Z_{in2} = 1/g_{m2}$

The emitter load on Tr_1 is Z_{in2} in parallel with R_E. In practice, $R_E \gg 1/g_{m2}$ and can be neglected. Therefore

$$V_E = e_1 \times \frac{1/g_{m2}}{1/g_{m1} + 1/g_{m2}}$$

The two transistors will be identical (very nearly) with equal collector currents and g_m values. Therefore

$$V_E = e_1 \times 0.5$$

and $V_{o2} = 0.5 g_{m2} R_L e_1$

For V_{o1}, we have a common emitter stage with an emitter resistance of $1/g_{m2}$. From equation [5.23]

$$V_{o1} = \frac{-h_{fe} R_L e_1}{h_{ie} + h_{fe}/g_{m2}}$$

but $h_{fe}/g_m = h_{ie}$

therefore

$$V_{o1} = \frac{-h_{fe} R_L e_1}{h_{ie} + h_{ie}} = -0.5 g_m R_1 e_1$$

The resulting differential output voltage V_o is

$$V_o = V_{o1} - V_{o2} = -0.5 g_m R_L e_1 - 0.5 g_m R_L e_1$$

$$= -g_m R_L e_1$$

Thus the voltage gain is the same as that for a single-stage common emitter amplifier. The input at the base of Tr_1 is given by equation [5.22]

$$Z_{in} = h_{ie} + 1/g_m \times h_{fe} = 2h_{ie} \qquad [5.35]$$

The above analysis would apply equally to inputs at e_2, although in this case, the differential output in the direction shown in Fig. 5.34 would be

$$V_o = +g_m R_L e_2$$

or by superposition, for both inputs together,

$$V_o = g_m R_L(e_2 - e_1) \qquad [5.36]$$

Example 5.14 *Determine V_o for the differential amplifier shown in Fig. 5.35. The input signals shown are a.c. with no d.c. component. The two transistors are identical.*

Fig. 5.35 Emitter coupled amplifier circuit for example 5.14.

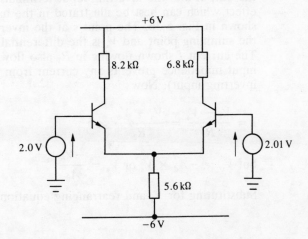

The only parameter required for the small-signal analysis is g_m and this is found from the d.c. conditions. Referring to the circuit,

$$V_E = -0.7\,V = -6 + 2I_C \times 5.6$$

Hence, $I_C = \dfrac{5.3}{2. \times 5.6} = 0.47\,mA$

$$g_{m1} = g_{m2} = 0.47 \times 40 = 18.5\,mA/V$$

V_{o1} due to $e_1 = \dfrac{-18.5 \times 8.2}{2} e_1$

V_{o1} due to $e_2 = \dfrac{+18.5 \times 8.2}{2} e_2$

$$V_{o1} = \frac{-18.5 \times 8.2}{2}(e_1 - e_2) = -75.9(e_1 - e_2)$$

Similarly, $V_{o2} = \dfrac{-18.5 \times 6.8}{2}(e_2 - e_1) = 62.9(e_1 - e_2)$

The differential output is thus given by

$$V_o = V_{o1} - V_{o2} = (-75.9 - 62.9)(e_1 - e_2)$$
$$= -138.8(e_1 - e_2)$$

Substituting values for e_1 and e_2,

$$V_o = -138.8(2.0 - 2.01) = 1.388\,V$$

You should now be able to attempt exercises 5.3 to 5.8.

Amplifiers using op-amps

The properties and equivalent circuit for the op-amp were discussed in Chapter 2. The important features to be included at this stage are a very high differential gain A_d, and input and output impedances which may be taken as infinity and zero respectively. For all amplifying applications, there are components connected between the output and the inverting input terminals. This has an important effect which can best be illustrated in the basic inverting amplifier shown in Fig. 5.36. The point s at the inverting input is known as the summing point and V_s is the differential input voltage ($v_- - v_+$). The current I shown flowing in R_1 also flows in R_2 (since the high input impedance prevents any current from flowing into the inverting input). Now

$$I = \frac{V_i - V_s}{R_1} = \frac{V_s - V_o}{R_2} \qquad [5.37]$$

but $V_o = -A_d \times V_s$ or $V_s = \dfrac{-V_o}{A_d}$ $\qquad [5.38]$

Substituting for V_s and rearranging equation [5.37]

Fig. 5.36 An op-amp
inverting amplifier.

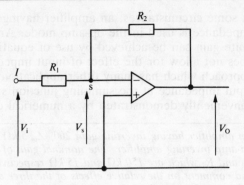

$$\frac{R_2}{R_1}\left(V_i + \frac{V_o}{A_d}\right) = -\left(V_o + \frac{V_o}{A_d}\right)$$

collecting terms, $\dfrac{R_2}{R_1}V_1 = -V_o\left(1 + \dfrac{1}{A_d} + \dfrac{R_2}{R_1 A_d}\right)$

Hence, voltage gain $A_v = \dfrac{V_o}{V_i} = \dfrac{-R_2/R_1}{1 + 1/A_d + \dfrac{R_2}{R_1} \times 1/A_d}$ [5.39]

In this last result, we can now remember that A_d is very large (typically 10^5) and thus the terms involving A_d in the denominator are negligible. The final result, therefore, is

$$A_v = \frac{-R_2}{R_1}$$ [5.40]

An alternative and very useful way of examining this result is as follows: if there is an output voltage V_o (within the range of the d.c. supplies), the terminal or summing junction input must be very small (from equation [5.38]). As A_d tends to infinity, this voltage V_s tends to zero and the point s is said to be a **virtual earth**. This is because it is at earth potential although no current flows to earth. In more general terms, whenever a circuit is connected between the output and the inverting input, the inverting and non-inverting inputs are held at virtually the same potential. In the inverting amplifier in Fig. 5.36, this concept allows a much simpler analysis as follows:

Since s is at 0 V, $I = V_i/R_1$
The voltage V_2 across R_2 is given by

$$V_2 = IR_2 = V_i\frac{R_2}{R_1}$$

But $V_o = -V_2$ (from the virtual earth to the output terminal)

therefore $V_o = -V_1 R_2/R_1$ and gain $A_v = \dfrac{-R_2}{R_1}$ [5.41]

This result is true provided that the open-loop gain A_d is very large.

<table>
<tr><td>

The inverting amplifier
with finite A_d and Z_{in}

</td><td>

In some circumstances, an amplifier having finite gain and input impedance is used in the op-amp mode. Analysis of the effect of finite gain can be achieved by use of equation [5.39] above but this does not allow for the effect of input impedance. A more direct approach which has many other applications starts from the total input impedance at the summing junction s. This can be conveniently demonstrated by a numerical example.

</td></tr>
</table>

Example 5.15
 An amplifier having inverting gain 200, Z_{in} 5 kΩ and Z_{out} 2 kΩ is used as an op-amp inverting amplifier. The nominal gain of 10 arises from the resistors R_2 and R_1 which are 150 kΩ and 15 kΩ respectively. Calculate the actual gain and comment on the relative effects of the three amplifier properties.

The required circuit is shown in Fig. 5.37. The current I is the sum of two currents,

$$I_i = V'/5 \text{ and } I_f = \frac{V' - (-200\,V')}{150 + 2}$$

therefore $I = V'\left(\dfrac{1}{5} + \dfrac{210}{152}\right)$ mA

and the input impedance Z'_{in} at s is given by

$$Z'_{in} = \frac{V'}{I} = \frac{1}{0.2 + 1.32} = 0.657 \text{ k}\Omega$$

(Note that this is the virtual earth point for an ideal amplifier.) Now, V' can be found by potential divider analysis (p.00).

$$V' = V_i \times \frac{0.657}{15.657} = 0.042 V_i$$

The open circuit output voltage would be this result multiplied by -200 but there is a small volt drop across the 2 kΩ output impedance. Referring to the circuit diagram,

$$V_o = -200 V' + 2I_f$$

$$= -200 V' + V'\frac{(1 + 200)2}{152} = -198 V' \qquad\qquad [5.42]$$

$$V_o = -198 \times 0.042 V_i$$

and $A_v = \dfrac{V_o}{V_i} = -8.28$

This result is appreciably smaller than the nominal gain value of 10 and the effects of the amplifier parameters may be considered separately.

Fig. 5.37 Inverting amplifier
circuit for example 5.15.

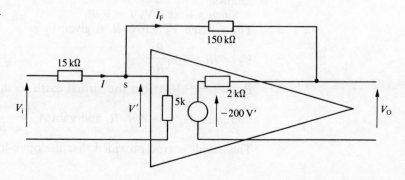

The output impedance of 2 kΩ has only the minor effect seen in equation [5.42], reducing the effective internal gain from 200 to 198 and in the expression for I_f when the feedback resistor is increased from 150 kΩ to 152 kΩ.

The finite gain results from an input impedance of 152/201 or 0.756 kΩ. If the amplifier Z_{in} was very large, the resulting system gain would be $0.756/15.756 \times 200$ or 9.6 but the shunting effect of the 5 kΩ Z_{in} causes the further reduction to the calculated 8.28. These effects should not really be considered separately as a larger internal gain would reduce the effect of Z_{in}. A gain of 2000, for example, results in a system gain of 9.8 compared with a figure of 9.95 resulting from a Z_{in} much greater than 5 kΩ.

The non-inverting amplifier

The circuit for the basic non-inverting amplifier is shown in Fig. 5.38. Notice that R_2 (between output and inverting input) again provides the conditions for zero voltage between v_+ and v_-. Since no current is taken by the inverting input to the op-amp,

$$v_- = V_o \times \frac{R_1}{R_1 + R_2}$$

but $V_i = v_+ = v_-$

therefore $V_o = V_i \frac{(R_1 + R_2)}{R_1}$

or $A_v = \frac{V_o}{V_i} = \left(1 + \frac{R_2}{R_1}\right)$ [5.43]

This result has no minus and is therefore non-inverting and the gain has a minimum value of 1.

Fig. 5.38 A non-inverting amplifier.

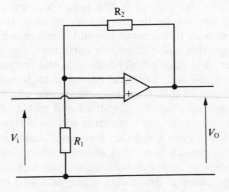

The voltage follower

The voltage follower is the op-amp equivalent to the BJT emitter follower and is a simplified version of the non-inverting amplifier in the last section. The circuit diagram is shown in Fig. 5.39 and can be regarded as an extreme case of Fig. 5.38 where R_2 is zero and R_1 is infinite. Alternatively

$$V_i = v_+ = v_- = V_o$$

therefore $V_o/V_i = 1$, a non-inverting gain of 1.

Fig. 5.39 A voltage
follower amplifier.

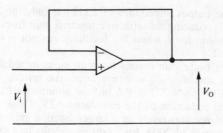

The input impedance is that of the op-amp itself and will be very
high, and the output impedance very low, for the op-amp is further
reduced by feedback making this circuit an ideal impedance
matching buffer.

A further and wider study of op-amp circuits is provided later in
this chapter and the concepts introduced in these basic amplifier
circuits are applied and extended in these sections.

You should now be able to attempt exercises 5.9 to 5.10.

FREQUENCY DISTORTION IN AMPLIFIERS

In the previous sections in this chapter, amplifier properties were
determined with the assumption that any frequency-sensitive
components can be neglected (usually as approximate short circuits).
These assumptions are usually valid for signals in the operating
frequency range. For direct coupled amplifiers, this would be from
d.c. to perhaps several hundred kilohertz while for capacitor coupled
circuits the bottom of the range may be about 100 Hz and in some
case the top of the range may be several megahertz. If signals are
used with frequencies beyond these mid-band ranges, the amplifier
properties will change. In particular, the gain and phase shift will
alter, the gain usually falling as the frequency departs further from
the mid-band. Practical analogue information often consists of a
mixture of different frequencies and amplitudes. Since some
frequencies can be amplified and phase shifted more than others,
the circuit output waveform will be different from that at the input.
This effect is known as **frequency distortion**.

Frequency distortion at low frequencies is usually due to
capacitors in series with signal paths, which have an increased
impedance at low frequencies, and also to decoupling capacitors
(particularly those in the emitter or source circuit) which fail to act
as effective short circuits under low frequency signal conditions. Low
frequency distortion is usually only applicable in capacitor coupled
amplifiers.

High frequency distortion, which applies to all amplifiers, is the
result of shunt capacitance which, as impedance reduces with
frequency, reduces the effective load impedance, and with it the
amplifier gain. The shunt capacitance is usually a combination of
input capacitance to devices and stray capacitance between the
circuit wires and to earth. In some very high frequency circuits,
series lead inductance can also contribute to the gain reduction.

In this section, three basic capacitance effects are analysed in detail using the techniques described in Chapter 3. The high frequency performance of the transistor is then examined with reference to the hybrid π and y-parameter equivalent circuits. Finally, the introduction of frequency sensitive components to op-amp circuits is investigated.

Shunt capacitance loading of amplifiers

The output of any of the amplifier circuits described in the section on small-signal circuit analysis (p.109) can be represented by the Norton equivalent of a current generator feeding the effective load resistance. In the case of common emitter and common source stages, the generator is simply $-g_m V_{in}$, for common base it is $+g_m V_{in}$, while for the common collector and common drain it is again $+g_m V_{in}$ but the effective load is $1/g_m$. In every case, there will be a capacitance in parallel with this arrangement. This may be simply the stray capacitance between wiring and components and earth or it may be the input capacitance to the next stage, which is discussed later in this section. The equivalent circuit including this capacitance is shown in Fig. 5.40. From the circuit,

$$V_o = \pm g_m V_{in} Z$$

where Z is the parallel combination of R and X_c

$$A_v = \frac{V_o}{V_{in}} = \frac{\pm g_m R/j\omega C}{R + 1/j\omega C}$$

Fig. 5.40 Shunt capacitance loading of an amplifier.

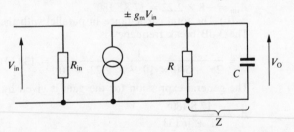

Multiplying numerator and denominator by $j\omega C$ and rearranging,

$$A_v = \frac{\pm g_m R}{1 + j\omega CR}$$

$$= \frac{A_{vm}}{1 + j\omega/\omega_1}$$

$$\text{or} \quad \frac{A_{vm}}{1 + jf/f_1} \tag{5.44}$$

where A_{vm} is the mid-band gain and $\omega_1 = 1/CR$ the 3 dB or break frequency. The Bode frequency response for this circuit is shown in Fig. 5.41 (see p.128). Thus as signal frequency increases above ω_1, the voltage gain falls at a constant rate of 20 dB/decade and the phase shift relative to the mid-band value increases to $-90°$ at frequencies much greater than ω_1.

Fig. 5.41 The Bode
frequency response for
shunt capacitance loading.

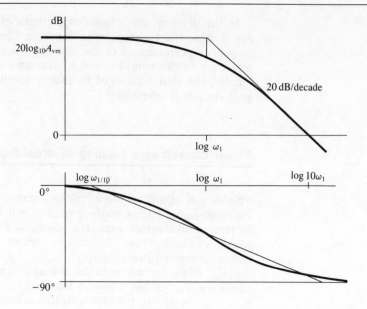

Example 5.16

A common source amplifier has a drain load of 5 kΩ and is coupled to an external load of 4 kΩ in parallel with 500 pF capacitance. Stray capacitance and the FET output capacitance account for another 40 pF. The g_m of the FET can be taken as 8 mA/V. Determine (a) the mid-band gain , (b) the gain when the signal frequency is 100 kHz, (c) the frequencies when the gain has fallen to half the mid-band value and to unity, and (d) the phase shift at these frequencies.

(a) For the mid-band calculation, the total load is the parallel combination $(5 \times 4)/9$ or 2.22 kΩ. This gain is therefore $A_{vm} = -8 \times 2.22 = 17.8\angle 180°$.

(b) The total capacitance in parallel with the load is $500 + 40 = 540$ pF. The 3 dB break frequency is

$$f_1 = \frac{1}{2\pi \times 540 \times 10^{-12} \times 2.22 \times 10^3} = 133 \text{ kHz}$$

The general expression for the gain is given by [5.44]

$$A_v = \frac{17.8/180°}{1 + \mathrm{j}f/133}$$

where the frequency is in kHz. Therefore, when f = 100 kHz,

$$A_v = \frac{\dfrac{17.8}{180}}{1 + \mathrm{j}\dfrac{100}{133}} = \frac{17.8/180}{1.25/37} = 14.2\angle -217°$$

(c) For this part, the modulus of the gain is required, so from the general expression for the gain, $|A_v| = \dfrac{17.8}{|1 + \mathrm{j}f/133|}$, which, for half gain, $= 17.8/2$

therefore $2 = |1 + \mathrm{j}f/133|$

and $4 = 1 + (f/133)^2$

so $f^2 = 133^2 \times 3$ and $f = 133/3 = 230$ kHz

Similarly, for unity gain,

$$\left|\frac{17.8}{1 + jf/133}\right| = 1$$

or $1 + (f/133)^2 = 17.8^2$

and $f = 133/17.8^2 - 1 = 2.36$ MHz

(d) The phase angle at the half gain frequency is obtained from

$$A_v = \frac{17.8 \,\underline{/180}}{1 + \dfrac{j230}{133}}$$

$$\underline{/A_v} = 180 - \tan^{-1}\left(\frac{230}{133}\right) = -240°$$

and at the unity gain frequency,

$$\underline{/A_v} = 180 - \tan^{-1}\left(\frac{2360}{133}\right) = -267°$$

These angles could have been given as 120° and 93° respectively. It is more convenient to express them as a further lag on top of the 180° due to the amplifier inversion which can be taken as plus or minus 180°. The effect will then be seen to be of the same form for non-inverting amplifiers, or to contrast with other circuits which give a leading phase angle.

Amplifiers with series capacitance

Capacitors are used in series with both input and output of amplifiers to avoid d.c. mismatch. The effect on the circuit frequency response in each case is the same. Figure 5.42 shows the equivalent circuit for a common emitter amplifier with signal source and a.c. coupled load.

Fig. 5.42 Series capacitance loading of an amplifier: (a) complete small-signal equivalent circuit; (b) modified equivalent circuit for the output circuit.

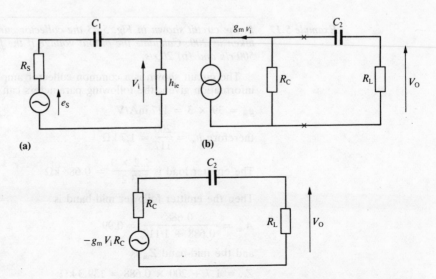

The series form of the input circuit is more convenient for this analysis (since C is in series with the final R in both circuits) so the output circuit is converted, by Thévenin's theorem, to the alternative equivalent shown in Fig. 5.42(b).

Analysing the output circuit by the potential divider relationship

$$V_o = -g_m V_{in} R_C \times \frac{R_L}{R_C + R_L + 1/j\omega C_2} \qquad [5.45]$$

Dividing numerator and denominator by $(R_C + R_L)$

$$A_v = \frac{-g_m \dfrac{R_C R_L}{R_C + R_L}}{1 + \dfrac{1}{j\omega C_2 (R_C + R_L)}}$$

or $A_v = \dfrac{A_{vm}}{1 - j\dfrac{\omega_1}{\omega}}$ \qquad [5.46]

since $\dfrac{R_C R_L}{R_C + R_L}$ is the total mid-band load and ω_1 is the frequency given by

$$\omega_1 = \frac{1}{C_2(R_C + R_L)} \qquad [5.47]$$

This result can be expressed in more general terms which can be applied to the input circuit as well. The numerator is the circuit gain when the series capacitor is disregarded. The frequency ω_1 is that due to the total series C and R. Thus for the input circuit,

$$A_v = \frac{\dfrac{h_{ie}}{R_S + h_{ie}}}{1 - j\dfrac{\omega_1}{\omega}} \qquad [5.48]$$

where $\omega_1 = 1/C_1(R_S + h_{ie})$

Example 5.17   *In the circuit shown in Fig. 5.43 the collector current is 3 mA and h_{fe} can be taken as 200. Calculate the output voltage if the frequency of the signal is (a) 500 r/s and (b) 20 r/s.*

The circuit shown is a common collector amplifier and, from the information given, the following parameters can be found.

$g_m = 39 \times 3 = 117 \text{ mA/V}$

therefore $h_{ie} = \dfrac{200}{117} = 1.7 \text{ k}\Omega$

The emitter load is $\dfrac{2.2 \times 1}{3.2} = 0.688 \text{ k}\Omega$

Then the emitter follower mid-band is

$A_{vm} = \dfrac{0.688}{0.688 + 1/117} = 0.99$

and the mid-band Z_{in} is

$Z_{in} = 1.7 + 200 \times 0.688 = 139.3 \text{ k}\Omega$

Fig. 5.43 Series capacitance
coupled transistor circuit
for example 5.17.

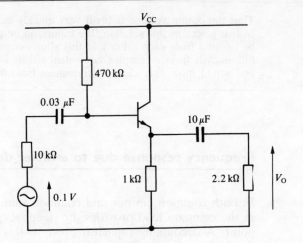

But this is parallel with 470 kΩ, therefore

$$\text{total mid-band } Z_{\text{in}} = \frac{139.3 \times 470}{139.3 + 470} = 107 \text{ k}\Omega$$

For the input circuit, the mid-band gain is given by

$$A_{\text{v}} = \frac{107}{107 + 10} = 0.91$$

Thus the overall mid-band gain $A_{\text{vm}} = 0.91 \times 0.99 = 0.9$.

The load circuit break frequency (equation [5.47]) is given by

$$\omega_2 = \frac{1}{10^{-5}(2200 + 58.5)} = 44.3 \text{ r/s}$$

The value of 58.5 here is $1/g'_{\text{m}}$, allowing for the 10 kΩ source. This would
be modified at low frequencies due to the impedance of C_2. This, however,
would only have a minor effect upon ω_2.

For the input circuit, the break frequency is given by

$$\omega_1 = \frac{1}{0.03 \times 10^{-6} \times (10 + 139.3) \times 10^3} = 223 \text{ r/s}$$

This result too is approximate since, at low frequencies, the $H_{\text{fe}}R_{\text{E}}$ term will
be increased and tend to 200×1 kΩ. In this case, the 10 μF capacitor
represents only 450 Ω and this has little effect upon the result for ω_1.

The overall low frequency gain is given by

$$A_{\text{v}} = \frac{0.9}{\left(1 - j\dfrac{44.3}{\omega}\right)\left(1 - j\dfrac{223}{\omega}\right)}$$

(a) At $\omega = 500$ r/s,

$$V_{\text{o}} = \frac{0.1 \times 0.9}{\left(1 - j\dfrac{44.3}{500}\right)\left(1 - j\dfrac{223}{500}\right)} = 0.082\angle +29° \text{ V}$$

(b) At $\omega = 20$ r/s,

$$V_{\text{o}} = \frac{0.1 \times 0.9}{\left(1 - j\dfrac{44.3}{20}\right)\left(1 - j\dfrac{223}{20}\right)} = 0.0033\angle +151° \text{ V}$$

Thus the output voltage falls off very quickly as signal frequency is reduced. In two places in this solution, the input and output circuits are assumed to be isolated from each other and this allows equation [5.46] to be used. A full analysis from a complex equivalent circuit would prove very complicated and would show only slight discrepancies from the above result.

Frequency response due to emitter decoupling

In both common emitter and common source amplifiers, resistance in the common lead provides d.c. feedback, stabilising the operating point. A decoupling capacitor prevents this feedback from reducing the mid-band gain. This analysis is necessary to see what order of capacitance is required for a particular gain frequency specification.

Figure 5.44 shows an equivalent circuit for the essential parts of the amplifier. The source resistance R_s is in series with h_{ie} and must be included in this analysis (as for the output impedance analysis of the common collector amplifier). This is included by using the modified g_m' given by

$$g_m' = \frac{h_{fe}}{h_{ie} + R_S} = \frac{h_{fe}}{R} \qquad [5.49]$$

where $R = R_S + h_{ie}$.

From the circuit, if $h_{fe} \gg 1$,

$$e = V + \frac{g_m'V}{Y} = V\left(1 + \frac{g_m'}{Y}\right)$$

where Y is the combined emitter circuit admittance and

$$V_o = -g_m'R_LV$$

The voltage gain, $\dfrac{V_o}{e} = \dfrac{-g_m'R_L}{1 + g_m'/Y}$ $\qquad [5.50]$

but $Y = \dfrac{1}{R_E} + j\omega C_E$

Substituting into equation [5.50]

$$\frac{V_o}{e} = \frac{-g_m'R_L}{1 + \dfrac{g_m'}{1/R_E + j\omega C_E}}$$

$$= \frac{-g_m'R_L(1/R_E + j\omega C_E)}{1/R_E + g_m' + j\omega C_E} \qquad [5.51]$$

For typical BJT circuits, $1/R_E \ll g_m'$.

Rearranging equation [5.51] and substituting for g_m' from equation [5.49]

$$\frac{V_o}{e} = \frac{\dfrac{-h_{fe}R_L}{R\,R_E}(1 + j\omega C_E R_E)}{\dfrac{h_{fe}}{R}\left(1 + j\omega\dfrac{C_E R}{h_{fe}}\right)}$$

Fig. 5.44 Equivalent circuit for amplifier with emitter decoupling capacitor and emitter resistance.

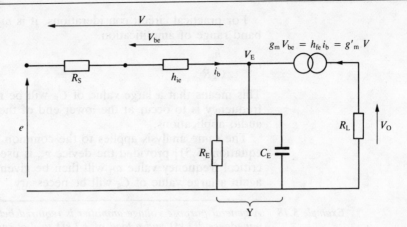

$$= \frac{-R_L}{R_E} \frac{(1 + j\omega/\omega_1)}{(1 + j\omega/\omega_2)} \qquad [5.52]$$

The gain term, $-R_L/R_E$, is the approximate result for the common emitter amplifier with emitter resistance equation [5.24]. ω_1 is $1/C_E R_E$ and is determined by the emitter circuit. ω_2 however is C_E combined with the resistance seen at the emitter looking back into the transistor

$$\left(\frac{h_{ie} + R_S}{h_{fe}}\right).$$

At high frequencies, ω/ω_1 and $\omega/\omega_2 \gg 1$; then

$$\frac{V_o}{e} = \frac{-R_L}{R_E} \times \frac{C_E R_E}{C_E R/h_{fe}}$$

$$= \frac{-h_{fe} R_L}{R} = -g'_m R_L \qquad [5.53]$$

which is the familiar mid-band gain under these circumstances.

Combining equations [5.52] and [5.53] in a Bode gain frequency response results in the graph shown in Fig. 5.45.

Fig. 5.45 Bode gain frequency response for amplifier with emitter decoupling.

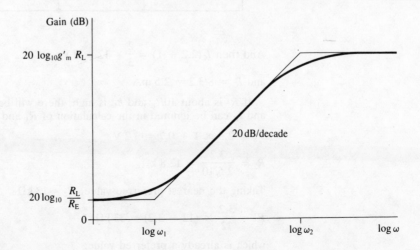

For practical circuit considerations, it is ω_2 that limits the midband range of amplification.

$$\omega_2 = \frac{h_{fe}}{C_E R} \qquad\qquad [5.54]$$

This means that a large value of C_E will be required if this break frequency is to occur at the lower end of the frequency range for audio applications.

The same analysis applies to the common source amplifier up to equation [5.51] provided the device g_m is used instead of g'_m. The critical frequency value ω_2 will then be given by $\omega_2 = g_m/C_S$ and again a large value of C_S will be necessary.

Example 5.18   *A general purpose voltage amplifier is required between a signal source of impedance 2.2 kΩ and a load of 2.2 kΩ to d.c. earth. Design a common emitter amplifier to operate from a 12 V supply using a transistor with a nominal h_{fe} of 250. The required lower 3 dB break frequency is 150 Hz.*

A circuit for the amplifier is shown in Fig. 5.46 with components labelled for identification. The starting point is the choice of collector load R_C and transistor Q point. A high R_C tends to increase the voltage gain but also necessitates a lower I_C which reduces the gain. A compromise choice is an R_C of 2.2 kΩ which will also provide maximum power transfer to the external load. A further guideline is to make V_{CE} about $V_{CC}/3$ and R_L about $2R_E$. Thus choose $R_E = 1$ kΩ.

Fig. 5.46 Common emitter amplifier circuit for example 5.18.

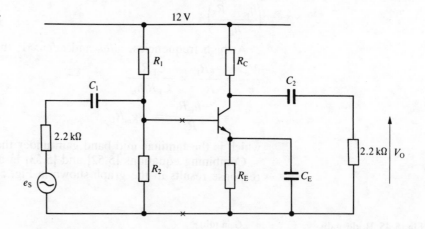

And then $I_C(2.2 + 1) = \frac{2}{3} \times 12 = 8$ V

and $I_C = 8/3.2 = 2.5$ mA

If R_2 is about $10R_E$ and h_{fe} is high, there will be adequate d.c. stability and I_B can be ignored in the calculation of R_1 and R_2.

$V_B = 2.5 \times 1 + 0.7 = 3.2$ V

$R_2 = \dfrac{3.2}{2.5/10} = 12.8$ kΩ

Taking the nearest preferred value, $R_2 = 12$ kΩ.

$R_1 = \dfrac{3.2}{12} \times (12 - 3.2) = 33$ kΩ

which is already a preferred value.

To investigate the performance and choose the capacitor, further information is required for the transistor.

$$g_m = 2.5 \times 39 = 97.5 \text{ mA/V}, \quad h_{ie} = \frac{250}{97.5} = 2.6 \text{ k}\Omega$$

It is convenient to work in terms of g_m' (see last section) for which we require the effective source components e_s' and R'.

$$R_B = \frac{R_1 R_2}{R_1 + R_2} = 8.8 \text{ k}\Omega$$

$$e_s' = e_s \times \frac{8.8}{11} = 0.8 e_s.$$

$$R' = \frac{2.2 \times 8.8}{11} = 1.76 \text{ k}\Omega$$

Now, $g_m' = \dfrac{h_{fe}}{h_{ie} + R'} = \dfrac{250}{1.76 + 2.6} = 57 \text{ mA/V}$

The mid-band gain is given by

$$A_{vm} = -g_m' R_L$$

where R_L is the two 2.2 kΩ resistors in parallel. Therefore

$$A_{vm} = -57 \times 11 = -63$$

or in terms of e_s rather than e_s',

$$A_{vm} = -63 \times 0.8 = -50.4$$

The first capacitor to choose is C_E with reference to [5.54]. The required 3 dB frequency ω_2 is given by

$$\omega_2 = 2\pi \times 150 = 942 \text{ r/s} = \frac{h_{fe}}{C_E R} = \frac{g_m'}{C_E}$$

Hence, $C_E = \dfrac{57 \times 10^{-3}}{942} = 60 \ \mu\text{F}$

The corresponding ω_1 is 16.6 r/s or 2.6 Hz and is of no interest in this problem. The remaining series capacitors are chosen to provide 'breaks' at frequencies well below the first break, say at $\omega = 100$ r/s. Applying equation [5.47] and [5.48] capacitor values of about 2 μF will be found to be adequate.

BJT performance at high frequencies

The simplified h-parameter equivalent circuit does not account for the observable high frequency behaviour. The more general hybrid π circuit is used to explain these effects and to define the parameter f_T. This circuit is described in Chapter 2 and shown in Fig. 2.25. Figure 5.47 shows a loaded BJT and the corresponding equivalent circuit.

There are two high frequency effects which may be analysed separately: (a) The capacitor C_{be}' may be determined in terms of the current-dependent g_m and f_T a basic parameter of the BJT; (b) the feedback Miller effect due to C_{bc}' and the loaded voltage gain. These combine to give an input capacitance resulting in a dominant break frequency on the input circuit (loading the preceding stage) at a

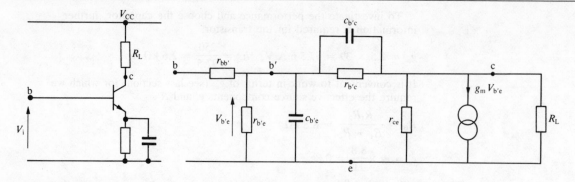

Fig. 5.47 A common emitter amplifier and its hybrid π equivalent circuit.

frequency much lower than any break due to the output circuit.

Consider the approximate equivalent circuit when the output is short circuit to earth so that the high frequency short-circuit current gain can be found. This is shown in Fig. 5.48.

Fig. 5.48 Approximate hybrid π equivalent circuit with output short circuit.

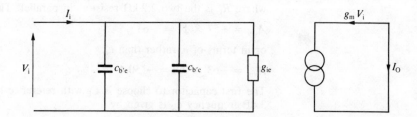

Since the output is short circuit to earth, c'_{bc} and r'_{bc} are in parallel with r'_{be} and c'_{be}. r'_{bc} is very large and can be neglected and r'_{bb} is small and can also be neglected (if required it could be included with the output impedance of the previous stage). r_{ce} and R_L are short circuited.

The input admittance Y_{in} is given by

$$Y_{in} = \frac{I_i}{V_i} = g_{ie} + j\omega(c'_{be} + c'_{bc})$$

The short circuit current gain is given by

$$h_{fe} = \frac{g_m V_i}{I_i} = \frac{g_m}{Y_{in}} = \frac{g_m}{g_{ie} + j\omega(c'_{be} + c'_{bc})} \qquad [5.55]$$

Now h_{fe} will fall with increased frequency and will have fallen by 3 dB when the real and imaginary parts of equation [5.55] are equal. The frequency at which this occurs can be defined as ω_{hfe} and found from

$$g_{ie} = \omega(c'_{be} + c'_{bc})$$

$$\text{or } \omega = \frac{g_{ie}}{c'_{be} + c'_{bc}} = \frac{1}{h_{ie}(c'_{be} + c'_{bc})}$$

For practical BJT values, $c'_{bc} \ll c'_{be}$, therefore

$$\omega_{hfe} \simeq \frac{1}{h_{ie}c'_{be}} \qquad [5.56]$$

or $c'_{be} \simeq \dfrac{1}{\omega_{hfe} h_{ie}}$ [5.57]

In general, $h_{fe} = \dfrac{h_{feo}}{1 + j\omega/\omega_{hfe}}$

where $h_{feo} = g_m h_{ie}$

and $|h_{fe}|^2 = \dfrac{|h_{feo}|^2}{1 + \left(\dfrac{\omega}{\omega_{hfe}}\right)^2}$

The transisition frequency ω_T is defined as the frequency at which the short-circuit current gain h_{fe} has fallen to unity. There at $\omega = \omega_T$,

$$1 = \dfrac{|h_{feo}|^2}{1 + \left(\dfrac{\omega_T}{\omega_{hfe}}\right)^2}$$

and as $1 \ll \left(\dfrac{\omega_T}{\omega_{hfe}}\right)^2$, $\quad h_{feo} = \dfrac{\omega_T}{\omega_{hfe}}$

and $\omega_T = h_{feo} \times \omega_{hfe}$ or $f_T = h_{feo} \times f_{hfe}$ [5.58]

This may be substituted in equation [5.57]

$$c_{be} = \dfrac{1}{h_{ie} \dfrac{\omega_T}{h_{feo}}} = \dfrac{g_m}{\omega_T}$$ [5.59]

Thus the short circuit input capacitance is determined from g_m or $39 I_C$ and the parameter f_T which is quoted by the device manufacturer.

For the second effect, the simplified equivalent circuit is redrawn under loaded conditions in Fig. 5.49. Referring to the circuit,

$$I_i = V_i(g_{ie} + j\omega c'_{be}) + (V_i - V_o)j\omega c'_{bc}$$

but $V_o = -g_m V_i Z_L$

(where Z_L includes r'_{bc}, r'_{ce}, c'_{bc} and R_L). Therefore

$$Y_{in} = \dfrac{I_i}{V_i} = \dfrac{1}{h_{ie}} + j\omega\left[c'_{be} + c'_{bc}(1 + g_m Z_L)\right]$$ [5.60]

This result is the admittance for a circuit consisting of h_{ie} in parallel with a capacitance

$$C_{in} = c'_{be} + c'_{bc}(1 + g_m Z_L).$$ [5.61]

This amplification of the feedback capacitance c'_{bc} (or any other feedback component in this position) is known as the **Miller effect**.

Fig. 5.49 Simplified hybrid π equivalent circuit on load.

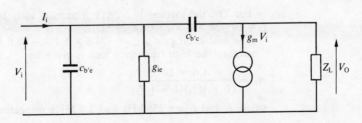

As Z_L is not a pure resistance, equation [5.60] will be further complicated by a j^2 term which results in a positive or a negative resistance in shunt with the other input admittance terms. With a mainly resistive load, this effect will not be important but it is taken advantage of, when an inductive load is used in some types of oscillator.

Example 5.19

In the two-stage amplifier shown in Fig. 5.50, both transistors have a nominal h_{fe} of 170 and operate at a d.c. collector current of 1.9 mA. The transition frequency f_T is quoted as 80 MHz and c_{bc}' is 2 pF. Determine the high frequency gain performance and illustrate your answer with a Bode gain plot.

From the information supplied,

$g_m = 76 \text{ mA/V}$ and $h_{ie} = 2.3 \text{ k}\Omega$

From equation [5.61],

$$c_{be}' = \frac{76 \times 10^{-3}}{2\pi \times 8 \times 10^7} = 150 \text{ pF}$$

The total input capacitance to Tr_2 is given from equation [5.59]

$C_{in} = 150 \text{ pF} + 76 \times 2.7 \times 2 \text{ pF} = 560 \text{ pF}$

This is the major part of the capacitance loading Tr_1 but, strictly, the 2 pF c_{bc}' for Tr_1 and any stray wiring capacitance should be included. Therefore estimate total C_s as 565 pF.

Fig. 5.50 Two-stage amplifier circuit for example 5.19.

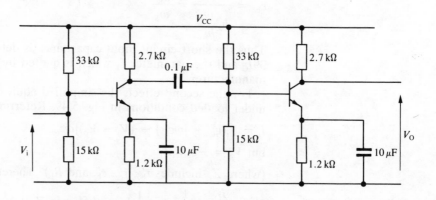

The resistive loading on Tr_1 is given by

$$\frac{1}{R} = \frac{1}{2.7} + \frac{1}{33} + \frac{1}{15} + \frac{1}{2.3} \text{ ms}$$

from which, R = 1.1 kΩ. Thus the break frequency f_1 is given by

$$f_1 = \frac{1}{2\pi \times 560 \times 10^{-12} \times 1100} = 258 \text{ kHz}$$

For Tr_2, the loading is 2.7 kΩ in parallel with 2 pF, making $f_2 = 10^{12}/2\pi \times 2 \times 2700$ or 29 MHz.

The mid-band gain $A_{vo} = -74.6 \times 1.1 \times -74.6 \times 2.7 = 1.6 \times 10^4$. Therefore the high frequency gain is given by

$$A_v = \frac{1.6 \times 10^4}{(1 + jf/f_1)(1 + jf/f_2)}$$

where f_1 and f_2 are 258 kHz and 2.9 MHz respectively.

For the Bode plot, the dB gain is $20 \log_{10} 1.6 \times 10^4$ or 84 dB. The frequency scale must extend from less than one-tenth of the lowest break to more than the highest break, say 10 kHz to to 10 MHz. The required response is shown in Fig. 5.51.

Fig. 5.51 Frequency response for example 5.19.

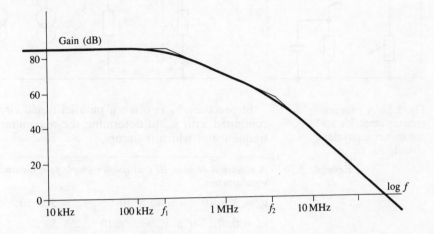

The use of y-parameters for BJT amplifiers

Some transistors are designed specifically for high frequency applications and their behaviour is more conveniently described by y-parameters than by h-parameters. These y-parameters are complex and frequency sensitive, but their use in the equivalent circuit method is no different from any other equivalent circuit based on a set of parameters.

Figure 5.52 shows a common emitter amplifier and its equivalent circuit using y-parameters. The load is specified as Y_L to correspond to the parameter form and that this too will usually be complex.

Following the usual equivalent circuit method, the variables are specified as V_i and V_o, which are respectively the dependent generator variables v_{be} and v_{ce}. Two nodal equations can be written

$$I_i - y_{re}V_o = V_i(Y_B + y_{ie}) \qquad [5.62]$$

$$-y_{fe}V_i = V_o(y_{oe} + Y_L) \qquad [5.63]$$

The voltage gain is obtained directly from equation [5.63] and is given by

$$A_v = \frac{V_o}{V_i} = \frac{-y_{fe}}{y_{oe} + Y_L} \qquad [5.64]$$

For the input admittance, this result is substituted into equation [5.62] and rearranged to give

$$Y_{in} = \frac{I_i}{V_i} = Y_B + y_{ie} - \frac{y_{fe}\,y_{re}}{y_{oe} + Y_L}$$

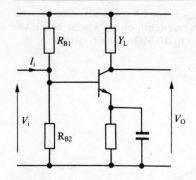

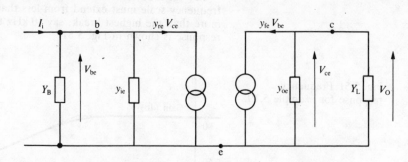

Fig. 5.52 A common emitter amplifier and y-parameter equivalent circuit.

In practice, Y_L is often a parallel tuned circuit which must be combined with y_{oe} to determine the maximum voltage gain and the frequency at which it occurs.

Example 5.20

A common emitter BJT amplifier employs a transistor with the following y-parameters

$y_{ie} = 3 \times 10^{-3} + j\omega 40 \times 10^{-12}$ S, $y_{oe} = 5 \times 10^{-5} + j\omega 1.3 \times 10^{-12}$ S

$y_{fe} = 0.01\angle 340°$ S, $y_{re} = 5 \times 10^{-6}\angle 268°$ S.

The collector load is a parallel tuned circuit with a coil of 0.22 mH and Q 33 and a capacitor C. The maximum gain of the amplifier is required to be at $\omega = 3 \times 10^7$ r/s. Calculate the required value of C, the maximum gain, the bandwidth and the input impedance at the resonant frequency.

Since all the components are in parallel, it is necessary to find the parallel equivalent admittance and susceptance for the coil.

At ω_o, $X_L = 0.22 \times 10^{-3} \times 3 \times 10^7 = 6600$ and $R = X_L/Q = 200\ \Omega$.

The equivalent admittance $Y = \dfrac{1}{200 + j6600} = (4.5 - j150)10^{-6}$ S.

The overall admittance in parallel with the y_{fe} generator will be

$Y = y_{oe} + j\omega C + (4.5 - j150)10^{-6}$

$= 5 \times 10^{-5} + j\omega 1.3 \times 10^{-12} + j\omega C + 4.5 \times 10^{-6} - j150 \times 10^{-6}$

At resonance, the imaginary terms are zero. Thus,

$3 \times 10^7(1.3 \times 10^{-12} + C) = 150 \times 10^{-6}$

or $39 + 30C = 150$ with C in pF

therefore $C = \dfrac{150 - 39}{30} = 3.7$ pF

The total admittance including y_{oe} is $Y = 5 \times 10^{-5} + 4.5 \times 10^{-6}$
$= 5.45 \times 10^{-5}$.

From equation [5.64]

$A_v = \dfrac{-0.001/340°}{5.45 \times 10^{-5}} = 183\angle 160°$

For the bandwidth, $Q = \dfrac{150 \times 10^{-6}}{5.45 \times 10^{-5}} = 2.75$

and the bandwidth $= \omega_o/2Q$

$= \dfrac{3 \times 10^7}{2 \times 2.75} = 5.45 \times 10^6$ r/s

The frequency response of basic op-amp amplifiers

The basic op-amp amplifiers are normally directly coupled to signals, loads and other amplifiers but, on occasion, a.c. capacity coupling may be required as shown in Fig. 5.53 for an inverting amplifer.

Fig. 5.53 An a.c. coupled inverting amplifier.

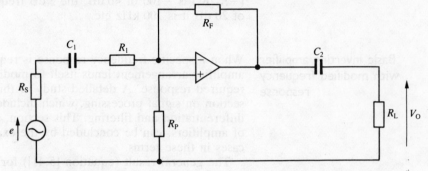

The effect of series C at the input and output will be the same as that for the BJT amplifier (p.361). Thus the overall gain will be given by

$$A_v = \frac{A_{vm}}{(1 + j\omega/\omega_1)(1 + j\omega/\omega_2)}$$

where $A_{vm} = \dfrac{R_1}{R_s + R_1} \times \dfrac{-R_F}{R_1} \times \dfrac{R_L}{R_o + R_L}$

The last term here approximates to unity as R_o for the op-amp is negligible. Therefore

$$A_{vm} = \frac{-R_F}{R_s + R_1}$$

$$\omega_1 = \frac{1}{C_1(R_1 + R_s)} \text{ and } \omega_2 = \frac{1}{C_2 R_L}$$

Fig. 5.54 The open and closed loop frequency reponses of an op-amp inverting amplifier.

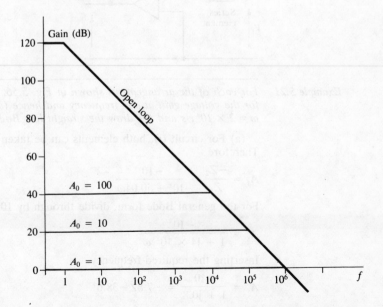

The high frequency performance is examined in detail later (p.481). It is useful to appreciate, however, that the upper 3 dB frequency for an internally compensated amplifier is inversely proportional to the gain. This is shown in the Bode plot in Fig. 5.54. The op-amp is shown as having an open-loop gain A_d of 10^6 and an open-loop 3 dB frequency of 1 Hz. If the circuit gain $(-R_F/R_1)$ is -100 or 40 dB, the 3 dB frequency is 10 kHz, for a gain of 20 dB it is 100 kHz etc.

Basic inverting amplifier with modified frequency response

When a specific frequency response is required, the inverting amplifier arrangement lends itself to modification to produce a required response. A detailed study of this area is found in the section on signal processing, which includes integration, differentiation and filtering. This section, on the frequency response of amplifiers, can be concluded by the examination of some simple cases in these terms.

The general result (equation [5.41]) for the gain of an inverting amplifier can be applied just as well for two impedances or two admittances or a combination of both. The result can then be expressed

$$A_v = \frac{-Z_2}{Z_1} = \frac{-Y_1}{Y_2} = \frac{-Z_2 Y_1}{1} = \frac{-1}{Z_1 Y_2}$$

for the circuit shown in Fig. 5.55. The choice of equation to use depends on the form of the series and feedback elements. For series combinations, an impedance Z form is best, while for parallel combinations the admittance from Y leads to a simpler result. This is best illustrated by an example.

Fig. 5.55 The feedback components of an op-amp.

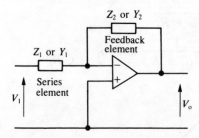

Example 5.21

For each of the arrangements shown in Fig. 5.56, derive a general expression for the voltage gain at any frequency and hence (a) find the gain as $\omega = 2 \times 10^4$ r/s and (b) draw the straight line Bode gain plot.

(a) For circuit (i), both elements can be taken in the series form. Therefore

$$A_v = \frac{-Z_2}{Z_1} = \frac{-10^4}{10^3 + j0.04\omega}$$

For the general Bode form, divide through by 10^3

$$A_v = \frac{-10}{1 + j4 \times 10^{-5}\omega} \tag{5.65}$$

Inserting the required frequency,

$$A_v = \frac{-10}{1 + j0.8} = -7.8\angle-38.6°$$

Fig. 5.56 Frequency
sensitive op-amp circuits for
example 5.21.

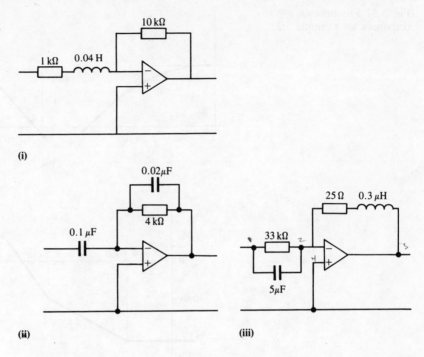

(i)

(ii)   (iii)

For circuit (ii), the parallel components in the feedback element make
the admittance form more suitable. Therefore

$$A_v = \frac{-Y_1}{Y_2} = \frac{-j10^{-7}\omega}{2.5 \times 10^{-4} + j2 \times 10^{-8}\omega}$$

and for the general form,

$$A_v = \frac{-j4 \times 10^{-4}\omega}{1 + j0.8 \times 10^{-4}\omega} \qquad [5.66]$$

which at $\omega = 2 \times 10^4 = \dfrac{-j8}{1 + j1.6} = 4.2\angle-148°$.

For circuit (iii), the combination of series and parallel elements makes
the form $-Y_1Z_2$ the best choice. Hence,

$$A_v = -Y_1Z_2 = -\left(\frac{1}{33 \times 10^3} + j\omega5 \times 10^{-9}\right)(25 + j3 \times 10^{-4}\omega)$$

which may be arranged for the general form to give

$$A_v = \frac{-25}{33 \times 10^3}(1 + j\omega1.65 \times 10^{-4})(1 + j1.2 \times 10^{-4}\omega) \qquad [5.67]$$

and when $\omega = 2 \times 10^4$,

$$A_v = -0.758 \times 10^{-3}(1 + j3.3)(1 + j\,0.24)$$

$$= 2.68 \times 10^{-3}\angle-93°$$

(b) The three Bode plots are shown in Fig. 5.57. For circuit (i) equation
[5.65] shows a low frequency constant gain of 10 or 20 dB and a single
break at $\omega = 2.5 - 10^4$.

For circuit (ii), equation [5.66] is rewritten

$$A_v = \frac{-j\omega 4 \times 10^{-4}}{1 + j\omega/1.25 \times 10^4}$$

Fig. 5.57 Frequency
responses for example 5.21.

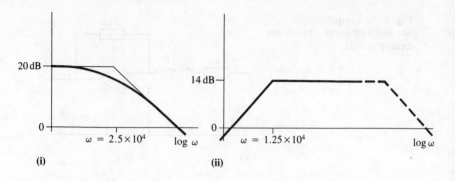

(i)   (ii)

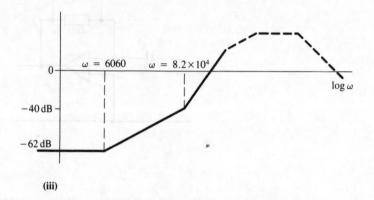

(iii)

and the Bode plot in this case has a constant gain at high frequencies given
by $A = 4 \times 10^{-4} \times 1.25 \times 10^4 = 5$, which is equivalent to 14 dB, and a low
frequency break at $\omega = 1.25 \times 10^4$ r/s.

For circuit (iii), equation [5.76] is rewritten

$$A_v = -0.758 \times 10^{-3}(1 + j\omega/6060)(1 + j\omega/8.3 \times 10^4).$$

In this case the low frequency constant gain is 0.756×10^{-3} or -62 dB with
two high frequency breaks at 6060 r/s and 8.3×10^4 r/s.

It is important to note that each of these frequency responses assumes an
ideal op-amp. In practice, the high frequency response of the op-amp (Fig.
5.54) will modify each of them as shown by the broken lines in Fig. 5.57.

You should now be able to attempt exercises 5.11 to 5.16.

LARGE SIGNAL AMPLIFIERS

Many audio amplifiers are required to deliver large amounts of
power to a load. Where this is the case then the use of small-signal
equivalent circuits is limited; the solution to problems with large
powers involved must be achieved graphically or by the use of
piecewise linear equivalent circuits.

When operated over a large dynamic range active devices are
notoriously non-linear and the large current and voltage swings

produced will produce large amounts of distortion. The undesirable effects of distortion must be minimised in a practical circuit and ways of achieving this will be discussed in this section. Feedback methods as discussed in the next section will play a large part in the reduction of distortion.

One problem with power amplifiers is caused by the relatively large amounts of heat that must be dissipated by the power device. This heat is generated largely at the base–collector junction and must be removed to the air, or some other cooling medium, to avoid damaging the device. Heat sinks are a popular method of assisting in the dissipation of the heat and methods of calculating the size of a heat sink to keep the junction temperature below a safe operating level will be discussed. Additionally, methods of improving the d.c. to a.c conversion efficiency will be investigated. When a large signal amplifier, or power amplifier as it is often called, is required to deliver large powers to a load the circuit is acting as a power converter. The input signal to the amplifier is a modulating signal which allows some of the available power from the power supply (input or d.c. power) to be converted to useful signal output power in the load (a.c. power). The effectiveness of this conversion is the measure of the efficiency of the amplifier. The conversion efficiency η, usually expressed as a percentage, is expressed as

$$\eta = \frac{\text{output a.c. power to load}}{\text{input d.c. power from supply}} \times 100\%$$

If P_{ac} is the output a.c. power delivered to the load and P_{dc} is the input d.c. power from the power supply, then

$$\eta = \frac{P_{ac}}{P_{dc}} \times 100\% \qquad [5.68]$$

It follows that the remainder of the input d.c. power (i.e. that part of the available input power *not* converted to useful signal power) is dissipated as heat in the output electrode of the device and the load. If P_{diss} represents this wasted power then

$$P_{diss} = P_{dc} - P_{ac} \qquad [5.69]$$

Dividing both sides of this equation by P_{dc} gives:

$$\frac{P_{diss}}{P_{dc}} = 1 - \frac{P_{ac}}{P_{dc}}$$

hence $\dfrac{P_{diss}}{P_{dc}} = 1 - \dfrac{\eta}{100}$ if η is a percentage

and $P_{diss} = P_{dc}\left(1 - \dfrac{\eta}{100}\right)$

Example 5.22 *A large signal amplifier has an efficiency of 15% and delivers an output power of 4.5 W to a resistive load. Calculate the d.c. power taken from the supply and the power dissipated under these signal conditions.*

Since $\eta = \dfrac{P_{ac}}{P_{dc}} \times 100\%$

then $P_{dc} = \dfrac{P_{ac}}{\eta} = 100\%$

$$P_{dc} = \frac{4.5 \times 100}{15} \text{ W}$$

$$= 30 \text{ W}$$

Also $P_{diss} = 30\left(1 - \dfrac{\eta}{100}\right)$

so $P_{diss} = 30\left(1 - \dfrac{15}{100}\right)$

$$P_{diss} = 30 \times 0.85 = 25.5 \text{ W}$$

This result can also be shown as follows: since

$$P_{dc} = 30 \text{ W} = P_{in}$$

and $P_{ac} = 4.5 \text{ W} = P_{out}$

then $P_{diss} = P_{dc} - P_{ac} = (30 - 4.5) \text{ W} = 25.5 \text{ W}$

Example 5.23   *If a large signal amplifier has an efficiency of 22% and a power dissipation at device output electrode of 1.5 W, calculate the output power of the amplifier.*

Since $P_{diss} = P_{dc} - P_{ac}$

and $\eta = \dfrac{P_{ac}}{P_{dc}} \times 100\%$

then $P_{dc} = \dfrac{P_{ac}}{\eta}$

and $P_{diss} = \dfrac{P_{ac}}{\eta} - P_{ac}$

so $P_{diss} = P_{ac}\left(\dfrac{1}{\eta} - 1\right)$

Rearranging this equation gives

$$P_{ac} = \frac{P_{diss}}{\dfrac{1}{\eta} - 1}$$

$$= \frac{1.5}{\dfrac{1}{0.22} - 1} \text{ W}$$

$$P_{ac} = 0.423 \text{ W}$$

The heat dissipation aspect of a power amplifier is one which must be taken into very careful consideration in the design of such circuits. Where very large powers are involved, special ventilation or cooling arrangements must be used to assist the device in dissipating the heat. This may involve the use of heat sinks with large surface area, provided by fins, to allow maximum rate of heat dispersion to the atmosphere. In more severe cases cooling fans or the use of pumped cooling water may be required. In addition it must be borne in mind that in the event of the input modulating signal being absent then the input d.c power must be dissipated in the output electrode of the active device and its load.

Consider a device which has a sinusoidal signal $v = V \sin \omega t$ superimposed upon a d.c. voltage of V_Q present at its output terminal. This voltage is assumed to be produced by a current having a sinusoidal value $i = I \sin \omega t$ superimposed upon a direct current of value I_Q flowing in the resistive load of the amplifier. For both the voltage and current waveforms $v(i)$ represents instantaneous values, $V(I)$ represents peak sinusoidal values while $V_Q(I_Q)$ represents the mean or d.c. value. The voltage and currrent waveforms are shown in Figs 5.58(a) and (b) respectively.

Fig. 5.58 Power device output waveforms: (a) voltage; (b) current. Sinusoidal signals are assumed.

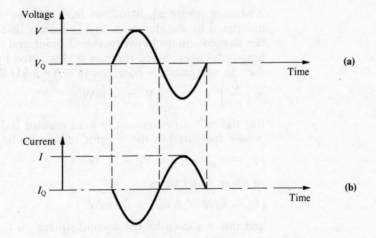

The amplifier is assumed to operate such that the alternating current output flows for 360° of the modulating input voltage cycle, i.e. class A bias, and that the amplifier is an inverting device.

The instantaneous power dissipated at the device output electrode is given by vi watts, where

$$v = V_Q + V \sin \omega t$$

and $i = I_Q - I \sin \omega t$

therefore

$$p_{\text{diss}} = vi = (V_Q + V \sin \omega t)(I_Q - I \sin \omega t)$$

$$= V_Q I_Q + I_Q V \sin \omega t - V_Q I \sin \omega t - VI \sin^2 \omega t \qquad [5.70]$$

Since the expression $\sin^2 \omega t$ can be replaced by $\dfrac{1 - \cos 2\omega t}{2}$ then equation [5.70] can be rewritten as

$$p_{\text{diss}} = V_Q I_Q + I_Q V \sin \omega t - V_Q I \sin \omega t - \frac{VI}{2} + \frac{VI}{2} \cos 2\omega t$$

Also, since the *average* power of all sine and cosine terms over a cycle is zero, then all sine and cosine terms will disappear, and

$$p_{\text{diss}} = V_Q I_Q - \frac{VI}{2} \text{ W}$$

The product of V_Q and I_Q represents the d.c. power at the device output electrode while $VI/2$ represents the a.c. power output. Since $V_{\text{rms}} = 0.707V$ and $I_{\text{rms}} = 0.707I$, then

a.c. power $= 0.707V \times 0.707I = 0.5VI = \dfrac{VI}{2}$ W

The resistive load will also dissipate power because of the mean current taken from the supply. This power can be taken as $I_Q^2 R$ watts where R is the load resistance in ohms. Consider the circuit of Fig. 5.59(a) which has a supply voltage of 12 V and a resistive load of 1 kΩ. The d.c. load line can be plotted as in Fig. 5.59(b). Assuming the Q point is selected in the centre of the load line, then

$$V_Q = 6\ \text{V},\ I_Q = 6\ \text{mA}$$

Assuming no signal, it follows from the above that the power dissipated by the device is $V_Q I_Q$ which is that power represented by the shaded square between the Q point and the vertical and horizontal axes. Since the power dissipated by the resistance is $I_Q^2 R$ then in this case the power is $(6\ \text{mA})^2\ 1\ \text{k}\Omega$ W

$$= \left(\frac{6}{10^3}\right)^2 \times 1 \times 10^3\ \text{W} = 36\ \text{mW}$$

But the volt drop across the load resistor is $V_S - V_Q = V_R$ so that power dissipated in the resistor could also be represented by

$$(V_S - V_Q) \times I_Q = V_R \times I_Q\ \text{W}$$

which in this case is

$$(12 - 6)\ \text{V} \times 6\ \text{mA} = 36\ \text{mW}$$

and this is shown by the second square on the output characteristic.

Fig. 5.59 Power bipolar transistor: (a) output circuit; (b) d.c. load line.

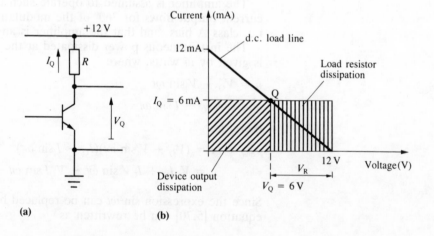

It follows from this that the total dissipation is given by the sum of the squares, or

$$V_Q I_Q + V_R I_Q\ \text{W}$$

$$= I_Q[V_Q + (V_S - V_Q)]\ \text{W}$$

$$= I_Q V_S\ \text{W}$$

This is the total power (P_{dc}) supplied from the power supply and in the absence of a signal this power must be dissipated as heat in the device and in the load resistance. Thus in the absence of a signal

$$P_{\text{diss}} = P_{\text{dc}}$$

However, the presence of a signal allows the active device to convert some of the d.c. power to a.c. output power and the amount dissipated is reduced by that amount, i.e.

$$P_{\text{diss}} = P_{\text{dc}} - P_{\text{ac}}$$

as shown in equation [5.69].

Maximum power dissipation at the device output electrode will be specified by a manufacturer and it is essential that this rated value is not exceeded when the device is in a practical circuit. The device output electrode power dissipation has already been specified as $V_Q I_Q$ under no-signal conditions and the design values for V_Q and I_Q must be such that their product $V_Q I_Q$ must be less than the manufacturer's rated dissipation value. Since P_{diss} is the product of instantaneous voltage and current at the device output electrode, a range of points can be overplotted on the graph of I_{out} against V_{out} to give a maximum power dissipation curve. As an example, consider a BJT device where the rated power dissipation at the collector must not exceed 20 W. Since $P_{\text{diss}} = V_{\text{ce}} \times I_{\text{c}}$ then for selected values of V_{ce}

V_{ce}(V)	I_{c}(mA)	
2.0	10.0	
5.0	4.0	
10.0	2.0	
20.0	1.0	etc.

and the curve plotted gives the result shown in Fig. 5.60.

Fig. 5.60 Power dissipation curve.

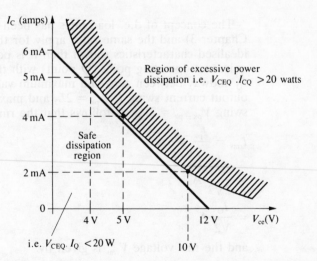

As can be seen from this example a d.c. load line plotted from $V_{\text{ce}} = V_{\text{cc}} = 12$ V to an intercept of 6 mA on the I_{c} axis (i.e. for a load resistance of 2 kΩ) allows a working point to be chosen which is within the safe dissipation region. Care must be taken however to ensure that, if an a.c. load line is involved, the choice of component values does not take the load line through the excessive power dissipation region. The concept of power dissipation will be discussed again later in this section since the power dissipation is a function of temperature.

The conversion efficiency of the active device is of prime importance in power amplifier applications. It follows that the greater the efficiency the smaller the value of input power required for a given required output power. Also, possible dissipation difficulties are resolved. The following sections will examine various circuit arrangements and classes of bias, and values for conversion efficiency under ideal and likely actual conditions will be evaluated.

Class A bias with direct resistive load

Class A bias with resistive load allows output current to flow in the device for all the input modulating cycle. For the resistive loaded amplifier stage refer to Fig. 5.61(a), which is a generalised circuit. The analysis is similar for all devices assuming their output characteristics have the idealised values shown in Fig. 5.61(b).

Fig. 5.61 Resistance loaded power amplifier: (a) circuit diagram; (b) output characteristics and d.c. load line.

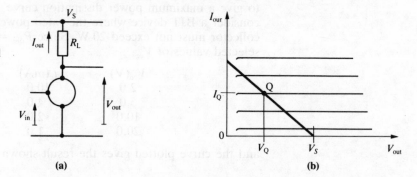

The concept of d.c. load lines has been discussed earlier (see Chapter 3) and the same rules apply for this circuit. Assuming the idealised characteristics shown then it is possible to select a quiescent operating point (Q point) with the d.c. values for V_Q and I_Q midway between zero and maximum values, so that maximum output current swing $I_{pk-pk} = 2I_Q$ and maximum output voltage swing $V_{pk-pk} = 2V_Q$. It follows that the rms current I_{rms} is given by

$$I_{rms} = \frac{I_{pk-pk}}{2\sqrt{2}} \text{ A}$$

$$= \frac{2I_Q}{2\sqrt{2}} \text{ A}$$

$$= \frac{I_Q}{\sqrt{2}} \text{ A}$$

and the rms voltage V_{rms} is given by

$$V_{rms} = \frac{V_{pk-pk}}{2\sqrt{2}} \text{ V}$$

$$= \frac{2V_Q}{2\sqrt{2}} \text{ V}$$

$$= \frac{V_Q}{\sqrt{2}} \text{ V}$$

so that $P_{ac} = V_{rms} \times I_{rms} = \dfrac{V_Q}{\sqrt{2}} \times \dfrac{I_Q}{\sqrt{2}}$

$$= \frac{V_Q I_Q}{2} \text{ W}$$

and since $P_{dc} = V_S \times I_Q$ then

$$\eta = \frac{P_{ac}}{P_{dc}} = \frac{V_Q I_Q}{2 V_S I_Q} \times 100\%$$

$$= \frac{V_Q I_Q}{4 V_Q I_Q} \times 100\%$$

or $\eta = 25\%$

Example 5.24　*Assuming idealised transistor characteristics, a supply voltage of +12 V and a resistive load of 2 kΩ as shown in Fig. 5.62(a), find the power dissipation levels for the load and the active device in the absence of a signal. What would be the maximum output power and efficiency of the amplifier?*

Fig. 5.62 Resistance loaded power stage: (a) circuit; (b) d.c. load line for example 5.24.

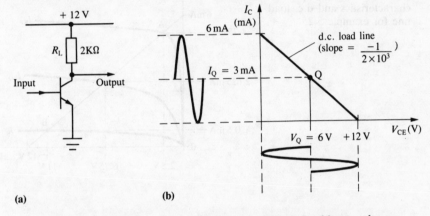

(a)　　　　　　　　(b)

The load line is shown on the output characteristic. In this case the a.c. and d.c. load lines are coincident. It follows that for maximum output power, assuming ideal characteristics, the Q point is midway along the load line so that:

$$V_Q = \frac{V_{cc}}{2} = 6 \text{ V}$$

$$I_Q = \frac{I_{c(max)}}{2} = 3 \text{ mA}$$

Power dissipated in load resistor $= P_{diss(R)} = I_Q^2 R = (3 \times 10^{-3})^2 \times 2 \times 10^3$ W

$$= 18 \text{ mW}$$

Power dissipated at the collector $= P_{diss(c)} = V_Q I_Q = 6 \times 3 \times 10^{-3}$ watts

$$= 18 \text{ mW}$$

Total input power $= P_{diss(R)} + P_{diss(c)} = 36 \text{ mW}$

For a maximum signal swing voltage can have a peak-to-peak value of 12 V while the peak-to-peak current swing is 6 mA as shown in Fig. 5.62(b).

Thus output power $= V_{out(rms)} \times I_{out(rms)}$

$$= \frac{V_{pk - pk}}{2\sqrt{2}} \times \frac{I_{pk - pk}}{2\sqrt{2}}$$

$$= \frac{12}{2\sqrt{2}} \times \frac{6 \times 10^{-3}}{2\sqrt{2}}$$

$$= \frac{72}{8} \text{ mW}$$

$$= 9 \text{ mW}$$

and efficiency

$$= \frac{P_{ac}}{P_{dc}} \times 100\%$$

$$= \frac{9 \times 10^{-3}}{36 \times 10^{-3}} \times 100\% = 25\%$$

Example 5.25   *Using the same data as in the previous example but using actual transistor characteristics which give maximum voltage and current swings as shown in Fig. 5.63 find the output power and efficiency of the amplifier.*

Fig. 5.63 Device output characteristics and d.c. load line for example 5.25.

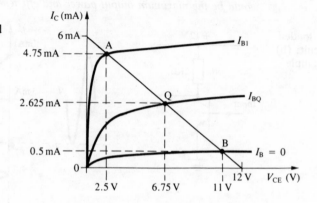

Because the actual characteristics prevent excursions from the Q point beyond points A and B under signal conditions, the voltage and current swings are reduced compared to the idealised case. Assuming class A operation, the bias point has been moved so that the Q point is now in the middle of the swing AB. This gives a value for V_Q of 6.75 volts and I_Q of 2.625 mA. Thus

$$P_{dc} = V_{cc} \times I_Q = 12 \times 2.625 \text{ mW}$$

$$= 31.5 \text{ mW}$$

and $V_{out(pk - pk)} = 11 - 2.5 \text{ V} = 8.5 \text{ V}$

$I_{out(pk - pk)} = 4.75 - 0.5 \text{ mA} = 4.25 \text{ mA}$

Hence $P_{ac} = \dfrac{V_{out(pk - pk)}}{2\sqrt{2}} \times \dfrac{I_{out(pk - pk)}}{2\sqrt{2}}$

$$= \frac{8.5 \times 4.25}{8} \text{ mW}$$

$$= 4.156 \text{ mW}$$

and efficiency $\eta = \dfrac{P_{ac}}{P_{dc}} \times 100\%$

$$= \frac{4.156}{31.5} \times 100\%$$

$$= 14.3\%$$

The analysis of example 5.24 showed that with ideal conditions assumed, a bigger voltage and current swing can be allowed than is normally the case in practice. In reality voltage and current swings would have to be reduced to avoid distortion produced by device saturation and cut-off effects. The result of allowing for practical reduction in voltage and current swings can be seen in example 5.25 is a reduction in the efficiency of the stage compared to the theoretical maximum.

Class A bias with transformer-coupled load

The low efficiency of a class A resistive loaded stage can be improved by using a transformer coupled load as shown in Fig. 5.64.

Fig. 5.64 Basic single-ended power amplifier stage.

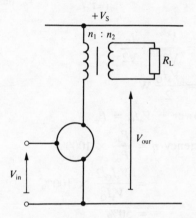

Since the d.c. resistance of the transformer primary is assumed to be zero then the d.c. load line is a vertical line passing through V_s on the V_{out} axis. The load under a.c. conditions is given by

$$R_L' = \left(\frac{n_1}{n_2}\right)^2 \times R_L$$

where n1/n2 is the transformer turns ratio.

This arrangement has the advantage that the use of a step-down transformer with $n_1 > n_2$ can give a value of $R_L' > R_L$ and provide a better power match to the device. Additionally, a larger output voltage swing can be accommodated. The a.c. load line intercepts the V_{out} axis at a value greater than V_s as Fig. 5.65 shows. If the position of the load line and the chosen Q point is such that

$$V_{out(max)} = 2V_s$$

and $I_{out(max)} = 2I_Q$

then the signal swing at the output can have a maximum possible value.

$$\text{Output power} = \frac{V_{out(max)}}{2\sqrt{2}} \times \frac{I_{out(max)}}{2\sqrt{2}}$$

and $P_{ac} = \dfrac{2V_s}{2\sqrt{2}} \times \dfrac{2I_Q}{2\sqrt{2}}$

Fig. 5.65 Idealised a.c. load line for single-ended stage showing that output voltage swing can have a maximum value of $2V_s$.

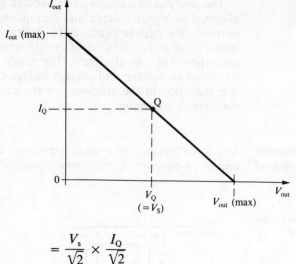

$$= \frac{V_s}{\sqrt{2}} \times \frac{I_Q}{\sqrt{2}}$$

$$= \frac{V_s I_Q}{2} \text{ W}$$

Input power $= V_s I_Q = P_{dc}$

and efficiency $\eta = \dfrac{P_{ac}}{P_{dc}} \times 100\%$

$$= \frac{V_s I_Q/2}{V_s I_Q} \times 100\%$$

$$= 50\%$$

In practice this efficiency would be reduced because of the need to allow for non-ideal characteristics.

The reason why in this arrangement the output voltage swing can exceed the supply voltage is because of the inductance of the transformer primary. Under a.c. conditions the reflected load is effectively in parallel with the primary inductance L_p and since L_p is large compared to R_L' then the effect of L_p has been ignored. However, the effect of L_p, once d.c. conditions have been established, is to maintain that direct current I_Q constant. Consider again a circuit with a supply voltage of $+12$ V and a quiescent output current of 6 mA but this time the load of the device is a large inductance L_p with parallel reflected resistance R_L'. Figure 5.66(a) shows the quiescent condition.

The a.c. load line is shown in Fig. 5.67. Since the a.c. load line is given by $-1/R_L'$, then $R_L' = 24$ V/12 mA $= 2$ kΩ.

From Fig. 5.66(b) it can be seen that when the current increases above its d.c. value the incremental value is passed through R_L' while L_p maintains I_Q at its quiescent value. Thus 12 volts is dropped across R_L' and device voltage falls to 0 V which is the idealised value for saturation.

When the current in the circuit tries to fall below the quiescent value then since the device is cut off the quiescent current in L_p is diverted via R_L' to give a volt drop across R_L' of polarity shown in Fig. 5.66(c). Thus voltage across the device must go to $+24$ V or twice the supply voltage, in this idealised case.

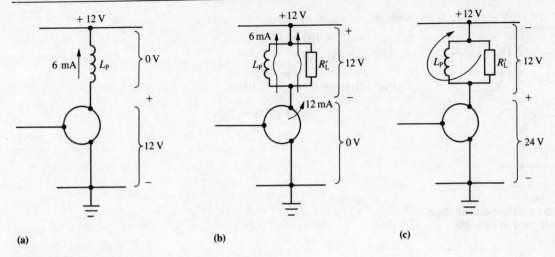

(a)   **(b)**   **(c)**

Fig. 5.66 Effect of inductance of single-ended stage transformer primary: (a) quiescent conditions; (b) maximum device current condition; (c) minimum device current condition.

Fig. 5.67 A.C. load line for a single-ended stage of supply voltage 12 V and reflected load 2 kΩ.

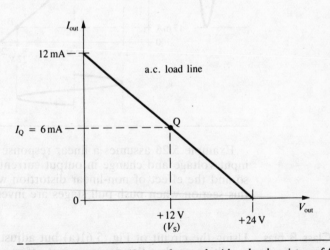

Example 5.26   The circuit of Fig. 5.68 is to be used with a load resistor of 8 Ω and a transformer of turns ratio $n_1/n_2 = 5$. If the peak voltage and current swings are as shown in Fig. 5.69, calculate the output power and amplifier efficiency.

From Fig. 5.69 the values for $V_{out(pk-pk)}$ and $I_{out(pk-pk)}$ are

$$V_{out(pk-pk)} = 22 - 4 = 18 \text{ V}$$

$$I_{out(pk-pk)} = 167 - 17 = 150 \text{ mA}$$

Therefore

$$P_{ac} = \frac{18 \times 150}{8} \text{ mW}$$

$$= 337.5 \text{ mW}$$

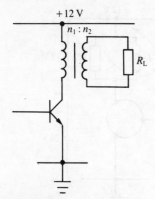

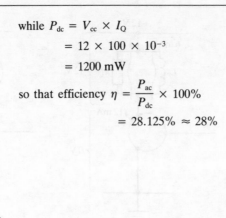

while $P_{dc} = V_{cc} \times I_Q$

$$= 12 \times 100 \times 10^{-3}$$

$$= 1200 \text{ mW}$$

so that efficiency $\eta = \dfrac{P_{ac}}{P_{dc}} \times 100\%$

$$= 28.125\% \approx 28\%$$

Fig. 5.68 Single-ended stage circuit used in example 5.26.

Fig. 5.69 A.C. load line for the circuit of example 5.26.

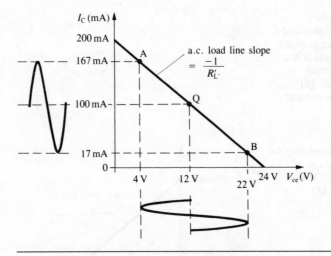

Example 5.26 assumes a linear response between the change in input voltage and change in output current. This is not necessarily so and the effect of non-linear distortion will be discussed later in this section when push-pull stages are investigated.

Class B bias   Using the circuit of Fig. 5.61(a) but adjusting the quiescent bias condition so that the device is just off produces class B bias. This class of bias is defined when output current flows for only 180° of the input cycle swing.

Figure 5.70 shows an idealised set of device characteristics for class B bias. It follows from the characteristics that the output current waveform consists of a half-cycle of sinusoidal swing with the second half-cycle giving zero current. The average, or d.c. component of this waveform is found by integrating the waveform over one cycle:

$$I_{dc} = \frac{1}{2\pi} \int_0^\pi I \sin \omega t \, dt = \frac{I}{\pi} \text{ amps} \qquad [5.71]$$

where I is the peak value of current swing. Thus d.c. power is supplied from the power supply is

$$P_{dc} = \frac{V_s I}{\pi} \text{ watts}$$

The output power is sinusoidal for half a cycle and has a peak voltage swing of V_s so that

$$P_{ac} = \frac{1}{2}\left(\frac{V_s}{\sqrt{2}} \times \frac{I}{\sqrt{2}}\right)$$

$$= \frac{V_s I}{4}$$

and efficiency $\eta = \dfrac{P_{ac}}{P_{dc}} \times 100\%$

$$= \frac{\dfrac{V_s I}{4}}{\dfrac{V_s I}{\pi}} \times 100\%$$

$$= 78.5\%$$

Fig. 5.70 Output current waveform for a class B bias stage.

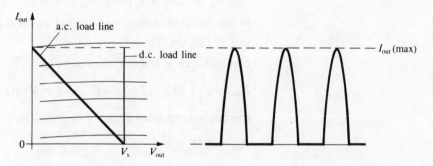

This is a considerable advantage over class A efficiency. However, the main advantage is gained at low power output levels. In the class A stage with low a.c. power levels, the device has to dissipate large powers at its output electrode since the input power is fixed. In class B, however, the d.c. value of the output current decreases with reduced signal hence reducing the possible dissipation. In the event of no-signal in class B then the standby dissipation is zero. The use of class B is obviously preferred in those situations where output power is likely to be small or even zero for long periods, especially if a battery supply is used.

Example 5.27 *If the circuit of Fig. 5.62(a) is biased at cut-off then assuming maximum output the waveforms of Fig. 5 71 result. Use these waveforms to determine circuit powers and the efficiency of the stage.*

From the waveforms of Fig. 5.71 it can be determined that:
With no signal the current drawn from the supply is 0.5 mA; with a signal the average current for the half-cycle of sinusoidal current is, from equation [5.71]

$$I_{dc} = \frac{I}{\pi}$$

In this case the maximum value of current swing is $4.75 - 0.5 = 4.25$ mA

Fig. 5.71 Output characteristics and load line for the class B circuit in example 5.27.

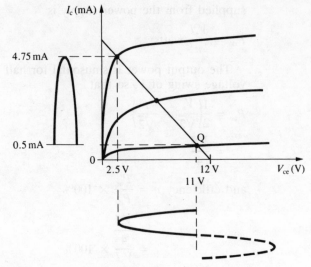

Hence mean current $= \dfrac{4.25}{\pi} + 0.5 = 1.85$ mA

and $P_{dc} = (1.85 \times 12)$ mW $= 22.2$ mW

In this circuit instantaneous transistor power is given by:

$(0.5 + I \sin \omega t).(V_{cc} - 1 - V \sin \omega t)$

over half a cycle and the mean power is

$$P_{diss} = \frac{1}{2\pi}\int_o^\pi (0.5 + I \sin \omega t).(V_{cc} - 1 - V \sin \omega t)$$

which when expanded gives:

$$P_{diss} = \underbrace{2.75 + \frac{(V_{cc} - 1)/V}{\pi R_L}}_{P_{dc}(\text{device})} \underbrace{- \frac{V}{2\pi} - \frac{V^2}{4R_L}}_{P_{ac}}$$

This gives P_{diss} in mW if R_L is measured in kilohms. I has been substituted for convenience since $I = V/R_L$. Substituting for $V_{cc} = 12$ V, $V = (11 - 2.5)$ $= 8.5$ V and $R_L = 2$ kΩ gives

$$P_{diss} = 2.75 + \frac{11 \times 8.5}{2\pi} - \frac{8.5}{2\pi} - \frac{8.5^2}{8}$$

or

$$P_{diss} = \underbrace{16.278}_{P_{dc}(\text{device})} \underbrace{- 9.03}_{P_{ac}}$$

P_{dc} for the device is not the same as P_{dc} (22.2 mW) since some power is dissipated in the load resistor. However, efficiency can be calculated as

$$\eta = \frac{P_{ac}}{P_{dc}} \times 100\%$$

$$= \frac{9.03}{22.2} \times 100\%$$

$$= 40.6\%$$

which is considerably less than the theoretical maximum value of 78.5% but nonetheless much better than the practical efficiencies for class A stages.

If class C bias is used then efficiency can be improved beyond that available for class B and could approach 100% in theory. Class C is not used for audio amplifiers because of the distortion it would introduce but it is widely used for tuned power amplifiers. Class C bias is defined as that bias where output current flows for less than half a cycle of the input cycle swing.

All of the efficiencies discussed in the previous sections have assumed sinusoidal input signals. Efficiencies will differ for other types of signals and can be higher than those obtained from sinusoidal signals. The circuit of Fig. 5.62(a) for example, with a square wave input, can give an efficiency approaching 100%. However, the majority of applications will involve sinusiodal input signals and no further discussion on square wave inputs will be considered.

Distortion in power amplifiers

At least three types of distortion may be produced in a power amplifier: **Frequency distortion**: this is produced when the amplitude of the output signal varies with variations of frequency of the input signal. **Phase distortion**: this is produced when the phase difference between input and output signals is non-linear with frequency. **Amplitude**, or **harmonic**, **distortion**: this is produced when the amplitude of the output signal is not directly related to the amplitude of the input signal by a constant factor over the complete 360° of an input cycle.

Amplitude distortion can be produced by non-linearity of the characteristic as Fig. 5.72 shows. The distorted output current contains a fundamental and several harmonics. The current and voltage variables are related by a power series of the form

$$I = A_0 + A_1.V_1 + A_2.V_1^2 + A_3.V_1^3 + \ldots + A_n.V_1^n \qquad [5.72]$$

Fig. 5.72 General mutual characteristic to show the effect of non-linear distortion.

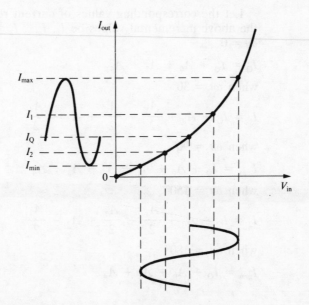

and by evaluating certain of the terms certain harmonics could be determined. For example, using only the first three terms would enable the fundamental and second harmonic terms to be evaluated. Using the fourth terms would enable the third harmonic to be determined etc.

Dividing the input variable into four equal parts as suggested in Fig. 5.72 allows five corresponding pairs of V and I to be obtained, giving five simultaneous equations. The constants can be found from these equations.

Let us assume the input variable is a voltage of the form:

$$V_I = V \sin \omega t \qquad [5.73]$$

Substituting equation [5.73] in equation [5.72] and expanding gives, for a specified value of V,

$$I = I_Q + A_0 + A_1 \sin \omega t + A_2 \cos 2\omega t + A_3 \sin 3\omega t + A_4 \cos 4\omega t + \cdots$$

$$[5.74]$$

Equation [5.74] assumes that the trigonometrical relationships such as

$$\sin^2 \omega t = \frac{1}{2}.(1 - \cos 2\omega t)$$

$$\sin^3 \omega t = \frac{1}{4}.(3 \sin \omega t - \sin 3\omega t)$$

etc., have been used.

For equation [5.74] I_Q is the quiescent d.c. output current, A_0 is an added d.c. value when the signal is present, and A_1 etc., are harmonic amplitudes. These coefficients may be evaluated by considering values of ωt that produce equal increments of change in the input voltage. For a sine function these values are:

0°, 30°, 90°, 150° and 180° or

$$0, \frac{\pi}{6}, \frac{\pi}{2}, \frac{5\pi}{6} \text{ and } \pi \text{ radians}$$

Let the corresponding values of current relating to the voltages at the above incremental values be I_1, I_{max}, I_2 and I_{min}. Then when $\omega t = 0°$

$$I_Q = I_Q + A_0 + A_2 + A_4$$

when $\omega t = 30°$

$$I_1 = I_Q + A_0 + \frac{A_1}{2} + \frac{A_2}{2} + A_3 - \frac{A_4}{2}$$

when $\omega t = 90°$

$$I_{max} = I_Q + A_0 + A_1 + A_2 - A_3 + A_4$$

when $\omega t = 150°$

$$I_2 = I_Q + A_0 + \frac{A_1}{2} + \frac{A_2}{2} + A_3 - \frac{A_4}{2}$$

when $\omega t = 180°$

$$I_{min} = I_Q + A_0 + A_2 + A_4$$

These equations may be solved to give amplitudes of the coefficients as follows:

$$A_0 = \frac{I_{max} + I_{min} + 2(I_1 + I_2)}{6} - I_Q \qquad [5.75]$$

$$A_1 = \frac{I_{max} - I_{min} + I_1 - I_2}{3} \qquad [5.76]$$

$$A_2 = \frac{I_{max} + I_{min} - 2I_Q}{4} \qquad [5.77]$$

$$A_3 = \frac{I_{max} - I_{min} - 2(I_1 - I_2)}{6} \qquad [5.78]$$

$$A_4 = \frac{I_{max} + I_{min} - 4(I_1 - I_2) + 6I_Q}{12} \qquad [5.79]$$

Example 5.28 *For the current-voltage relationship shown in Fig. 5.73 find the mean value of the output current when an input voltage centred at 0.7 V swings 0.1 V peak.*

Fig. 5.73 Mutual characteristic for the determination of non-linear distortion in example 5.28.

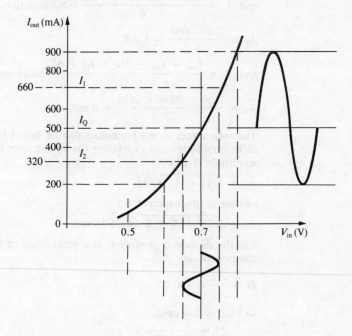

From the characteristics the value of I_Q is 470 mA and this would be the output current in the absence of a signal. However, the signal causes the d.c. level to rise to a value of $(I_Q + A_0)$ mA, where A_0 is obtained from equation [5.75] as

$$A_0 = \frac{I_{max} + I_{min} + 2(I_1 + I_2)}{6} - I_Q$$

Hence $A_0 = \dfrac{900 + 200 + 2(660 + 320)}{6} - 470$

$$= \frac{1100 + 1960}{6} - 470$$

$$= \frac{3060}{6} - 470$$

$$= 510 - 470 = 40 \text{ mA}$$

Hence mean current in the presence of a signal is 510 mA.

Example 5.29

For the transfer characteristic shown in the previous example find the peak value of the fundamental, second, third and fourth harmonics. Express total distortion, produced by the non-linearity of the characteristics, as a percentage.

Since $A_1 = \dfrac{I_{max} - I_{min} + I_1 - I_2}{3}$ from equation [5.76], then

$$A_1 = \frac{900 - 200 + 660 - 320}{3} \approx 313 \text{ mA}$$

and $A_2 = \dfrac{I_{max} + I_{min} - 2I_Q}{4}$ from equation [5.77], then

$$A_2 = \frac{1100 - 940}{4} = 40 \text{ mA}$$

and $A_3 = \dfrac{I_{max} - I_{min} - 2(I_1 - I_2)}{6}$ from equation [5.78], then

$$A_3 = \frac{700 - 680}{6} = 3 \text{ mA}$$

and $A_4 = \dfrac{I_{max} + I_{min} - 4(I_1 + I_2) + 6I_Q}{12}$ from equation [5.79], then

$$A_4 = \frac{1100 - 3920 + 2820}{12} = 0 \text{ mA}$$

The total effect of the harmonics can be found by summing the square of their amplitudes and expressing the square root of this sum as the harmonic amplitude A_T; i.e.

$$A_T = \sqrt{A_2^2 + A_3^2 + A_4^2}$$

so that in this example:

$$A_T = \sqrt{25^2 + 3^2 + 0^2} = 25.18 \text{ mA}$$

and the distortion produced, as a percentage of the fundamental component, is

$$D = \frac{A_T}{A_1} \times 100\%$$

so that in this case:

$$D = \frac{25.18}{313} \times 100\% = 8.04\%$$

Push-pull operation

The device used in a power amplifier stage has a non-linear characteristic liable to give rise to distortion of the output waveform. Consider Fig. 5.74 which shows the mutual characteristic for a bipolar junction transistor.

For class A bias, current flows for 160° of input voltage waveform but the curvature of the mutual characteristic gives rise to a current output waveform which is distorted compared to the input voltage

Fig. 5.74 BJT mutual characteristic showing the output current distortion due to non-linearity of the characteristic. Class A bias is used.

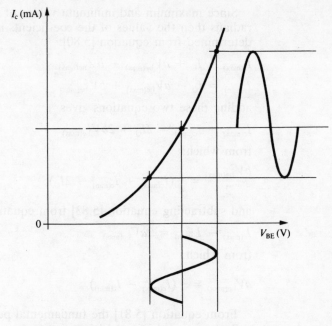

waveform. For simplicity the characteristic can be assumed to obey the relationship:

$$I_c = I_Q + aV_{be} + bV_{be}^2 + \ldots \qquad [5.80]$$

where a and b are constants.

If $V_{be} = V_{be(max)} \sin \omega t$ then:

$$I_c = I_Q + aV_{be(max)} \sin \omega t + bV_{be(max)}^2 \sin^2 \omega t + \ldots$$

$$= (I_Q + \frac{1}{2}bV_{be(max)}^2) + aV_{be(max)} \sin \omega t - \frac{1}{2}bV_{be(max)}^2 \cos 2\omega t \qquad [5.81]$$

The first term in equation [5.81] represents the d.c. component of the output current. The second term represents the fundamental frequency component, while the third term represents the second harmonic component.

The three components and the resultant added waveform have been drawn for one cycle in Fig. 5.75. The resultant waveform is exaggerated because some of the higher order terms in the original equation have been ignored.

Fig. 5.75 Effect of second harmonic term contribution to output current distortion.

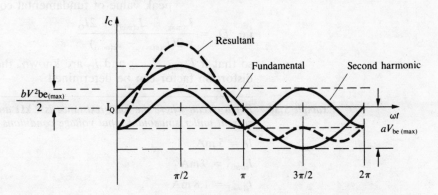

Since maximum and minimum values of I_c occur at $\pi/2$ and $3\pi/2$ radians then the values of the coefficients a and b can be determined from equation [5.80]:

$$I_{c(max)} = I_Q + aV_{be(max)} + bV_{be(max)}^2 \qquad [5.82]$$

$$I_{c(min)} = I_Q - aV_{be(max)} + bV_{be(max)}^2 \qquad [5.83]$$

adding these two equations gives

$$I_{c(max)} + I_{c(min)} = 2I_Q + 2bV_{be(max)}^2$$

from which:

$$\frac{bV_{be(max)}^2}{2} = \frac{1}{4}(I_{c(max)} + I_{c(min)} - 2I_Q)$$

and subtracting equation [5.83] from equation [5.82] gives

$$I_{c(max)} - I_{c(min)} = 2aV_{be(max)}$$

from which:

$$aV_{be(max)} = \frac{1}{2}(I_{c(max)} - I_{c(min)})$$

From equation [5.81] the fundamental peak component of collector current is given by:

$$I_{c1} = aV_{be(max)}$$

or $I_{c1} = \dfrac{1}{2}(I_{c(max)} - I_{c(min)})$

Hence useful power delivered to the resistive load is given by the product of the square of the fundamental current component and load resistance R_L (or reflected load resistance R_L' in the case of a transformer coupled load); i.e.

$$P_i = \left(\frac{\frac{1}{2}(I_{c(max)} - I_{c(min)})}{\sqrt{2}}\right)^2 R_L$$

or $P_i = \dfrac{(I_{c(max)} - I_{c(min)})^2 R_L}{8}$ W

In this case the factor which measures non-linear distortion associated with the resultant current waveform is the distortion factor (D) for the second harmonic, where

$$D_2 = \frac{\text{peak value of second harmonic component of current}}{\text{peak value of fundamental component}}$$

i.e. $D_2 = \dfrac{I_{c(max)} + I_{c(min)} - 2I_Q}{2(I_{c(max)} - I_{c(min)})}$

so that if $I_{c(max)}$, $I_{c(min)}$ and I_Q are known, the second harmonic distortion factor can be determined.

Example 5.30   *For a circuit where the load resistance is 2 kΩ and the values of collector current under sinusoidal input voltage conditions are:*

$I_Q = 5\,\text{mA}$

$I_{c(max)} = 9\,\text{mA}$

$I_{c(min)} = 1.8\,\text{mA}$

find the fundamental a.c. power output and the second harmonic distortion factor.

$$I_{c(max)} + I_{c(min)} = 10.8 \, \text{mA}$$

$$I_{c(max)} - I_{c(min)} = 7.2 \, \text{mA}$$

Fundamental a.c. power output:

$$P_i = \frac{(I_{c(max)} - I_{c(min)})^2 R_L}{8} \, \text{W}$$

$$= \frac{(7.2 \times 10^{-3})^2 \times 2 \times 10^3}{8} \, \text{W}$$

therefore $P_i = 12.96 \, \text{mW}$

Second harmonic distortion factor D is

$$D_2 = \frac{I_{c(max)} + I_{c(min)} - 2I_Q}{2(I_{c(max)} - I_{c(min)})}$$

$$= \frac{10.8 - 10}{2 \times 7.2} = 0.055$$

D_2 is usually expressed as a percentage, i.e. D_2 is 5.5% in this case.

Class A push-pull stage

The a.c. signal output from a class A single-ended amplifier will suffer from the distortion mentioned in the previous section. Since the worst distortion contribution is from the second harmonic content of the overall signal, the use of push-pull (double-ended) operation can reduce the second harmonic content and give a reasonable output at the same efficiency level as for a single-ended stage.

Figure 5.76 shows a possible circuit arrangement for a class A

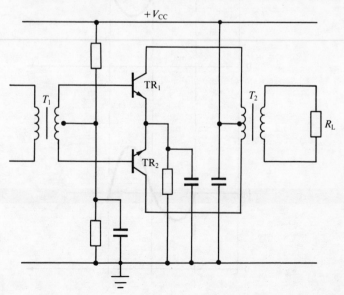

Fig. 5.76 Class A push-pull circuit using BJTs.

push-pull amplifier using bipolar junction transistors. Under steady-state conditions the current through each active device is *additive* at the centre tap of T_2. Under the steady-state (d.c) conditions the transistors act as if they are in *parallel*. The advantage of this arrangement is that the direct current flowing in each half of the primary of T_2 is in opposition; thus the flux in the transformer core produced by one current is minimised by the other current, reducing the possibility of d.c. saturation of the transformer core. For a balanced circuit the direct currents through each transistor would be equal thus completely cancelling the two magnetic fluxes.

Under signal conditions the effect of T_1 is to act as a phase-splitter so that TR_1 and TR_2 are driven in antiphase. The alternating currents produced in TR_1 and TR_2 collectors under these conditions are shown in Fig. 5.77. This means that as the signal current due to TR_1 is increasing, say, then the signal current to TR_2 is decreasing. The signal currents are therefore *additive* in the primary of T_2, i.e. as far as signal currents are concerned the two transistors appear to be in *series*. Since the two signal collector currents are added in T_2 primary (i.e. T_2 acts as a phase adder) the distortion effect is minimised. Using the value of signal collector current given in equation [5.80], i.e.

$$i_c = I_Q + aV_{be} + bV_{be}^2 + \ldots$$

then it follows that

$$i_{c1} = I_Q + aV_{be} + bV_{be}^2 \ldots \tag{5.84}$$

$$i_{c2} = I_Q - aV_{be} + bV_{be}^2 \tag{5.85}$$

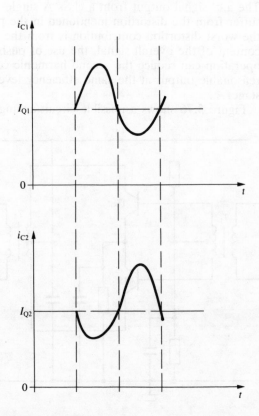

Fig. 5.77 One cycle of output current from each transistor of the class A push-pull BJT circuit.

Since i_{c2} is produced by an *inverted* V_{be}. The equations assume identical matched transistors.

The additive signal current in T_2 primary is produced by:

$$i_{c(total)} = i_{c1} + (-i_{c2})$$

therefore $i_{c(total)} = I_Q + aV_{be} + bV_{be}^2 \ldots - I_Q + aV_{be} - bV_{be}^2 + \ldots$

so that $i_{c(total)} = 2aV_{be}$ [5.86]

This shows that the second harmonic effect has been eliminated. The same could be said of all *even* harmonics, although as stated earlier the second harmonic distortion effect is the biggest contributor to the overall distortion. The elimination of second harmonic distortion can be seen by considering a *dynamic characteristic* obtained by using a *transfer characteristic* for each device. As Fig. 5.78 shows, the same input signal is used to produce i_{c1} and i_{c2} and matched devices are assumed.

Fig. 5.78 Transfer characteristics for each BJT in the push-pull circuit. Using the second device characteristic inverted, as shown, allows each device output current to be plotted directly. The output current in the phase-adding transformer primary is the sum of these two currents.

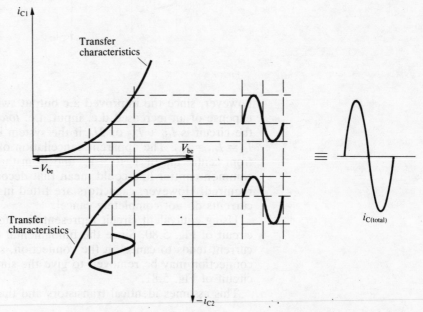

The two currents shown in Fig. 5.77 are the same as shown in Fig. 5.78 although i_{c2} is inverted (i.e. $-i_{c2}$). Inspection shows that summing the instantaneous values of i_{c1} and $-i_{c2}$ gives the resultant current $i_{c(total)}$. However, taking the sum of instantaneous currents for TR_1 and TR_2 of the transfer characteristics gives the dynamic characteristic of Fig. 5.79.

Inspection of Figs 5.78 and 5.79 shows that the resultant current is the same in both cases. The 'linearity' of the dynamic characteristic is instrumental in the elimination of even-order harmonics and their distortion effect in the output of each individual transistor.

As Figs 5.78 and 5.79 also show there is a larger current output swing for the push-pull circuit than for either of the individual circuits and with reduced distortion. The efficiency is *not* improved,

Fig. 5.79 The dynamic characteristic can be plotted directly by summing device currents for all voltage values. The total output current can then be found directly from the dynamic characteristic.

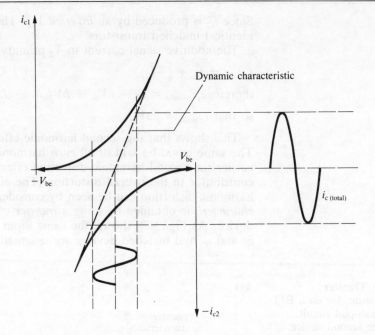

however, since the improved a.c output swing is produced at the expense of an increased d.c. input, i.e. *total* direct current input to the circuit is $I_{Q1} + I_{Q2}$ or $2I_Q$ if the system is balanced with $I_Q = I_{Q1} = I_{Q2}$. The apparent cancellation of signal current flowing from centre tap of T_2 to $+V_{cc}$ and from the earth line to the common emitter line could mean that decoupling capacitors are not required. However, capacitors are fitted in practice since the signal currents do not completely cancel.

Using equivalent circuit representation for TR_1 and TR_2 gives the circuit of Fig. 5.80. Since the fundamental component of collector current tends to cancel in the connection, shown as AB, then this connection may be removed to give the simplified output equivalent circuit of Fig. 5.81.

This assumes identical transistors and that:

Fig. 5.80 Equivalent circuit representation of a class A push-pull stage.

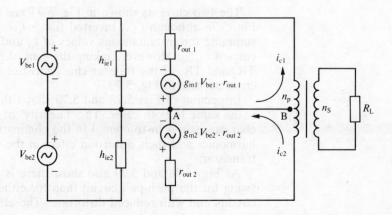

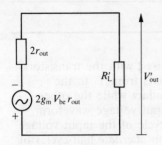

Fig. 5.81 Simplified equivalent circuit representation of a class A push-pull stage.

$$g_m = g_{m1} = g_{m2}$$

$$V_{be} = V_{be1} = V_{be2}$$

$$r_{out} = r_{out1} = r_{out2}$$

also the reflected load $R'_L = \left(\dfrac{2n_p}{n_s}\right) \times R_L$

or $R'_L = 4\left(\dfrac{n_p}{n_s}\right)^2 \times R_L$ [5.87]

It follows that the voltage developed across the primary of T_2 is V'_{out} where

$$V'_{out} = 2g_m . r_{out} . V_{be} . \left(\frac{R'_L}{R'_L + 2r_{out}}\right) \qquad [5.88]$$

and actual voltage output V_{out} is given by

$$V_{out} = \left(\frac{n_s}{2n_p}\right) V'_{out} \qquad [5.89]$$

or $V_{out} = \left(\dfrac{n_s}{2n_p}\right) \times 2g_m . r_{out} . V_{be} . \left(\dfrac{R'_L}{R' + 2r_{out}}\right)$ V   [5.90]

Example 5.31   *For a class A push-pull amplifier stage $R_L = 4\ \Omega$, $g_m = 120\ mS$ and $r_{out} = 400\ \Omega$ at the operating point. Assuming that transformer T_1 delivers voltage V_{be} of 0.2 V rms to each device, find (a) turns ratio of T_2 to achieve maximum power to the load, and (b) the power output to the load.*

(a) Since $R'_L = \left(\dfrac{2n_p}{n_s}\right)^2 . R_L$

and since for maximum power transfer

$$R'_L = 2 . r_{out}$$

then $\left(\dfrac{2n_p}{n_s}\right)^2 = \dfrac{2 . r_{out}}{R_L} = \dfrac{800}{4} = 200$

and $\dfrac{2n_p}{n_s} = \sqrt{200} = 14$

so that $\dfrac{n_p}{n_s} = 7:1$

The effective a.c. resistance seen by each device will be $\dfrac{800}{4} = 200\ \Omega$.

(b) Since $V'_{out} = 2 . g_m . r_{out} . V_{be} . \left(\dfrac{R'_L}{R'_L + 2 . r_{out}}\right)$

then $V'_{out} = 2 \times 120 \times 10^{-3} \times 400 \times 0.2\left(\dfrac{800}{1600}\right)$ volts

$$= 240 \times 10^{-3} \times 80 \times 0.5\ \text{V}$$

$$= 9.6\ \text{V (rms)}$$

but $V_{out} = \left(\dfrac{2n_p}{n_s}\right) = \dfrac{9.6}{14} = 0.68\ \text{V (rms)}$

and $P_{out} = \dfrac{V^2_{out}}{R_L} = \dfrac{(0.68)^2}{4}$ W

$$= 115\ \text{mW}$$

Class B push-pull stage

The same circuit as for Fig. 5.76 could be used with the transistors biased to cut-off. The antiphase voltage inputs from T_1 to the transistor bases will cause one device to conduct while the other device is cut off during half-cycles of the input voltage waveform. Thus each device conducts for only half a cycle of the input voltage waveform but, since each device conducts on *alternate* half-cycles of the input waveform, the summing transformer T_2 will recombine the individual current half-cycle pulses into a sinusiodal output current. This is shown in Fig. 5.82.

Fig. 5.82 Currents and voltages in a class B push-pull stage.

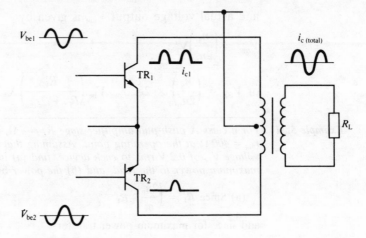

The circuit of a class B push-pull stage should be represented by an equivalent circuit similar to that of the class A stage.

Since the current in the primary of transformer T_2 is given by the difference in the collector currents then, assuming turns ratio of n_p/n_s for each half of the circuit as before, the load current I_L is given by

$$I_L = \frac{n_p}{n_s}(i_{c1} - i_{c2}) \tag{5.91}$$

The voltage developed across each half of the primary winding (V_1 or $-V_2$ as the case may be) produces an induced e.m.f. in the secondary of T_2 which, assuming no transformer losses, is given by

$$V_1 = -V_2 = \frac{n_p}{n_s}.V_L$$

where V_L is the voltage across the load resistance R_L.

but $V_1 = I_L R_L$

so that

$$V_1 = \left(\frac{n_p}{n_s}\right)^2 R_L(i_{c1} - i_{c2})$$

and $V_2 = \left(\dfrac{n_p}{n_s}\right)^2 R_L(i_{c2} - i_{c1})$

and since $\left(\dfrac{n_p}{n_s}\right)^2 R_L = R_L'$, the reflected load resistance, then

$$V_1 = R_L'(i_{c1} - i_{c2}) \tag{5.92}$$

and $V_2 = R_L'(i_{c2} - i_{c1})$ $\hspace{2cm}$ [5.93]

It follows that the collector voltage for TR_1 and TR_2 must be

$$V_{ce1} = V_{cc} - V_1$$
$$= V_{cc} - R_L'(i_{c1} - i_{c2}) \tag{5.94}$$

$$V_{ce2} = V_{cc} - V_2$$
$$= V_{cc} - R_L'(i_{c2} - i_{c1}) \tag{5.95}$$

Since the class B mode of operation assumes one transistor is off when the other conducts then i_{c1} or i_{c2} must always be zero. Assuming that a positive input voltage causes i_{c2} to be zero then equation [5.94] reduces to

$$V_{ce1} = V_{cc} - i_{c1}.R_L'$$

which is the normal equation for the response of a single transistor.

When the input voltage is negative, i_{c1} is zero and the operational path for TR_1 is a horizontal straight line to the right of the Q point. As V_{in} goes negative then, as equation [5.94] shows, the value of V_{ce1} depends on i_{c2}, which is now not zero, since i_{c2} flowing in the bottom half of the primary winding will induce a voltage in the top half. Since $V_1 = -V_2$ at all instants of time it follows that the peak value of V_{ce1} must reach $2V_{cc}$ assuming ideal conditions. The i_c versus v_{ce} relationship for either TR_1 or TR_2 is shown in Fig. 5.83.

Fig. 5.83 Output characteristic plot for each transistor in a class B push-pull stage.

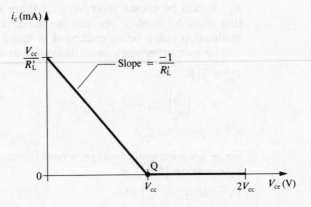

Using the principle of a composite load line then the two identical transistors can have the combined characteristic shown in Fig. 5.84.

Fig. 5.84 Composite output
characteristic for a class B
push-pull stage.

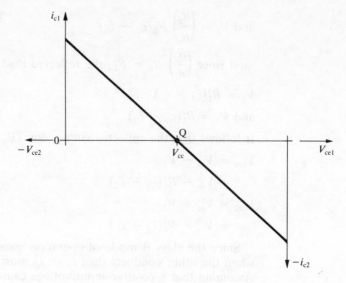

The output power possible from the stage is limited by the rated
maximum voltage, current and power dissipation values for the
transistors. Maximum voltage limit is allowed for by choosing the
supply voltage V_{cc} to be equal to, or less than, half the maximum
rated collector voltage. Maximum current limit may then be set to
prevent excessive waveform distortion at large values of I_c. With
regard to power dissipation there is a major difference between this
circuit and the class A push-pull circuit. For class A operation
maximum collector dissipation occurs under quiescent conditions and
reduces for signal conditions. Thus it is a simple matter to safely
locate the quiescent operating point within the allowable collector
dissipation area, i.e. to the left of the rated power dissipation curve.
For a class B stage the quiescent dissipation is zero and dissipation
increased with signal level. The value of dissipation is a function of
R'_L, as will be shown later, so that there is a minimum value of R'_L
that must be used to prevent the maximum possible collector
dissipation value being exceeded at some intermediate signal value.

The instantaneous power delivered to the load R_L is given by

$$P_{ac} = i_L^2 R_L$$

$$= R_L \left(\frac{n_p}{n_s}\right)^2 (i_{c1} - i_{c2})^2$$

$$= R'_L (i_{c1} - i_{c2})^2$$

Since instantaneous voltage across the transformer primary is given
by equation [5.92] as

$V_1 = R'_L (i_{c1} - i_{c2})$ then

$$P_{ac} = V_1 \times (i_{c1} - i_{c2})$$

From the output characteristic and assuming ideal characteristics it
can be seen that the maximum peak value of V_1 is V_{cc} while the
maximum peak value of current is I_{max}, so that maximum P_{ac} is

$$P_{ac} = \frac{V_{cc}}{\sqrt{2}} \frac{I_{max}}{\sqrt{2}} = \frac{V_{cc} I_{max}}{2} \text{ W}$$

$$= \frac{V_{cc}^2}{2R_L} \text{ W} \qquad [5.96]$$

Since the input d.c. power is the mean value of current over one cycle multiplied by the supply voltage V_{cc} then for a sine wave

$$\text{maximum } P_{dc} = V_{cc} \times I_{max} \times \frac{2}{\pi}$$

$$P_{dc} = \frac{2.V_{cc}^2}{\pi R_L'} \text{ W} \qquad [5.97]$$

The maximum collector conversion efficiency is

$$\eta = \frac{P_{ac(max)}}{P_{dc(max)}} \times 100\%$$

$$\text{or } \eta = \frac{V_{cc}I_{max}/2}{V_{cc}I_{max}2/\pi} \times 100\%$$

$$\text{or } \eta = \frac{\pi}{4} \times 100\% = 78.5\%$$

which is the same theoretical maximum efficiency as for a single transistor operated in class B bias. The advantage with push-pull of course if that the complete output voltage waveform is now possible instead of simply half a cycle as was the case for the single transistor circuit. Each of the two push-pull transistors contributes half a cycle to the overall output waveform.

The power dissipation at device collectors is given by

$$2P_{diss} = P_{dc} - P_{ac}$$

where P_{diss} is the average dissipation of each device.

Considering values of P_{ac} and P_{dc} other than maximum values

$$P_{diss} = \frac{1}{2}\left[\left(V_{cc} \times I \times \frac{2}{\pi}\right) - \left(\frac{V \times I}{2}\right)\right]$$

where I is the peak value of $(i_{c1} - i_{c2})$, which could be less than I_{max}, and V is the peak value of V_1, which could be less than V_{cc}. Therefore

$$P_{diss} = \frac{1}{2}\left[\left(V_{cc} \times \frac{V}{R_L'} \times \frac{2}{\pi}\right) - \left(\frac{V \times V}{2R_L'}\right)\right]$$

since $I = \dfrac{V}{R_L'}$

$$\text{and } P_{diss} = \frac{V_{cc}^2}{4R_L'}\left[\left(\frac{2V}{V_{cc}} \times \frac{2}{\pi}\right) - \left(\frac{V}{V_{cc}}\right)^2\right]$$

$$\text{or } P_{diss} = \frac{1}{2}P_{ac(max)}\left[\left(\frac{4}{\pi} \cdot \frac{V}{V_{cc}}\right) - \left(\frac{V}{V_{cc}}\right)^2\right] \qquad [5.98]$$

This is the equation of a parabola, which can be obtained by plotting P_{diss} against possible peak values of V up to the maximum value of V_{cc}. The value for P_{diss} is zero if

$$\frac{V}{V_{cc}} = \frac{4}{\pi} \text{ or if } \frac{V}{V_{cc}} = 0$$

i.e. if $V = \dfrac{4V_{cc}}{\pi}$ or 0

The maximum value for P_{diss} can be found by differentiating equation [5.98] with respect to V and equating the result to zero.

$$\frac{dP_{diss}}{dV} = \frac{1}{2}P_{ac(max)}\left[\frac{4}{\pi V_{cc}} - \frac{2V}{V_{cc}^2}\right] = 0$$

or $\dfrac{4}{\pi V_{cc}} = \dfrac{2V}{V_{cc}^2}$

or $\dfrac{V}{V_{cc}} = \dfrac{2}{\pi}$

or $V = \dfrac{2V_{cc}}{\pi}$

Substituting this value V in equation [5.98] gives

$$P_{diss(max)} = \frac{1}{2}P_{ac(max)}\left[\frac{4}{\pi}\cdot\frac{2}{\pi} - \left(\frac{2}{\pi}\right)^2\right]$$

$$P_{diss(max)} = \frac{1}{2}P_{ac(max)}\left[\frac{8}{\pi^2} - \frac{4}{\pi^2}\right]$$

$$P_{diss(max)} = \frac{2}{\pi^2}P_{ac(max)}$$

Thus the maximum dissipation rating of a transistor in class B push-pull can be considerably less than the maximum power it can deliver to a load. This is a vast improvement over the use of a single stage.

Since the signal level may be such as to produce the rated collector dissipation then this should be allowed for and, if the rated collector dissipation value is $P_{c(max)}$, then

$$P_{c(max)} \geqslant \frac{2}{\pi^2}P_{ac(max)} \tag{5.99}$$

or $P_{c(max)} \geqslant \dfrac{2}{\pi^2}\dfrac{V_{cc}^2}{2R_L'}$

or $P_{c(max)} \geqslant \dfrac{V_{cc}^2}{\pi^2 R_L'}$

giving $R_L' \geqslant \dfrac{V_{cc}^2}{\pi^2 P_{ac(max)}} \tag{5.100}$

This value of R_L' would be the threshold value above which the rated collector dissipation is exceeded as Fig. 5.85 shows.

If a circuit is constructed using the value $R_L' = V_{cc}^2/\pi^2 P_{ac(max)}$ it transpires that the load line crosses the hyperbola of device rated maximum allowable dissipation. The instantaneous dissipation may thus be very high, even though the average value may be low, and this could destroy the device.

A useful way of determining a suitable value for R_L' which does not allow the load line to stray into the region of excess dissipation is to place the load line tangentially to the dissipation hyperbola as shown in Fig. 5.86. If the point of contact is chosen so that it lies half-way along the load line it follows that the instantaneous collector dissipation at that point is

$$P_{c(max)} = \frac{V_{cc}}{2} \times \frac{I}{2}$$

Fig. 5.85 Curves of power dissipation plotted against voltage for various values of effective load resistance.

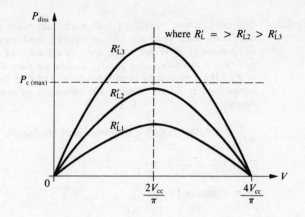

Figure 5.85

Fig. 5.86 Determination of effective load resistance by positioning the a.c. load line tangentially to the power dissipation curve at the quiescent point.

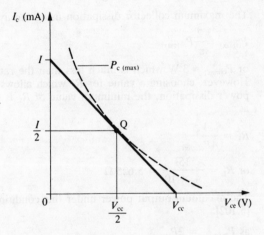

or $P_{c(max)} = \dfrac{V_{cc} \times I}{4}$ W

or $P_{c(max)} = \dfrac{V_{cc}^2}{4R_L'}$

$\left(\text{since } R_L' = \dfrac{V_{cc}}{I} \right)$

Hence the value of R_L' from equation [5.100] is

$$R_L' = \frac{V_{cc}^2}{4P_{c(max)}}$$ [5.101]

and from equation [5.96] the corresponding maximum output signal power is

$$P_{ac(max)} = 2P_{c(max)}$$ [5.102]

Compare this to equation [5.99] which gave $P_{ac(max)} = (\pi^2/2)P_{c(max)}$. This latter expression is for the case where the *average* dissipation is considered rather than the instantaneous dissipation. Even though, for the case where instantaneous power dissipation is considered, two push-pull transistors operating in class B bias can still deliver four times as much power as one device operating in class A.

Example 5.32   *Bipolar junction transistors with maximum rated collector current of 2 A, maximum rated collector voltage of 30 V and maximum rated collector dissipation of 10 W are to be used in a push-pull class B amplifier stage. Estimate a suitable value for R'_L and hence find the maximum possible value of P_{ac}. What would be a suitable turns ratio of the output transformer if the actual load is 3.5 Ω? What is the efficiency of the stage? Assume sinusoidal signals.*

For greatest possible output power choose $V_{cc} = \frac{1}{2}$ rated maximum value.

$$R'_L = \frac{V_{cc}}{I_{max}} = \frac{15}{2} = 7.5 \ \Omega$$

from equation [5.96]:

$$P_{ac(max)} = \frac{V^2_{cc}}{2R'_L} = = \frac{15^2}{2 \times 7.5} = 15 \ W$$

The maximum collector dissipation using the average value equation is

$$P_{c(max)} = \frac{2}{\pi^2} P_{ac(max)}$$

or $P_{c(max)} \approx 3 \ W$ which is much less than the rated maximum value of 10 W. However, choosing a value for R'_L which allows for instantaneous values of power dissipation, the minimum value of R'_L is given, from equation [5.101] as

$$R'_L = \frac{V^2_{cc}}{4P_{c(max)}}$$

or $R'_L = \dfrac{15^2}{4 \times 10} = 5.625 \ \Omega$

The maximum output power under this condition is given by equation [5.102]

as $P_{ac(max)} = 2P_{c(max)}$

or $P_{ac(max)} = 2 \times 10 = 20 \ W$

However, although the value of $P_{ac(max)}$ is higher in this case, the use of the minimum value of R'_L (5.625 Ω) operating with a V_{cc} of 15 volts (i.e. half the rated maximum value of 30 V) would cause the maximum rated collector current to be exceeded since $I_{c(max)} = V_{cc}/R'_L = 15/5.625 = 2.67$ A. Finally, using the average collector dissipation value of 10 W, the minimum value of R'_L necessary could be found from equation [5.100] where

$$R'_L = \frac{V^2_{cc}}{\pi^2 P_{c(max)}}$$

or $R'_L = \dfrac{15^2}{\pi^2 \times 10} = 2.28 \ \Omega$

and the maximum power output under this condition is

$$P_{ac(max)} = \frac{\pi^2}{2} . P_{c(max)}$$

or $P_{ac(max)} = 50 \ W$

Obviously the latter two values for $P_{ac(max)}$ are better than the first value. However, in both cases the rated maximum collector current has been exceeded, in the last calculation by a figure of 4.58 A. The use of maximum collector currents greater than the rated value could lead to difficulties and would certainly increase distortion.

Choosing the first value of R'_L, i.e. 7.5 Ω then since

$$\left(\frac{2n_p}{n_s}\right)^2 R_L = 4R_L'$$

a suitable turns ratio is

$$\frac{2n_p}{n_s} = \sqrt{\frac{4R_L'}{R_L}} = \sqrt{\frac{30}{3.5}} = 1.87{:}1$$

Also, because ideal characteristics have been assumed, efficiency is 78.5%

with $P_{ac(max)} = 15$ W and

$$P_{dc} = \frac{2}{\pi} \times \frac{V_{cc}^2}{R_L} \approx 19.1 \text{ W}$$

Example 5.33   *Identical transistors to be used in a class B push-pull power output stage have input and output characteristics shown in Fig. 5.87. Assuming the input voltage swing is such that the maximum value of I_b is 20 mA peak find: (a) fundamental output power, (b) value of mean output current under signal conditions; (c) total distortion and (d) conversion efficiency.*

Harmonic analysis: for this arrangement the collector current has the following values:

$I_{max} = 700$ mA

$I_{max} = 0$ mA

$I_1 = 100$ mA

$I_2 = 0$ mA

$I_Q = 0$ mA

Thus from equations [5.75], [5.76] and [5.78]:

$$A_0 = \frac{700 + 200}{6} = 150 \text{ mA}$$

$$A_1 = \frac{700 + 100}{3} = 266.67 \text{ mA}$$

$$A_3 = \frac{700 - 200}{6} = 83.34 \text{ mA}$$

Values for A_2 and A_4 have not been found since the even harmonic terms will cancel.

(a) Fundamental output power for each transistor is given by

$$P_{ac} = \frac{A_1 \times V_{ce(pk)}}{2} \text{ watts}$$

$$= \frac{266.67 \times 10^{-3} \times 17}{2} \text{ watts}$$

$$\approx 2.26 \text{ W}$$

Hence total output power = $2.P_{ac} = 4.52$ W

(b) Value of mean output current under signal conditions is

$$I_Q + A_0 = 150 \text{ mA}$$

(c) Total distortion. Since even harmonics cancel the only distortion produced is due to the third harmonic term A_3. Hence

$$D = \frac{A_3}{A_1} \times 100\%$$

$$\text{or } D = \frac{83.34}{266.67} \times 100\% \approx 31\%$$

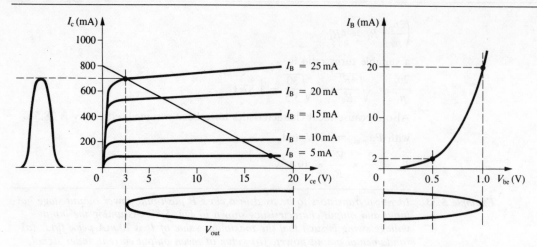

Fig. 5.87 Characteristics for a class B push-pull stage used for example 5.33.

The value for distortion can be evaluated for a single transistor as shown above since the total contribution of A_1 and A_3 terms for the push-pull arrangement is simply twice the value for each transistor considered alone.

(d) Conversion efficiency. Since

$P_{dc} = (I_Q + A_0).V_{cc}$ for one transistor,

$P_{dc} = 150 \times 10^{-3} \times 20$ watts

$P_{dc} = 3.0$ W

Conversion efficiency $\eta = \dfrac{P_{ac}}{P_{dc}} \times 100\%$

$$= \dfrac{2.26}{3} \times 100\% \approx 75.3\%$$

A major disadvantage of class B operation is **crossover distortion**. As the transfer characteristic for a typical silicon bipolar junction transistor in Fig. 5.88 shows, the transistor does not conduct when the base-emitter voltage is less than approximately 0.5 V. With the device biased at cut-off there will be considerable distortion at the 'bottom-end' of the current waveform. This type of distortion produces odd harmonics that are not minimised by the push-pull action.

Once again a dynamic characteristic could be produced from the device transfer characteristics as shown in Fig. 5.89.

Class AB operation

A class AB circuit can be produced by biasing each transistor so that in the absence of a signal input each device is on the threshold of conduction with a small idling current flowing. Figure 5.90 shows the effect of this on the dynamic characteristic. Class AB bias is such that current will flow in the output for just slightly greater than half a cycle of input voltage swing.

Fig. 5.88 Output current 'pulse' for a class B device showing the effect of 'crossover distortion'.

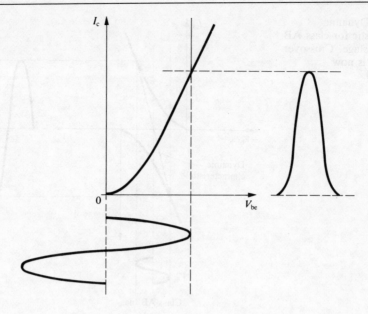

Fig. 5.89 Dynamic characteristic for class B push-pull stage showing the effect of crossover distortion.

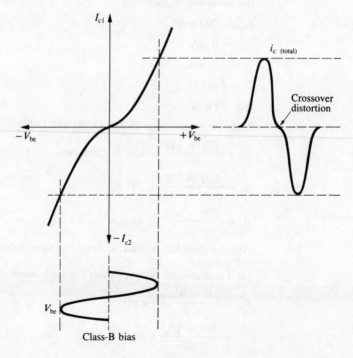

Fig. 5.90 Dynamic
characteristic for class AB
push-pull stage. Crossover
distortion is now
minimised.

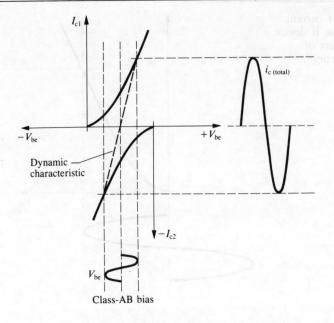

Class-AB bias

Example 5.34   *The transistors used in Example 5.33 are to be biased in class AB. Assume the same load line is to used and that the value of V_{cc} is adjusted to occur at 18 V. Find: (a) fundamental output power, (b) value of mean output current under signal conditions, (c) total distortion and (d) conversion efficiency. Assume the same value of peak I_c as in the previous example.*

The input conditions are altered by the new value of bias as shown in Fig. 5.91. Harmonic analysis: for this arrangement the collector current has the following values:

$I_{max} = 700 \, \text{mA}$

$I_{min} = 0 \, \text{mA}$

$I_1 = 220 \, \text{mA}$

$I_2 = 0 \, \text{mA}$

$I_3 = 75 \, \text{mA}$

Thus from equations [5.75], [5.76] and [5.78]:

$$A_0 = \frac{700 + 440}{6} - 100 = 115 \, \text{mA}$$

$$A_1 = \frac{700 + 220}{3} = 307 \, \text{mA}$$

$$A_3 = \frac{700 - 440}{6} = 43 \, \text{mA}$$

Again values for A_2 and A_4 have not been calculated since even harmonics will cancel.

(a) Fundamental output power for each transistor is given by

$$P_{ac} = \frac{A_1 \times V_{ce(pk)}}{2} \, \text{watts}$$

$$= \frac{307 \times 15.5}{2} \, \text{watts}$$

$$= 2.38 \, \text{W}$$

Fig. 5.91 Input
characteristic and signal
levels for example 5.34.

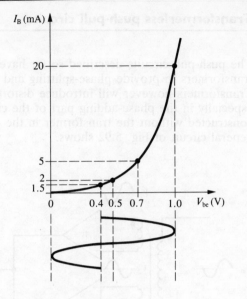

Hence total output power

$$= 2.P_{ac}$$

$$= 4.76 \text{ W}$$

(b) Value of mean output current under signal conditions is

$$I_Q + A_0 = 75 + 115 = 190 \text{ mA}$$

(c) Total distortion. Only distortion present is due to the third harmonic term A_3 so that

$$D = \frac{A_3}{A_1} \times 100\%$$

or $D = \dfrac{43}{307} \times 100\% = 14.0\%$

The value of distortion has been calculated for a single transistor since the values for A_1 and A_3 for both transistors is simply twice the value for an individual transistor. Compare the percentage distortion using class AB with the value obtained in example 5.33. The effect of a standing (mean) current has markedly reduced the distortion.

(d) Conversion efficiency. Since

$$P_{dc} = (I_Q + A_0)V_{cc} \text{ for one transistor}$$

$$P_{dc} = 190 \times 10^{-3} \times 18 \text{ W}$$

$$= 3.42 \text{ W}$$

conversion efficiency $\eta = \dfrac{P_{ac}}{P_{dc}} \times 100\%$

$$\eta = \frac{2.38}{3.42} \times 100\% = 70$$

Compare the efficiency of this arrangement with the value obtained in Example 5.33. Some advantage in terms of efficiency is thus traded off for improved distortion in class AB compared to class B operation.

Transformerless push-pull circuits

The push-pull circuits described so far have used centre-tapped transformers to provide phase-splitting and phase-adding circuits. Transformers however will introduce distortion themselves, especially in the phase-adding part of the circuit. Circuits can be constructed without the transformer in the output stage as the general circuit of Fig. 5.92 shows.

Fig. 5.92 Block diagram of a class A push-pull stage without a phase-adding transformer.

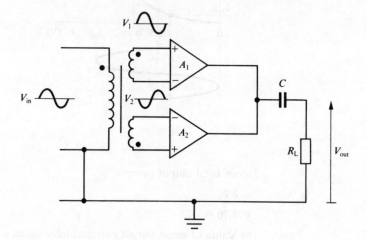

In this circuit amplifiers A_1 and A_2 are operated in class B (or AB) so that A_2 is cut off (or just conducting) when A_1 is fully conducting and vice versa. A possible circuit arrangement is shown in Fig. 5.93.

Since in this arrangement TR_1 only conducts during the positive half-cycle (assuming class B bias) of V_{in} and TR_2 only conducts during the negative half-cycle of V_{in} the circuit action can be easily explained by considering half-cycle of input voltage at a time, i.e.

Fig. 5.93 Possible circuit arrangement with BJTs in a class A push-pull stage without a phase-adding transformer.

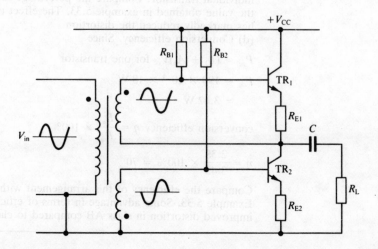

First half-cycle-V_{in} positive

TR_2 cuts off presenting a high impedance (Z) so circuit effectively becomes as shown in Fig. 5.94.

Since $Z \gg R_L$ its effect may be neglected and an output voltage is developed across R_L that is *in phase* with the input voltage since the circuit is a common-collector (emitter-follower) stage.

Fig. 5.94 Circuit modification for the first half-cycle of input voltage.

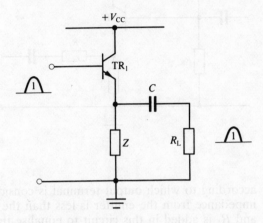

Second half-cycle-V_{in} negative

TR_1 cuts off presenting a high impedance (Z) so the circuit effectively becomes as shown in Fig. 5.95.

The output voltage is again across R_L but is now inverted because the circuit is effectively a common-emitter stage.

Fig. 5.95 Circuit modification for the second half-cycle of input voltage.

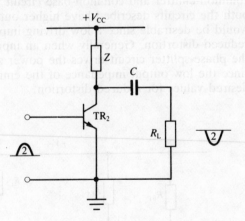

In this arrangement if the left-hand plate of C is kept at $+V_{cc}/2$ V then a maximum swing of $\pm V_{cc}/2$ volts can be obtained at the output. The disadvantage of the arrangement is that the output impedance of a common-collector stage is different from that of a common-emitter stage and also that the voltage gain of the circuit is less than unity.

The need for a phase-splitter transformer could be eliminated by the use of phase-splitting circuits, one of which is shown in Fig. 5.96. This is a split load phase inverter and, like the previous circuit, presents different output impedances to the following circuits

Fig. 5.96 A typical phase-splitter circuit using a BJT.

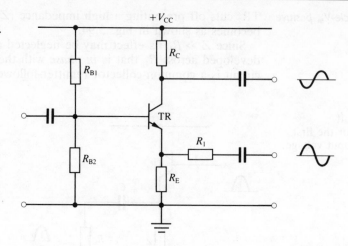

according to which output terminal is considered. The output impedance from the emitter is less than the value from the collector and R_1 is added in this circuit to equalise the two output impedances. Again the voltage gain of the circuit is less than unity.

A second circuit which overcomes the above problems is shown in Fig. 5.97. This is an emitter-coupled phase-splitter. In this arrangement TR_1 is a common-emitter circuit giving an output which is antiphase with V_{in}. TR_2 is arranged as a common-base stage giving an output which is in phase with V_{in}. The output impedance of a common-emitter and common-base circuit is similar and high. In fact both the circuits described have higher output impedances than would be desirable since a low driving impedance is preferred for reduced distortion. Generally when an input transformer is not used the phase-splitter circuit drives the power stage via emitter-followers since the low output impedance of the emitter-followers gives the desired values for reduced distortion.

Fig. 5.97 Emitter-coupled phase-splitter circuit.

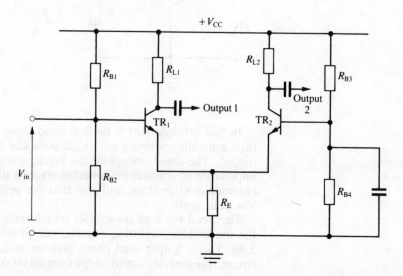

An alternative transformerless push-pull circuit uses the general circuit of Fig. 5.98. In this arrangement the same input signal is applied to both A_1 and A_2 simultaneously and both devices operate in the same mode, i.e non-inverting in this case.

Fig. 5.98 Block diagram of a transformerless push-pull circuit.

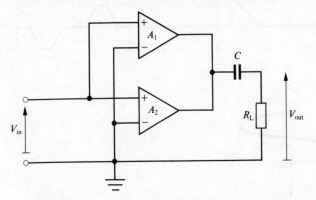

Obviously to produce a class B push-pull stage either A_1 or A_2 must be cut off at any instant while the other device conducts. The seemingly odd state of affairs where both devices are the same and receive the same input and yet only one device conducts at a given instant is resolved when matched pairs of power transistors (i.e. n-p-n/p-n-p for bipolar junction transistors) are used in what is known as **complementary symmetry**. Matching suggests that apart from conduction being a process of electrons or holes depending on the device, the output characteristics of each device are the same.

Consider the basic circuit as illustrated in Fig. 5.99. For this circuit a value of V_{in} with polarity shown would turn on TR_1 (n-p-n device) and turn off TR_2. As the polarity of V_{in} reverses the effect on the transistors also reverses and TR_2 conducts with TR_1 off. When V_{in} is absent both TR_1 and TR_2 will conduct equally under d.c. conditions and the voltage at the emitter junction will be $+V_{cc}/2$ volts. Thus when a signal is present and, say, TR_1 conducts, it acts as an emitter-follower with load of R_L and effective supply voltage of $+V_{cc} - V_{cc}/2$ or $+V_{cc}/2$ volts. Similarly when TR_2 conducts it too acts as an emitter-follower with a load of R_L and a supply voltage of $+V_{cc}/2$ volts which in this case is the charge across C. For this reason C should be large (at least 500 μF) in order to keep the supply voltage steady.

Fig. 5.99 Basic complementary circuit using BJTs.

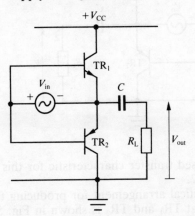

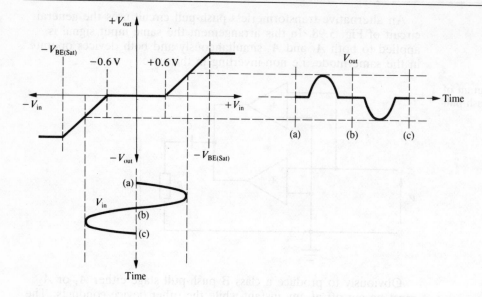

Fig. 5.100 Transfer
characteristic for the circuit
of Fig. 5.99.

The disadvantage of the circuit shown is that both TR_1 and TR_2 will not conduct until the voltage input between base and emitter exceeds approximately 0.6 V for a silicon device. Thus for the transfer characteristic shown in Fig. 5.100 nothing happens between 0 V and ±0.6 V while above this level conduction occurs. Certain values of V_{in} will cause V_{be} to saturate and limit the output but this has been ignored on this diagram.

As can be seen from Fig. 5.100 the output waveform suffers from the crossover distortion effect mentioned earlier for class B push-pull stages. What is needed is a bias circuit which keeps TR_1 and TR_2 on the threshold of conduction when the signal is absent and responds instantly when the signal appears. A possible arrangement is shown in Fig. 5.101.

Fig. 5.101 Modified
complementary circuit using
BJTs.

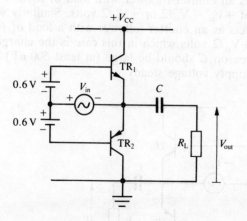

A revised transfer characteristic for this arrangement is shown in Fig. 5.102.

A practical arrangement for producing the required no-signal bias voltages at TR_1 and TR_2 is shown in Fig. 5.103.

Fig. 5.102 Transfer
characteristic for the circuit
of Fig. 5.101.

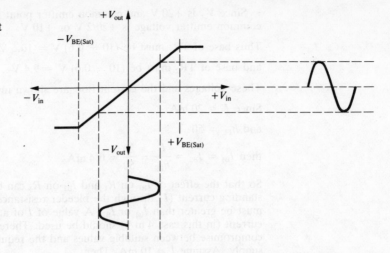

In this circuit when V_{in} is absent the common emitter voltage is $+V_{cc}/2$ volts as before and the biasing current flowing through $R_1 R_2$ and R_3 will produce base voltages of $((+V_{cc}/2) + 0.6)$ V and $((+V_{cc}/2) - 0.6)$ V for TR_1 and TR_2 respectively.

Since the volt drop across R_2 (i.e. the voltage difference between TR_1 base and TR_2 base voltage) must equal 1.2 V it follows that this could be achieved using two diodes instead of R_2 (since each diode has a forward conduction of 0.6 V across it). There are advantages to be gained in the use of diodes since temperature variations affecting TR_1 and TR_2 would also affect the diodes in a like manner. Thus there would be compensation for the variation in TR_1 and TR_2 collector current that could occur between the bases especially in integrated circuits where the fabrication of a diode is preferable to that of a resistor.

Example 5.35   *For the circuit shown in Fig. 5.103 calculate the values of R_1 R_2 and R_3 required to bias TR_1 and TR_2 for the threshold of conduction. Assume V_{cc} is +20 V and that transistors are identical with h_{FE} of 50 and as expected direct current of 20 mA will flow at each collector.*

Fig. 5.103 Practical
complementary circuit
arrangement using BJTs.

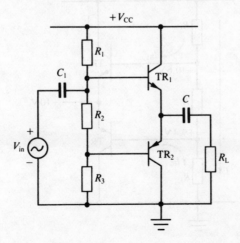

Since V_{cc} is +20 V and common emitter point is at $+V_{cc}/2$ volts then common emitter voltage is +20/2 V or +10 V.

Thus base of TR_1 must be $(10 + 0.6)$ V = 10.6 V

and base of TR_2 must be $(10 - 0.6)$ V = 9.4 V

These voltages and the current flow are shown in Fig. 5.104.

Since I_c = 20 mA

and h_{FE} = 50

then $I_{B1} = I_{B2} = \dfrac{I_c}{h_{FE}} = \dfrac{20}{50} = 0.4$ mA

So that the effect of I_{B1} on R_1 and I_{B2} on R_3 can be neglected, the value of standing current (I) through the bleeder resistance network R_1 R_2 and R_3 must be greater than I_{B1} or I_{B2}. A value of I of at least 10 times the base current (in this case 4 mA) should be used. There must be of course a compromise between suitable values and the requirements of the power supply. Assume I = 10 mA. Then:

$$R_1 = \left(\frac{20 - 10.6}{10}\right) k\Omega = 940 \; \Omega$$

$$R_2 = \left(\frac{10.6 - 9.4}{10}\right) k\Omega = 120 \; \Omega$$

$$R_3 = \left(\frac{9.4 - 0}{10}\right) k\Omega = 940 \; \Omega$$

Consider the circuit of Fig. 5.103 using the values for R_1 R_2 and R_3 calculated in Example 5.52. If the signal voltage increases positively to a peak value of, say, 5 V, the effect on the voltages at the various points can be examined. For example, the base of TR_1 will rise from 10.6 V to 15.6 V and the peak 5 V change will result in a voltage rise at TR_2 base given by

$$\left[9.4 + \left(\frac{R_3}{R_2 + R_3}\right)5\right] \text{ volts}$$

$$= \left[9.4 + \left(\frac{940}{1060}\right)5\right] \text{ volts}$$

$$= [9.4 + 4.43] \text{ V}$$

$$= 13.83 \text{ V}$$

Fig. 5.104 Circuit used to calculate the bias resistance values for example 5.35.

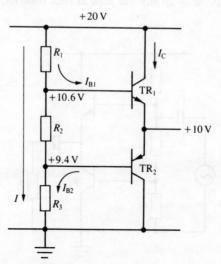

This means that the volt drop across R_2 is now $(15.6 - 13.83)$ V or 1.77 V. Since TR$_1$ acts as an emitter-follower the full 5 V increase is apparent at the left-hand plate of capacitor C so that the emitter junction rises to +15 V. Because C is so large the full 5 V is passed directly to the load R_L so that in this case V_{out} rises from 0 V to +5 V.

Because the emitter voltage is +15 V with TR$_1$ base at +15.6 V and TR$_2$ base at +13.83 V it follows that both TR$_1$ and TR$_2$ are conducting. TR$_2$ should be non-conducting under these circumstances since it would otherwise remove current from the load. This can be achieved by adding small resistors, typically 1 Ω in series with the transistor as Fig. 5.105 shows.

Fig. 5.105 Modified complementary circuit with emitter resistors added to prevent both devices conducting simultaneously.

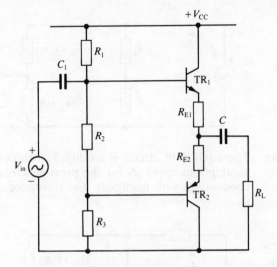

Now as V_{in} goes positive by 5 V peak this voltage change must be passed to the emitter of TR$_1$ and its 'load' of R_{E1} in series with R_L. If it is assumed that R_{E1} is 1 Ω and R_L is 5 Ω then the volt drop across each resistor is

$$V_{R_{E1}} = \left(\frac{1}{1+5}\right). \ 5 \text{ V} \approx 0.83 \text{ V}$$

$$V_{R_L} = V_{out} = \left(\frac{5}{1+5}\right). 5 \text{ V} \approx 4.17 \text{ V}$$

The no signal volt drop across R_{E1} and R_{E2} is very small (20 mV in the example shown) and will not affect the bias conditions.

Now the left-hand plate of capacitor C is at 14.17 V at peak input signal level and with the base voltage of TR$_2$ at 13.83 V (base voltages TR$_1$ and TR$_2$ are unaffected by adding R_E to the emitter circuit) then TR$_2$ is well off under all conditions of a positive input swing if V_{in}.

Using the circuit of Fig. 5.105 and analysing circuit response in two stages (i.e. for positive half-cycle of input voltage when TR$_1$ conducts, TR$_2$ OFF and for negative half-cycle of input voltage when TR$_2$ conducts and T$_1$ is OFF) gives a simplified equivalent circuit.

Positive half-cycle input   The equivalent circuit is as shown in Fig. 5.106. This assumes that the forward biased base-emitter junction of TR$_1$ has negligible resistance and that the bias resistance R_1 can be neglected (this will be explained shortly).

The resistances R_{E1} and R_L appear in this model to be values $(1 + h_{FE})$ times greater because TR$_1$ as an emitter-follower with 100% negative feedback will have an apparent input resistance of

$(1 + h_{FE})R'_L$ where R'_L is total emitter load, in this case, $R_{E1} + R_L$.

Since the voltage V_{in} is across both resistive combinations it follows that the output voltage for this half-cycle, V_{out1} is given by the ratio:

$$V_{out} = V_{in}\left(\frac{(h_{FE} + 1)R_L}{(h_{FE} + 1)R_L + (h_{FE} + 1)R_{E1}}\right)$$

$$\text{or } V_{out1} = V_{in}\left(\frac{R_L}{R_L + R_{E1}}\right) \qquad [5.103]$$

Fig. 5.106 Equivalent circuit for the complementary circuit of Fig. 5.105 for positive half-cycle of input voltage.

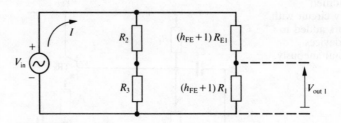

Negative half-cycle input   The equivalent circuit is modified as shown in Fig. 5.107. The same assumptions apply as for the previous circuit. Now TR_2 is conducting, with negligible a.c. resistance and TR_1 is off.

Fig. 5.107 Equivalent circuit for the complementary circuit of Fig. 5.105 for negative half-cycle of input voltage.

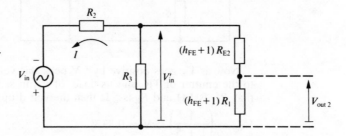

From the equivalent circuit it can be deduced that the voltage across the parallel combination of R_3 and $(h_{FE} + 1)(R_{E2} + R_L)$ is a fraction of V_{in} according to the values of R_2 and R_3, i.e.

$$V'_{in} = V_{in}\left(\frac{R_3}{R_2 + R_3}\right)$$

$$\text{and } V_{out2} = V'_{in}\left(\frac{(h_{FE} + 1)R_L}{(h_{FE} + 1)R_L + (h_{FE} + 1)R_{E2}}\right)$$

$$\text{or } V_{out2} = V_{in}\left(\frac{R_3}{R_2 + R_3}\right)\left(\frac{R_L}{R_{E2} + R_L}\right) \qquad [5.104]$$

It follows from equations [5.103] and [5.104] that the values for V_{out1} and V_{out2} are not the same but differ by a value determined by the ratio $R_3/(R_2 + R_3)$. The following example should make this clear.

Example 5.36   *Using the same circuit as for the previous example and assuming matched transistors with h_{FE} of 50 and peak input voltage of 5 V, find output voltage swings for positive and negative half-cycles of the input.*

From equation [5.103]:

$$V_{\text{out1}} = V_{\text{in}}\left(\frac{R_L}{R_L + R_{E1}}\right)$$

$$V_{\text{out1}} = 5\left(\frac{5}{5 + 1}\right) \approx 4.17 \text{ V}.$$

while from equation [5.104]:

$$V_{\text{out2}} = V_{\text{in}}\left(\frac{R_3}{R_2 + R_3}\right)\left(\frac{R_L}{R_L + R_{E2}}\right)$$

$$V_{\text{out2}} = 5\left(\frac{940}{940 + 120}\right)\left(\frac{5}{5 + 1}\right) \approx 3.7$$

The disparity in voltage levels in Example 5.36 is obviously a source of distortion and the only way of minimising it is to reduce the value of R_2 while still maintaining the required volt drop across it for biasing purposes. As mentioned earlier, a diode, or in this case a pair of diodes, would keep the required voltage difference between the two transistor bases but at the same time would offer a low forward reistance thus minimising the difference between V_{out1} and V_{out2}.

Another disadvantage of the circuit of Fig. 5.105 is due to the signal loss in the bias circuit. Considering again the equivalent circuit of Fig. 5.106 for the positive half-cycle of input voltage it can be seen that the signal current through the bias resistance network R_2 and R_3 is

$$i_1 = \frac{V_{\text{in}}}{R_2 + R_3} \text{ amps} \qquad [5.105]$$

while the signal current through the output resistances is

$$i_2 = \frac{V_{\text{in}}}{(h_{FE} + 1)(R_{E1} + R_L)} \text{ amps} \qquad [5.106]$$

Ideally the value of i_2 should be very much greater than i_1 to minimise signal loss through the bias network.

Example 5.37 *Using the same circuit as for the previous example, find the peak values of signal current through the bias network and the load during positive half-cycle of V_{in}.*

From equation [5.105]

$$i_1 = \frac{V_{\text{in}}}{R_2 + R_3}$$

hence $i_1 = \dfrac{5}{940 + 120}$ amps

$$= 4.72 \text{ mA}$$

and from equation [5.106]

$$i_2 = \frac{V_{\text{in}}}{(h_{FE} + 1)(R_{E1} + R_L)}$$

hence $i_2 = \dfrac{5}{(50 + 1)(1 + 5)}$

$$= 16.3 \text{ mA}$$

and signal current supplied $(i) = i_1 + i_2 \approx 21 \text{ mA}$.

During the negative half-cycle of input voltage the signal current through the bias resistance R_3 is

$$i_1 = \frac{V_{in}\left(\dfrac{R_3}{R_2 + R_3}\right)}{R_3}$$

$$\text{or } i_1 = \frac{V_{in}}{R_2 + R_3} \text{ A} \qquad\qquad [5.107]$$

which is the same as equation [5.105], while the signal current through the output resistance is

$$i_2 = \frac{V_{in}\left(\dfrac{R_3}{R_2 + R_3}\right)}{(h_{FE} + 1)(R_{E2} + R_L)} \qquad\qquad [5.108]$$

Example 5.38   *Using the same circuit as for the previous example find the peak values of signal current through R_3 and the load during negative half-cycle of V_{in}.*

From equation [5.107]

$$i_1 = \frac{V_{in}}{R_2 + R_3}$$

$$= \frac{5}{1060} = 4.72 \text{ mA}$$

while from equation [5.108]

$$i_2 = \frac{V_{in} \cdot \left(\dfrac{R_3}{R_2 + R_3}\right)}{(h_{FE} + 1)(R_{E2} + R_L)}$$

$$= 14.4 \text{ mA}$$

Again signal current supplied $(i) = i_1 + i_2 = 19.12 \text{ mA} \approx 19 \text{ mA}$

A method of reducing the signal drain through the bias network, by making the resistance of the bias circuit appear larger than its actual value under signal conditions, is known as 'bootstrapping'. An example of the use of bootstrapping in the complementary stage is shown in Fig. 5.108.

In this circuit the bottom-end of R_3 is connected to earth via R_L instead of directly as before. The same value of bias current I is assumed since the effect of R_L on the bias circuit resistance is negligible. There will be a quiescent volt drop across R_L but again this is negligible compared to the expected voltage swing under signal conditions.

Bearing in mind that the equivalent circuit of an emitter-follower has an input impedance of $(h_{FE} + 1)R_L'$ where R_L' is the total emitter load and assuming h_{ie} can be neglected, the bootstrapped complementary circuit can be redrawn in equivalent circuit form for both half-cycles of V_{in}. Thus the peak signal voltage swing across R_L can be found for each half-cycle of the input as before.

Positive half-cycle input   Equivalent circuit becomes as shown in Fig. 5.109. From the equivalent circuit

$$V_{in} = I[(R_2 + R_3)/\!/(h_{FE} + 1)R_{E1}] + IR_L + h_{FE}I_B R_L \qquad\qquad [5.109]$$

(the sign $/\!/$ means 'in parallel with') but

Fig. 5.108 A bootstrapped complementary stage.

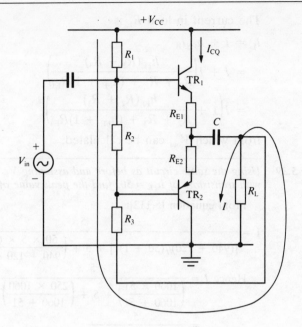

Fig. 5.109 Equivalent circuit representation of the bootstrapped complementary circuit for positive half-cycle of input voltage.

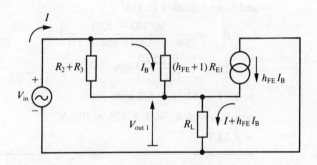

$$I_{\text{B}} = I\left[\frac{R_2 + R_3}{R_2 + R_3 + (h_{\text{FE}} + 1)R_{\text{E1}}}\right] \qquad [5.110]$$

and substituting for I_{B} in equation [5.109] gives

$$V_{\text{in}} = I[(R_2 + R_3)/\!/(h_{\text{FE}} + 1)R_{\text{E1}}] + IR_{\text{L}} + I\left(\frac{h_{\text{FE}}R_{\text{L}}(R_2 + R_3)}{R_2 + R_3 + (h_{\text{FE}} + 1)R_{\text{E1}}}\right) \qquad [5.111]$$

from which

$$I = \frac{V_{\text{in}}}{((R_2 + R_3)/\!/(h_{\text{FE}} + 1)R_{\text{E1}}) + R_{\text{L}} + \left(\dfrac{h_{\text{FE}}R_{\text{L}}(R_2 + R_3)}{R_2 + R_3 + (h_{\text{FE}} + 1)R_{\text{E1}}}\right)} \qquad [5.112]$$

The current in load R_L is

$$I_L = I + h_{FE}I_B$$

$$= I + I\left(\frac{h_{FE}(R_2 + R_3)}{R_2 + R_3 + (h_{FE} + 1)R_{E1}}\right)$$

$$= I\left(1 + \frac{h_{FE}(R_2 + R_3)}{R_2 + R_3 + (h_{FE} + 1)R_{E1}}\right) \qquad [5.113]$$

from which V_{out} can be calculated.

Example 5.39   *Using the same circuit as before and assuming V_{in} has a peak value of 5 V with transistors of $h_{FE} = 50$, find the peak value of V_{out}.*

From equation [5.112]:

$$I = \frac{5}{[(940 + 120)//(50 + 1)1] + 5 + \left(\dfrac{50 \times 5 \times (940 + 120)}{940 + 120 + (50 + 1)1}\right)}$$

Hence $I = \dfrac{5}{\left(\dfrac{1060 \times 51}{1060 + 51}\right) + 5 + \left(\dfrac{250 \times 1060}{1060 + 51}\right)}$

$$= \frac{5}{48.67 + 5 + 238.5}$$

$$= 17 \, \text{mA}$$

and from equation [5.113]

$$I_L = I\left(1 + \frac{50(940 + 120)}{940 + 120 + (50 + 1)1}\right)$$

$$= 17\left(1 + \frac{50 \times 1060}{1111}\right) \text{mA}$$

$$= 17(1 + 47.7) \approx 828 \, \text{mA}$$

and $V_{out} = I_L R_L = 5 \times 828 \times 10^{-3} \, \text{V}$

$$= 4.14 \, \text{V}$$

Negative half-cycle input.   Equivalent circuit becomes as shown in Fig. 5.110. From the equivalent circuit:

$$V_{in} = I\left[R_2 + \frac{R_3(h_{FE} + 1)R_{E2}}{R_3 + (h_{FE} + 1)R_{E2}} + R_L\right] + h_{FE}I_B R_L \qquad [5.114]$$

while $I_B = I\left[\dfrac{R_3}{R_3 + (h_{FE} + 1)R_{E2}}\right]$ \qquad [5.115]

Hence substituting for I_B in equation [5.114] gives

$$V_{in} = I\left[R_2 + R_L + \frac{R_3(h_{FE} + 1)R_{E2}}{R_3 + (h_{FE} + 1)R_{E2}}\right] + \frac{h_{FE}IR_LR_3}{R_3 + (h_{FE} + 1)R_{E2}}$$

or $V_{in} = I\left[R_2 + R_L + \dfrac{R_3(h_{FE} + 1)R_{E2} + h_{FE}R_LR_3}{R_3 + (h_{FE} + 1)R_{E2}}\right]$ \qquad [5.116]

from which I can be calculated. Also since current in the load $I_L = I + h_{FE}I_B$ then

Fig. 5.110 Equivalent circuit representation of the bootstrapped complementary circuit for negative half-cycle of input voltage.

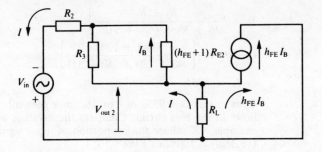

$$I_L = I + \frac{h_{FE}IR_3}{R_3 + (h_{FE} + 1)R_{E2}}$$

$$\text{or } I_L = I\left(1 + \frac{h_{FE}R_3}{R_3 + (h_{FE} + 1)R_{E2}}\right) \tag{5.117}$$

from which V_{out2} can be calculated.

Example 5.40 *Using the circuit of example 5.39 with the same values for $V_{in(pk)}$ and h_{FE} for the transistors, find the peak value of V_{out2}.*

From equation [5.116]

$$I = \frac{5}{120 + 5 + \left(\dfrac{940(50 + 1)1 + 50 \times 5 \times 940}{940 + (50 + 1)1}\right)}$$

hence $I = \dfrac{5}{120 + 5 + \left(\dfrac{47940 + 235000}{940 + 51}\right)}$

$I = \dfrac{5}{125 + 285.5} \approx 12.2 \text{ mA}$

and from equation [5.117]

$$I_L = I\left(1 + \frac{50 \times 940}{940 + (50 + 1)1}\right)$$

$$= 12.2\left(1 + \frac{50 \times 940}{991}\right) \text{ mA} = 591 \text{ mA}$$

so that $V_{out2} = I_L R_L = 5 \times 591 \times 10^{-3} = 2.95 \text{ V}$

The values for V_{out1} and V_{out2} for the bootstrapped complementary amplifier still show a disparity proving that bootstrapping does not improve the value of V_{out2}. Only a reduction in the value of R_2 will do that. The significance of bootstrapping is the improvement in the proportion of input signal current I that flows through the device input compared to that amount that flows through the input bias circuit. Consider the following example.

Example 5.41 The bootstrapped complementary amplifier of Fig. 5.108 has the values for input signal current for the postive and negative half-cycles of input voltage respectively. Calculate the proportion of signal current through the input of the device and through the bias circuit for each half-cycle of input.

(a) Positive half-cycle of input. Since from equation [5.110]

$$I_B = I\left[\frac{R_2 + R_3}{R_2 + R_3 + (h_{FE} + 1)R_{E1}}\right]$$

then $I_B = I\left[\dfrac{940 + 120}{940 + 120 + (50 + 1)1}\right]$

or $I_B = 0.95I$

Thus in this case 95% of I or 16.2 mA goes into the device while 0.8 mA flows in the bias circuit. Compare these values with the values obtained in example 5.35 where the proportion of input signal current to the device was $(16.3/20) \times 100\%$ or 81%

(b) Negative half-cycle of input. From equation [5.115]

$$I_B = I\left[\frac{R_3}{R_3 + (h_{FE} + 1)R_{E1}}\right]$$

$$I_B = I\left[\frac{940}{940 + (50 + 1)1}\right]$$

or $I_B = 0.95I$

Since in this case $I \approx 12.2$ mA then $I_B \approx 11.6$ mA so that 0.6 mA flows into the bias circuit.

Compare the values of example 5.41 with those obtained in example 5.38 where the proportion of input signal current to the device was $(14.4/19.12) \times 100\%$ or 75%

The reductions in signal current through the bias circuit in a bootstrapped complementary amplifier as explained in example 5.41 is caused by making the bias resistance under signal conditions appear to be larger than it actually is. Referring again to Fig. 5.108 when the input signal goes positive, say, then by emitter-follower action via TR_1 the output also goes positive by an amount slightly less than the input variation. (Example 5.36 showed that for an input voltage swing of $+5$ V peak, V_{out} increased from 0 V to $+4.17$ V peak.) Thus the 'bottom-end' of R_3 connected to the right-hand plate of capacitor C means that instead of the $+5$ V peak causing a signal current to flow to earth via R_2 and R_3, the actual signal voltage across R_2 and R_3 is the difference between V_{in} and V_{out} and signal current via R_2 and R_3 must fall. The $+5$ V peak applied to the base of TR_1 with a *reduced* signal current through R_2 and R_3 suggests that the value of the bias resistors is higher than the actual figure. Using the signal voltage levels of example 5.36 together with the calculated bias resistors' share of signal current from example 5.41(a) suggests the bias resistance $(R_2 + R_3)'$ has a value of:

$$(R_2 + R_3)' = \frac{\text{volts applied}}{\text{current taken}} = \left(\frac{5.0}{0.8}\right) \text{k}\Omega = 6.25 \text{ k}\Omega$$

Bootstrapping therefore gives the advantage of high a.c. bias resistance while leaving the d.c. bias conditions unaffected.

The driver stage

Examples 5.37 and 5.39 have shown that with circuit component, and active device parameter, values used the input signal current for a +5 V positive input voltage swing is 20 mA and 17 mA respectively. In either case the input resistance of the stage as seen by the source is very low, i.e. $5/20 \times 10^{-3}\,\Omega$ or 250 Ω and $5/17 \times 10^{-3}\,\Omega$ or 300 Ω respectively.

To drive such a stage directly with a voltage amplifier would result in the gain of the driver being severely reduced by the shunting action of the input resistance of the power stage on the collector load of the driver. An emitter-follower however could be used to buffer the voltage amplifier and the power amplifier and this can be done by substituting an emitter-follower circuit for the bias resistor R_1 as shown in Fig. 5.111.

Fig. 5.111 Driver stage for the complementary output circuit.

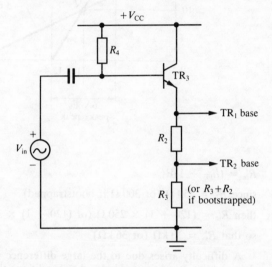

Example 5.42 *For the circuit of Fig. 5.111 assuming TR₃ collector current has a quiescent value of 10 mA and $R_2 + R_3$ (or $R_2 + R_3 + R_L$, if bootstrapped) is 1060 Ω, calculate the value of R_4 and effective input resistance of the stage. Assume h_{FE} for TR₃ is 120 and that bias voltages for TR₁ and TR₂ base are 10.6 V and 9.4 V respectively. Take V_{cc} as 20 V.*

For the circuit of Fig. 5.111, if I_{CQ} for TR₃ is 10 mA and h_{FE} is 120 then

$$I_{BQ} = \frac{10}{120}\,\text{mA}$$

and the volt drop across $R_4 = V_{CC} - V_{BB}$

$$= 20 - 11.2 = 8.8\,\text{V}$$

(since $V_{BB} = 0.6\,\text{V}$ above TR₃ emitter which is 10.6 V under quiescent conditions).

Therefore

$$R_4 = \frac{8.8}{\left(\dfrac{10}{120} \times 10^{-3}\right)} = 106\,\text{k}\Omega$$

By emitter-follower action effective input resistance is

$$R'_{in} = h_{ie} + (h_{FE} + 1)R'_L$$

where R'_L is the effective emitter load under a.c. conditions. This can be simplified by neglecting h_{ie} since its value will be small compared to $(h_{FE} + 1)R'_L$ so that

Fig. 5.112 D.C. and a.c.
load lines plotted on the
output characteristics to
show the large variation in
load value between the d.c.
and a.c. conditions.

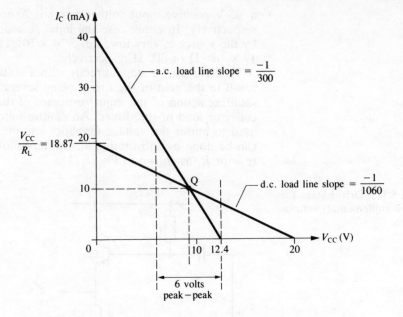

$R'_{in} = (h_{FE} + 1)R'_{L}$

since $R'_{L} = 250 \, \Omega$ (or $300 \, \Omega$ if bootstrapped)

then $R'_{in} = (120 + 1) \times 250 \, \Omega$ (or $(120 + 1) \times 300 \, \Omega$)

so that $R'_{in} = 30 \, k\Omega$ (or $36 \, k\Omega$)

A difficulty arises due to the large difference in emitter resistance under
d.c. and a.c. conditions. Under d.c. conditions the load R_{L} is $1060 \, \Omega$ while
under a.c. conditions the best value for R'_{L} is $300 \, \Omega$ for the bootstrappèd
circuit. The effect of the large difference between the d.c. and a.c. values of
R'_{L} can be best explained with reference to the characteristic of Fig. 5.112.

Little accuracy has been lost by showing I_{c} on the characteristic since
$I_{c} \approx I_{E}$ in practice. The d.c. load line for TR$_{3}$ will intercept the V_{ce} axis at
V_{cc} and the I_{c} axis at V_{cc}/R_{L} as shown. The Q-point occurs for I_{CQ} of
$10 \, mA$. Since the a.c. load is $300 \, \Omega$ the a.c. load line must pass through the
Q-point with a slope of $-1/300$ and intercept the V_{ce} axis at $12.4 \, V$. This
will only give a possible $6 \, V$ peak-peak or $3 \, V$ peak voltage swing at the
output.

To avoid restricting the a.c. swing at TR$_{3}$ output the a.c. and d.c.
resistances must be made nearly as equal as possible. Since for the
bootstrapped circuit

$R_{L} = R_{2} + R_{3} + R_{L}$

and $R_{L} = [(R_{2} + R_{3})/(h_{FE} + 1)R_{E1}] + R_{L} + \left(\dfrac{h_{FE}R_{L}(R_{2} + R_{3})}{R_{2} + R_{3} + (h_{FE} + 1)R_{E1}} \right)$

from equation [5.111] then for $R_{L} = R'_{L}$

$R_{2} + R_{3} + R'_{L} =$
$$\dfrac{(R_{2} + R_{3})(h_{FE} + 1)R_{E1}}{R_{2} + R_{3} + (h_{FE} + 1)R_{E1}} + R'_{L} + \dfrac{h_{FE}R_{L}(R_{2} + R_{3})}{R_{2} + R_{3} + (h_{FE} + 1)R_{E1}}$$

Hence $[(R_{2} + R_{3}) + (h_{FE} + 1)R_{E1}] = (h_{FE} + 1)R_{E1} + h_{FE}R_{L}$

and $(R_{2} + R_{3}) + R_{E1} = R_{E1} + h_{FE}R_{L}$

or $R_{2} + R_{3} = h_{FE}R_{L}$

In the examples 5.35 to 5.42 it has been assumed that $R_2 + R_3 = 1060\ \Omega$, while $h_{FE} = 50$ and $R_L = 5\ \Omega$ so that $h_{FE}R_L = 250\ \Omega$. Thus to improve matters the values of R_2 and R_3 should be reduced although this does mean a higher standby current from the power supply for the power amplifier biasing.

Power output Generally the same rules apply as for the previous class B explanation except in this case the maximum peak output voltage swing is restricted to a smaller value than previously. Since TR_3 acts as a driver input stage the emitter of TR_3 (and the base of TR_1) can change between 0 V and $+V_{cc}$. Thus the voltage across R_L can vary from a maximum value of $V_{cc}/2 - V_{BE1}$ to a minimum value of $-V_{cc}/2 + V_{BE2}$. Thus the peak-peak swing across

$$R_L = \frac{V_{cc}}{2} - V_{BE1} - \left(-\frac{V_{cc}}{2} + V_{BE2}\right)$$

i.e. $V_{pk-pk} = V_{cc} - V_{BE1} - V_{BE2}$

However this ignores R_E and including this resistance output voltage swing to a peak-peak value of

$$V_{pk\text{-}pk} \times \left(\frac{R_L}{R_E + R_L}\right)\ \text{V}$$

thus $$P_{ac} = \left[\frac{V_{pk-pk}}{2\sqrt{2}} \times \frac{R_L}{R_E + R_L}\right]^2 \times \frac{1}{R_L} \qquad [5.118]$$

so that under maximum conditions and ignoring the effects of V_{BE1}, V_{BE2} and R_E

$$P_{ac} \approx \frac{V_{cc}^2}{8R_L} \qquad [5.119]$$

Compare this with equation [5.96] for the conventional class B push-pull amplifier.

Since the mean d.c. current drawn from the supply depends on the peak value of output voltage, which can have a maximum value of $\dfrac{V_{cc}}{2}$ ignoring V_{BE1} or V_{BE2}, then

$$I_{dc} = \frac{1}{\pi}\left(\frac{V_{pk}}{R_L}\right)$$

and $$P_{dc} = \frac{V_{pk}V_{cc}}{\pi R_L}\ \text{W} \qquad [5.120]$$

and maximum value of P_{dc} occurs when $V_{pk} = \dfrac{V_{cc}}{2}$ so that

$$P_{dc(max)} = \frac{V_{cc}^2}{2\pi R_L}\ \text{W} \qquad [5.121]$$

Compare this with equation [5.97] for the conventional class B push-pull amplifier.

Maximum efficiency $\eta = \dfrac{P_{ac}}{P_{dc}} \times 100\%$

$$= \frac{\left(\dfrac{V_{cc}^2}{8R_L}\right)}{\left(\dfrac{V_{cc}^2}{2\pi R_L}\right)} \times 100\%$$

$$= \frac{\pi}{4} \times 100\% = 78.5\%$$

which is the same as for the conventional class B push-pull circuit.
Power dissipated in transistor $P_{diss} = P_{dc} - P_{ac}$

and for each transistor, $P_{diss} = \dfrac{V_{pk}V_{cc}}{2\pi R_L} - \dfrac{V_{pk}^2}{4R_L}$ [5.122]

and differentiating equation [5.122] with respect to V_{pk} gives
maximum power dissipation when $V_{pk} = V_{cc}/\pi$. Hence

$$P_{diss(max)} = \frac{V_{cc}^2}{2\pi^2 R_L} - \frac{V_{cc}^2}{4\pi^2 R_L}$$

or $P_{diss(max)} = \dfrac{V_{cc}^2}{4\pi^2 R_L}$ [5.123]

Example 5.43 *For a complementary push-pull circuit where V_{cc} is 20 V, R_L is 5 Ω what is
(a) maximum power output, (b) power drawn from supply at maximum
output and (c) maximum transistor dissipation? Ignore the effects of V_{be} and
R_E for each transistor output stage.*

(a) From equation [5.119] $P_{ac(max)} = \dfrac{V_{cc}^2}{8R_L} = \dfrac{400}{40} = 10$ W.

(b) From equation [5.121] $P_{dc(max)} = \dfrac{V_{cc}^2}{2\pi R_L} = \dfrac{400}{10\pi} = 12.72$ W.

(c) From equation [5.123] $P_{diss(max)} = \dfrac{V_{cc}^2}{4\pi^2 R_L} = \dfrac{400}{20\pi^2} = 2$ W.

Example 5.44 *Using the same circuit as for example 5.41 but assuming that the peak-peak
voltage swing at the base of TR_1 of the circuit of Fig. 5.108 is limited to
6.0 V, find the output power, d.c. power drawn from the supply and stage
efficiency. Assume that the effects of V_{be1} and V_{be2} and R_{E1} and R_{E2} must be
taken into account and that $V_{be1} = V_{be2} = 0.6$ V and $R_{E1} = R_{E2} = 1$ Ω.*

From equation [5.118], $P_{ac} = \left[\dfrac{V_{pk-pk}}{2\sqrt{2}} \times \left(\dfrac{R_L}{R_L + R_E}\right)\right]^2 \times \dfrac{1}{R_L}$ watts

where $V_{pk-pk} = 6.0 - 0.6 - 0.6 = 4.8$ V

$$P_{ac} = \left[\frac{4.8}{2\sqrt{2}} \times \frac{5}{6}\right]^2 \times \frac{1}{5} \text{ watts}$$

$$P_{ac} = 0.4 \text{ W.}$$

From equation [5.120], $P_{dc} = \dfrac{V_{pk}V_{cc}}{\pi R_L}$ watts

$$P_{dc} = \frac{2.4 \times 20}{5\pi} \times \frac{5}{6} \text{ watts}$$

$$P_{dc} = 2.55 \text{ W.}$$

Efficiency $= (P_{ac}/P_{dc}) \times 100\% = (0.4 \times 100)/2.55 = 15.7\%$

This poor efficiency is due to the large discrepancy between a.c. load
resistance and d.c. load resistance as mentioned in this section under the

heading of the driver stage. If it can be arranged that the a.c. and d.c. loads are the same for this driver stage then the peak-peak voltage swing at the base of TR_1 in Fig. 5.108 can be 20 V. However, if V_{be1} and V_{be2} and R_{E1} and R_{E2} are still taken into account, the revised value of P_{ac} is

$$P_{ac} = \left[\frac{18.8}{2\sqrt{2}} \times \frac{5}{6}\right]^2 \times \frac{1}{5} = 6.136 \text{ W}$$

and since $P_{dc} = \dfrac{9.6 \times 20}{5} \times \dfrac{5}{6} = 10.19 \text{ W}$

then efficiency $\eta = (6.136/10.19) \times 100\% = 60\%$

So that efficiency is much improved compared to the previous value but is still much less than the theoretical maximum possible value.

Including a voltage amplifier the complete complementary amplifier takes the form shown in Fig. 5.113.

Fig. 5.113 A possible complementary stage with voltage amplifier and driver stage.

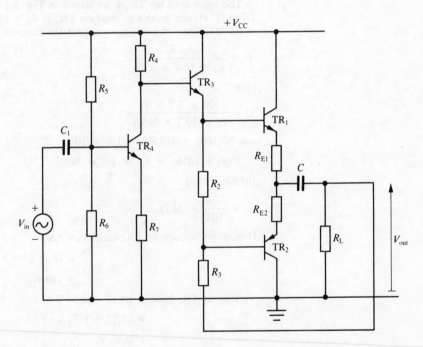

When considering the design of the buffer stage TR_3 in example 5.42, the value of R_4 was found to be 106 kΩ. This will now have to be modified to allow for the fact that R_4 is now also the collector load for the voltage amplifier TR_4. This can be explained best by considering the following example.

Example 5.45 *Assuming the criteria regarding quiescent voltages and currents for TR_3 are the same as described in example 5.42, find suitable values for R_4, R_5, R_6 and R_7. Assume standby current for TR_4 is 5 mA and its h_{FE} is 100. Voltage gain of TR_4 is to be 50 and h_{ie} is 1 kΩ.*

Base voltage of TR_3 (and collector voltage of TR_4) = 11.2 V (from example 5.42). Base current of TR_3 flowing in R_4 is ≈ 85 μA (from example

5.42), and since standby current for TR_4 is 5 mA, the effect of I_B for TR_3 can be neglected. Thus value of $R_4 = \dfrac{V_{cc} - V_c \text{ of } TR_4}{I_{CQ}}$

$$= \left(\frac{20 - 11.2}{5}\right) k\Omega = 1.76 \ k\Omega$$

From equation [5.116]. the a.c. resistance in the emitter of TR_3 is given by

$$\frac{V_{in}}{I} = \left[R_2 + R_L + \frac{R_3(h_{FE} + 1)R_{E2} + h_{FE}R_LR_3}{R_3 + (h_{FE} + 1)R_{E2}}\right]$$

and using values from example 5.40 the a.c. resistance is 410.5 Ω.

TR_3 is in the common-collector mode so that its effective input resistance is given by $h_{ie} + (h_{FE} + 1)R$ where R is the a.c. emitter resistance, h_{FE} the forward current gain and h_{ie} the input resistance for TR_3. For $h_{FE} = 120$ and $R = 410.5 \ \Omega$, effective input resistance is $1000 + (120 \times 410.4) = 50 \ k\Omega$ (assuming that h_{ie} is 1 kΩ).

The total load on TR_4 is as shown in Fig. 5.114, 50 kΩ in parallel with 1.76 kΩ. Hence total a.c. load on TR_4 is $R'_L = (1.76 \times 50)/(1.76 + 50)$ ≈ 1.7 kΩ. Since the voltage gain for the stage is

$$A_v = \frac{h_{FE}R'_L}{h_{ie} + (h_{FE} + 1)R_7}$$

then

$$50 = \frac{100 \times 1.7 \times 10^3}{(1 \times 10^3) + 101R_7}$$

and $50(1000 + 101R_7) = 170 \times 10^3$

or $1000 + 101R_7 = 3.4 \times 10^3 = 3400$

$101R_7 = 2400$

$$R_7 = \frac{2400}{101} = 24 \ \Omega.$$

Hence the voltage at TR_4 emitter $\approx I_cR_7 \qquad (I_c \approx I_e)$

$$= (5 \times 10^{-3} \times 24) \ V$$

$$= 120 \ mV$$

Voltage at TR_4 base $= (0.12 + V_{be}) \ V$

$$= (0.12 + 0.7) \ V$$

$$= 0.82 \ V$$

Base currrent for TR_4 is $\dfrac{I_c}{h_{FE}} = \dfrac{5}{100}$ mA $= 50 \ \mu A$ and a suitable bleeder current would be $10 \times I_B = 0.5$ mA. Thus $R_5 + R_6 = \dfrac{20 \ V}{0.5 \ mA} = 40 \ k\Omega$ and since $R_6 \times$ bleeder current $= 0.82$ V

$R_6 = 0.82/0.5 \ k\Omega = 1.64 \ k\Omega$

and $R_5 = 38.36 \ k\Omega$

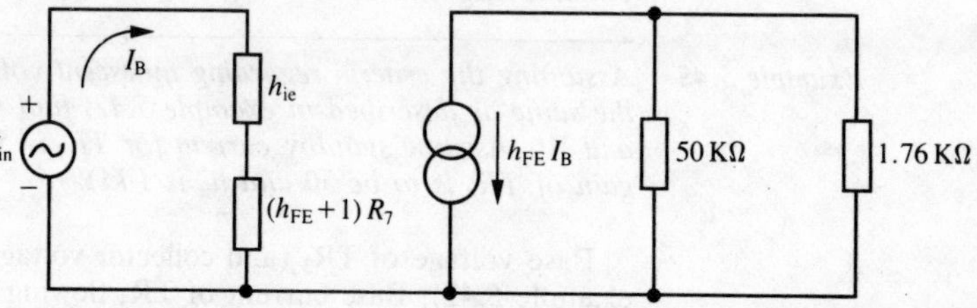

Fig. 5.114 Equivalent circuit for the voltage amplifier of Fig. 5.113.

A modification to the circuit of Fig. 5.113 which allows operating point stability because of negative feedback is shown in Fig. 5.115.

Fig. 5.115 Complementary circuit with d.c. negative feedback to give operating point stability.

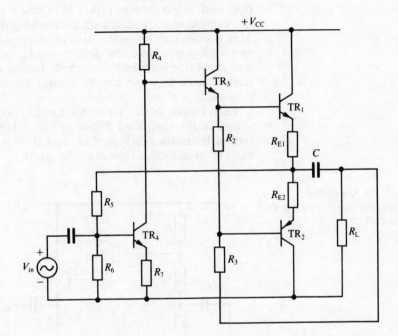

The mid-point of the output stage is connected to the bias circuit R_5 and R_6 of TR$_4$. The values of R_5 and R_6 are thus modified to give the same bias conditions as before. Should the operating point, nominally at $+V_{cc}/2$ for quiescent conditions, attempt to rise, say, then the base voltage of TR$_4$ will increase and TR$_4$ conducts more. The increased quiescent collector current from TR$_4$ will cause a larger volt drop across R_4 so that the base voltage of TR$_3$ must fall. It follows that base voltage of TR$_1$ must fall as does the base voltage of TR$_2$. Thus TR$_1$ conducts less and TR$_2$ conducts more. Reduced conduction makes for larger resistance and vice versa so that the volt drop across TR$_1$ collector-emitter will rise while that across TR$_2$ will fall. The mid-point voltage will therefore fall and this compensates for the original rise at the mid-point.

Example 5.46   *For the modified circuit of Fig. 5.115 calculate the values of bias resistors R_5 and R_6 to achieve the same bias conditions of example 5.45.*

The voltage at the mid-point of the output circuit is $+V_{cc}/2$ or $+10$ V. The voltage at the base of TR$_4$ = 0.83 V (from example 5.45). If the bleeder current for R_5 and R_6 is to be 0.5 mA as before then

$$R_5 + R_6 = \frac{10\ \text{V}}{0.5\ \text{mA}} = 20\ \text{k}\Omega$$

and since $R_6 = \dfrac{0.82\ \text{V}}{0.5\ \text{mA}} = 1.64\ \text{k}\Omega$

then $R_5 = 18.36\ \text{k}\Omega$

Quasicomplementary symmetry push-pull amplifiers

It is often advantageous to use identical devices in the output stage (i.e. both n-p-n devices rather than one n-p-n and one p-n-p as in the complementary stage) since matching of devices is made easier. This is especially so in the case of integrated circuit complementary amplifier stages since the devices are fabricated at the same time and should be identically matched. It is also preferable to use n-p-n devices since they can handle greater amounts of power than p-n-p devices.

Each output power transistor has its own driver transistor as shown in the simplified circuit of Fig. 5.116. Both the driver and output transistor are biased in class B with a very small idling current to minimise crossover distortion.

Fig. 5.116 Simplified quasicomplementary class B push-pull stage.

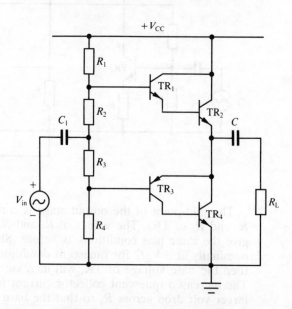

The arrangement of TR_1 and TR_2 forms a Darlington Pair. The two transistors could be combined into a 'single' transistor as shown in Fig. 5.117. The effect of connecting the two transistors in this fashion can be explained with the aid of the simplified hybrid-parameter equivalent shown in Fig. 5.118. From Fig. 5.118 since

$$I_{B2} = h_{FE1}I_B + I_B$$
$$= I_B(1 + h_{FE1})$$

then $h_{FE2}I_{B2} = h_{FE2}(1 + h_{FE1})I_B$

and total collector current $I_c = h_{FE1}I_B + h_{FE2}I_{B2}$

$$= h_{FE1}I_B + h_{FE2}(1 + h_{FE1})I_B$$
$$= I_B(h_{FE1}h_{FE2} + h_{FE1} + h_{FE2})$$

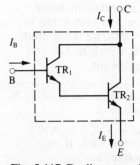

Fig. 5.117 Darlington pair configuration of TR_1 and TR_2 of the circuit of Fig. 5.116.

so that *overall* forward gain is I_c/I_B or $h_{FE1}h_{FE2} + h_{FE1} + h_{FE2}$ which is much greater than the individual forward current gain of either TR_1 or TR_2 alone.

Fig. 5.118 Equivalent
circuit of the Darlington
pair TR_1 and TR_2.

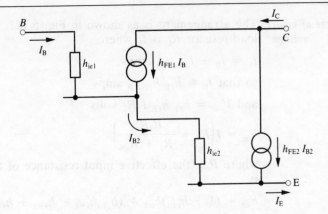

Thus if $h_{FE1} = h_{FE2} = 100$, say, then $h_{FE(overall)} = 10\ 200 \approx 10\ 000$ (or $h_{FE1}h_{FE2}$)

The arrangement for TR_3 and TR_4 can be similarly treated as shown in Fig. 5.119. Using h-parameter equivalent circuits in Fig. 5.120, similar reasoning as for TR_1 and TR_2 shows that in this case the total collector current is given by

$$I_c = h_{FE2}I_{B2} + h_{FE1}I_B$$

while $I_{B2} = h_{FE1}I_B$

so that $I_c = h_{FE2}h_{FE1}I_B + h_{FE1}I_B$

$$= I_B(h_{FE1}.h_{FE2} + h_{FE1})$$

In this case $h_{FE1} = h_{FE2} = 100$

so that $h_{FE(overall)} = 10\ 100 \approx 10\ 000$ (or $h_{FE1}h_{FE2}$)

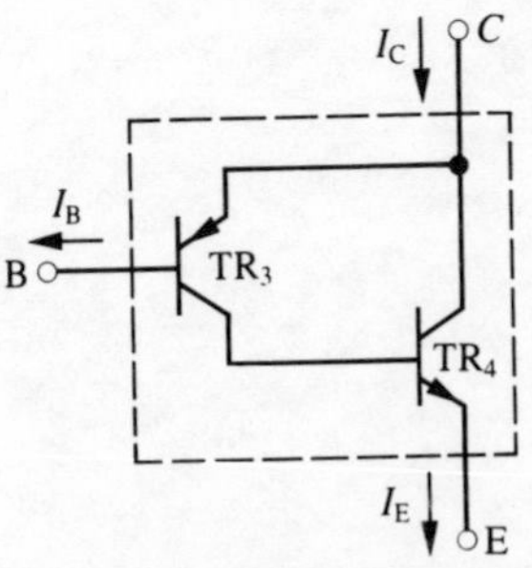

Because of the direction of the quiescent current the composite arrangement of TR_3 and TR_4 acts as a single p-n-p transistor.

Let us now consider the action of each of the composite transistors for each half-cycle of input voltage. It is assumed that since TR_1 and TR_2 act as a composite n-p-n device these transistors conduct while TR_3 and TR_4 switch off (since TR_3 and TR_4 also represent a composite device).

Fig. 5.119 Arrangement of
TR_3 and TR_4 of Fig. 5.116
as a single transistor.

Fig. 5.120 Equivalent
circuit of the 'single'
transistor of Fig. 5.119.

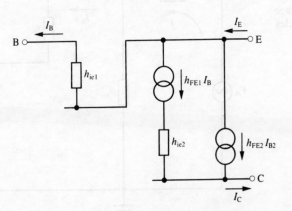

Positive half-cycle of input voltage

The arrangement is as shown in Fig. 5.121. The total current in the load resistor R_L is I_E where

$$I_E = I_B + I_{c(overall)} \text{ amps}$$

so that $I_E \approx h_{FE1}h_{FE2}I_B$ amps

and $V_{out} = h_{FE1}h_{FE2}I_BR_L$ volts

$$V_{in} = I\left(R_2 + \frac{R_1 R_{eff}}{R_1 + R_{eff}}\right)$$

where R_{eff} (the effective input resistance of the composite transistor) is

$$h_{ie1} + (1 + h_{FE1})h_{ie2} + (h_{FE1}h_{FE2} + h_{FE1} + h_{FE2} + 1)R_L$$

thus $R_{eff} \approx h_{FE1}h_{FE2}R_L$

and since $R_{eff} \gg R_1$ then

$$V_{in} \approx I(R_1 + R_2)$$

$$I_B = I\left(\frac{R_1}{R_1 + R_{eff}}\right)$$

so that $V_{in} = \left(\dfrac{R_1 + R_{eff}}{R_1}\right)(R_1 + R_2)I_B$

$$\approx \frac{R_{eff}}{R_1}(R_1 + R_2)I_B$$

$$V_{in} \approx R_{eff}\left(1 + \frac{R_2}{R_1}\right)I_B$$

$$\approx h_{FE1}h_{FE2}R_L\left(1 + \frac{R_2}{R_1}\right)I_B$$

Voltage gain $= A_v = \dfrac{V_{out}}{V_{in}}$

and $A_v = \dfrac{h_{FE1}h_{FE2}I_BR_L}{h_{FE1}h_{FE2}R_L\left(1 + \dfrac{R_2}{R_1}\right)I_B}$

or $A_v \approx 1$

Fig. 5.121 Equivalent circuit for the quasicomplementary stage during positive half-cycles of input voltage.

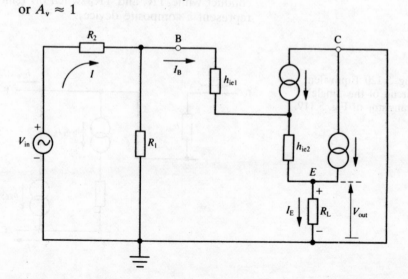

Negative half-cycle of input voltage

The arrangement is as shown in Fig. 5.122. Similar reasoning applies as for positive half-cycle of input voltage. With

$$V_{out} \approx h_{FE1}h_{FE2}I_B R_L \text{ volts}$$

and $V_{in} = I\left(R_3 + \dfrac{R_4 R_{eff}}{R_4 + R_{eff}}\right)$

where in this case, $R_{eff} = h_{ie1} + (h_{FE1} + h_{FE2}h_{FE1})R_L$

or $R_{eff} = h_{FE1}h_{FE2}R_L$

and since $R_{eff} > R_4$ then

$$V_{in} \approx I(R_3 + R_4)$$

$$I_B \approx I\left(\frac{R_4}{R_4 + R_{eff}}\right)$$

$$\approx I\frac{R_4}{R_{eff}}$$

so that $V_{in} \approx I_B\dfrac{R_{eff}}{R_4}(R_3 + R_4)$

$$\approx I_B R_{eff}\left(\frac{R_3}{R_4} + 1\right)$$

$$\approx I_B h_{FE1}h_{FE2}\left(1 + \frac{R_3}{R_4}\right)$$

and voltage gain $A_v = \dfrac{V_{out}}{V_{in}}$

$$= \frac{h_{FE1}h_{FE2}R_L I_B}{I_B h_{FE1}h_{FE2}\left(1 + \dfrac{R_3}{R_4}\right)} \approx 1$$

Fig. 5.122 Equivalent circuit for the quasicomplementary stage during negative half-cycles of input voltage.

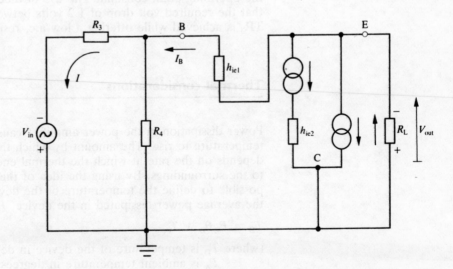

If $R_3 = R_2$ and $R_4 = R_1$ the voltage gain can be the same for positive and negative input swings with the value of output peak voltage very nearly equal to the input peak voltage, provided the peak-peak input voltage does not exceed the supply voltage value.

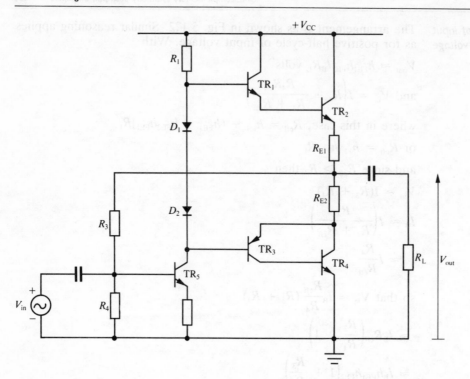

Fig. 5.123 Possible circuit arrangement for a quasicomplementary stage.

A possible quasicomplementary stage is shown in Fig. 5.123. Many of the principles already established have been incorporated in this circuit. For example, the use of R_{E1} and R_{E2} to ensure that the composite transistor TR_1/TR_2 turns off on negative half-cycles (TR_3/TR_4 on positive half-cycles) and the feedback from the junction of R_{E1} and R_{E2} to the base bias circuit of TR_5 in order to maintain the operating point constant. The use of diodes D_1 and D_2 ensures that the required volt drop of 1.2 volts between the base of TR_1 and TR_3 is achieved while offering a low a.c. resistance.

Thermal considerations

Power dissipation in the power amplifier causes the device temperature to rise. The amount by which the temperature rises depends on the rate at which the thermal energy can be transferred to the surroundings. By using the idea of **thermal resistance** θ it is possible to define the temperature of the device in terms of θ and the average power dissipated in the device, P_D, i.e.

$$T_D = P_D\theta_D + T_A \qquad [5.124]$$

(where T_D is temperature of the device in degrees Celsius
T_A is ambient temperature in degrees Celsius and
θ_D is thermal resistance in degrees Celsius/W)

In practice θ_D consists of many resistances in series. There is resistance between device and case θ_{DC}, resistance between case and heat sink, where fitted, θ_{CS} and resistance between heat sink and

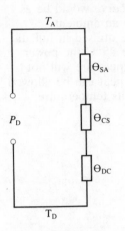

Fig. 5.124 Thermal resistances of a power device with a heat sink.

ambient temperature θ_{SA}. The thermal resistance can be treated in a similar manner to ohmic resistance as Fig. 5.124 shows.

In this equivalent circuit equation [5.124] is modified to $T_D = P_D(\theta_{SA} + \theta_{CS} + \theta_{DC}) + T_A$. The circuit shows that P_D is analogous to current flow which causes a temperature drop across the resistance so that temperature is analogous to voltage.

Resistance θ_{DC} depends on device construction and should be small for high-power transistors.

Resistance θ_{CS} depends on many factors. In the circumstances where the transistor collector is in electrical contact with the case while the heat sink is at ground potential an electrical insulator must be placed between the case and the heat sink. Mica is the material most often used and its thickness and area introduce a variation to the value of θ_{CS}. The use of silicon grease between the mica washer, case and heat sink also introduces a factor of variation which tends to reduce θ_{CS}. Typical values of θ_{CS} vary from about 0.1 to 2° C/W.

Resistance θ_{SA} depends on such functions as temperature variations (temperature of heat sink in the region of the transistor will be greater than the remaining area) and the area of the heat sink. The dependence of θ_{SA} on heat sink area is inverse but not linearly so since the removal of heat from the sink is an unequal process. A curve showing possible values of θ_{SA} against area for a 4 mm thick aluminium sheet is shown in Fig. 5.125. Values for θ_{SA} are given by the manufacturer and may be reduced to values less than 1° C/W by forced cooling.

Fig. 5.125 Curve of θ_{SA} against area (in cm²) for a 4 mm thick aluminium sheet.

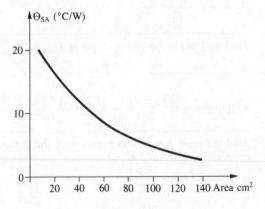

Some manufacturers give information regarding the effects of thermal reistance in terms of what is called a **power derating curve**. Since from equation [5.124]

$$P_D = \frac{T_D - T_A}{\theta_D}$$

and assuming a maximum operating temperature for silicon devices is quoted (this is 200° C in Fig. 5.126) then the maximum collector power dissipation at any ambient temperature T_A is given by

$$P_{D(max)} = \frac{T_{D(max)} - T_A}{\theta_D}$$

Assuming the manufacturer's information gave $P_{D(max)}$ at 2 W up to 25° C ambient temperature then the derating curve would be as shown in Fig. 5.126. The curve shows that up to an ambient temperature of 25° C the rated dissipation can be utilised in full. If the ambient temperature is allowed to rise above 25° C the power dissipation must be decreased so that device temperature will not exceed 200° C. At 200° C it can be seen that no heat can be allowed to be dissipated from the device without raising its temperature above the limit of 200° C.

Fig. 5.126 Derating curve for a power transistor device.

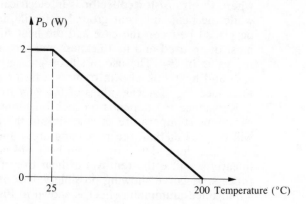

Note from Fig. 5.126 that the value of θ_D can be found from the slope of the curve between the T_A and $T_{D(max)}$ points since

$$\theta_D = \frac{T_{D(max)} - T_A}{P_D}$$

Example 5.47   *Find θ_D from the derating curve of Fig. 5.126.*

$$\theta_D = \frac{T_{D(max)} - T_A}{P_D}$$

so that $\theta_D = \dfrac{200 - 25}{2} = 87.5°$ C/W

Example 5.48   *Find the rated power dissipation from the curve of Fig. 5.126 if the ambient temperature rises to 100° C.*

$$P_{D(max)} = \frac{T_{D(max)} - T_A}{\theta_D}$$

so that $P_{D(max)} = \dfrac{200 - 100}{87.5} \approx 1.14$ W

The results from examples 5.47 and 5.48 suggest a device operating without a heat sink since in practice θ_D would be less than 87.5° C/W with a heat sink. Without a heat sink θ_D reduces to

$$\theta_D = \theta_{DC} + \theta_{CA}$$

and it would be expected that for a value of 87.5° C/W for θ_D the majority of this value would come from θ_{CA}. Possible values are:

$$\theta_{DC} = 12.5° \text{ C/W}$$

$$\theta_{CA} = 75.0° \text{ C/W}$$

Fig. 5.127 The effect on thermal resistances of adding a heat sink. The thermal resistance θ_{CA} is reduced to a lower value because of the shunting effect of θ_{SA} and θ_{CS}.

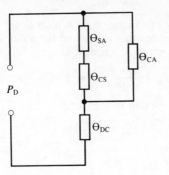

The analogy with electrical resistance can also be applied here as the thermal resistance diagram of a transistor with a heat sink shows in Fig. 5.127.

Thus the very much lower values of θ_{SA} and θ_{CS} effectively shunt the very much higher value of θ_{CA}.

Example 5.49 *Assume a 2.5 W rated transistor is to be used with a heat sink at an ambient temperature of 100° C. If the value for θ_{DC} is taken at 12.5° C/W and if θ_{CS} is 2° C/W find the area of 4 mm thick aluminium heat sink required.*

If $T_{D(max)} = 200°$ C as before then

$$\theta_D = \frac{T_{D(max)} - T_A}{P_{D(max)}}$$

or $\theta_D = \dfrac{200 - 100}{2.5} = 40°$ C/W

but $\theta_D = \theta_{DC} + \theta_{CS} + \theta_{SA}$

or $\theta_{SA} = \theta_D - \theta_{DC} - \theta_{CS}$

$$= 40 - 12.5 - 2.0$$

$$= 25.5°\text{ C/W}$$

From Fig. 5.125 it can be deduced that for $\theta_{SA} = 25.5°$ C/W a heat sink area of less than 10 cm² is required.

Different heat sinks can give different values of θ_{SA} and information can be obtained from manufacturer's data.

Example 5.50 If the transistor of example 5.49 is used with a heat sink giving a value for θ_{SA} of 2° C/W, find the maximum possible dissipation.

$$P_{D(max)} = \frac{200 - 100}{12.5 + 2 + 2}$$

$$= \frac{100}{16.5} \approx 6 \text{ W}$$

For given values of temperature and thermal resistance it is often useful to plot $P_{D(max)}$ on the $V_{ce} - I_c$ output characteristic to give an indication of safe operating values for V_{CEQ} and I_{CQ} without exceeding thermal limitations.

Example 5.51   If the transistor of example 5.50 is used at 25° C find $P_{D(max)}$. Plot the dissipation curve for this value, and the value at 100° C found in example 5.50, on the graph of V_{ce} against I_c.

At 25° C, $P_{D(max)} = \dfrac{200 - 25}{16.5} \approx 10.5$ W.

The curves are plotted in Fig. 5.128.

Fig. 5.128 Power dissipation curves used in example 5.51.

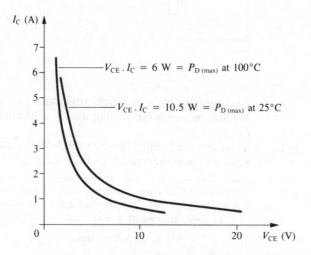

From the results of example 5.51, a choice of V_{CEQ} and I_{CQ} may be made to give satisfactory operating conditions while still allowing for a temperature rise to 100° C without exceeding required dissipation levels.

An integrated circuit audio power amplifier

The LM380 is an audio power amplifier with an internally fixed gain of 50 (34 dBs) and an output which automatically centres at one half of the supply voltage. The input may be ground referenced or a.c. coupled as required. The output stage is protected with circuit current limiting and a thermal shutdown facility. The device can be worked at a maximum supply voltage of 22 V, take a peak current of 1.3 A and sustain a package dissipation, assuming an infinite heat sink, of 10 W. The circuit diagram and the block diagram representation are shown in Figs 5.129(a) and (b) respectively.

Referring to the circuit diagram the input stage is a p-n-p emitter-follower (Q_1 or Q_2) driving a p-n-p differential amplifier (Q_3 and Q_4) with a constant current source load. Q_6, in association with the 'diode' Q_5 is the constant current source for Q_3 and Q_4. The p-n-p input is chosen to reference the input to ground enabling the input transducer to be directly coupled. The second stage is a common-emitter voltage gain amplifier (Q_{12}) with a current source load. Q_{11}, in association with the 'diode' Q_{10} provides the constant current source. Internal compensation is provided by the pole-splitting

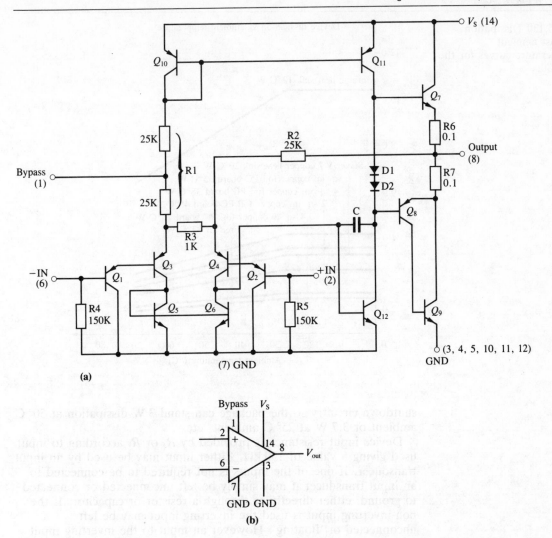

Fig. 5.129 LM380 audio
power amplifier: (a) circuit
diagram; (b) block diagram.

capacitor C'; this compensation being designed to maintain wide
bandwidth (100 kHz at 2 W; 8 Ω). The final stage is the output stage
provided by the quasicomplementary pair Q_7 and Q_8/Q_9 acting as
emitter-followers. The diodes D_1 and D_2 are to prevent crossover
distortion in the output waveform and the resistors R_6 and R_7 also
assist in this aim (see under quasicomplementary symmetry push-pull
amplifiers).

The output is biased to half the supply voltage by the resistor
ratio R_1/R_2. D.c. negative feedback through resistor R_2 balances the
differential stage with the output at half the supply voltage. The
gain is fixed at 50 by the internal feedback network R_2–R_3. This
gives a ratio of 25:1 but the gain is doubled due to the current
source providing full differential gain for the input stage.

The output current of the LN380 is rated at 1.3 A peak and the
package is rated at 35° C/W when soldered into a printed circuit
board with 6 square inches of 2 ounce copper foil. This is shown on
the dissipation versus ambient temperature curves of Fig. 5.130. The
device junction temperature is limited to 150° C via the thermal

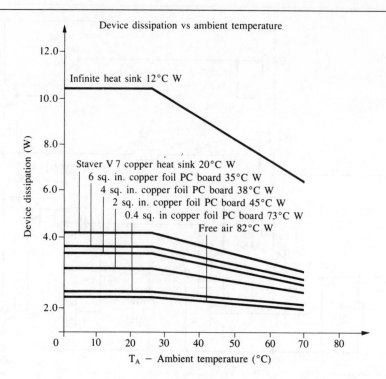

Fig. 5.130 Dissipation against ambient temperature curves for the LM380.

Device dissipation vs ambient temperature

Infinite heat sink 12°C W

Staver V 7 copper heat sink 20°C W
6 sq. in. copper foil PC board 35°C W
4 sq. in. copper foil PC board 38°C W
2 sq. in. copper foil PC board 45°C W
0.4 sq. in copper foil PC board 73°C W
Free air 82°C W

Device dissipation (W)

T_A – Ambient temperature (°C)

shutdown circuitry so the package can stand 3 W dissipation at 50° C ambient or 3.7 W at 25° C ambient, etc.

Device input resistance is provided by R_4 or R_5 according to input used giving a value of 150 kΩ. Either input may be used by an input transducer. If one of the inputs is not required to be connected to an input transducer it may simply be left unconnected or connected to ground, either directly or through a resistor or capacitor. If the non-inverting input is used the inverting input may be left unconnected or 'floating'. However an input to the inverting input with the non-inverting input left floating may result in stray coupling providing positive feedback via the floating input. This may be avoided by

(a) A.c. grounding the unused input with a small capacitor. Preferable when using high source impedance transducers.

(b) Grounding the unused input with a resistor. Preferable when using low to moderate d.c. source impedance transducers and when output offset from half supply voltage is critical.

(c) Shorting the unused input to ground. Used with low d.c. source impedance transducers or when output offset is not critical.

Figure 5.131 shows a simple application of the LM380 where apart from the loadspeaker the only external component required is an output coupling capacitor. The possible connection of R_c and C_c shown dotted in this circuit is for suppression of a 5 MHz to 10 MHz small amplitude oscillation which can occur during the negative swing into a load which draws high current. The oscillation is at too high a frequency to affect a speaker output but should be borne in mind when radio frequency (r.f.) oscillations may cause problems in nearby circuits.

Fig. 5.131 A possible
LM380 circuit application.

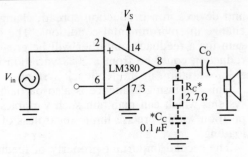

*For stability with high current loads

You should now be able to attempt exercises 5.17 to 5.55.

THE PRINCIPLES OF FEEDBACK

A **feedback system** is one for which the input signal is modified in some way by the system output before the internal processing by the system itself. This general arrangement is illustrated in Fig. 5.132.

Fig. 5.132 The block
diagram arrangement for a
feedback system.

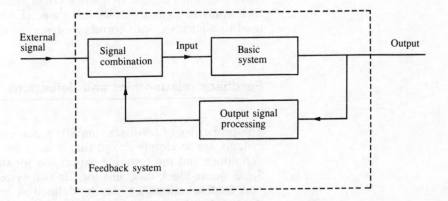

The signal combination shown is usually a simple addition or subtraction but multiplication, logical or other combinations may also be used. The output signal processing is often simple attenuation of amplitude reduction but this too may be combined with differentiation or filtering in various ways.

Most electronic circuits and systems employ feedback in one way or another and since this can affect the system behaviour to a marked degree, it is important to have a clear appreciation of the principles involved.

Feedback can affect the overall system properties in many ways some of which may or may not be desirable. With correctly applied feedback, the most important effect is a stabilising influence. For example, the gain of an amplifier without feedback is subject to modification due to a range of external factors. These include change of d.c. supply levels, change of component values of device parameters (due to ageing or replacement), variation in components and devices in a production spread, change in external loading and change in environmental conditions. The corresponding changes in gain for a feedback amplifier will be greatly reduced. A typical reduction could be from a 20% variation in the basic system gain to a 1% variation in the feedback system.

This stabilising influence is also particularly important in control systems which can maintain such variables as temperature, speed or position within precise limits regardless of variation in external loading.

The second important property of feedback systems is concerned with the bandwidth or frequency response. In general, if the system without feedback correctly processes only signals within a particular frequency range or bandwidth, the equivalent feedback system will satisfactorily process signals over a wider frequency range, i.e. it will have a greater bandwidth,

Further properties of feedback which are of particular interest in amplifiers include the reduction of harmonic distortion due to non-linearities within the amplifier and modification of both input and output impedances.

All the above remarks are applicable to correctly designed feedback systems: under certain circumstances, converse effects can occur resulting in less stability, more distortion, less bandwidth etc. In a particular system, the advantage of feedback can apply to signals within a certain frequency range and the opposite effects then apply to signals outside this range. Even this concept can be used to advantage with correct design.

Feedback relationships and definitions

The properties of feedback amplifiers and closed loop control systems are so closely linked that it is useful to compare the basic definitions and the resulting expression for the system gain. Figure 5.133 shows block diagrams for the two system types.

A **feedback amplifier** may be defined as an amplifier for which the **terminal input signal** is the sum of an external signal and a signal proportional to the output signal. This is shown in Fig. 5.133a

Fig. 5.133 Feedback systems: (a) a feedback amplifier; (b) a closed loop control system.

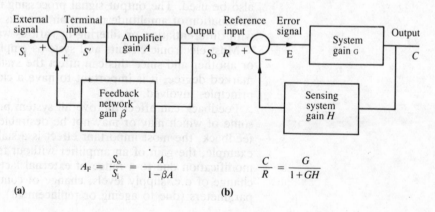

$$A_F = \frac{S_o}{S_i} = \frac{A}{1-\beta A}$$

(a)

$$\frac{C}{R} = \frac{G}{1+GH}$$

(b)

where the signal proportional to the output signal is βs_o and this is *added* to the external signal s_i at the the summing junction. The resulting terminal input signal s' is then amplified by the gain A to give the output s_o. Signals s_i, s_o and s' are used at this stage as in practical systems signals may be considered as either voltage signals or current signals according to the method of connection. In the equivalent control system for Fig. 5.133b the **error signal** is the *difference* between a reference input and the processed output. The use of the difference instead of the sum used for the feedback amplifier causes a sign change between the formula for system gain for the amplifier and that for the control system.

From the definition of the feedback amplifier which is shown in Fig. 5.133a,

$$s' = s_i + \beta s_o$$

But $s_o = As'$

$$\therefore s' = s_i + \beta As'$$

collecting terms, $s'(1 - \beta A) = s_i$

or $s' = \dfrac{s_i}{1 - \beta A}$

hence, $s_o = As' = \dfrac{As_i}{1 - \beta A}$ and the system gain

or gain with feedback, $A_f = \dfrac{s_o}{s_i} = \dfrac{A}{1 - \beta A}$ [5.125]

Both gain A and feedback factor β may be complex quantities having both modulus and angle each of which depend upon signal frequency. A more general form of equation [5.125] includes these as $A \angle \theta$ and $\beta \angle \phi$. The resulting expression is then,

$$A_f = \frac{A\angle\theta}{1 - \beta A\angle(\theta + \phi)}$$ [5.126]

For a particular amplifier, the modulus of the denominator $|1 - \beta A \angle (\theta + \phi)|$ may be either greater than one or less than one depending on the signal frequency. These two conditions in turn result in a decrease or an increase respectively in the gain compared with the gain without feedback. For convenience, the two conditions are known as **negative and positive feedback** and may be defined in the following way:- The feedback is said to be negative if the gain with feedback $|A_f|$ is less than the gain without feedback $|A|$.

The feedback is said to be positive if the gain with feedback $|A_f|$ is greater than the gain without feedback $|A|$.

It is important to note that the term negative feedback amplifier is incorrect since the nature of the feedback changes with the signal frequency. In general, an amplifer having negative feedback for signals of medium frequency will have positive feedback for both high and low signal frequencies. This change in the nature of the feedback is in fact responsible for the bandwidth improvement mentioned in the introduction to this section.

Many of the properties due to feedback are most marked in the frequency ranges for which $\beta A \angle (\theta + \phi) = \beta A \angle 180°$. The resulting gain equation [5.126] becomes

$$A_f = \frac{A}{1 + \beta A} \qquad\qquad [5.127]$$

and the condition can be referred to as **simple negative feedback**, SNFB.

The different types of feedback can be illustrated by an example.

Example 5.52 *The gain and feedback factor of an amplifier have the different values at different frequencies listed in the table below. In each case, determine the nature of the feedback and the system gain.*

	f_1	f_2	f_3	f_4
A	$5000 \angle 180°$	$4500 \angle 160°$	$1000 \angle 65°$	$500 \angle 20°$
β	$0.02 \angle 0°$	$0.018 \angle -5°$	$1.148 \times 10^{-3} \angle -10°$	$0.001 \angle -15°$

At frequency f_1, the gain A is $5000 \angle 180°$ and the feedback factor β is $0.02 \angle 0°$. Applying the equation [5.126],

$$A_f = \frac{5000 \angle 180°}{1 - 100 \angle 180°} = \frac{5000 \angle 180°}{101} = 49.5 \angle 180°.$$

This is an example of simple negative feedback since $\theta + \phi = 180°$.

At f_2,

$$A_f = \frac{4500 \angle 160°}{1 - 81 \angle 155°} = \frac{4500 \angle 180°}{1 - (-73.4 + j34.2)}$$

$$= \frac{4500 \angle 160°}{81.9 \angle -24.7°} = 54.9 \angle 185°.$$

The feedback is again negative since the gain with feedback, 54.9 is less than the gain without feedback 4500. It is however larger than that at f_1; this does occur in practical systems and is considered in a later section on frequency response with feedback.

At f_3,

$$A_f = \frac{1000 \angle 65°}{1 - 1.148 \angle 55°} = \frac{1000 \angle 65°}{1 - (0.658 + j0.94)} = \frac{1000 \angle 65°}{1 \angle -70°} = 1000 \angle 135°.$$

In this case, the gain modulus is unchanged so the feedback is neither positive nor negative. It has however, had the effect of changing the phase angle.

At f_4,

$$A_f = \frac{500 \angle 20°}{1 - 0.5 \angle 5°} = \frac{500 \angle 20°}{1 - (0.498 + j0.044)} = \frac{500 \angle 20°}{0.503 \angle -5°} = 922 \angle 25°.$$

Thus at f_4, the feedback is positive since 922 is greater than the gain of 500 without feedback at this frequency.

Feedback and gain stability

A typical amplifier will be subjected to changes of environment such as temperature and d.c. supply levels and there will also be internal changes due to ageing. Such changes can all result in variation in the gain of the amplifier. In a production run of such amplifiers, component and parameter tolerances will cause an even greater sample to sample variation in gain. In many applications, such variation would be quite unacceptable for correct and reliable system operation and feedback is employed to reduce the spread in gain to within acceptable limits.

Consider the feedback equation [5.127] for simple negative feedback. If the **loop gain** βA is much larger than one (the loop gain is the product of the gain around the loop)

$$A_f \simeq \frac{A}{\beta A} = \frac{1}{\beta} \qquad [5.128]$$

The value of β, the feedback factor, is commonly determined by the ratio of two resistors so if these are chosen to be high stability, low tolerance components, the gain is apparently independent of the changes mentioned above.

Equation [5.128] is an approximation and in practice it may not be valid as large changes in A can occur. It is useful to examine some practical problems by means of some numerical examples.

Example 5.53 *The gain of a particular amplifier is found to be 750 on test but under worst conditions of reduced d.c. supply and external loading, this figure is reduced to 400. Compare this percentage variation for the basic amplifier with the corresponding variation if SNFB is applied using a β factor of (a) 0.005, (b) 0.03 and (c) 0.1.*

The required percentage variation is in terms of the nominal value. For the amplifier without feedback, this is given by

$$\frac{750 - 400}{750} \times 100 = 46.7\%.$$

(a) The two values of gain for the first feedback factor are respectively,

$$A_{max} = \frac{750}{1 + 0.005 \times 750} = 157.9 \text{ and } A_{min} = \frac{400}{1 + 0.005 \times 400} = 133.3.$$

$$\text{Percentage variation} = \frac{157.9 - 133.3}{157.9} = 15.6\%$$

Similar calculations for other values of β are left for the reader but the resulting maximum gain and percentage variation are respectively

(b) 31.9 and 3.5%

(c) 9.87 and 1.1%

From the results in example 5.53 we can see that the use of feedback for stabilisation of gain is a 'trade-off' situation. Very tight tolerances are easily obtained but at the expense of a large reduction in gain.

Example 5.54 The production spread in gain for an amplifier is found to be from 3000 to 12 500. Negative feedback is to be employed to make the percentage variation better than $\pm 10\%$ on a nominal figure. Determine the required value of feedback factor β (assuming simple negative feedback) and the resulting nominal gain.

The minimum gain (-10%) with feedback is

$$A_{1f} = \frac{3000}{1 + 3000\beta} = 0.9A_n \text{ where } A_n \text{ is the nominal gain.}$$

Maximum gain ($+10\%$) with feedback is

$$A_{2f} = \frac{12\,500}{1 + 12\,500\beta} = 1.1A_n$$

From each of these equations, an expression for A_n can be obtained and these two expressions can then be equated.

$$A_n = \frac{1}{0.9} \times \frac{3000}{1 + 3000\beta} = \frac{1}{1.1} \times \frac{12\,500}{1 + 12\,500\beta}$$

Cross multiplying,

$$3000 \times 1.1(1 + 12\,500\beta) = 0.9 \times 12\,500(1 + 3000\beta)$$

$$3300 + 4.125 \times 10^7\beta = 11\,250 + 3.375 \times 10^7\beta$$

Collecting terms and rearranging,

$$\beta = \frac{11\,250 - 3300}{(4.125 - 3.375) \times 10^7} = 1.06 \times 10^{-3}.$$

Taking an expression for the nominal gain A_n from the equation for A_{1f} and substituting for this value of β,

$$A_n = \frac{3000}{0.9(1 + 3000 \times 1.06 \times 10^{-3})} = 797.$$

Checking for maximum and minimum values,

$$A_{1f} = A_{min} = 717.7 = 0.9 \times 797.$$

$$A_{2f} = A_{max} = 877.2 = 1.1 \times 797.$$

Thus the required limits of $\pm 10\%$ have been achieved with reduction in average gain level by a factor of about 10.

A gain desensitising factor

It is sometimes useful to have a general factor which shows how effective the feedback is in making the final gain insensitive to changes in the original open loop gain.

For a constant feedback factor β, the equation [5.127],

$$A = \frac{A}{1 + \beta A} \text{ can be differentiated to give}$$

$$\frac{dA_f}{d_A} = \frac{(1 + \beta A) - A\beta}{(1 + \beta A)^2} = \frac{1}{(1 + \beta A)^2}$$

and $dA_f = \dfrac{dA}{(1 + \beta A)^2}$

Dividing this equation by equation [5.127],

$$\frac{dA_f}{A_f} = \frac{\dfrac{dA}{(1 + \beta A)^2}}{\dfrac{A}{(1 + \beta A)}} = \frac{dA}{A} \times \frac{1}{(1 + \beta A)} \qquad [5.129]$$

The two terms $\dfrac{dA_f}{A_f}$ and $\dfrac{dA}{A}$ are respectively, the changes in gain with and without feedback which can be expressed as percentages.

Equation [5.129] shows that the percentage variation is reduced by a factor $1/(1 + \beta A)$, which is exactly the same as the factor reducing the gain itself.

The numerical examples 5.53 and 5.52 do not show exactly this result since in each case we are concerned with large changes in gain. Equation [5.129] is obtained by differentiation and is thus only valid for small changes.

Feedback and harmonic distortion

All amplifiers are, to some extent, non-linear due to the curvature of the characteristics of the active devices. The non-linearity is more pronounced if large signals take a particular device near to the saturation and cut-off regions of operation. In all cases, the non-linearity results in harmonic distortion which is caused by the effective generation of harmonic signals within the amplifier. These signals are not present in the input but their frequencies are multiples of the input signal frequency. The amplitude of these distorting signals is related to the amplitude of the output signal. The effect of feedback on such amplifiers can be investigated by considering the distorting amplifier shown in Fig. 5.134.

Fig. 5.134 A distorting voltage amplifier with feedback.

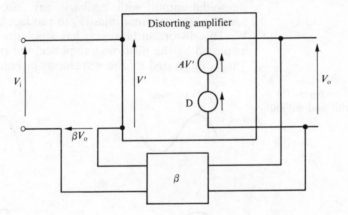

Consider the case with no feedback: a particular value of V_o will result in a corresponding value of D. The relationship between these two variables is non-linear so we can only make comparisons if, in our complete system, we return to the same value of V_o and thus, to the same value of D.

From the diagram, and the feedback definition,

$$V' = V_i + \beta V_o$$

but $V_o = AV' + D$ [5.130]

$$\therefore \ V' = V_i + \beta(AV' + D)$$

and $V'(1 - \beta A) = V_i + \beta D$

hence $V' = \dfrac{V_i + \beta D}{1 - \beta A}$ [5.131]

Substituting in equation [5.130],

$$V_o = A\frac{(V_i + \beta D)}{1 - \beta A} + D$$

Combining over a common denominator,

$$V_o = \frac{A}{1 - \beta A}V_i + \frac{\beta AD}{1 - \beta A} + \frac{D(1 - \beta A)}{1 - \beta A}$$

and $V_o = \underbrace{\frac{A}{1 - \beta A}V_i}_{\substack{\text{amplified} \\ \text{signal}}} + \underbrace{\frac{D}{1 - \beta A}}_{\substack{\text{actual distortion present} \\ \text{in the output } V_o}}$ [5.132]

This result shows that the signal is amplified by the familiar gain with feedback and the actual distortion in the output is the distortion D introduced by the amplifier *reduced* by a factor $1/(1 - \beta A)$. Thus if simple negative feedback is used, the output distortion is $D/(1 + \beta A)$. However, the value of D is only known if the original output signal level V_o is maintained by increasing V_i to compensate for the fall in signal gain. If this is done, the actual output distortion will be less for the same output than it was in the original amplifier. This provides a very real improvement as most of the distortion occurs in the final stage of amplification where the signal is largest. This effect can also be described in the following way: with no feedback, a sinusoidal signal produces a distorted sinusoidal output; with feedback, any distortion in the output is added to the external input V_i to produce a *distorted* terminal input V'. This distortion however, has a polarity such that when it is amplified by the distorting amplifier, the result is more sinusoidal. This is illustrated by the waveforms in Fig. 5.135.

Fig. 5.135 Waveforms for amplifiers with and without feedback.

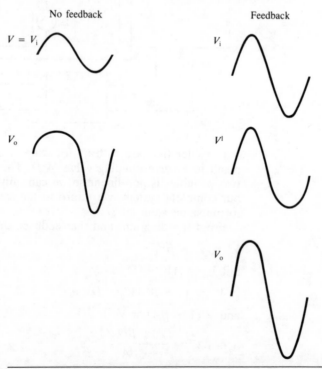

No feedback

$V = V_i$

V_o

Feedback

V_i

V'

V_o

Example 5.55   *An amplifier is tested with an input signal of 10 mV at f = 2 kHz. The output is found to include 4 V at 2 kHz and 0.8 V at 4 kHz. If SNFB is applied with a β of 0.006, find the required new level of input signal to provide the same output at 2 kHz and the change in percentage distortion.*

From the information provided, the gain A of the amplifier is 4 V/10 mV or 400 and the percentage distortion is $\dfrac{0.8}{4} \times 100$ or 20%.

The gain with feedback is given by $A_f = 400/(1 + 2.4) = 117.6$.

Thus to maintain the output signal level, the input must be increased by a factor of 400/117.6 to 34 mV. The 4 kHz component in the output would then still be 0.8 V but for the feedback which reduces this figure to 0.8/3.4 or 0.235 V. The percentage distortion with feedback is therefore $(0.235/4) \times 100$ or 5.9%. This is of course the same figure as $20\%/(1 + \beta A) = 5.9\%$.

Example 5.56   *A two stage amplifier consists of a pre-amp with a gain of 250 and an output stage with gain of 80. At the specified maximum output level of 10 V, 12% harmonic distortion is present. Design a feedback system to reduce this figure to 2% and state how the pre-amp performance must be modified.*

Feedback will be applied to the output stage to reduce the distortion and the pre-amp gain must be increased to restore the specified output level. To find β,

$$2 = \frac{12}{1 + \beta A}$$

$$\therefore \ 1 + \beta A = 6 \text{ and } \beta = \frac{6 - 1}{A} = \frac{5}{80} = 0.0625$$

New gain of output stage $= \dfrac{80}{6} = 13.3$.

Required new gain of the pre-amp $= 250 \times 6 = 1500$.

Amplifier frequency response with feedback

In general, amplifiers have a frequency range for which the gain and phase shift are nearly constant. For direct coupled amplifiers, this range, or bandwidth will be from d.c. to 1 MHz or more or from d.c. to a few Hz depending on the type. A.c. coupled amplifiers also have a limiting lower frequency, typically of 100 Hz or so and an upper limit. For frequencies outside this mid-band range, the gain falls and the phase shift changes. If feedback is applied, it will be designed to be negative feedback over the mid-band frequency range and, to a first approximation, equation [5.128], $A_f = 1/\beta$ will apply as the gain falls. This suggests that there will be an improvement in bandwidth when feedback is applied. This is correct, but the situation is more complex as the phase change can result in the feedback becoming positive at extreme signal frequencies resulting in a further improvement in bandwidth.

Analysis to investigate the increase in bandwidth with feedback is best based upon a graphical representation of the frequency response which includes both gain and phase shift, i.e. the polar or Nyquist plot. We shall consider first how such a representation can be combined with the general feedback formula $A_f = A/1 - \beta A$ at a single frequency.

Figure 5.136a shows the Nyquist plot for the gain of a two stage direct coupled amplifier. Notice that the two stages result in zero

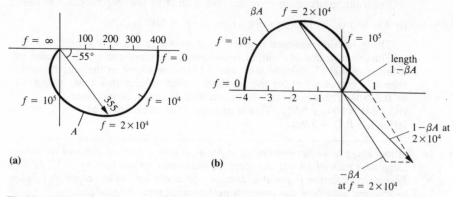

Fig. 5.136 Nyquist plots for a two stage direct coupled amplifier: (a) plot for A; (b) plot for βA.

phase shift at d.c. $(f = 0)$ and that as the gain falls at high frequencies, the phase shift changes to $-180°$ (a variation between $180°$ and $0°$ respectively would also be possible if one of the stages was non-inverting). Notice also that the d.c. gain is 400 and that there is no appreciable fall in gain until the signal frequency exceeds 10^4 Hz. Now if feedback, negative at d.c. is applied with a constant β of $-1/100$, we can draw the Nyquist plot for the loop gain βA as shown in Fig. 5.136. The different scale now shows βA at d.c. to be $4 \angle 180°$ and thus $A_{\text{fo}} = \dfrac{400}{1 - 4 \angle 180°} = 80$.

The shape and frequency scale of the βA plot is the same as that for the original A plot since β, in this case, is a simple fraction with a constant $180°$ phase shift. If the feedback formula is to be applied at any frequency other than d.c., we must consider how $1 - \beta A$ can be represented. The gain phasor for βA at $f = 2 \times 10^4$ Hz is shown on Fig. 5.136b and directly opposite is the phasor for $-\beta A$ at the same frequency. The point $+1$ is also shown on the positive real axis. If these two are added by phasor construction, we have the phasor for $1 - \beta A$ at $f = 2 \times 10^4$ Hz. Finally, taking the gain from the gain plot at the same frequency as $A = 322 \angle -55°$ we can obtain $|A_f| = 83$. If this construction had to be done for each frequency, it would be a very tedious business; however, examination of Fig. 5.136b shows the same length as the $1 - \beta A$ phasor for that frequency.

In practice, unless β is frequency dependent, a single polar plot can be drawn to represent both A and βA by using two separate scales. This can now be demonstrated by means of an example.

Example 5.57 An amplifier has the gain and phase shift values given in the table below. Feedback, negative at d.c. is applied with a β of 0.005. Determine the frequency response with feedback and plot the information as a Bode gain plot. Estimate the frequency range for which the feedback is positive and find the maximum value of β for normal amplifier operation.

$f(\times 10^5)$ Hz	0	0.2	0.5	1	2	3	4	5	7
$|A|$	800	794	761	666	437	269	167	108	50.3
$\angle A°$	180	168	151	124	79.5	47.7	24.7	7.8	-15

The polar plot for A is shown in Fig. 5.137 with a scale of 1 cm = 100. Since β is a simple fraction, the same plot represents βA if a new scale is used. In this case, the d.c. value for βA is, $800 \times 0.005 = 4$. Thus a scale of

Fig. 5.137 Nyquist plot for example 5.57.

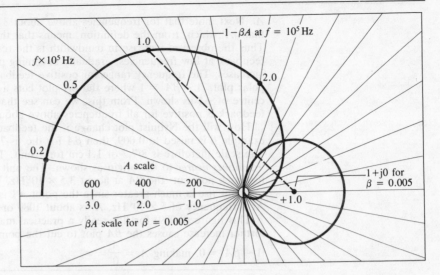

2 cm = 1 unit will allow the same plot to be used for βA. We can also mark the point +1 on the graph as shown.

Now, for each frequency, the value of $1 - \beta A$ can be measured (using the scale for βA). For example, at $f = 10^5$ Hz, the line for $1 - \beta A$ is shown and measured to be 8.1 cm, equivalent to a factor of 4.05. The resulting gain with feedback A_f, is given by 666/4.05 or 164. This result, and those for the other signal frequencies are shown in the following table together with the dB equivalents for the gain with feedback.

$f(\times 10^5)$ Hz	0	0.2	0.5	1	2	3	4	5	7
$|A_f|$	160	159	160	164	193	299	358	216	67.1
$20 \log_{10}|A_f|$	44	44	44	44.3	45.7	49.5	51	46.7	36.5

The two frequency responses A and A_f are shown as Bode gain plots in Fig. 5.138. The first point to observe is that while A falls continuously for frequencies above 10^5 Hz, A_f rises to a maximum (with a resonant rise) at 6.3×10^5 Hz and then falls steadily in the same way as A. The net result is an increase in bandwidth from about 1.5×10^5 Hz to about 5.9×10^5 Hz for

Fig. 5.138 Bode plots for example 5.57: frequency response with and without feedback.

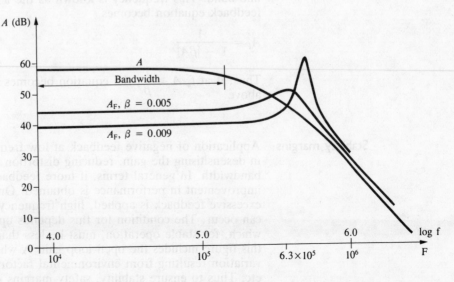

A_f. Next, note that for frequencies above about 3×10^5 Hz, $|A_f|$ is greater than $|A|$ which, from the definition, means that the feedback is positive. Thus the observed increase in bandwidth is the result of the negative feedback at low frequencies, gradually becoming positive as frequency increases. The frequency range for positive feedback is easily seen on the polar plot. $|1 - \beta A| = 1$ where the βA plot cuts a unit circle drawn on a centre of $+1$ as shown. From this, we can see that in this example, the feedback is positive for all frequencies above about 3×10^5 Hz.

How will the Nyquist plot change if the feedback is increased? If, for example β is raised to 0.009, then βA for d.c. is 7.2 represented by 8 cm. The scale therefore is 8/7.2 or 1.1 cm for 1 unit. The new scale point for $+1$ *moves nearer* to the origin as shown. The unit circle for positive feedback now cuts the βA at about 3.5×10^5 Hz. The maximum gain on the resonant rise is much greater since the point $+1$ is so close to the βA plot. At a frequency of 5×10^5 Hz, A_f is about 1080 or 61 dB.

This figure (0.009) for β is really a practical maximum value since a small further increase causes the βA plot to cut the point $+1$ and at this point $|1 - \beta A| = 0$, making $A_f = \dfrac{A}{0} = \infty$.

Stability of feedback amplifiers

The last example introduces a very important aspect of feedback. Increase of negative feedback makes the gain less subject to variation, reduces the distortion and increases the bandwidth, but with excessive feedback, the resulting high frequency positive feedback can cause the amplifier to become unstable and oscillate without any input. The rule for instability is

If the locus of the loop gain βA cuts or encloses the point $+1$, the system will be unstable.

This condition will always occur at the frequency where the phase has changed by 180° or π radians from that at low frequency or mid-band. This frequency is known as the π frequency f_π. At f_π, the feedback equation becomes

$$A_f = \frac{A}{1 - |\beta A|} \qquad [5.133]$$

Thus, if at f_π $A \geqslant \dfrac{1}{\beta}$, the equation becomes $A/0$ as mentioned above.

Stability margins Application of negative feedback at low frequencies is advantageous in desensitising the gain, reducing distortion and increasing the bandwidth. In general terms, if more feedback is applied, more improvement in performance is obtained. On the other hand, if excessive feedback is applied, high frequency instability or oscillation can occur. The condition for this depends upon the loop gain βA which, for stable operation, must be less than 1 at f_π. Unfortunately, this figure includes the open loop gain A which is subject to variation resulting from environmental factors, parameter spread, etc. Thus to ensure stability, safety margins must be allowed for in any design. Two margins are commonly used, the gain margin and the phase margin.

The gain margin is measured at f_π the frequency at which the angle of the loop gain βA is $0°$. At f_π, the gain margin is the difference between 0 dB and the actual dB gain for βA.

The phase margin is the angle of the loop gain βA at the frequency at which $\beta A = 1$.

The gain and phase margins are most easily measured and calculated from Bode gain and phase plots, but they can also be obtained from the polar plot in which case, the gain margin may be expressed as a fraction. The principles of calculation and use of gain and phase margins are best illustrated by means of examples.

Example 5.58   *The amplifier described in Example 5.57 has a gain given by*

$$A = \frac{800}{(1 + jf/f_1)(1 + jf/f_2)(1 + jf/f_3)}$$

where $f_1 = 2 \times 10^5$, $f_2 = 3 \times 10^5$ and $f_3 = 5 \times 10^5$.

(a) Determine the gain and phase margins if the feedback factor β is 0.005.
(b) Compare these results with those obtained in Examples 5.57.
(c) Estimate the permissible percentage variation in the d.c. gain A_o.

The Bode gain and phase plots for βA constructed using the techniques discussed in Chapter 3 (p.00). The gain plot is shown in Fig. 5.139 and the phase plot in Fig. 5.140.

Fig. 5.139 The Bode gain plot for example 5.58.

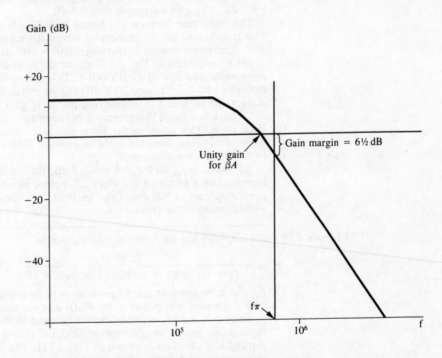

(a) The gain plot starts from $+12$ dB since $\beta A_o = 800 \times 0.005 = 4$ and 20 $\log_{10}4$ is 12 dB. The gain is constant up to the first break frequency at 2×10^5 Hz and finishes with a final slope of -60 dB/decade above 5×10^5 Hz. The phase plot shows the *relative phase* since the signs of β and A will always ensure SNFB at low frequencies. The phase plot therefore starts at $0°$ and finishes at $-270°$ ($-90°$ for each frequency break). In terms of the relative phase, the π frequency is that frequency at which the relative phase of the loop gain is $180°$ or π radians.

Fig. 5.140 The Bode phase plot for example 5.58.

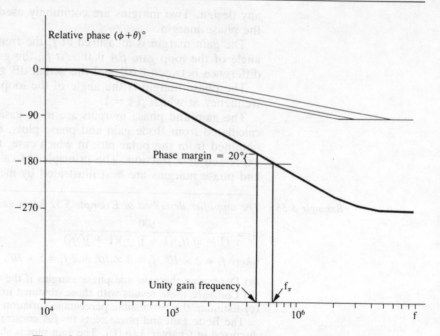

The π frequency is found from Fig. 5.140 to be 6.4×10^5 Hz. A line at this frequency is drawn on the gain plot to find that the loop gain at f_π is -6.5 dB. The gain margin is thus 6.5 dB.

The unity gain frequency is found from Fig. 5.139 to be 4.8×10^5 Hz. This is indicated on the phase plot where the corresponding phase shift is $160°$. The phase margin is therefore $180° - 160°$ or $20°$.

(b) Examination of Fig. 5.137 shows the f_π to have approximately the same value and that at f_π, $\beta A \simeq 0.5$. This means that βA could be multiplied by 2 (increased by 6 dB) before instability occurs. The phase margin can be found by identifying the unity gain point where $\angle\beta A$ is about $30°$. There is a small disagreement between the results from the polar and Bode plots. This arises as the Bode plot is a straight line approximation. Bode plots taken from the table in example 5.58 would give identical results to those from the polar plot.

(c) Since, at f_π, $20 \log_{10}\beta A = -6.5$ dB, $\beta A = 0.47$. This could be increased by a factor of $1/0.47$ or 2.1 before instability occurs. Thus the open loop gain of 800 could rise to 1690, an increase of more than 100% without undesirable oscillation.

Example 5.59

An amplifier has an open loop gain given by

$$A = \frac{2550}{(1 + j\omega/10^4)(1 + j\omega/10^5)^2(1 + j\omega/3 \times 10^5)}$$

Feedback, negative at low frequencies is to be applied. Determine the value of β if the gain is required to be 10 dB and the resulting phase margin.

Since β is not known, βA_0 cannot be used to draw the Bode gain plot. Instead, we ignore the gain factor 2250 and simply construct the gain plot starting at 0 dB. This is shown in Fig. 5.141. The phase plot in Fig. 5.142 starts at $0°$ and falls to $-360°$ (four break frequencies).

Inspection shows the π frequency (ω_π in this case) to be 7.4×10^4 and this is added to the gain plot. Since the gain margin is to be 10 dB, $20 \log_{10}\beta A$ at ω_π must be -10 dB. The 0 dB line for the loop gain can thus be drawn and projected to find βA_0 is 7.5 dB or 2.37. As A_0 is 2550,

$$\beta = \frac{2.37}{2550} = 9.3 \times 10^{-4}$$

Fig. 5.141 The Bode gain plot for example 5.59.

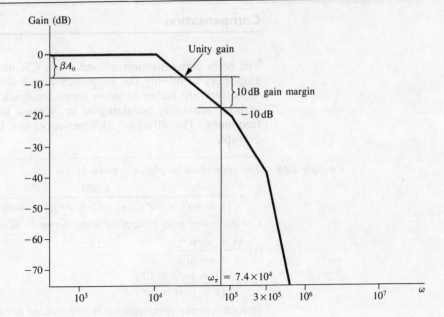

Fig. 5.142 The Bode phase plot for example 5.59.

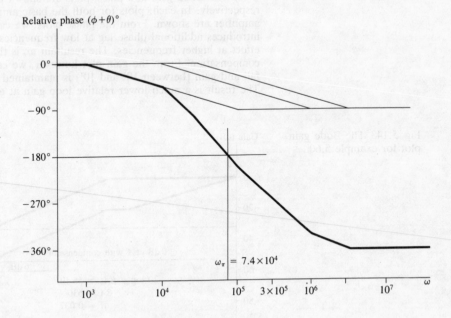

Unity gain for βA is at $= 2.4 \times 10^4$ r/s and the resulting phase margin is about $85°$.

Compensation

The term compensation is used when CR networks are added to amplifiers to modify the frequency response with feedback. The objectives are either to allow more feedback at low frequencies without instability occurring or to increase the bandwidth at high frequency. The effect of such networks can be illustrated by an example.

Example 5.60   *The gain of an amplifier is given by*

$$A = \frac{17000}{(1 + j\omega/2 \times 10^4)(1 + j\omega/5 \times 10^5)(1 + j\omega/2 \times 10^6)}$$

Compensation is to be applied using networks having transfer functions of

(i) $\dfrac{(1 + j\omega/10^4)}{(1 + j\omega/10^3)}$

or (ii) $\dfrac{(1 + j\omega/3 \times 10^5)}{(1 + j\omega/4 \times 10^6)}$.

In each case the compensation is to be added in cascade with the amplifier, and β is to be adjusted to give a gain margin of 6 dB. For each of the three situations (no compensation, network (i), network (ii)) find (a) the low frequency loop gain βA_o and (b) the π frequency. Discuss the results in terms of the effectiveness of the feedback.

(i) The Bode gain and phase plots are shown in Figs 5.143 and 5.144 respectively. In each, plots for both the basic amplifier and the compensated amplifier are shown. From the phase plots, we can see that the compensator introduces additional phase lag at low frequencies (10^2 to 10^5) but has no effect at higher frequencies. The resultant ω_π is the same with and without compensation. From the gain plot however, we can see that an additional fall and gain (between 10^3 and 10^4) is maintained for all higher frequencies. The result is a much lower relative loop gain at ω_π. Allowing for the

Fig. 5.143 The Bode gain plot for example 5.60i.

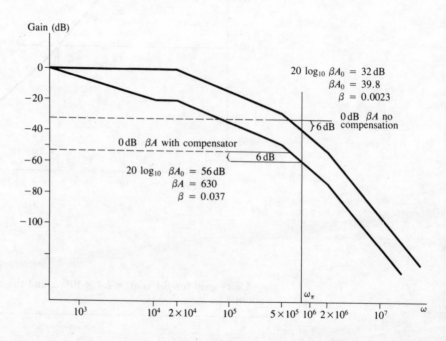

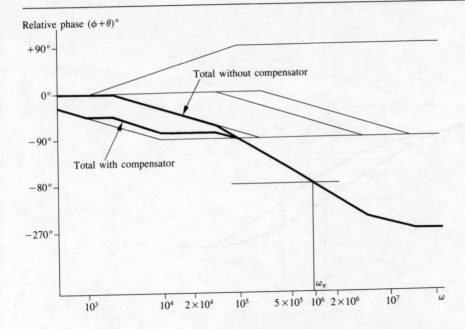

Fig. 5.144 The Bode phase plot for example 5.60i.

specified gain margin of 6 dB, the two absolute 0 dB levels have been added. The corresponding low frequency dB gains are 32 dB and 56 dB respectively. Finally, the equivalent βA_0 gain values of 39.8 and 630 are calculated and the corresponding values of feedback factor β are found to be 0.0023 (uncompensated) and 0.037 (compensated). The ratio in the loop gain figures of about 16 suggests that a considerable improvement in gain stabilisation and reduction in distortion can be achieved by the use of this type of compensation.

(ii) The Bode gain and phase diagrams for the second arrangement are shown in Figs 5.145 and 5.146 respectively. In this case, there is little difference between the compensated and uncompensated responses at low

Fig. 5.145 The Bode gain plot for example 5.60ii.

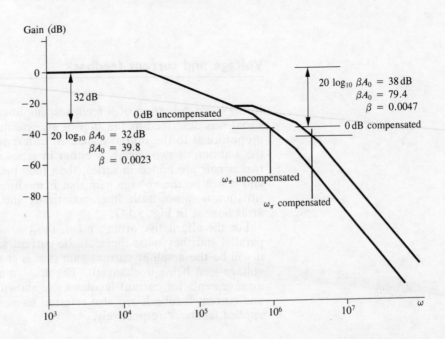

Relative phase $(\phi + \theta)°$

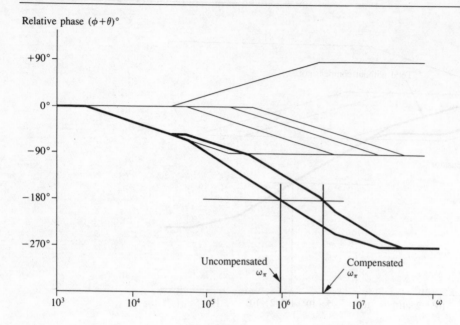

Fig. 5.146 The Bode phase plot for example 5.60ii.

frequencies. At high frequencies however, the leading phase angle of the compensator has moved ω_π from 9.6×10^5 to 3.5×10^6. This suggests that a similar improvement in bandwidth will be obtained. referring to the gain plot, the relative 0 dB levels have been added, allowing for the 6 dB gain margin. The corresponding low frequency loop gains are 32 dB (39.8) and 38 dB (79.4) with β values of 0.0023 (uncompensated) and 0.0047 (compensated). This shows a small increase in low frequency loop gain ($\times 2$) but not as much as with the first type of compensation. The use of this 'lead' compensation is for the improvement in bandwidth.

Voltage and current feedback

In the basic definition of a feedback amplifier, the terminal input signal was described as the *sum* of an external signal and a signal proportional to the output signal. In different circuit arrangements, the addition of two signals is either in series or in parallel. If the two signals are added in series, then they must be voltage signals and it will be the voltage gain that is modified by the feedback. This situation is shown both diagrammatically and in a practical circuit arrangement in Fig. 5.147.

For the alternative arrangement, the two signals are added in parallel and they must therefore be current signals. This means that it will be the amplifer current gain that is changed by feedback (the voltage gain being unchanged). The diagrammatic and practical arrangements for current feedback are shown in Fig. 5.148. Voltage and current feedback are also referred to as *series applied* and *shunt applied* feedback respectively.

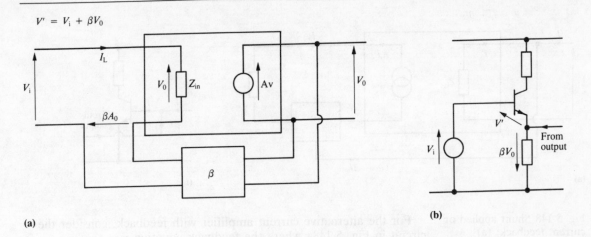

$V' = V_i + \beta V_0$

(a)

(b)

Fig. 5.147 Series applied or voltage feedback: (a) diagramatically; (b) transistor circuit arrangement.

The input impedance of feedback amplifiers

The input impedance of any circuit is obtained by dividing the applied voltage by the current flowing into the circuit. With feedback amplifiers, either the applied voltage is the sum of two voltages (series applied feedback) or the applied current is the sum of two currents (shunt applied feedback). In the first case, a larger voltage is required for the same current and this represents an increase in input impedance. If the input current is the sum of two currents, the increased input current demonstrates a reduction of input impedance. In each case, the analysis comes from the basic feedback definition and equation.

Consider Fig. 5.147a;

$V' = V_i + \beta V_o$

or $V' = V_i + \beta A V'$

and $V_i = V'(1 - \beta A)$

Now dividing both sides of this equation by the input current I_i,

$$\frac{V_i}{I_i} = \frac{V'}{I_i}(1 - \beta A)$$

But $\dfrac{V'}{I_i} = Z_{in}$, the input impedance of the amplifier without feedback

and $\dfrac{V_i}{I_i} = Z_{inf}$, the input impedance of the feedback amplifier.

Therefore

$$Z_{inf} = Z_{in}(1 - \beta A) \qquad [5.134]$$

Note that with SNFB the result becomes $Z_{in}(1 + \beta A)$ showing that the input impedance is *increased* by the same amount as the voltage gain is reduced. At extremes of frequency however, where the feedback becomes positive, Z_{in} will be reduced.

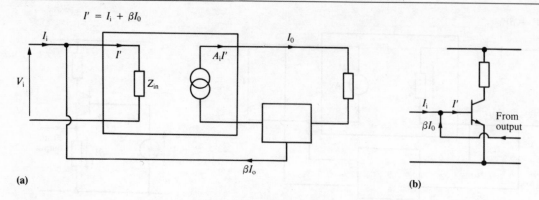

(a)

(b)

Fig. 5.148 Shunt applied or current feedback: (a) diagramatically; (b) transistor circuit arrangement.

For the alternative current amplifier with feedback, consider the circuit in Fig. 5.148a where the feedback equation is

$$I' = I_i + \beta I_o \qquad [5.135]$$

Rearrangement in the same way results in

$$I_i = I'(1 - \beta A)$$

This is now divided by the voltage to give

$$\frac{I_i}{V_i} = \frac{I'}{V_i}(1 - \beta A)$$

This is an admittance equation involving the input admittance of the amplifer without feedback I'/V_i and the input admittance with feedback, Y_{inf},

and $Y_{inf} = Y_{in}(1 - \beta A)$

or $Z_{inf} = \dfrac{Z_{in}}{1 - \beta A} \qquad [5.136]$

In this case, the input impedance is reduced by SNFB by the same amount as the current gain is reduced.

Derivation of feedback signals

For both voltage and current feedback, the feedback signals can be obtained in two ways. The feedback circuit of β network can be connected in parallel with the output, in which case the feedback signal will be proportional to the output voltage. Alternatively, the β network can be connected in series with the output and the resulting signal is proportional to the output current. These two situations can be referred to as **voltage derived feedback** and **current derived feedback** respectively or simply as **shunt output feedback** and **series output feedback**.

In every case, the calculated feedback factor β must apply to the appropriate signal form, i.e. as a voltage ratio β_v for a voltage amplifier having gain A_v or a current ratio β_i for a current amplifier with gain A_i.

These feedback networks are connected in series or in parallel with the amplifier output so it is not surprising that the amplifier output impedances will be modified and that the amount of modification depends on the amount of feedback and on the amplifier gain.

Voltage derived voltage feedback amplifiers

Consider the voltage amplifier shown in Fig. 5.149; this is an example of shunt output, series input (or voltage derived voltage feedback). The amplifier is shown here with its output impedance Z_{out} and the gain A_v will be the normal loaded voltage gain. The impedance of the feedback network is usually too large to change the loading appreciably, but if this is not the case, Z_{out} and A_v can be modified accordingly. The terminal input voltage V' will depend upon the feedback network and upon Z_{in} and Z_s the input source impedances. The source e is not shown as for the calculation of output impedance any external generators are suppressed. Thus V' may be written as βV_o. (In general β will not have the same value as that used in calculating Z_{inf} and A_{vf} as Z_s and Z_{in} will load the feedback network and further attenuate V'.)

Fig. 5.149 The output impedance of a voltage derived voltage feedback amplifier (shunt output, series input).

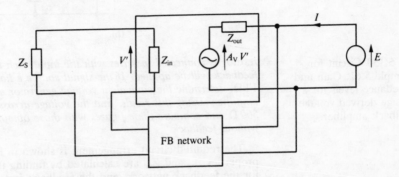

To find Z_{outf}, the output impedance with feedback, a voltage E is applied to the output terminals, the resulting I is calculated and the ratio E/I is the required impedance. From the circuit in Fig. 5.149,

$$E = IZ_{out} + A_v V'$$
$$= IZ_{out} + A_v \beta E \text{ since } V_o = E$$
$$\therefore E(1 - A_v\beta) = IZ_{out}$$

and $\dfrac{E}{I} = Z_{outf} = \dfrac{Z_{out}}{1 - \beta A_v}$

or, with SNFB,

$$Z_{outf} = \frac{Z_{out}}{1 + \beta A_v} \tag{5.137}$$

and the output impedance is *reduced* with feedback. The direction of change in output impedance is that which might be expected since the feedback network is in parallel with the amplifier output.

Example 5.61 *An amplifier has a voltage gain of 1500 and input and output impedances respectively of 2 kΩ and 3 kΩ. Feedback is provided using a potential divider in shunt with the output, the values being 100 kΩ and 400 Ω. The smaller*

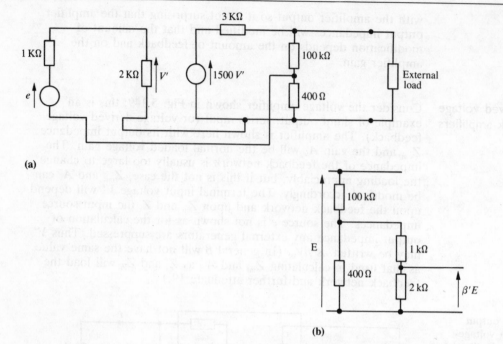

(a)

(b)

Fig. 5.150 Circuit for example 5.61. Gain and impedance levels for a voltage derived voltage feedback amplifier.

resistor is connected in series with the input such that simple negative feedback will be applied. If the signal source e has internal resistance of 1 kΩ, determine the equivalent voltage generator at the output terminals of the amplifier. Using this result, find the voltage across external loads of 1 kΩ and 500 Ω and compare these figures with those obtained with the same amplifier without feedback.

The required circuit arrangement is shown in Fig. 5.150a. The forward properties A_{vf} and Z_{inf} are calculated by finding the loaded gain, allowing for the feedback network, and the feedback factor β. Since 100 kΩ $\gg$ 3 kΩ, the gain will be taken as 1500. The β factor is simply 0.4/100.4 or 1/251. Hence, for SNFB,

$$1 + \beta A = (1 + 1500/251 = 6.98.$$

Thus $A_{vf} = 1500/6.98 = 215$, and $Z_{inf} = 2 \times 6.98 = 13.96$ kΩ.

For the output impedance with feedback, Z_{outf}, the β factor is modified by the input network as shown in Fig. 5.150b.

$$\beta' = \frac{\dfrac{0.4 \times 3}{3.4}}{100 + \dfrac{0.4 \times 3}{3.4}} \times \frac{2}{3} = 2.34 \times 10^{-3}$$

Thus $Z_{outf} = \dfrac{3000}{1 + 1500 \times 2.34 \times 10^{-3}} = 664$ Ω.

The various results from above are combined in the equivalent circuit for the complete feedback amplifier shown in Fig. 5.151. For the 1 kΩ external load,

$$V_o = e \times \frac{13.96}{14.96} \times 215 \times \frac{5}{5.664} = 188.1e.$$

and for the 500 Ω load,

$$V_o = e \times \frac{13.96}{14.96} \times 215 \times \frac{5}{5.664} = 177.1e \text{ a 6\% fall.}$$

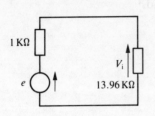

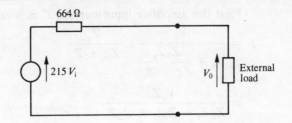

Fig. 5.151 Equivalent circuit for the feedback amplifier in example 5.61.

The same load changes for the amplifier without feedback would have resulted in the following output voltages; for the 1 kΩ external load,

$$V_o = e \times \frac{2}{3} \times 1500 \times \frac{10}{13} = 769e.$$

and for 500 Ω,

$$V_o = e \times \frac{2}{3} \times 1500 \times \frac{5}{8} = 625e, \text{ a 19\% fall.}$$

These results demonstrate the effect of a reduction in output impedance on the variation of gain due to loading effects. This reduction is the direct effect of the negative feedback applied to the amplifier and the same results could have been obtained by applying the feedback formulae directly to the results for loaded gain without feedback except for the additional effect due to change in input impedance.

Voltage derived current feedback

The circuit shown in Fig. 5.152 is a current amplifier with a terminal input signal which is the sum of an external signal I and a feedback signal βI_o. For the output impedance calculation the signal current I is suppressed. The feedback network is in parallel with the output of the amplifier and is thus derived from the output voltage. The gain A is the short circuit current gain of the amplifier and Z_{out} is the output impedance of the amplifier without feedback (including the final load). The amplifier input current can be calculated directly in terms of the output voltage V_o. To calculate the Z_{outf}, the output impedance with feedback, a current signal I is applied at the output and the resulting voltage V_o is determined.

Fig. 5.152 The output impedance of a voltage derived current feedback amplifier (shunt output, shunt input).

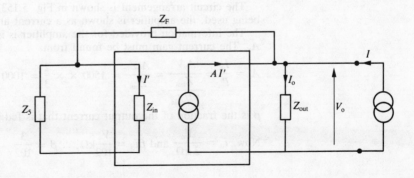

First the amplifier input current I' is found,

$$I' = \frac{V_o}{Z_f + \dfrac{Z_{in}Z_s}{Z_{in} + Z_s}} \times \frac{Z_s}{Z_{in} + Z_s}$$

$$= \frac{V_oZ_s}{Z_{in}Z_f + Z_sZ_f + Z_{in}Z_s}$$

$$= \frac{V_o}{Z_f + Z_{in} + \dfrac{Z_{in}Z_f}{Z_s}} = \frac{V_o}{Z_{eff}}$$

Writing an equation at the output node,

$$I + \frac{A \times V_o}{Z_{eff}} = \frac{V_o}{Z_{out}}$$

rearranging, $I = V_o\left(\dfrac{1}{Z_{out}} - \dfrac{A}{Z_{eff}}\right)$

and $Z_{outf} = \dfrac{V_o}{I} = \dfrac{1}{\dfrac{1}{Z_{out}} - \dfrac{A}{Z_{eff}}}$

For SNFB, the gain A will be negative and the resulting output impedance is given by

$$Z_{outf} = \frac{1}{\dfrac{1}{Z_{out}} + \dfrac{A}{Z_{eff}}}$$ [5.138]

which is the parallel combination of Z_{out} and Z_{eff}/A.

The widest application of this type of feedback (shunt, shunt) is in op-amp circuits and it is usually more convenient to analyse these using an alternative approach which is discussed in the section on op-amp circuits (p.482). However, this traditional feedback approach does demonstrate some useful principles which are illustrated in the following example.

Example 5.62   *The amplifier described in Example 5.61 ($A = 1500$, $Z_{in} = 2\ k\Omega$, $Z_{out} = 3\ k\Omega$) is to used with voltage derived current feedback provided by a 100 kΩ resistor between input and output terminals. Assuming SNFB, calculate the equivalent Norton generator at the output terminals. The external signal source has internal resistance of 1 kΩ. Hence find the output voltage for an external load of 1 kΩ.*

The circuit arrangement is shown in Fig. 5.153. Since current feedback is being used, the amplifier is shown as a current amplifier.

The information provided for the amplifier is in terms of the voltage gain A_v. The current gain must be found from,

$$A_i = \frac{I_o}{I'} = \frac{A_vV'}{\dfrac{Z_{out}}{V'/Z_{in}}} = \frac{A_vZ_{in}}{Z_{out}} = 1500 \times \times \frac{2}{3} = 1000$$

β is the fraction of the output current that is fed back to the input.

Now, $I_o = \dfrac{V_o}{3\ k\Omega}$ and $\beta I_o = \dfrac{V_o}{102}k\Omega,\ \therefore\ \beta = \dfrac{3}{102}$

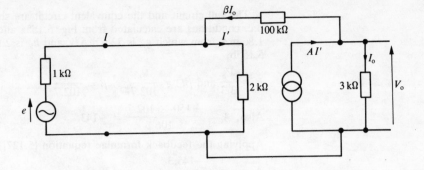

Fig. 5.153 Circuit for example 5.62. Gain and impedance levels for a voltage derived current feedback amplifier.

Applying equations [5.127] and [5.132],

$$A_{if} = \frac{1000}{1 + \dfrac{3000}{102}} = 32.9$$

and $Z_{inf} = \dfrac{2\ \text{k}\Omega}{1 + \dfrac{3000}{102}} = 65.8\ \Omega.$

For the output impedance with feedback, we apply equation [5.138]

$$Z_{outf} = \frac{1}{\dfrac{1}{3\ \text{k}\Omega} + \dfrac{1000}{Z_{eff}}}$$

where $Z_{eff} = 100\ \text{k}\Omega + 2\ \text{k}\Omega + \dfrac{2 \times 100}{1}\ \text{k}\Omega = 302\ \text{k}\Omega.$

$$\therefore Z_{outf} = \frac{3 \times 302/1000}{3 + 302/1000}\ \text{k}\Omega = 274\ \Omega.$$

The final circuit arrangement is shown in Fig. 5.154. From this circuit,

$$I' = \frac{e}{1065.8}$$

and $V_o = \dfrac{32.9e}{1065.8} \times \dfrac{274}{1274} \times 1000 = 6.64e.$

This value is much smaller than the final output in the previous example (188e). This is the result of the impedance mismatch at the input. This type of feedback results in a very low input impedance which is unsuitable if direct voltage amplification is required.

Fig. 5.154 Equivalent circuit for the solution of example 5.62.

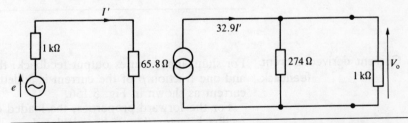

Example 5.63   *A transistor has* h_{fe} *150 and is employed in a common emitter amplifier with collector load 3.3 kΩ and a decoupled emitter resistor of 1 kΩ. Base bias is provided by a 100 kΩ resistor between collector and base and the d.c. supply is 10 V. Calculate the input impedance, current gain and voltage gain for the circuit.*

The full circuit and the equivalent circuit are shown in Fig. 5.155. The d.c. conditions are calculated from Fig. 5.155a and I_c is found to be 1.86 mA from which g_m is 72.6 mA/V and h_{ie} is 2.07 kΩ. Referring to Fig. 5.155b,

$$I_o = \frac{V_o}{3.3} \quad \text{and} \quad \beta I_o = \frac{V_o}{102.7}, \quad \therefore \beta = \frac{3.3}{102.7}$$

Also, $A_i = \dfrac{-150 \times 102.7}{106} = -143.$

Applying the feedback formulae (equation [5.127])

$$A_{if} = \frac{-145.3}{1 + \dfrac{145.3 \times 3.3}{102.7}} = -25.6.$$

and $Z_{inf} = \dfrac{2.07}{1 + \dfrac{145.3 \times 3.3}{102.7}} = 0.365 \text{ kΩ}.$

For the voltage gain,

$$V_i = I_i \times 0.365, \quad V_o = -25.6 I_i \times 3.3$$

$$\therefore A_v = \frac{-25.6 \times 3.3}{0.365} = -231.5.$$

Fig. 5.155 Circuits for example 5.63: (a) full circuit; (b) equivalent circuit.

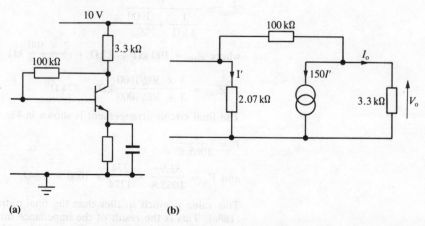

(a)   (b)

Note that a direct calculation of A_v from $g_m R_{Leff}$ results in virtually the same answer, demonstrating that current feedback does not effect the voltage gain.

Current derived current feedback

For shunt input, series output feedback, the output current is split and one fraction β of the current is added to the external signal current as shown in Fig. 5.156.

For the forward properties, the loaded current gain A_i' and the feedback factor β must be calculated.

If $R_1 \ll R_L$, $A_i' = \dfrac{A_i Z_o}{R_L + Z_o}$

and $R_2 \gg Z_{in}$, $\beta = \dfrac{R_1}{R_1 + R_2}$

Fig. 5.156 A current derived current feedback amplifier (series output, shunt input).

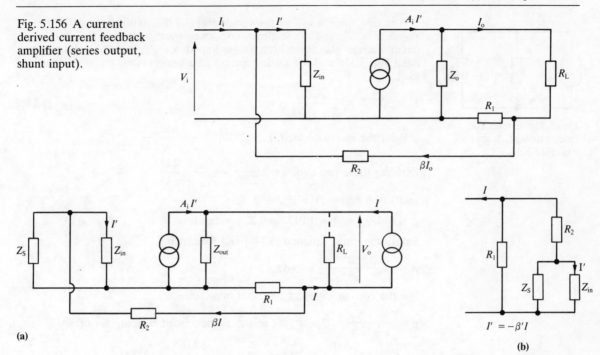

Fig. 5.157 The output impedance of a current feedback amplifier: (a) full circuit; (b) β network.

In extreme cases, the additional loading effects may be calculated numerically. The results for A_{if} equation [5.125] and Z_{inf} equation [5.136] can then be applied.

For the output impedance with feedback, the suppressed source must be included and a signal applied at the output terminals as shown in Fig. 5.157a. The external load resistor R_L is shown unconnected as this can be included afterwards if necessary. The effective β factor is calculated from the part circuit shown in Fig. 5.157b.

Summing the currents at the output,

$$I + A_i I' = \frac{V_o}{Z_{out}}$$

and $I(1 - \beta' A_i) = \dfrac{V_o}{Z_{out}}$

$$Z_{outf} = \frac{V_o}{Z_{out}} = Z_{out}(1 - \beta' A_i) \qquad [5.139]$$

and for SNFB, the output impedance is increased to $Z_{out}(1 + \beta' A_i)$. In most applications, the external effect of this will be small as the load R_L will be in parallel with Z_{outf} and this will dominate the effective output impedance.

Example 5.64   *A two-stage common emitter amplifier has a current gain into the final 1 kΩ load of 3000, but if the load is increased to 2 kΩ, the gain falls to 2750. The amplifier input impedance is 5 kΩ. Feedback is provided with a 1 kΩ resistor in the emitter lead of the output transistor and a 270 kΩ resistor from this point to the base of the first transistor. Calculate the resulting current gain into the 1 kΩ load, the input impedance and the output impedance assuming a source impedance of 10 kΩ.*

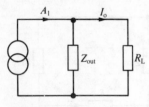

Fig. 5.158 Circuit for calculation of A_i and Z_{out} in example 5.64.

The first stage is to find the components of the amplifier as shown in Fig. 5.157a. Z_{in} and the feedback components are given, but the short circuit current gain A_i and the output impedance Z_{out} must be determined from Fig. 5.158 and the loaded current gain information. For the original load,

$$I_o = 3000I' = \frac{A_i Z_{out}}{Z_{out} + 1\text{ k}\Omega} \times I' \qquad [5.140]$$

and for the increased load, $I_o = 2750I' = \dfrac{A_i Z_{out}}{Z_{out} + 2\text{ k}\Omega} \times I'$

Dividing these two equations, $\dfrac{3000}{2750} = \dfrac{Z_{out} + 2\text{ k}\Omega}{Z_{out} + 1\text{ k}\Omega}$

and $1.017(Z_{out} + 1) = Z_{out} + 2$

$\therefore 0.017 Z_{out} = 2 - 1.017$, and $Z_{out} = 58\text{ k}\Omega$.

Substituting into equation [5.140] and rearranging,

$$A_i = \frac{3000(58 + 1)}{58} = 3052.$$

For the current gain and the input impedance,

$$\beta = \frac{1}{1 + 276} = 3.6 \times 10^{-3} \text{ and } A_i \text{ is the loaded current gain of } 3000.$$

$$\therefore A_{if} = \frac{3000}{1 + \dfrac{3000}{277}} = 254 \text{ and } Z_{inf} = \frac{5\text{ k}\Omega}{1 + \dfrac{3000}{277}} = 423\ \Omega.$$

For the output impedance, the circuit for β' is that shown in Fig. 5.158b with $R_1 = 1\text{ k}\Omega$, $R_2 = 250\text{ k}\Omega$, $Z_{in} = 5\text{ k}\Omega$ and $Z_s = 10\text{ k}\Omega$.

Hence, $\beta' = \dfrac{1}{1 + 250 + \dfrac{5 \times 10}{15}} \times \dfrac{10}{15} = 2.62 \times 10^{-3}$

and $Z_{outf} = 58(1 + 2.62 \times 10^{-3} \times 3052) = 522\text{ k}\Omega$.

But the overall Z_{out} includes the parallel 1 kΩ load and is thus virtually 1 kΩ.

Current derived voltage feedback

For this type of feedback, the feedback signal is obtained from the output current (series output) and is added in series with the external signal. The equivalent circuit is shown in Fig. 5.159.

As with the last example, the open circuit gain and output impedance are required since R_f is in series with the load.

From the circuit,

$$V_o = IR_L \text{ and } V_o = IR_F$$

$$\therefore = \frac{R_F}{R_L} \qquad [5.141]$$

and the loaded voltage gain is given by

$$A_v' = \frac{A_v R_L}{Z_{out} + R_L + R_F} \qquad [5.142]$$

The usual voltage feedback equations can then be applied to the calculation of A_{vf} and Z_{inf}.

Fig. 5.159 Current derived
voltage feedback (series
output, series input).

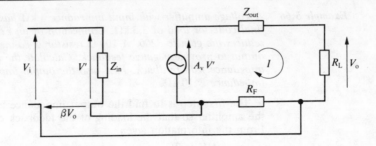

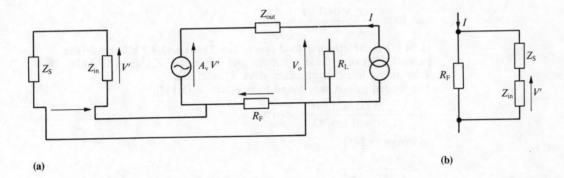

(a)   **(b)**

Fig. 5.160 The output
impedance for current
derived voltage feedback:
(a) full circuit; (b) feedback
network.

For the output impedance, a current I is injected as shown in Fig.
5.160a and the signal source is suppressed. The value of V' in terms
of I is found from the feedback circuit shown in Fig. 5.160b but if
Z_s is small and $Z_{in} \gg R_F$, then $V' \simeq -IR_F$. Writing an equation for
the output circuit,

$$V_{out} = A_vV' + I(Z_{out} + R_F) = -A_vIR_F + I(Z_{out} + R_F)$$

$$\therefore Z_{out} = \frac{V_o}{I} = Z_{out} + R_F(1 - A_v) \qquad [5.143]$$

For SNFB, $Z_{outf} = Z_{out} + R_F(1 + A_v)$ $\qquad [5.144]$

but, once again, this is in parallel with R_L so the overall output
impedance will not be increased by very much.

Example 5.65

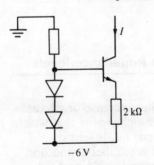

Fig. 5.161 Constant current
tail circuit for example
5.65.

*A single stage transistor amplifier is used to provide the 'constant current' tail
for an emitter coupled amplifier. The arrangement is shown in Fig. 5.161, the
transistor having h_{fe} 180 and h_{oe} 25×10^{-6}S. Calculate the value of the
constant current and the output impedance for the current source.*

For the d.c. analysis, the two diodes hold the base of the transistor at
1.4 V above the negative supply rail. The emitter is thus at 0.7 V above the
rail giving an emitter current of 0.7/2 or 0.35 mA. Since h_{fe} is large, I_B can
be neglected and the collector current $I_C = I = I_E$. The constant current is
therefore 0.35 mA. From this value, we find $g_m = 14$ mA/V, and
$h_{ie} = 12.9$ kΩ.

To calculate the output impedance, the open circuit voltage gain A_v' is
required. In this case, the open circuit loading on the transistor is the
internal h_{oe} of 25 μS or the equivalent of 40 kΩ. Thus $A_v' = -g_mR_L = -560$.
Applying equation [5.144],

$$Z_{outf} = 40 \text{ kΩ} + 2 \text{ kΩ}(1 + 560) = 1162 \text{ kΩ}.$$

Example 5.66   *A voltage amplifier with input impedance 4 kΩ has a gain of 5000 with the final collector load of 3.3 kΩ. Connection of an external load of 4.7 kΩ reduces the gain to 3200. A 100 Ω resistor is connected in series with the output to provide negative feedback. Calculate the resulting gain and input impedance with feedback. Find also the output impedance if the source impedance is 10 kΩ.*

The first step is to find the output impedance and open circuit gain of the amplifier so that the loading of the feedback circuit can be included. From the information given

$$5000 = \frac{A_v'V \times 3.3}{Z_{out} + 3.3}$$

and $$3000 = \frac{A_v'V \times 1.94}{Z_{out} + 1.94}$$

(1.94 kΩ is the effective load due to the 33 kΩ and 4.7 kΩ resistors in parallel.) Dividing these equations and solving for Z_{out} gives 13.4 kΩ. Substituting in either equation gives $A_v' = 25\ 302$.

The loaded gain is now found from equation [5.142],

$$A_v = \frac{25302 \times 1.94}{1.94 + 0.1 + 13.4} = 3179$$

and from [5.141],

$$\beta = \frac{0.1}{1.94} = \frac{1}{19.4}$$

Applying the results for series applied feedback (equation [5.134])

$$A_{vf} = \frac{3179}{1 + \dfrac{3179}{19.4}} = 19.28 \text{ (approximately } 1/\beta)$$

$$Z_{inf} = 4\left(1 + \frac{3179}{19.4}\right) = 659 \text{ kΩ}.$$

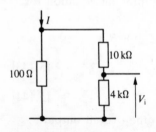

Fig. 5.162 Circuit for output impedance calculation in example 5.66.

For the output impedance calculation, we have to express V_{in} in terms of I allowing for the effects of Z_S, Z_{in} and R_F. The circuit is shown in Fig. 5.162. Since $R_F \ll (Z_{in} + Z_S)$, the voltage across R_F is simply 100 Ω × I. This voltage is reduced by the series $Z_{in} + Z_S$ to make $V_{in} = (4/14) \times 0.1$ kΩ × I. Now applying equation [5.143],

$$Z_{outf} = 13.4 + \frac{4}{14} \times 0.1(1 + 25302) = 736 \text{ kΩ}.$$

The overall output impedance will now include the effective load of 1.94 kΩ and is hardly changed.

Summary of the effects of feedback on impedance levels

We have seen that negative feedback can either increase or decrease input and output impedances. In each case, the change is that which would be expected from the type of connection, i.e. a series connection increases the impedance level and a parallel connection reduces it. In all cases, the change is by a factor $1 + \beta A$ and the only problem is to determine exactly which values of β and A are applicable. For the forward properties, gain and input impedance, the loaded gain, current or voltage as appropriate is used and the source impedance is not relevant. For the reverse effect, output impedance, the source impedance is involved in the determination of β but the external load is not involved in the feedback equations. For practical purposes, feedback methods provide quick and easily applicable approximate results if second order effects are neglected. If more precise calculations are required, a general analysis using mesh or nodal equations is probably just as convenient as trying to determine the precise values of β and A that should be used. For easy reference, a summary of the results is given in Table 5.1.

Table 5.1 Gain and impedance levels with SNFB

Type of Feedback	Voltage gain	Current gain	Input impedance	Output impedance
Series shunt Voltage derived voltage feedback	$\dfrac{A}{1 + \beta A}$ loaded A_v incuding FB network usual β $\dfrac{R_2}{R_1 + R_2}$	unchanged	$Z_{in}(1 + \beta A)$ same factor as for A_v	$\dfrac{Z_{out}}{1 + \beta A}$ o/c gain β modified by R_2
Series series Current derived voltage feedback	$\dfrac{A}{1 + \beta A}$ A_v loaded by R_L and R_F usual $\beta = \dfrac{R_F}{R_T}$ unchanged	$Z_{in}(1 + \beta A$ same factor as for A_v	$Z_{out} + R_F(1 + A)$ open circuit A_v	
Shunt series Current derived current feedback	unchanged	$\dfrac{A}{1 + \beta A}$ loaded A_i including FB network usual $\beta = \dfrac{R_1}{R_1 + R_2}$ $\dfrac{Z_{in}}{1 + \beta A}$ same factor as for A_i	$Z_{out}(1 + \beta A)$ s/c/ gain β modified by R_s	
Shunt shunt† Voltage derived current feedback	unchanged	$\dfrac{A}{1 + \beta A}$ A_i loaded by R_L and R_F usual $\beta = \dfrac{R_L}{R_F}$	$\dfrac{Z_{in}}{1 + \beta A}$ same factor as for A_i	s/c gain Z_{out} in parallel with $Z_{F/A}$

† Shunt shunt feedback is usually more conveniently analysed by op-amp methods.

General summary of the properties of feedback

Amplifiers employing negative feedback with the resulting fall in gain will be less sensitive to variation of parameters and environment, will add less distortion and noise to the signal and will have an improved frequency response. If choice of circuit arrangement allows, the above advantages can be combined with modification of impedance levels as desired. Excessive feedback can result in unwanted oscillation at extremes of frequency.

You should now be able to attempt exercises 5.56 to 5.62.

OP-AMP CIRCUITS

Op-amps and feedback

We have seen how feedback modifies all the properties of amplifiers. An op-amp can be regarded as the ideal amplifier for use with feedback and it is in practice useless as an amplifer if feedback is *not* used. Many different amplifier arrangements are used and these are discussed in the following sections but as feedback is applied in all cases, we should first examine the effect of this feedback upon some of the basic op-amp properties. In particular, the properties of voltage offset and frequency response should be considered. In each case these can be examined for the basic inverting amplifier discussed in Chapter 5 (p.354).

Voltage offset with feedback

The inverting amplifier shown in Fig. 5.163 has a differential gain A_d and an input impedance which tends to infinity. The equivalent circuit consists of an ideal op-amp together with a voltage generator representing the input voltage offset V_{OI}. Remember that a typical voltage offset is about 1 mV and that if both input are earthed, the high A_d amplifying this voltage drives the amplifier into saturation.

Fig. 5.163 Op-amp voltage off-set with feedback.

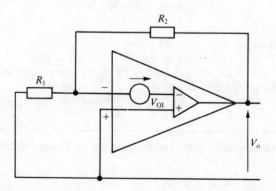

From the circuit shown we can write an equation for the differential input voltage for the 'ideal' amplifier,

$$V_{\text{diff}} \doteq \frac{V_o R_1}{R_1 + R_2} + V_{\text{OI}}$$

V_o is obtained by multiplying V_{diff} by A_d. Therefore

$$V_o = A_d\!\left(\frac{V_o R_1}{R_1 + R_2} + V_{\text{OI}}\right)$$

rearranging, $V_o = \dfrac{A_d V_{\text{OI}}}{1 - \dfrac{A_d R_1}{R_1 + R_2}} = \dfrac{A_d(R_1 + R_2)V_{\text{OI}}}{R_2 + R_1(1 - A_d)}$ [5.145]

Since A_d is very large, and negative, R_2 is negligible compared with $A_d R_1$ and the terms A_d terms cancel. Therefore

$$V_o = \left(1 + \frac{R_2}{R_1}\right)V_{\text{OI}}$$

For typical circuit values of R_1 and R_2, the output offset will be between two and ten times the internal V_{OI}, i.e. a few millivolts. Some op-amps have a provision for additional external circuitry to 'balance out' any residual output offset voltage.

Example 5.67

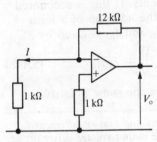

12 kΩ

I

1 kΩ

1 kΩ

V_o

Fig. 5.164 Circuit for example 5.67.

An op-amp using ±12 V d.c. supplies has an open loop gain A_d of 1000 and if both inputs are earthed, the output voltage is 5 V. Determine the output offset if it is used in the circuit shown in Fig. 5.164.

From the open loop information, the input offset V_{IO} can be found:

$$V_{\text{IO}} = \frac{5}{1000} = 5 \text{ mV}$$

From first principles, if no other signals are present and the input currents are negligible, V_+ is earthed and therefore V_- will be at ±5 mV. The circuit current I shown will therefore be

$$I = \frac{\pm 5 \text{ mV}}{1 \text{ k}\Omega} = \pm 5 \text{ μA}.$$

Since this current flows through the 12 kΩ resistor, the output voltage is given by,

$$V_o = \pm 5 \text{ mV} \pm 5 \text{ μA} \times 12 \text{ k}\Omega = \pm 65 \text{ mV}.$$

Alternatively, from equation [5.145]

$$V_o = \left(1 + \frac{12}{1}\right)5 \text{ mV} = 65 \text{ mV}.$$

Off-sets due to bias currents

For all op-amps, there will be very small bias currents flowing into the inverting and non-inverting input terminals. For FET input op-amps, these bias currents are so small that their effect upon any voltage offset can be neglected, but with BJT inputs, problems can arise.

The op-amp in the circuit in Fig. 5.165 has BJT input stages and to examine the effect of bias currents, the voltage offset V_{IO} will be assumed to be zero. The various currents including the equal bias currents I_B are indicated on the circuit diagram.

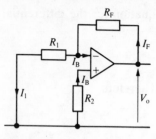

Fig. 5.165 Op-amp off-sets resulting from bias currents.

Since $V_{IO} = 0$, $V_- = V_+ = -I_B R_2$

and $\therefore I_1 = \dfrac{-I_B R_2}{R_1}$

and $I_F = -I_B \dfrac{R_2}{R_1} + I_B = I_B \left(1 - \dfrac{R_2}{R_1}\right)$

$V_o = V_- + I_F R_F = -I_B R_2 + I_B R_F \left(1 - \dfrac{R_2}{R_1}\right)$

which can be arranged to give,

$$V_o = I_B \left[R_F - R_2 \left(\frac{R_F}{R_1} + 1 \right) \right] \qquad [5.146]$$

Thus, if the effect of any bias currents is to be minimised, let

$R_F = R_2 \left(\dfrac{R_F}{R_2} + 1 \right)$ and this can be rearranged to give

$$R_2 = \frac{R_1 R_F}{R_1 + R_F} \qquad [5.147]$$

This result shows that the combined d.c. resistance connected to the two input terminals should be the same if input offsets due to bias currents are to be avoided.

In practical op-amps, there will also be a current offset I_{IO} which is the difference between the two bias currents. If this is accounted for, equation [5.146] must be modified by the addition of a term $I_{IO} R_F$. The total output offset due to V_{IO}, I_B and I_{IO} is given by

$$V_{OO} = V_{IO} \left(1 + \frac{R_F}{R_1}\right) + I_B \left(R_F - R_2 \left(\frac{R_F}{R_1} + 1\right)\right) + I_{IO} R_F \qquad [5.148]$$

where the three terms may or may not have the same polarity.

Example 5.68

The op-amp in Fig. 5.166 has the following offsets and bias requirements; V_{IO} 2 mV, I_B 0.1 μA, I_{IO} 0.02 μA. Determine the output voltage offset (a) as shown and (b) with the optimum value of R_2.

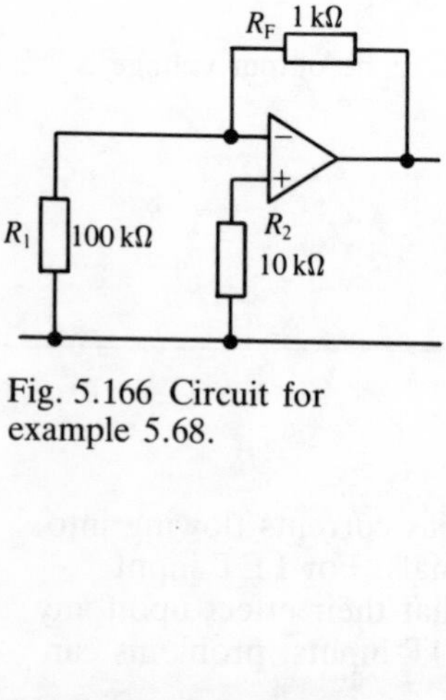

For the circuit as shown, equation [5.148] can be applied directly.

$V_{OO} = \pm 2(1 + 10)$ mV $+ 0.1(1 - 0.01 \times 11)$ V $\pm 0.02 \times 1$ V.

The off-set effects may act in the same direction as those due to the bias currents or they can act in the opposite way. Taking the worst case,

$V_{OO} = 0.131$ V.

For minimum bias current effects, apply equation [5.147],

$R_2 = \dfrac{1 \times 0.1}{1.1}$ MΩ $= 91$ kΩ.

The output voltage offset then becomes

$V_{OO} = \pm 2(1 + 10)$ mV $\pm 0.02 \times 1$ V $= 42$ mV or 2 mV.

Fig. 5.166 Circuit for example 5.68.

Bandwidth and slew rate with feedback

Op-amps are multistage amplifiers and for each stage, there will be a high frequency break resulting in the gain falling and a limiting phase change of 90° per stage. As discussed in Chapter 5 (p.458), if such an amplifier is used in a feedback arrangement, the feedback, which is negative at low frequencies, becomes positive and at the π frequency the system becomes unstable if the loop gain is greater than one. Op-amps are usually used with very large loop gain and compensation must be used to avoid instability. For general purpose op-amps, internal compensation is provided to give an *additional* break at a few Hz. Then as frequency increases the gain falls from a very high value to a value less than one before the π frequency is reached. Thus, for resistive feedback circuits, the system cannot become unstable. The effect upon the bandwidth can easily be analysed as follows.

The forward gain is given by,

$$A = \frac{A_o}{1 + j\omega/\omega_1}$$

where A_o is the differential gain at d.c. when $\omega = 0$ and ω_1 is the frequency of the additional break (the dominant break frequency).

If feedback, negative at low frequencies is applied, the standard feedback formula can be used.

$$A_f = \frac{\dfrac{A_o}{1 + j\omega/\omega_1}}{1 + \dfrac{\beta A_o}{1 + j\omega/\omega_1}}$$

$$= \frac{A_o}{1 + j\omega/\omega_1 + \beta A_o}$$

$$\text{rearranging} = \frac{A_o}{1 + \beta A_o + j\omega/\omega_1}$$

and dividing by $1 + \beta A_o$,

$$A_f = \frac{\dfrac{A_o}{1 + \beta A_o}}{1 + \dfrac{j\omega}{\omega_1(1 + \beta A_o)}}$$

This result shows a low-frequency gain of $\dfrac{A_o}{1 + \beta A_o}$ and a break frequency of $\omega_1(1 + \beta A_o)$. Thus as the gain is reduced by the factor $1 + \beta A_o$, the bandwidth is increased by the same factor. This is illustrated for a number of β factors in Fig. 5.167.

From the diagram, the forward gain is seen to fall at 20 dB/ decade from ω_1 until at very high frequencies, it falls more steeply due to the higher internal breaks. Increasing feedback reduces the gain and increases the bandwidth until for typical gain figures the bandwidth is 10^4 or 10^5 times the open loop figure of ω_1.

The slew rate is the limiting rate in volts/second at which the output voltage of an op-amp can change. This is not modified by feedback and for large signals will limit high-frequency performance before bandwidth limitations have any effect. Details of slew rate limitation are given (see p.59) but the combined effect on frequency response can now be illustrated by an example.

Example 5.69   *An op-amp has an open loop gain of 2×10^4 and the dominant pole lies at 3 Hz. The slew rate is 0.5 V/μs. It is used in an inverting amplifier with a low frequency gain of 3. Determine the 3 dB bandwidth of the amplifier and the maximum input signal voltage when operating at this frequency.*

Reference to Fig. 5.167 shows that the ratio of closed loop to open loop bandwidth is the same as that for the gain. In this case,

$$\frac{A_d}{A} = \frac{2 \times 10^4}{3} = 6667 = \frac{\omega_1}{3 \times 2\pi}$$

The 3 dB frequency $\omega_1 = 6\pi \times 6667 = 1.26 \times 10^5$.

Slew rate is the maximum rate of change of output voltage, which, for a sinusoidal signal is given by

$$\frac{dV}{dt} = \omega \hat{V} \cos \omega t \quad \text{which is a maximum when } \cos \omega t = 1$$

$$\therefore \text{ maximum } \frac{dV}{dt} = \omega \hat{V}$$

Thus $0.5 \text{ V}/\mu\text{s} = 1.26 \times 10^5$

or $\hat{V} = 4$ V and $V_{\text{rms}} = 2.8$ V

The gain at this frequency (the 3 dB frequency) is $3/\sqrt{2}$.

The maximum input signal is given by

$$V_{\text{in}} = \frac{2.8}{3/\sqrt{2}} = 1.33 \text{ V}$$

Analysis by the 'op-amp method'

All amplifying arrangements using op-amps include negative feedback through a shunt feedback circuit. Analysis can be by first principles (p.354) or by using feedback theory; the feedback approach is not convenient as the shunt input arrangement is concerned with the current gain of the amplifier. Op-amps are nearly ideal voltage amplifiers (high Z_{in}, low Z_{out}) and another more useful approach is to consider the input impedance, due to this feedback arrangement on a voltage amplifier.

In the circuit in Fig. 5.168, no current flows into the amplifier inverting terminal and thus the currents I and I_F are equal.

$$I_F = \frac{(V_d - V_o)}{Z_F}$$

$$I = I_F = \frac{V_d(1 - A_d)}{Z_F} \quad \text{where } A_d \text{ is the differential gain of the}$$
amplifier. The input impedance

$$\frac{V_d}{I_F} = \frac{Z_F}{1 - A_d}$$

But for signals applied to the inverting input, A_d is negative and very large (10^5 to 10^7).

Fig. 5.167 Op-amp frequency response with feedback.

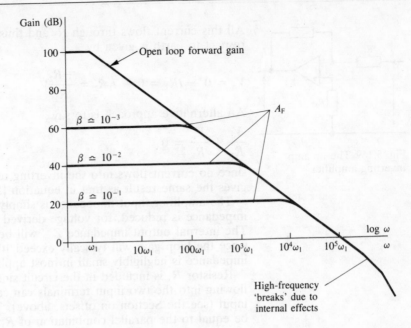

∴ $Z_{in} = \dfrac{Z_F}{1 + |A_d|} \simeq 0$

There is no connection between the inverting terminal and earth but the impedance is approximately zero which holds V_d at earth potential. Thus, in this arrangement, the summing point at the inverting input is said to be a virtual earth.

An alternative argument is to say that if V_o is finite (and not limited by saturation) and A_d tends to infinity, then the differential input must be zero. This suggests a virtual short circuit between the + and − terminals. Any rise in input voltage causes a rise in V_o which, through Z_F forces V_d back to zero.

Whichever way the circuit is considered, there are two rules which may be applied in all cases when A_d is very large and there is a circuit connecting V_o to the inverting terminal. These are

(a) The potentials V_+ and V_- at the non-inverting and inverting terminals respectively will be identical.

(b) No current flows into either the inverting of the non-inverting terminals.

These rules will be applied to the analysis of all the following op-amp circuits. The first group of circuits are ideally non-frequency sensitive as the only external components are resistors. However, the bandwidth and slew rate effects described above apply in all cases.

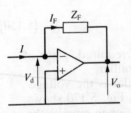

Fig. 5.168 The 'virtual earth' at an op-amp input.

The inverting amplifier

The circuit for the inverting amplifier is shown in Fig. 5.169. Since no current flows in R_3, V_+ and V_- are both at earth potential. The input current is given by

$I = \dfrac{V_i - 0}{R_1}$

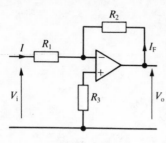

Fig. 5.169 The op-amp inverting amplifier.

All this current flows through R_2 and thus $I_F = \dfrac{-V_i}{R_1}$ and V_o, with respect to earth is given by

$$V_o = 0 - IR_2 = 0 + I_F R_2 = \frac{-R_2}{R_1}V_i \qquad [5.149]$$

An alternative approach is to say

$$\frac{V_i}{R_1} + \frac{V_o}{R_2} = 0$$

since no current flows into the inverting terminal. Rearranging then gives the same result as that in equation [5.149].

The amplifier input impedance is simply R_1 and the output impedance is reduced, for voltage derived feedback to $Z_{out}/(1 + \beta A)$. The internal output impedance Z_{out} will be a few tens of ohms and since the loop gain will typically exceed 10^4, the resulting output impedance is negligibly small in most applications.

Resistor R_3 is included in the circuit since the d.c. bias currents flowing into the two input terminals can cause a voltage offset at the input (see the Section on offsets, above). To avoid this, R_3 should be equal to the parallel combination of R_1 and R_2.

The non-inverting amplifier

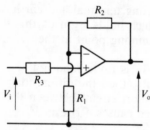

Fig. 5.170 The op-amp non-inverting amplifier.

A non-inverting amplifier has the same feedback as the inverting amplifier described in the last section. The change in input power however, does modify the gain equation. With reference to Fig. 5.170, negative feedback holds the differential input at zero and so,

$$V_i = V_+ = V_-$$

But $V_+ = V_o \times \dfrac{R_1}{R_1 + R_2} = V_i$

Rearranging, $A = \dfrac{V_o}{V_i} = \dfrac{R_1 + R_2}{R_1} = 1 + \dfrac{R_2}{R_1} \qquad [5.150]$

The minimum gain is one in this case while, if required, gains of less than one are available in the inverting amplifier.

A special version of the non-inverting amplifier is the voltage follower shown in Fig. 5.171. Comparison with Fig. 5.170 shows that, in this case, R_2 is a short circuit and R_1 is open circuit; the resulting gain from equation [5.150] is simply +1. The feedback arrangement results in the lowest possible output impedance of Z_{out}/A_d and the highest possible input impedance of $Z_{id}A_d$. Z_{out} and Z_{id} are the op-amp output impedance and differential input impedance respectively and A_d is the open loop differential gain.

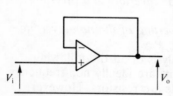

Fig. 5.171 A voltage follower amplifier.

The basic forms of op-amp amplifiers discussed above are sometimes modified in detail. The op-amp method of analysis can still be applied in all cases and this can usefully be illustrated by some examples.

Example 5.70   *Assuming that the op-amps in the circuits shown in Fig. 5.172 are ideal, determine from first principles the output quantities indicated.*

Circuit (a). Since there is a feedback path from output to inverting input, virtual earth conditions apply. The input current I is given by

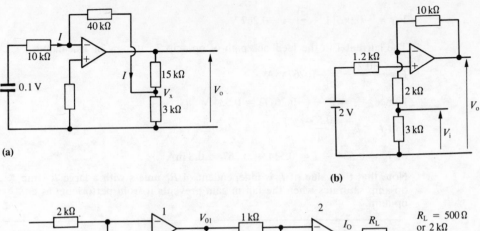

(a)

(b)

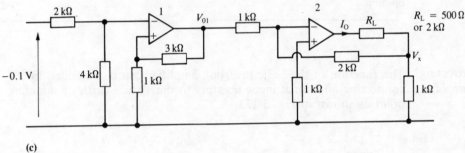

(c)

Fig. 5.172 Op-amp circuits for example 5.70.

$$I = \frac{0.1}{10} = 0.01 \text{ mA}$$

and this current flows in the 40 kΩ resistor. The node voltage $V_x = 0 - 40I = -0.4$ V.
Writing a nodal equation,

$$I + I_o = V_x/3 \text{ or } I_o = V_x/3 - I$$

and $I_o = \dfrac{-0.4}{3} - 0.01 = -0.143$ mA

$$\therefore V_o = V_x + 15I_o = -0.4 - 0.143 \times 15 = -2.55 \text{ V}.$$

Circuit (b). Virtual short circuit conditions apply, so,

$$V_- = V_+ = -2 \text{ V}.$$

(There is no volt drop across the 1.2 kΩ resistor as the input bias current is negligible.)
The indicated voltage V_1 is given by

$$V_1 = \frac{-2 \times 3}{2 + 3} = -1.2 \text{ V}$$

The current in the 3 kΩ resistor also flows in the 10 kΩ feedback resistor.

$$\therefore V_o = -2 + \frac{-2}{5} \times 10 = -6 \text{ V}.$$

Circuit (c). For op-amp 1, $V_+ = -0.1 \times \dfrac{4}{6} = -0.067$ V.

This stage is connected as a non-inverting amplifier and V_{o1} is therefore given by

$$V_{o1} = -0.067\left(1 + \frac{3}{1}\right) = -0.267 \text{ V}.$$

The current I in the feedback path of op-amp 2 can now be calculated;

$$I = \frac{-0.267}{1} = -0.267 \text{ mA}$$

Hence, $V_x = 0 - (-0.267)2 = 0.534$ V

and $I + I_o = \dfrac{0.534}{1}$,

or $I_o = 0.534 - I = 0.534 + 0.267 = 0.8$ mA.

Note that the value of I_o is independent of R_L unless with a large R_L, the op-amp saturates when the fall in gain prevents it from performing as an op-amp.

The summing inverting amplifier

The function of the basic inverting amplifier can be extended by connecting additional input resistors to the virtual earth of summing point as shown in Fig. 5.173.

Fig. 5.173 A summing inverting amplifier.

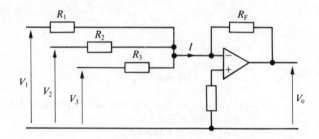

Since V_- is held at zero by the feedback, the current I shown is given by

$$I = \frac{V_1}{R_1} + \frac{V_2}{R_2} + \frac{V_3}{R_3}$$

V_o is, as before given by $-IR_F$,

$$\therefore V_o = \frac{R_F}{R_1}V_1 - \frac{R_F}{R_2}V_2 - \frac{R_F}{R_3}V_3 \qquad [5.151]$$

or with equal resistors throughout,

$$V_o = -(V_1 + V_2 + V_3)$$

This circuit thus provides, not only a summing facility, but also the choice of the scaling factor to be applied to each input. The input resistance to each signal is simply the resistance connected between that signal and the virtual earth.

Example 5.71   *The op-amp in the circuit shown in Fig. 5.174 may be assumed to be ideal. Calculate the indicated output voltage V_o.*

$$V_{o1} = -1 \times \frac{-10}{5} + 2 \times \frac{-10}{20} = 1 \text{ V}$$

Fig. 5.174 Summing
amplifier circuit for
example 5.71.

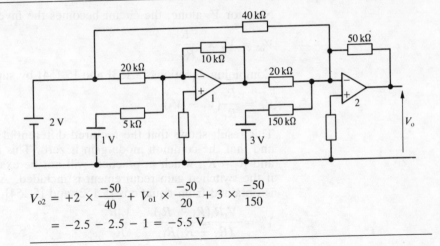

$$V_{o2} = +2 \times \frac{-50}{40} + V_{o1} \times \frac{-50}{20} + 3 \times \frac{-50}{150}$$

$$= -2.5 - 2.5 - 1 = -5.5 \text{ V}$$

The differential amplifier The basic op-amp is a differential amplifer, but the differential gain is much too high for practical applications. A common requirement is for the difference between two signals to be amplified while their average value, the common mode signal should be rejected, that is amplified by a gain of 0. Additional requirements for instrumentation include high input impedances and switched variation of differential gain A_d. A single op-amp arrangement for such an amplifier is shown in Fig. 5.175. This circuit is a composite, combining the inverting and non-inverting amplifiers discussed earlier. Analysis by application of superposition is most convenient.

Fig. 5.175 An op-amp
differential amplifier.

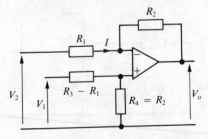

Consider first the output V_{o1} due to V_1 alone with the other input earthed. This circuit is simply the non-inverting amplifier with the input given by

$$V_1 \times \frac{R_4}{R_3 + R_4}.$$

Therefore

$$V_{o1} = \frac{V_1 R_4}{R_3 + R_4} + \frac{R_1 + R_2}{R_1} \qquad [5.152]$$

but $R_1 + R_2 = R_3 + R_4$.

Therefore

$$V_{o1} = V_1 \times \frac{R_4}{R_1} = V_1 \times \frac{R_2}{R_1} \qquad [5.153]$$

Now for V_2 alone, the circuit becomes the inverting amplifier and

$$V_{o2} = V_2 \times \frac{-R_2}{R_1} \qquad [5.154]$$

Combining equations [5.153] and [5.154] by superposition,

$$V_o = \frac{R_2}{R_1}(V_1 - V_2) \qquad [5.155]$$

This result shows that the required differential facility is provided and that the common mode gain is zero. This is only true if $R_1 = R_3$ and $R_2 = R_4$, which in practice, will not be exactly true, particularly if the switched gain requirement is included. A more general result is obtained from equations [5.152] and [5.154]

$$V_o = \frac{V_1 R_4(R_1 + R_2)}{(R_3 + R_4)R_1} - \frac{V_2 R_2}{R_1} \qquad [5.156]$$

For the common mode gain, $V_1 = V_2 = V_c$,

$$\text{and } A_c = \frac{V_o}{V_c} = \frac{R_4(R_1 + R_2) - R_2(R_3 + R_4)}{R_1(R_3 + R_4)}$$

$$= \frac{R_4 R_1 - R_2 R_3}{R_1 R_3 + R_1 R_4} \qquad [5.157]$$

Example 5.72 *The circuit shown in Fig. 5.176 is a differential amplifier with a nominal gain of 10. Resistor tolerances result in the actual values shown. Determine (a) the output voltage V_o, (b) the common mode gain, (c) the CMRR and (d) the differential input impedance.*

Fig. 5.176 Differential amplifier circuit for example 5.73.

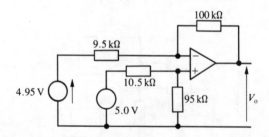

For the ideal case, $V_o = A_d(V_1 - V_2) = 10 \times 0.05 = 0.5$ V. From equation [5.156],

$$V_o = \frac{5(95 \times 109.5)}{105.5 \times 9.5} - 4.95 \times \frac{100}{9.5} = -0.21 \text{ V}.$$

Which shows a considerable error compared with the ideal 0.5 V. Now apply equation [5.15] to find A_c the common mode gain;

$$A_c = \frac{95 \times 9.5 - 100 \times 10.5}{9.5 \times 10.5 + 9.5 \times 95} = -0.147$$

The total output V_o will be given by

$$V_o = A_c \frac{(V_1 + V_2)}{2} + A_d(V_1 - V_2)$$

$$= -0.147 \times \frac{9.95}{2} + 10(5 - 4.95) = -0.23 \text{ V}$$

This not only agrees with the result obtained above but shows that the common mode gain has produced a larger output component than that due to the differential gain. The CMRR = 10/0.147 = 68 or 37 dB, which is obviously not sufficient. This poor result arises from the use of the 5% tolerance resistors. In practice, much tighter tolerances would be used and trimmed to minimise the common mode gain but this solution is not satisfactory when switched gains are required.

The differential input impedance is calculated by noting the virtual short circuit between the inverting and non-inverting terminals. This makes the differential Z_{in} simply 10.5 kΩ + 9.5 kΩ or 20 kΩ which is not very large (the impedances to earth are (10.5 + 95) kΩ and 9.5 kΩ).

An instrumentation amplifier

The limitations of the differential amplifier described in the last section and example can be overcome by the use of three op-amps as shown in Fig. 5.177. Op-amps A_1 and A_2 are modified voltage followers with the corresponding very high input impedance. The combined differential gain of these two stages is controlled by the single resistor R_1 which can thus be adjusted without involving resistor matching problems. The differential outputs are combined in the last stage A_3. This is the standard differential amplifier discussed in the previous section, but in this case the components can be carefully matched as no switching of components is required and the low input impedance is of no consequence as the output impedances of A_1 and A_2 are negligibly small.

Fig. 5.177 An instrumentation amplifier: (a) full circuit; (b) gain controlling network.

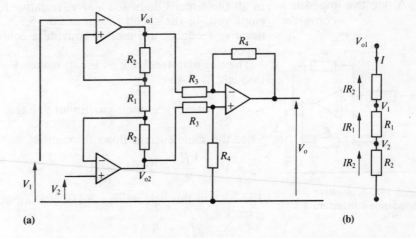

(a)   **(b)**

For the analysis of the circuit, consider the resistor chain R_2, R_1, R_2 between the outputs of A_1 and A_2; this is shown in Fig. 5.177b. The top and bottom of the chain are V_{o1} and V_{o2} respectively (the input signals) since feedback holds the inverting and non-inverting terminals at the same potential. The current I is thus determined by the difference between the input signal voltages.

$$I = \frac{V_1 - V_2}{R_1}$$

The volt drops across the two equal R_2 resistors can now be found in the direction shown in the circuit diagram. In each case,

$$V = \left(\frac{V_1 - V_2}{R_1}\right)R_2$$

The outputs of A_1 and A_2 can now be written in terms of the input signals V_1 and V_2.

$$V_{o1} = V_1 + (V_1 - V_2)\frac{R_2}{R_1}$$

$$V_{o2} = V_2 - (V_1 - V_2)\frac{R_2}{R_1}$$

The final output V_{o3} for the differential stage is given by

$$V_{o3} = \frac{R_4}{R_3}(V_{o2} - V_{o1})$$

$$= \frac{R_4}{R_3}\left[\left\{V_2 - (V_1 - V_2)\frac{R_2}{R_1}\right\} - \left\{V_1 + (V_1 - V_2)\frac{R_2}{R_1}\right\}\right]$$

collecting terms, $V_{o3} = \dfrac{R_4}{R_3}\left[\left(1 + \dfrac{2R_2}{R_1}\right)(V_2 - V_1)\right]$ [5.158]

As stated above, the single resistor R_1 can be used to select the required gain. A further advantage of this circuit is that low resistors can be used for A_3 minimising d.c. offset effects due to bias currents and with a low gain for A_3, the offset voltage effects are also minimised.

A negative impedance converter

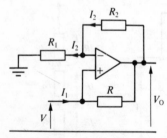

Fig. 5.178 A negative impedance converter.

In all the circuits discussed above, negative feedback only has been employed. In the circuit shown in Fig. 5.178, both positive and negative feedback are used to provide a controlled negative input impedance.

The negative feedback as usual, makes $V_- = V_+$ and the current through R_1 is given by

$$I_2 = \frac{V}{R_1} \text{ where } V \text{ is the signal input voltage.}$$

Since the same current flows through R_2,

$$V_o = V + \frac{VR_2}{R_1} = V\left(1 + \frac{R_2}{R_1}\right)$$

The current through resistor R is I_1 the input current and this is given by

$$I_1 = \frac{V - V\left(1 + \dfrac{R_2}{R_1}\right)}{R} = \frac{V}{R}\left(\frac{-R_2}{R_1}\right) \qquad [5.159]$$

Finally, the input impedance is $\dfrac{V}{I_1}$ and rearranging equation [5.159]

$$Z_{in} = \frac{-R_1}{R_2}R. \qquad [5.160]$$

Thus the negative impedance converter (NIC) circuit produces negative impedance combined with impedance scaling. This can have applications in oscillator circuits and in the constant current generator shown in the next section.

A voltage to current converter

In a voltage to current converter a current is caused to flow in a load in such a way that the value of the current is independent of the load resistance. The circuit shown in Fig. 5.179a is the same as the NIC circuit with the addition of an external resistance R and the load Z_L.

The ratio $\dfrac{R_1}{R_2}$ is also set to unity.

Since the input impedance to the NIC is simply $-R$, the equivalent circuit in Fig. 5.179b can be drawn. The voltage generator V_s and the

external R can be combined as a current generator $\dfrac{V_s}{R}$ in parallel with

R by use of Norton's theorem. The rearranged equivalent in Fig. 5.179c shows R and $-R$ in parallel and the

combined $R_p = \dfrac{-R \times R}{R - R} = \infty$.

Fig. 5.179 A voltage to current amplifier: (a) full circuit; (b) equivalent circuit; (c) Norton equivalent circuit.

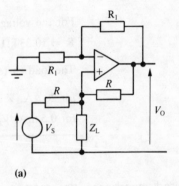

(a)

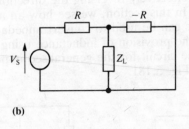

(b)

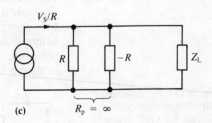

(c)

All that remains is the constant current $\dfrac{V_s}{R}$ that must flow in Z_L.

The resulting voltage across Z_L cannot usually be used directly as this is a high impedance point in the circuit. However, this voltage is effectively applied to the input of the non-inverting amplifier arrangement and the final V_o will simply be twice this voltage.

Example 5.73   *The circuit shown in Fig. 5.180 is to be used as a voltage to current converter between V_s and the current I in the 5 kΩ load. Determine the required value of R and the range of V_s if I is to be controlled between 20 µA and 0.5 mA.*

Fig. 5.180 The voltage to current converter circuit for example 5.73.

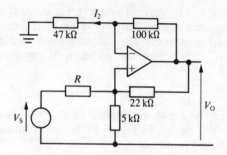

The first step is to analyse the negative impedance section of the circuit taken at V with the resistor missing. Following the steps in the previous section

$$I_2 = V/47 \text{ and } V_o = V + \frac{V}{47} \times 100 = 3.13 \text{ V}.$$

The current into the 22 kΩ resistor $I_{in} = \dfrac{V - 3.13 \text{ V}}{22}$

$$\therefore Z_{in} = \frac{V}{I_{in}} = \frac{22}{1 - 3.13} = -10.33 \text{ k}\Omega$$

For the voltage to current conversion, (see Fig. 5.179)

R = 10.33 kΩ.

The load current is given by $\dfrac{V_s}{R}$, so the required range of V_s is

for 20 μA, $V_s = 0.02 \times 10.33 = 0.207$ V,

for 0.5 mA, $V_s = 0.5 \times 10.33 = 5.17$ V.

A generalised impedance converter

The NIC circuit described above can only change the sign of an impedance (effectively reversing the direction of current flow through it). In this section, we see how an arrangement involving two op-amps can be used to convert impedance forms. A particular example is the provision of inductance using only capacitors and resistors. The circuit for the generalised impedance converter (GIC) is shown in Fig. 5.181.

Fig. 5.181 A generalised impedance converter. Circuit for GIC example 5.74.

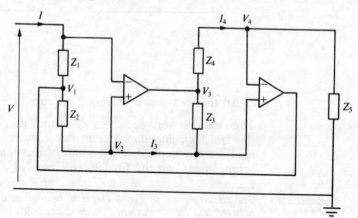

Analysis depends only on the two op-amp principles, the virtual short circuit due to negative feedback and zero current into either input terminal. From the first principle,

$$V_2 = V_4 = V$$

Now, by Ohm's law,

$$I_4 = \frac{V_4}{Z_5}$$

and $V_3 = V_4 + I_4 Z_4 = V + \dfrac{V Z_4}{Z_5}$

$$I_3 = \frac{V_2 - V_3}{Z_3} = \frac{V - V\left(1 + \dfrac{Z_4}{Z_5}\right)}{Z_3}$$

$$= \frac{-V Z_4}{Z_3 Z_5}$$

but $V_1 = V_2 + I_3 Z_2 = V - \dfrac{V Z_4 Z_2}{Z_3 Z_5}$

Finally, the input current I is given by

$$I = \frac{V - V_1}{Z_1} = \frac{V - V\left(1 - \dfrac{Z_4 Z_2}{Z_3 Z_5}\right)}{Z_1} = V \frac{Z_4 Z_2}{Z_1 Z_3 Z_5}$$

The circuit input impedance is thus given by

$$Z_{in} = \frac{V}{I} = \frac{Z_1 Z_3 Z_5}{Z_2 Z_4} \qquad [5.161]$$

Now, suppose an inductance is required and Z_2 (or Z_4) is a capacitor C and the remaining components are resistors:

$$Z_{in} = \frac{R_1 R_3 R_5}{\dfrac{1}{j\omega C} R_4} = j\omega C \frac{R_1 R_3 R_5}{R_4}$$

which is the impedance of an inductance of CR^2 henry if equal resistors are used. For example, four 10 kΩ resistors and a 0.001 μF capacitor provide an equivalent inductance of $10^{-9} \times 10^8$ H or 0.1 H.

Example 5.74   *The circuit in Fig. 5.181 is used with Z_1, Z_3 and Z_5 all 2 kΩ resistors and Z_4 is a 0.02 μF capacitor in parallel with a 69 kΩ resistor. (a) If Z_2 is a 1 kΩ resistor and a 0.1 μF capacitor is connected in series with the input, determine the resonant frequency and the bandwidth of the equivalent tuned circuit. (b) If Z_2 is a 0.02 μF capacitor, determine the input impedance at an angular frequency of 10^4 r/s.*

(a) Taking Z_{in} from equation [5.161]

$$Z_{in} = \frac{8 \times 10^9}{10^3 \times \dfrac{1}{1/69 \times 10^3 + j\omega 2 \times 10^{-8}}}$$

$$= 8 \times 10^6 (1.45 \times 10^{-5} + j\omega 2 \times 10^{-8})$$

$$= (116 + j\omega 0.16)\,\Omega$$

which is an effective series R and L circuit. When the 0.1 μF capacitor is added in series, resonance occurs at $\omega = \dfrac{1}{\sqrt{LC}}$, thus

$$\omega = \frac{1}{\sqrt{0.16 \times 10^{-7}}} = 7906 \text{ r/s}$$

The Q factor is given by $Q = \dfrac{\omega L}{R} = \dfrac{7906 \times 0.16}{116} = 10.9$. and the bandwidth is $\dfrac{\omega_o}{Q} = \dfrac{7906}{10.9} = 725$ r/s.

(b) When Z_2 is a 0.02 μF capacitor, the result for Z_{in} above is modified to,

$$Z_{in} = 8 \times 10^9 \times j\omega 2 \times 10^{-8}(1.45 \times 10^{-5} + j\omega 2 \times 10^{-8})$$

$$= j\omega 2.3 \times 10^{-3} - \omega^2\, 3.2 \times 10^{-6}$$

which at $\omega = 10^4$, becomes

$$Z_{in} = j23 - 320 \ \Omega$$

i.e. an inductance in series with a negative resistance.

You should now be able to attempt exercises 5.63 to 5.70.

Frequency sensitive op-amp circuits

All the circuits discussed in Section 5.8 have gains which are apparently independent of signal frequency. In practice, this will not be true and at higher frequencies the gain will, in all cases, fall as signal frequency increases. The op-amp approach to analysis assumes infinite differential gain A_d. In practice, with general purpose op-amps, the gain falls off with frequency as shown in Fig. 5.167 and for all circuits the bandwidth with feedback follows a similar pattern. If a wider bandwidth is required, op-amps without internal compensation can be used and external compensation circuits should be connected according to the manufacturer's instructions.

The various amplifiers arrangements can, however be made frequency sensitive if some or all of the resistors are replaced by complex impedances. In practice, this is usually limited to the basic inverting amplifier circuit when R_1 and R_2 can be replaced by impedances, as discussed earlier in this chapter. Two special examples of this are the operational integrator and differentiator which are discussed in the following sections. When signal filtering is required, alternative arrangements are necessary and these too are discussed later in this section. In all cases, the op-amp bandwidth limitation will override any effects produced by the external circuitry.

The operational integrator

The circuit for the basic inverting integrator is shown in Fig. 5.182 and the arrangement is obviously that of an inverting amplifier with the feedback resistor replaced by a capacitor.

If V_i is a constant or d.c. signal, the current i will be a constant

Fig. 5.182 An op-amp inverting integrator.

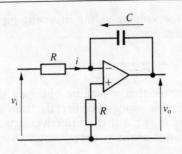

V_i/R into the virtual earth. This constant current charges the capacitor to give a ramp output voltage. The level of output voltage at any time is proportional to the current and to the time for which it has been flowing. If the current stops as V_i is reduced to zero, the capacitor charge stays constant and with it v_o. This is precisely the behaviour that should be expected with integration since,

if $v_o = \int v_i dt$ and v_i is a constant V_i,

then $v_o = \int V_i dt = V_i \times t$ which is a ramp voltage of slope V_i V/s.

For a.c. signals, the standard result can be used to give

$$\frac{V_o}{V_i} = -\frac{Z_F}{Z_1} = \frac{\dfrac{-1}{j\omega C}}{R} = \frac{-1}{j\omega CR} = \frac{+j}{\omega CR} \qquad [5.162]$$

This indicates a 90° phase shift at all frequencies (i.e. a sine wave input produces a cosine output) and a gain that falls from infinity at $\omega = 0$ (d.c.) at 20 dB/decade.

Now,

$$\int \sin \omega t dt = -\frac{1}{\omega} \cos \omega t$$

showing that equation [5.162] represents integration for a sine wave input with a sign change from the inverting amplifier and a 'gain' of $1/CR$.

For a more general analysis, we must consider instantaneous values of v and i as follows. Since the feedback provides a virtual earth,

$$i = \frac{v_i}{R}$$

This current flows in C producing a volt drop $v = \dfrac{1}{C}\int i dt$. Therefore

$$v_o = -\frac{1}{CR}\int v_i dt \qquad [5.163]$$

Thus, whatever the waveform of v_i, the output is the inverted integral of the input with a gain of $1/CR$.

The virtual earth is of particular interest in this case and can be analysed as follows: the voltage across the capacitor, v is given by

$$v = \frac{1}{C}\int i dt = v'(1 - A_d)$$

where v' is the voltage at the inverting input to the op-amp and thus A_d is negative.

Therefore $V' = \dfrac{1}{C(1 + |A_d|)} \displaystyle\int i\,dt$ [5.164]

This is the result that would be obtained if a capacitor of $C(1 + |A_d|)$F was connected directly from this point to earth. As A_d tends to infinity, so will the effective capacitance. For any frequency a.c., this is once again, a short circuit with no connection i.e. the virtual earth.

One problem that can occur results from the lack of any d.c. feedback which allows amplifier offset effects to cause the output to drift. This may be overcome by connecting a large resistance in parallel with the feedback capacitance. The resulting a.c. gain is given by

$$\frac{V_o}{V_i} = \frac{-1}{R(j\omega C + 1/R_F)}$$
$$= \frac{-R_F}{R(1 + j\omega CR)}$$

This result appears to vary somewhat from the required integration, but if the frequency response is compared with that for equation [5.162] as shown in Fig. 5.183 we can see that the only difference is at very low frequency signals. An a.c. gain frequency response falling at 20 dB per decade is equivalent to integration.

Fig. 5.183 The frequency response of an integrator.

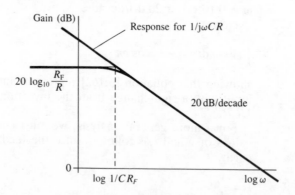

The differentiator
The circuit for the basic differentiator is similar to that for the integrator with the R and C interchanged as shown in Fig. 5.184. With a d.c. input, the current will be zero and thus v_o will be the same as the virtual earth i.e. zero. The capacitance C only passes a current when v_i is changing; this current then produces an output voltage proportional to the rate of change of input voltage v_i. An output equivalent to the rate of change is equivalent to the differential of the input.

As with the integrator, analysis can be for either a.c. signals or for signals expressed as time varying functions of the instantaneous variables v and i. For a.c.,

Fig. 5.184 An op-amp differentiator circuit.

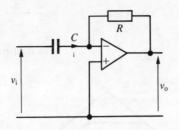

$$\frac{V_o}{V_i} = \frac{-R}{1/j\omega C} = -j\omega CR$$

which is the inverse of the integrator gain. There is a 90° phase shift once again, but the gain now increases with frequency. For instantaneous values,

$$i = C\frac{dv}{dt} \tag{5.165}$$

$$v_o = -iR = -CR\frac{dv}{dt} \tag{5.166}$$

Thus, whatever the input waveform (within op-amp limitations) the output v_o is given by the differential of the input with a sign change and a gain of CR.

There is however a problem associated with differentiators. The feedback through R and C adds a phase shift to the loop gain. As a result, the π frequency can occur where the gain is still greater than one producing an unstable system. This is avoided by the addition of a capacitor C_f in the feedback path and a resistor R_s in series with the input capacitor. A resistor R_p is also included to balance any offset due to bias current flowing in R. The complete circuit is shown in Fig. 5.185 and the associated frequency response plots are given in Fig. 5.186.

Fig. 5.185 An 'improved' differentiator circuit.

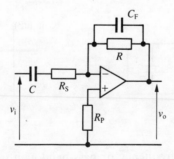

For a.c. signals, the gain is now given by

$$\frac{V_o}{V_i} = \frac{-1}{Z_1 Y_F} = \frac{-1}{(R_s + 1/j\omega C)(1/R + j\omega C_f)}$$

$$= \frac{-j\omega CR}{(1 + j\omega C_f R)(1 + j\omega CR_s)}$$

The basic differentiation function is still provided by the numerator at low frequencies (below $\omega = 1/CR_s$), but at high frequencies, the 20 dB roll-off indicates integration. Another

Fig. 5.186 Frequency
responses for
differentiators.

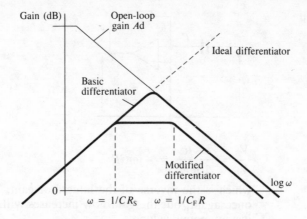

advantage of the modified circuit is that high frequency noise does
not receive excessive amplification.

Low-pass filters

In many electronic and communication applications, it is necessary
to select or reject signals according to their frequency. Circuits
performing this selection are known as filters and they are classified
according to the band of frequencies which is 'passed' or 'stopped'.
An ideal low-pass filter has a gain of one for all signals at
frequencies less than f_c, the cut-off frequency and a gain of zero for
signals at frequencies greater than f_c. The quality of a practical filter
is assessed by two factors; the roll-off and the gain at f_c. Consider
the simple first order low-pass filter and its frequency response
shown in Fig. 5.187

$$\frac{V_o}{V_i} = \frac{1/j\omega C}{R + 1/j\omega C} = \frac{1}{1 + j\omega CR} = \frac{1}{1 + j\omega/\omega_c}$$

Fig. 5.187 A low-pass filter
network and its frequency
response.

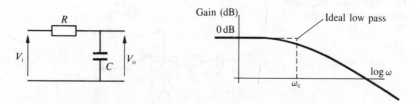

The low-frequency or pass-band gain is unity as required, the
roll-off is 20 dB/decade and the gain at ω_c is -3 dB or 0.707. For
most applications, this cut-off is not sharp enough. A number of
similar RC networks could be used with isolating voltage followers
between them. The resulting roll-off would be 20 dB/decade for
each section with a practical requirement of 60 or 80 dB/decade
being provided by three or four sections respectively. Unfortunately,
the gain at ω_c will also fall with more sections, by a factor of $\sqrt{2}$ for
each section. This problem can be overcome by the use of feedback
to produce a resonant rise which will cancel the excess fall in gain at
ω_c due to other sections.

In the circuit diagram in Fig. 5.188. the op-amp is connected as a voltage follower with a gain of one and no phase shift. The $R_2 C_2$ network provides first order filtering as shown in Fig. 5.187. R_1 and C_1 ensure that the combined roll-off will be 40 dB/decade, but C_2 also leads to positive feedback and a resonant rise near to f_c, the cut-off frequency.

Fig. 5.188 An op-amp low-pass second order filter.

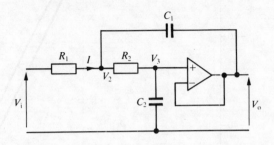

Since $V_o = V_3$, $V_o = V_2 \times \dfrac{1/j\omega C_2}{R_2 + 1/j\omega C_2} = \dfrac{1}{1 + j\omega C_2 R_2}$

The current I is now given by

$$I = \frac{V_2}{R_2 + 1/j\omega C_2} + (V_2 - V_o)j\omega C_1$$

$$= \frac{j\omega C_2 V_2}{1 + j\omega C_2 R_2} + V_2\left(1 - \frac{1}{1 + j\omega C_2 R_2}\right)j\omega C_1$$

$$= V_2\left[\frac{j\omega C_2 + (1 + j\omega C_2 R_2 - 1)j\omega C_1}{1 + j\omega C_2 R_2}\right]$$

$$= V_2\left[\frac{j\omega C_2 - \omega^2 C_1 C_2 R_2}{1 + j\omega C_2 R_2}\right]$$

but $V_1 = V_2 + IR_1$, therefore

$$V_1 = V_2\left[1 + \frac{(j\omega C_2 R_1 - \omega^2 C_1 C_2 R_1 R_2)}{1 + j\omega C_2 R_2}\right]$$

$$= V_2 \frac{1 + j\omega(C_2 R_2 + C_2 R_1)\ \omega^2 C_2 C_1 R_2 R_1}{1 + j\omega C_2 R_2}$$

and $V_o = \dfrac{1}{(1 + j\omega C_2 R_2)} \times V_2$

$$= \frac{1}{(1 + j\omega C_2 R_2)} \times \frac{(1 + j\omega C_2 R_2)V_1}{(1 + j\omega(C_2 R_2 + C_2 R_1) - \omega^2 C_1 C_2 R_1 R_2)}$$

thus $\dfrac{V_o}{V_1} = \dfrac{1}{1 - \omega^2 C_1 C_2 R_1 R_2 + j\omega(C_2 R_2 + C_2 R_1)}$ [5.167]

In this expression, the term $\omega^2 C_1 C_2 R_1 R_2$ results from the positive feedback and can cause a resonant rise. Figure 5.189 shows the response for two second order filters with a design cut-off frequency of 1 kHz. The first, known as a Butterworth response, passes

Fig. 5.189 Frequency
response for second order
low-pass filters.

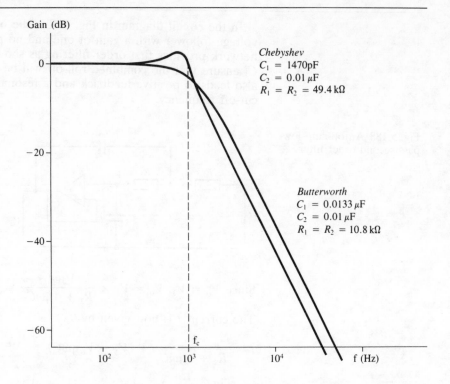

through −3 dB at f_c (regardless of the order). The second, a
Chebyshev response, passes through 0 dB at f_c but allows the pass-
band gain to be greater than unity. In each case, the roll-off is
40 dB/decade. Higher order filters are constructed with additional
simple *CR* sections or additional second order op-amp arrangements.
In each case, the design of all components must be precise if the
desired response is to be obtained.

High-pass filters

A high-pass filter is obtained in the same way as a low-pass filter
with the resistors and capacitors interchanged. Figure 5.190 shows
the circuit for a third order high-pass filter and the corresponding
frequency response. The values of the components must be precisely
designed to obtain the desired form of response and f_c. The values
shown provide a Butterworth response and an f_c of 500 Hz.

The additional third section is added to the front of the filter.
This allows the op-amp low-output impedance to be retained,
avoiding external loading effects. The additional stage will interact
with the second order circuit but this is allowed for in the
component design. Also, the op-amp bandwidth will limit the high-
frequency end of the pass band.

Band-pass and band-stop filters

Simple forms of band-pass and band-stop filters are similar to active
resonant circuits. An example of a band-pass filter in this class is
shown in Fig. 5.191. Negative feedback through R_3 provides a
virtual earth and all the current I which is given by $j\omega C_2 V_2$ flows in
R_3. Therefore

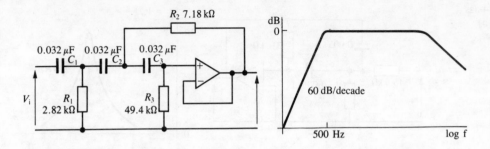

Fig. 5.190 A third order high-pass filter and its frequency response.

Fig. 5.191 An active band-pass filter.

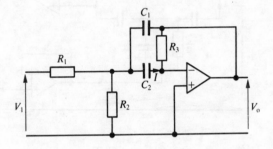

$$V_o = -IR_3 = -j\omega C_2 R_3 V_2$$

Writing a nodal equation,

$$\frac{V_1 - V_2}{R_1} = V_2(1/R_2 + j\omega C_2 + j\omega C_1) - j\omega C_1 V_o$$

Rearranging and substituting for V_2,

$$\frac{V_o}{V_1} = \frac{-j\omega C_3 R_3}{R_1(1/R_1 + 1/R_2 + j\omega(C_1 + C_2) - \omega^2 C_1 C_2 R_3)}$$

$$= \frac{-j\omega C_3 R_3}{\left(1 + \dfrac{R_1}{R_2}\right) - \omega^2 C_1 C_2 R_1 R_3 + j\omega(C_1 + C_2)R_1} \qquad [5.168]$$

This result can be compared with that for the low-pass filter in equation [5.167]. The denominator has a familiar form denoting a possible resonant rise. There is also a $j\omega$ term in the numerator which means that the low-frequency gain falls to zero as will the high-frequency gain with the ω^2 term in the denominator.

Figure 5.192a shows a practical filter circuit designed to provide a resonant frequency of 2 kHz and a Q factor of 10. Figure 5.192b shows the resulting frequency response.

The band stop filter circuit shown in Fig. 5.193 employs a twin T circuit which has a gain of 0 at $\omega = 1/CR$. The feedback through the two voltage followers allows the Q factor of the response to be controlled by the gain of the second op-amp and input potentiometer.

Nodal analysis results in an expression for the overall gain given by

Fig. 5.192 A band-pass
filter: (a) circuit; (b)
frequency response.

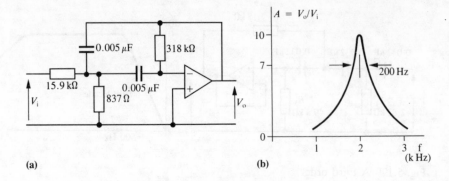

(a)

(b)

Fig. 5.193 An active band-
stop or notch filter.

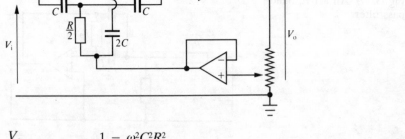

$$\frac{V_o}{V_i} = \frac{1 - \omega^2 C^2 R^2}{1 - \omega^2 C^2 R^2 + j4\omega(1 - A)CR}$$

[5.169]

In this result, at $\omega = 1/CR$, the numerator is zero. The
denominator is once again of a resonant form with the $j\omega$ term
controlled by the potentiometer setting A. Figure 5.194 shows some
typical response curves taking f_o as 1 kHz and various values of A.

You should now be able to attempt exercise 5.71.

Fig. 5.194 The frequency
response of an active band-
stop filter.

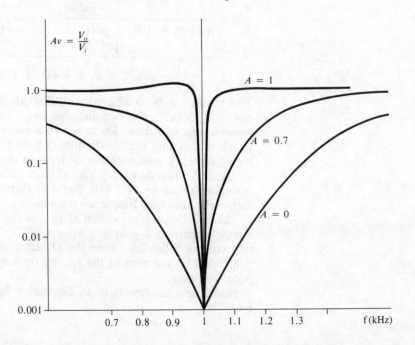

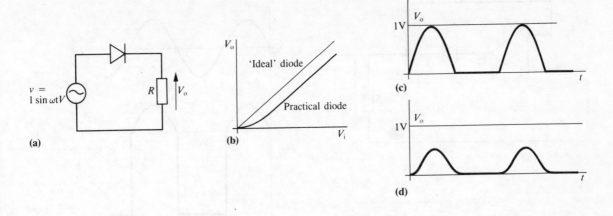

Fig. 5.195 Diode
rectification of a sine wave:
(a) circuit; (b) circuit
transfer characteristic; (c)
output waveform with
'ideal' diode; (d) output
waveform with practical
diode.

Signal-processing circuits

In many applications, it is necessary to process the waveform of a.c.
signals in various ways. Half-cycles or smaller parts may be 'clipped'
off, alternate half-cycles may be inverted to produce full wave
rectification or the peak of a sine wave may be 'clamped' to a
particular d.c. level. All these processes can be provided by diode
circuits but the resulting process is not always as 'precise' as may be
required as the diode characteristics are 'non-ideal'. This is a
particular problem in the forward biased region where for low
voltages, the resistance is very high. The effect of this on simple
rectification is shown in Fig. 5.195.

With the ideal diode, on positive half-cycles, the diode is a short
circuit for all positive voltages. On negative half-cycles, it is open
circuit and no voltage appears across the load. With the practical
diode, although current is passed during the positive half-cycle, part
of the voltage is dropped across the diode resistance. For higher
current values, the resistance is lower and a greater proportion of
the applied voltage appears across the load. This effect will be more
noticable with silicon diodes than germanium, but on the other
hand, the reverse leakage current is greater with germanium than
silicon. These diode properties can affect all of the processes
mentioned above, but circuits combining diodes with op-amps can
eliminate the problem. Analysis of such circuits involves only the
principle of op-amp operation described in the previous sections.

The precision diode The combination of a diode and an op-amp shown in Fig. 5.196
behaves very like the 'ideal' diode. The op-amp is connected as a
voltage follower with the diode in the feedback path. When the
input is positive, the diode is forward biased and feedback results in
the familiar virtual short circuit between inverting and non-inverting
terminals. V_o thus follows V_i exactly. This is true for all positive
inputs as even if the diode has high resistance, feedback still holds
the input conditions as described. With negative inputs, the diode is
reverse biased and no current can flow in the load R. This results in
zero V_o during negative half cycles.

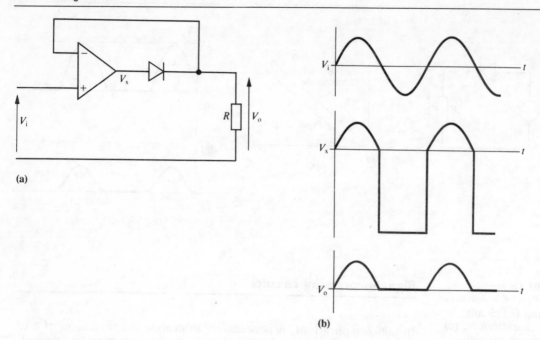

(a)

(b)

Fig. 5.196 The precision
diode: (a) circuit; (b)
waveforms.

Problems can occur with this circuit as with negative input
voltages, the whole of the input appears across the differential input
of the op-amp itself as shown in Fig. 5.196b. If it is not protected,
the op-amp can be damaged and in any case, the output will be
driven into saturation. This can then cause some distortion with
higher-frequency signals as the output change will be limited by slew
rate.

The precision half-wave rectifier

This circuit overcomes the limitations of the precision diode and, if
required allows for a gain on the conducting half-cycles. The circuit
arrangement is shown in Fig. 5.197.

During positive half-cycles of input, the amplifier output goes
negative, D_1 is turned off and D_2 is on providing negative feedback
and a path for the input current. Both input and output of the
amplifier are held at earth potential and V_o is thus zero.

On negative half-cycles, D_1 is on and D_2 is off. The feedback path
is now through R_2 and the usual inverting operation results in a V_o
of $-R_2/R_1 \times V_i$. Perfect rectification will occur if R_1 and R_2 are well
matched. The op-amp is not saturated in either half-cycle and this
improves the performance at higher-frequency operation.

The precision full-wave rectifier

The half-wave circuit can be adapted in various ways to provide
precision full-wave rectification. A common form is shown in Fig.
5.198. Amplifier A_1 provides half-wave rectification blocking the
negative half-cycles of the input and inverting the positive half-cycles
with a gain of unity. Amplifier A_2 is a summing inverting amplifier
with a gain of -1 for the input signal V_i and a gain of -2 for the
rectified output V_{o1}. The waveforms are shown in Fig. 5.199.

Waveforms (c) and (d) show the output of A_2 resulting from the
two inputs taken separately. Output (c) would be that if R_4 was
opened while output (d) would result if R_5 was opened. The total

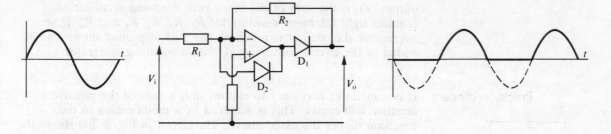

Fig. 5.197 The precision
half-wave rectifier.

Fig. 5.198 The precision
full-wave rectifier.

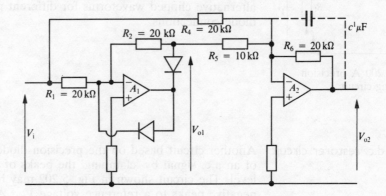

Fig. 5.199 Waveforms for
the precision full-wave
rectifier.

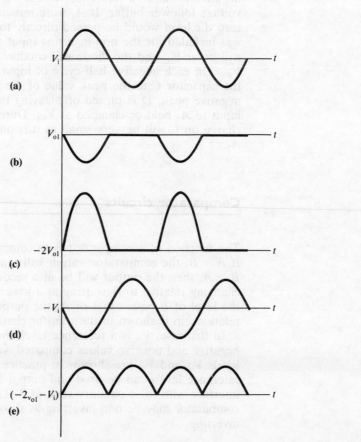

output (e), is the sum of the other two. Precision rectification requires tight tolerance resistors for R_1, R_2, R_4, R_5 and R_6. If an average or d.c. output is required, the $1\ \mu$F capacitor shown can be added to the circuit converting A_2 into a summing integrator.

Precision clippers

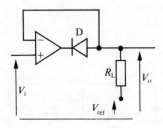

Fig. 5.200 A precision clipping circuit.

It is sometimes necessary to clip off only a part of the positive or negative half-cycles. This is achieved by a modification to the precision diodes discussed above. The circuit in Fig. 5.200 shows the load R_L of the precision diode returned to a reference level V_{ref}. Diode D will be conducting whenever the amplifier output and therefore V_i is less than V_{ref}. The wave forms in Fig. 5.201 show alternative clipped waveforms for different polarities of V_{ref} and diode connections.

A d.c. restorer circuit

Another circuit based on the precision diode restores the d.c. level of an a.c. signal by 'clamping' the peaks of the a.c. output to a d.c. level. The circuit shown in Fig. 5.202 may be used to clamp the negative peaks to a reference voltage V_{ref}. Amplifier A_2 is simply a voltage follower buffer. If A_1 were missing, the a.c. input with a zero d.c.level would be passed directly to V_o (provided a bias path was included for the non-inverting input to A_2). The output of A_1 will be at V_{ref} and the diode will conduct whenever V_{o1} is less than V_{ref}. On each negative half-cycle of input, D will conduct, charging the capacitor C to the peak value of the input. As V_i rises from the negative peak, D is turned off, leaving the d.c. level of the a.c. input to A_2 held or clamped at V_{ref}. During each cycle, the loss of charge on C will be very small as it is only into the inputs of A_1 and A_2.

Comparator circuits

The function of a comparator is to compare two voltages A and B. If $A > B$, the comparator output will remain at one voltage level; if $B > A$, then the output will be at a second level. Thus if B is changing relative to A starting at a lower value and increasing, as the level of B passes that of A, the output level will switch. This relationship is shown in the transfer characteristic in Fig. 5.203.

In this case, V_B is a reference level and V_A varies between negative and positive values compared with V_B. Two arbitrary output levels V_{o1} and V_{o2} are shown. In practice, positive, negative or zero reference levels can be used and output levels can be chosen to interface with logic circuitry such as TTL or CMOS as required. The comparator may be non-inverting as shown in Fig. 5.203 or inverting.

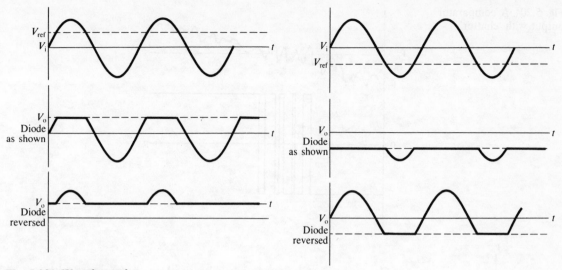

Fig. 5.201 Waveforms for
the precision clipping
circuit.

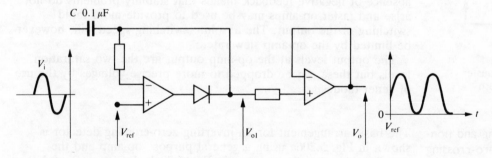

Fig. 5.202 A precision
clamping circuit with input
and output waveforms.

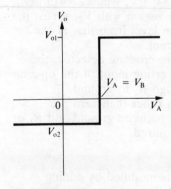

Fig. 5.203 The transfer
characteristic of a
comparator.

One disadvantage of a basic comparator is that noise
superimposed on a slowly changing input signal can cause the total
input to pass backwards and forwards past the reference level. The
resulting output will 'chatter', oscillating between the two levels as
shown in Fig. 5.204. This problem may be eliminated if hysteresis is
incorporated in the characteristic as shown in Fig. 5.205.

There are now effectively two reference levels V_1 and V_2. When
V_i starts from a negative value going positive, the output switches
level as the input passes V_1. When V_i returns from a positive level,
the output will not switch back until the second lower level is
reached. If the peak to peak noise amplitude is less than the width
of the hysteresis loop, there will be no chatter. The width of the
hysteresis loop can also be referred to as the noise immunity of the
system.

Fig. 5.204 A comparator output with 'chatter'.

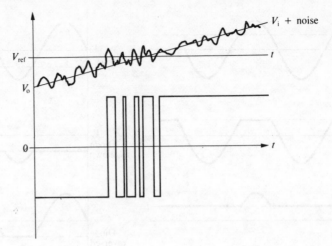

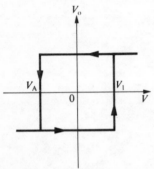

Fig. 5.205 A transfer characteristic with hysteresis.

General purpose op-amps can be used as comparators with considerable flexibility in design. Alternatively, special purpose comparator op-amps may be used. In either case, negative feedback is *not* used and the basic configuration is an open-loop arrangement. When hysteresis is required, positive feedback will be applied. The absence of negative feedback means that stability problems do not arise and faster op-amps may be used to provide more rapid switching of the output. The absolute switching speed will, however be limited by the op-amp slew rate.

The output levels at the op-amp output are the two saturation levels but these can be dropped to more precise voltages by the use of zener diodes.

Inverting and non-inverting zero-crossing detectors

The basic arrangement for an inverting zero-crossing detector is shown in Fig. 5.206a using a general purpose op-amp and the transfer characteristic is shown in Fig. 5.206b. Resistors R_1 and R_2 are equal to avoid off-set effects due to bias currents and are included to balance out temperature variation in the input voltage offset. Since the amplifier is open loop, if V_i is a millivolt above zero, the output will be saturated at the negative saturation level. Zener Z_2 will be forward biased, dropping 0.7 V, while Z_1 will be reverse biased and operating in the Zener region with V_{Z1} across it. Resistor R_o is included to drop the voltage level from that at the op-amp output and to limit the zener current.

For the non-inverting version of the zero-crossing detector, the input is applied through R_2 to the non-inverting input of the op-amp and the inverting input is connected through R_1 to ground.

With either circuit, if comparator levels other than zero are required, the reference level is connected between ground and R_1 or R_2 according to the mode of operation required.

The inverting comparator with hysteresis

If hysteresis is required, the basic circuit is modified by adding positive feedback between the output and the non-inverting input as shown in Fig. 5.207. Consider first the case when V_{ref} is zero and V_i is at first at a very negative value. Since the amplifier is inverting,

Fig. 5.206 An inverting
zero-crossing detector: (a)
circuit; (b) transfer
characteristic.

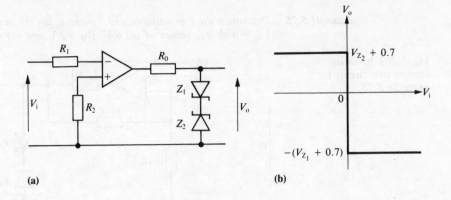

(a)   **(b)**

Fig. 5.207 An inverting
comparator with hysteresis.

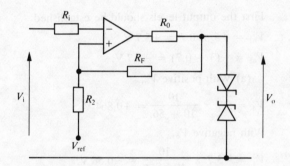

the output will be at the positive level of $V_{o2} = V_{Z2} + 0.7$. No
current flows into the op-amp non-inverting input, so the voltage
here V_+ will be $+V_{o2}R_2/R_F + R_2$. The output can only change if V_i
becomes more positive than this value. V_o then changes to
$V_{o1} = -(V_{Z1} + 0.7)$. The new level of V_+ is $V_{o1}R_2/R_F + R_2$ which is a
negative level. V_i must now be reduced to below this value before
the outut switches back. These levels are shown in Fig. 5.208.

If the zener voltages are equal, the **trip voltages** V_{T1} and V_{T2} will
be equal and of opposite polarity. The resulting hysteresis loop will
then be symmetrical about zero. The effect of V_{ref} can be
demonstrated by an example.

Fig. 5.208 Hysteresis loop
and trip levels for an
inverting comparator.

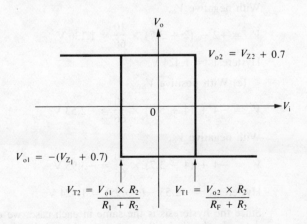

Example 5.75   *Determine the trip voltages and hysteresis for the comparator shown in Fig. 5.209 with V_{ref} values of (a) 0 V, (b) +2 V and (c) −4 V.*

Fig. 5.209 Inverting comparator circuit for example 5.75.

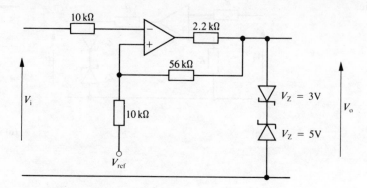

First the output levels should be established.

$V_{o1} = +5 + 0.7 = +5.7$ V.

$V_{o2} = -(3 + 0.7) = -3.7$ V.

(a) With positive V_o,

$$V_+ = 5.7 \times \frac{10}{10 + 56} = +0.864 \text{ V}$$

With negative V_o,

$$V_+ = -3.7 \times \frac{10}{10 + 56} = -0.56 \text{ V}.$$

Hysteresis $= 0.864 - (-0.56) = 1.424$ V.

(b) With positive V_o,

$$V_+ = +2 + (5.7 - 2) \times \frac{10}{66} = +2.56 \text{ V}.$$

or by superposition,

due to V_{ref}, alone $V_+ = \dfrac{2 \times 56}{66} = 1.697$ V,

due to V_o, alone $V_+ = \dfrac{5.7 \times 10}{66} = 0.864$ V,

and the combined value is, $V_+ = 1.697 + 0.864 = 2.56$ V.

With negative V_o,

$$V_+ = +2 - (2 + 3.7) \times \frac{10}{66} = 1.136 \text{ V}.$$

Hysteresis $= 1.424$ V.

(c) With positive V_o,

$$V_+ = -4 + (4 + 5.7) \times \frac{10}{66} = -2.53 \text{ V}.$$

With negative V_o,

$$V_+ = -4 + (4 - 3.7) \times \frac{10}{66} = -3.95 \text{ V}.$$

Hysteresis $= -2.53 - (-3.95) = 1.424$ V.

Since the hysteresis is the same in each case, we can see that this is determined only by the output levels and the resistor ratio.

Hysteresis $= [5.7 - (-3.7)] \times 10/66 = 1/424$ V.

The non-inverting comparator with hysteresis	The circuit for the non-inverting comparator is the same basic arrangement as the inverting circuit but the input and reference points are interchanged. This is shown in Fig. 5.210. The trip points will occur, in each direction, when $V_+ = V_{ref}$. For this condition, the current is determined by V_o and V_{ref}. This current can only flow through R_2 thus enabling V_i to be calculated.

Fig. 5.210 A non-inverting comparator with hysteresis.

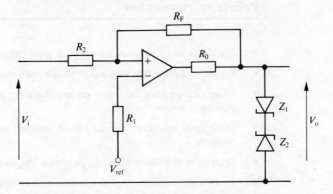

Example 5.76 *In the circuit shown in Fig. 5.211, both zener diodes have a V_Z of 4.3 V. The required lower-trip voltage is 0 V. Calculate the required value of R_F and the resulting upper-trip voltage.*

Fig. 5.211 Circuit and trip levels for example 5.76.

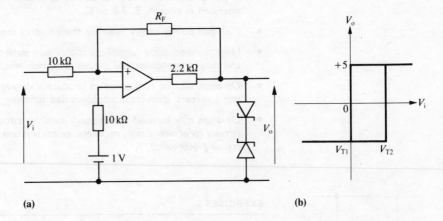

(a)   (b)

The two output levels are $\pm(4.3 + 0.7)$ or ± 5 V. At the lower trip level of 0 V, the current flowing in the 10 kΩ resistor will be $\dfrac{1-0}{10}$ mA (since $V_+ = 1$ V at the trip point). The same current flows from V_o, which is +5 V at this time, to V_+ which is at 1 V. Therefore

$$R_F = \frac{(5-1)}{1/10} = 40 \text{ k}\Omega.$$

For the second condition, V_o is -5 V; the current now flows in the opposite direction towards V_o and is given by

$$I = \frac{+1 - (-5)}{40} \text{ mA} = 6/40 \text{ mA}.$$

This current must flow from V_i to V_+ making,

$$V_i = +1 + \frac{6}{40} \times 10 = 2.5 \text{ V}.$$

You should now be able to attempt exercises 5.72 and 5.73.

Points to remember

- Transistor amplifiers may be analysed graphically or by suitable equivalent circuits.

- Amplifier applications with op-amps use inverting and non-inverting circuits.

- Effect of frequency distortion on amplifiers can be analysed using high-frequency equivalent circuit models.

- Small signal equivalent circuits can be analysed using electrical networks and methods.

- Analysis of amplifiers where large signals are involved requires graphical analysis

- Analysis of amplifiers provides information on the voltage gain, the current gain and the input and output impedance levels

- Variation of gain with frequency is best demonstrated graphically e.g. by Bode plots

- A power amplifier is used to deliver large powers to a load and may be operated in class A, B, AB or C.

- Push-pull power stages minimise the effect of non-linear distortion

- Negative feedback in amplifiers affects gain stability, frequency response, input and output impedance and output distortion levels.

- Op-amps may be configured to produce a variety of circuits including inverters, non inverters, differential amplifiers and summing amplifiers

- Op-amps may be used in frequency sensitive circuits (filters), signal processing circuits (precision diode, rectifiers, restorers) and comparator circuits (zero-crossing detectors)

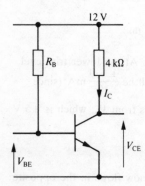

Fig. 5.212 Transistor circuit for exercise 5.1.

EXERCISES 5

5.1 The transistor shown in the circuit in Fig. 5.212 has the d.c. characteristics given in Fig. 5.5. The signal input base current is given by $i_b = 10 \sin \omega t \, \mu A$. (a) If R_B is 758kΩ, determine the current gain, the voltage gain and the input impedance. (b) Find the maximum and minimum values of R_B if the output signal is to be 'undistorted'.
$(97.5, -63, 4.75 \text{ k}\Omega, 1137 \text{ k}\Omega, 406 \text{ k}\Omega)$

5.2 The transistor shown in the circuit in Fig. 5.213 has the d.c. characteristics shown in Fig. 5.5. Determine (a) the output voltage and voltage gain V_o/V_{be} and (b) the value of R_2 giving the maximum undistorted output voltage and the corresponding voltage gain.
$(1.75 \text{ V}, -39, 4.9 \text{ k}\Omega, -59)$

Fig. 5.213 Circuit for exercise 5.2.

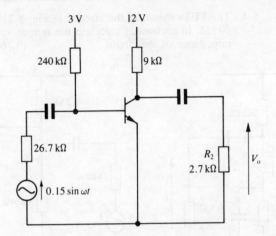

5.3 Determine for each of the transistor circuits shown in Fig. 5.214, the voltage gain, the input impedance, and assuming negligible source impedance, the output impedance.

$(-110, 53 \text{ k}\Omega, 27 \text{ k}\Omega; -151, 131 \Omega, 248 \Omega; 0.97, 19 \text{ k}\Omega, 12 \Omega; 95,$
$59 \Omega, 5.6 \text{ k}\Omega)$

Fig. 5.214 Transistor circuits for exercise 5.3.

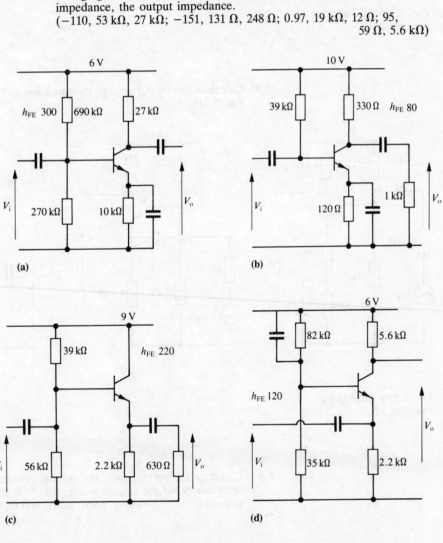

5.4   The FETs shown in the circuits in Fig. 5.215 have g_m 6 mA/V, and r_o 20 kΩ. In each case, calculate the output voltage and output impedance of the circuit.   (0.164 V, 2.8 kΩ; 0.26 V, 155 Ω)

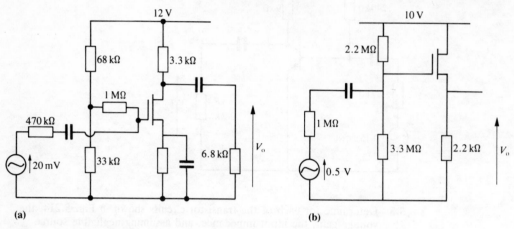

(a)   **(b)**

Fig. 5.215 FET circuits for exercise 5.4.

5.5   Calculate the output voltage indicated for each of the circuits shown in Fig. 5.216.   (1.8, 0.287, 0.6)

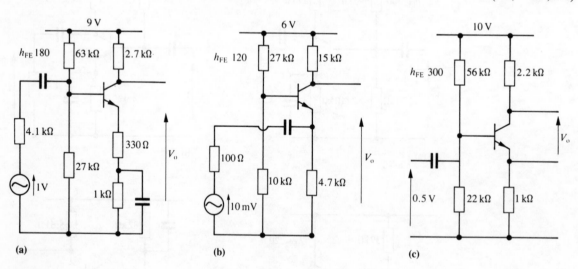

(a)   **(b)**   **(c)**

Fig. 5.216 Circuits for exercise 5.5.

5.6   Calculate the voltage gain, the current gain and the input and output impedances for the circuits in Fig. 5.217. A Z_s of 10 kΩ may be assumed.   (−181.5, 6.95, 383 Ω, 600 Ω; 0.987, 90, 137 kΩ, 85 Ω)

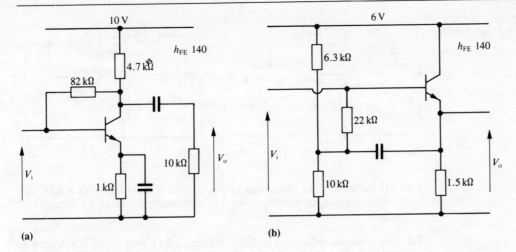

(a)

Fig. 5.217 Circuits for
exercise 5.6.

(b)

5.7 Two BJTs are Darlington connected and the combination is used as (a) common emitter or (b) as common collector. The relevant h parameters are

T_{r1}	h_{ie} 65 kΩ	h_{fe} 85	h_{oe} 2 × 10⁻⁶S
T_{r2}	800 Ω	110	2 × 10⁻⁴S

Determine in each case the input impedance to the circuit and the voltage gain if the external load is 2 kΩ.

$$(-83.9, 117 \text{ k}\Omega; 0.99, 18.8 \text{ M}\Omega)$$

5.8 The transistors in the differential amplifier circuit shown in Fig. 5.218 are 'identical'. Determine V_o, (a) if $e_1 = 0$ and $e_2 = 0.02$ V and (b) if $e_1 = 1.03$ V and $e_2 = 1.01$ V. $(-2.71, +2.71)$

Fig. 5.218 Emitter coupled amplifier circuit for exercise 5.8.

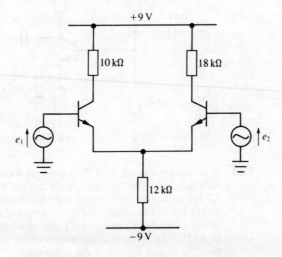

5.9 Assuming 'ideal' amplifiers, calculate the voltage gain for the circuit arrangements shown in Fig. 5.219. $(-4.68, +5.68)$

Fig. 5.219 Op-amp circuits
for exercises 5.9 and 5.10.

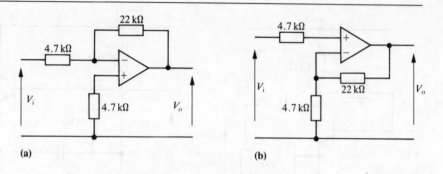

(a)   (b)

5.10 If the op-amp in the circuit in Fig. 5.219a is non-ideal with a gain of
40 and an input impedance of 5 kΩ, calculate the resulting voltage
gain.

(-4.1)

5.11 A common collector amplifier includes a BJT with g_m 40 mA/V and a
d.c. emitter load of 2 kΩ in parallel with a 0.01 μF capacitor.
Calculate (a) the low-frequency voltage gain, (b) the 3 dB bandwidth
and (c) the frequency at which the gain is 0.1. Zero source
impedance may be assumed. (0.988, 636 kHz, 6.26 MHz)

5.12 In a two-stage common emitter amplifier, both transistors have g_m
65 mA/V, h_{ie} 4 kΩ and collector loads of 2.7 kΩ. The effect of bias
components is negligible. The interstage and output shunt capacitance
are respectively 450 pF and 60 pF. Calculate the mid-band dB gain
and the two break frequencies. Hence calculate the gain and phase
shift at 400 kHz and 2 MHz.

(85, 220 kHz, 982 kHz, 78, $-83°$, 58.8, $-147°$)

5.13 Calculate the output voltage for the circuit shown in Fig. 5.220 if the
signal frequency is (a) 5 kHz or (b) 500 Hz. (1.92 V, 1.45 V)

Fig. 5.220 Circuit for
exercise 5.13.

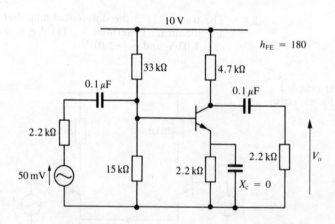

5.14 A transistor having h_{fe} 150 is used as a common emitter amplifier
with a 1.5 kΩ emitter resistor decoupled by a capacitor C_E. The
4.7 kΩ collector load is coupled by a capacitor C_C to an external
10 kΩ load. The signal source of internal resistance 2500 Ω is coupled
to the input by a capacitor C_S and the input impedance can be taken
as 3 kΩ. The minimum signal frequency will be 100 Hz. Calculate
suitable values for the three capacitors. (100 μF, 0.22 μF, 0.47 μF)

5.15 Manufacturer's data for a BJT includes the following: h_{FE} 300, f_T
100 MHz, $C_{b'c}$ 2 pF. It is operated as a common emitter amplifier at a
d.c. collector current of 3 mA with a collector load of 2.2 kΩ. The
signal source has internal impedance of 5 kΩ. Calculate the upper
three dB frequency for the system. (132 kHz)

5.16 An op-amp (which may be assumed to be ideal) is used as an inverting amplifier with the following series and feedback impedances.

	Z_1	Z_F
(a)	a 1 kΩ resistor	a 0.2 μF capacitor in parallel with a 3 kΩ resistor
(b)	a 20 kΩ resistor in parallel with a 0.1 μF capacitor	a 47 kΩ resistor in series with a 25 μF capacitor.

In each case, state without calculation the high- and low-frequency limiting gains and calculate the gain at a frequency of 150 Hz.

(3, 0, 2.7; 0.004, 2.35, 1.99)

5.17 A power amplifier stage has an efficiency of 20% and dissipates a power of 3.2 W at the device output electrode. What is the output power of the stage? Is the stage likely to be operated at class A or class B? (0.8 W, class A)

5.18 Using the circuit of Fig. 5.61 with $R_L = 1.8$ kΩ and $V_{cc} = 20$ V and assuming idealised device characteristics, find the output power and efficiency of the stage. What would be the maximum power dissipation for such a stage? (13.89 mW, 25%, 41.67 mW)

5.19 A class A single-ended power amplifier stage is required to deliver a maximum power of 20 W to a 5 Ω load. The supply voltage is 20 V. Calculate the required transformer turns ratio and collector efficiency. What would be the value of the collector quiescent current for maximum power transfer? (1.414:1. 50%, 2 A)

5.20 A class A single-ended power amplifer has a quiescent collector current of 2 A and can accommodate a peak swing of 2.7 A in the load. If the supply voltage is 25 V and assuming ideal characteristics find the actual load resistance for a circuit efficiency of 18%. What is the value of signal output power under these conditions? (2.5 Ω, 9 W)

5.21 The circuit of Fig. 5.68 has base bias resistances R_1 (base to $+V_{cc}$ line) of 10 kΩ and R_2 (base to ground) of 1 kΩ and has an emitter resistor R_e of 5 Ω, $R_L = 8$ Ω and $V_{cc} = 25$ V. If the transformer turns ratio is 5:1 and the primary d.c. resistance is 5 Ω, find: (a) maximum output power to load; (b) circuit efficiency; (c) power dissipation at collector output. Assume h_{FE} of the device is 50.

(455.6 mW; 27%; 1.665 W)

5.22 What reasons can you give for the improved efficiency of a class B power amplifier compared to a class A stage? What disadvantage, if any, occurs with a class B stage? If a power amplifier is required to deliver 250 W of signal power what would be the power rating for the transistor in: (a) class-A, (b) class-B bias? (250 W; 68.3 W)

5.23 A class B power amplifier uses a transformer with an effective primary resistance of 5 Ω. The resistive load is 15 Ω and the supply voltage is 20 V. Assuming maximum output power what is the collector efficiency and the value of the output power? (78.5%, 6.67 W)

5.24 The circuit used in example 5.23 is to have its output power reduced to allow for the effects of device saturation, with a maximum voltage peak swing at the output of 17 V. Calculate the new value of output power and collector efficiency. What is the value of device power dissipation under these conditions? (4.82 W. 66.75%, 2.4 W)

5.25 Design a class B push-pull amplifier to achieve a maximum power output in a 10 Ω load. Identical transistors are to be used with a maximum collector current of 1 A and a maximum power rating of 5 W. Assuming a supply voltage of 20 V calculate the power output and efficiency. ($R'_L = 20$ Ω, $2n_p/n_s = 2.828$:1; 10 W, 78.5%)

5.26 Using the current-voltage curve of Fig. 5.73 determine values for the coefficients A_0, A_1, A_2, A_3 and A_4 if an input voltage centred at 0.6 V swings 0.05 V peak. What would be the percentage distortion produced in this case?

(7.5 mA, 100 mA, 8.75 mA, 2.5 mA, 1.25 mA, 9.19%)

5.27 A circuit has a load resistance of 1.8 kΩ and gives collector current values under sinusoidal input conditions of: I_{max} = 12 mA, I_{min} = 0.8 mA, I_Q = 6.2 mA. Find the fundamental a.c. power output and second harmonic distortion factor.   (28.2 mW, 1.78%)

5.28 A class A push-pull power amplifier has a load resistance of 20 Ω. If identical transistors with g_m = 80 mS, r_{out} = 500 Ω are used and a transformer of turns ratio of 10:1, find the output voltage and power to the load. Is the power delivered to the load a maximum value? If not, to what value would R_L need to be changed to achieve maximum power and what is the value of that power? Assume V_{be} to each device is 0.3 V rms.   (1.6 V rms, 128 mW, no, 10 Ω, 144 mW)

5.29 What do you understand by the curve of power dissipation against voltage for a class B push-pull stage as plotted in Fig. 5.85? Explain the shape of the curve. What is the advantage of a class B push-pull stage compared to a single-ended class B stage? What disadvantage occurs in the class B push-pull stage if the rated collector dissipation is given by

$$P_{c(max)} = \frac{2}{\pi^2} P_{ac(max)} \ ?$$

5.30 A class B push-pull amplifier is required to deliver 12 W of power to a 4 Ω loudspeaker. If the supply voltage is 20 V find the turns ratio of the transformer and the minimum voltage, current and power ratings of the stage transistors.   (6.4:1, 0.5 A, 40 V, 6 W)

5.31 A class B push-pull stage uses transistors with the following maximum rated values: collector current 1 A, collector voltage 20 V and collector dissipation 20 W. Suggest a suitable value for R_L if the transformer turns ratio is 3.34:1. Under these conditions what would be the value of P_{ac}?   (3.58 Ω, 5 W)

5.32 Using the characteristics of Fig. 5.87, but assuming an input current swing of 15 mA peak, find: (i) fundamental output power; (ii) value of mean output current under signal conditions; (iii) total distortion; (iv) conversion efficiency.   (1.39 W; 118 mA; 31%; 58.9%)

5.33 Repeat the previous question but using a peak input current swing of 10 mA.   (435 mW; 87 mA; 54%; 25%)

5.34 Draw the circuit diagram of a transformer-coupled push-pull stage and explain how the stage produces an output signal under class B bias conditions. Why is it necessary to use matched transistors in such a circuit?

5.35 What is meant by cross-over distortion? Explain how class AB biasing may be used to minimise the effect of cross-over distortion.

5.36 What do you understand by a transformerless push-pull circuit? Describe one transformerless push-pull circuit and show how your circuit produces one cycle of output voltage waveform. Assume class B bias. What are the advantages to be gained in *not* using transformers in a push-pull circuit?

5.37 A complementary symmetry push-pull stage uses the circuit of Fig. 5.104. If the transistors are identical with h_{FE} = 50 and supply voltage of 30 V, find suitable values of R_1, R_2 and R_3. Assume I_{CQ} = 25 mA.   (1.44 kΩ, 120 Ω, 1.44 kΩ)

5.38 A complementary symmetry push-pull stage uses the circuit of Fig. 5.105. Assuming matched transistors with h_{FE} of 30 and component values for R_1, R_2 and R_3 of 1 kΩ, 100 kΩ and 1 kΩ respectively, $R_{e1} = R_{e2}$ = 1 Ω and R_L = 10 Ω, find output voltage swings for positive and negative half-cycles of the input of 5 V peak. What effect is the value of supply voltage going to have on the value of the output voltage swing? Assume V_{CC} = 20 V.   (4.55 V, 4.13 V)

5.39 Using the circuit of Fig. 5.105 and the component and transistor values of example 5.43, find the peak values of signal current through the bias network and the load during positive and negative half-cycles of input voltage swing.   (4.5 mA, 14.67 mA, 4.5 mA, 13.33 mA)

5.40 A bootstrapped complementary symmetry amplifier has $R_1 = R_3 = 1$ kΩ, $R_2 = 100$ kΩ, $R_{e1} = R_{e2} = 1$ kΩ and $R_L = 10$ Ω. Identical transistors with $h_{FE} = 30$ are used. Find peak values of output voltage if V_{in} changes by ± 5 V peak. (4.53 V, 4.31 V)

5.41 Why is the bootstrapped complementary symmetry amplifier of Fig. 5.108 to be preferred to that of an ordinary complementary symmetry amplifier? Using the circuit and component values of example 5.45 show by calculation how the improvement in the use of bootstrapping is to be gained.

5.42 Why is it necessary to provide a driver stage shown by TR$_3$ in Fig. 5.111? What configuration is the transistor TR$_3$ connected in? What would be the effect on the bias to TR$_1$ and TR$_2$ if the h_{FE} for TR$_3$ is 150 and the quiescent value of TR$_3$ collector current is 8 mA. Assume supply voltage of +20 V and $R_2 = 100$ Ω while $R_3 = 1$ kΩ.

5.43 A complementary symmetry class B push-pull amplifier is shown in Fig. 5.105, If V_{CC} is +30 V, $R_L = 10$ Ω find: (i) maximum power output, (ii) power drawn from supply at maximum output, (iii) maximum transistor dissipation. (11.25 W, 14.32 W, 2.28 W)

5.44 For a complementary symmetry class B push-pull stage it is required to deliver 20 W to an 8 Ω load. Find a suitable value for the supply voltage. What would be the required power rating of the transistors (assumed identical)? (36 V, 4 W)

5.45 A complementary symmetry push-pull stage is shown in Fig. 5.105. R_L is 10 Ω, R_e is 1 Ω and $V_{CC} = +30$ V. If $V_{BE} = 0.6$ V and the peak to peak voltage swing at TR$_1$ base is 10 V find: (i) output power, (ii) d.c. power and (iii) stage efficiency. (800 mW, 3.82 W, 21%)

5.46 Using the circuit of Fig. 5.105 but assuming the a.c. and d.c. loads are identical for the driving stage, repeat example 5.43 allowing for $V_{BE} = 0.6$ V. (8.57 W, 12.5 W, 68.5%)

5.47 Draw the circuit diagram of a quasi-complementary stage. What is the major advantage of such a stage compared to a complementary symmetry stage?

5.48 A transistor has a thermal resistance $\theta_{DC} = 0.75°$ C/W and $T_{D(max)} = 120°$ C. Find: (i) the power dissipated by the transistor if the case is maintained at $60°$ C, (ii) the power that can be dissipated with $\theta_{CA} = 1.5°$ C/W at an ambient temperature of $60°$ C. (80 W, 26.67 W)

5.49 Find the heat sink requirement which would allow 5 W dissipation for a transistor. Assume that $\theta_{CS} = 0.35°$ C/W, $T_{D(max)} = 200°$ C, $\theta_{DC} = 10°$ C/W and the ambient temperature is $85°$ C. ($\theta_{SA} = 12.65°$ C/W)

5.50 The curve of Fig. 5.125 may be used to determine the size of a square aluminium heat sink for a power transistor. If it is required that there should be 5 W dissipation at an ambient temperature of $50°$ C/W, find the size of heat sink required. Assume $\theta_{DC} = 0.8°$ C/W, $\theta_{CS} = 0.2°$ C/W, $T_{D(max)} = 100°$ C. (85 cm^2)

5.51 If the transistor of example 5.49 is used with a large finned heat sink with $\theta_{SA} = 0.75°$ C/W, what would be the maximum allowable power dissipation if the transistor is mounted directly on the sink? (28.57 W)

5.52 The transistor of example 5.51 is to be electrically insulated from the sink such that θ_{CS} is raised to $5°$ C/W. What is the effect on the maximum allowable power dissipation? (7.63 W)

5.53 A silicon transistor has a derating curve with $P_D = 100$ W up to an ambient temperature of $25°$ C and a maximum case temperature of $150°$ C. Plot the curve and find: (i) the value of θ_D, (ii) the rated power dissipation if the temperature should rise to $110°$ C. (1.25° C/W, 32 W)

5.54 Plot the power dissipation curve for the power transistor where P_D at $25°$ C = 100 W. If the input power to the device, used as a power amplifier, is 40 W and the maximum and ambient temperatures are $100°$ C and $40°$ C respectively and $\theta_{DC} = 0.6°$ C/W find a suitable value for θ_{CA}. (0.15° C/W)

5.55   Manufacturer's data for a transistor gives $T_{D(max)} = 200°$ C, $P_{D(max)} = 20$ W at 25° C case temperature and $P_{D(max)} = 2.5$ W at 25° C ambient temperature. Determine the derating curves for both ambient and case temperatures. Find θ_{DC}, θ_D and θ_{CA}.

$$(8.75° \text{ C/W}. \ 70° \text{ C/W}, 61.25° \text{ C/W})$$

5.56   Three feedback amplifiers have the following gain and feedback factors:

	A	β
(a)	$5000 \ \angle 120°$	$3 \times 10^{-4} \ \angle 25°$
(b)	$75 \ \angle -45°$	$0.01 \ \angle 60°$
(c)	$450 \ \angle -116.7°$	$0.002 \ \angle 180°$

In each case, calculate the gain with feedback and state whether the feedback is positive or negative.

$$(2092 \ \angle 141°, \text{ negative}; 220.5 \ \angle -10.3°. \text{ positive}, 450 \ \angle 170°, \text{ neither})$$

5.57   An amplifier having a nominal gain of 2500 is found to have a production spread in gain of $\pm 30\%$. Simple negative feedback is to used to reduce the spread to within $\pm 4\%$. Calculate the required β factor and the resulting maximum and minimum values of gain.

$$(2.86 \times 10^{-3}, 316, 291.7)$$

5.58   A two stage amplifier has gains of 20 and 120 respectively. If the input is adjusted to give the required output level of 5 V, the output is found to include 15% harmonic distortion. Negative feedback is applied to the second stage to reduce this figure to to 3%. Calculate the required β factor and the necessary new gain for the first stage.

$$(0.033, 100)$$

5.59   The gain frequency response of an amplifier is given by the following table:

$\omega(\text{r/s} \times 10^4)$	0	0.5	1	2	5	10
$\|A\|$	5000	3984	2450	926	136	22.1
$\angle A°$	0	-59	-101	-148	-202	-232

Feedback is provided with $\beta = 0.0015 \ \angle 180°$. Determine the frequency response of the resulting feedback amplifier and find (a) the low-frequency gain, (b) the maximum gain and the frequency at which it occurs, (c) the frequency range for which the feedback is positive, (d) the maximum value to which β could be increased before the system becomes unstable.

$$(588, 1867, 3 \times 10^4, 1.9 \times 10^4, 3.3 \times 10^{-3})$$

5.60   An amplifier has a gain specified by

$$A = \frac{5000}{(1 + j\omega/10^4)^2(1 + j\omega/5 \times 10^4)}$$

Feedback is applied with a negative real β factor of 0.0015. By means of a Bode plot determine (a) the gain margin, (b) the β value if a gain margin of 10 dB was required, (c) the resulting phase margin, (d) the resulting low-frequency gain. $(6 \text{ dB}, 10^{-3}, 35°, 833)$

5.61   An amplifier has an open circuit gain of 18 000 and input and output impedances respectively of 2 kΩ and 3 kΩ and an external load of 4 kΩ. Feedback is applied using the following feedback networks: (a) a potential divider in parallel with the output of 20 kΩ and 22 Ω, or (b) a 10 Ω resistor in series with the output. In each case the feedback signal is added in series with the input such that simple negative feedback results. For each arrangement, calculate the voltage gain and input impedance and assuming a source impedance of 1 kΩ the output impedance. $(830, 22.8, 0.199, 385, 53.4, 4)$

5.62 The amplifier described in exercise 5.61 is used with a current feedback signal (a) derived from the output current with a current splitting network of 200 Ω and 100 kΩ or (b) derived from the output voltage with a 220 kΩ resistor between the input and output. If the external load is 1 kΩ and SNFB is assumed, determine in each case, the current gain and input impedance with feedback and the resulting voltage gain. (473, 110, 4280, 214.5, 4285)

5.63 An op-amp is to used as an inverting amplifier with a voltage gain of 8. With no feedback and both inputs earthed, the output off-set is 6 V. Determine the necessary open-loop gain A_d if the circuit off-set is not to exceed 10 mV. (5400)

5.64 An op-amp is specified as having input bias current of 0.2 μA and input voltage and current off-sets of 1.5 mV and 0.01 μA. It is used as an inverting amplifier with a feedback resistor of 1 MΩ and a gain of 5. If the non-inverting input is connected to earth through a 10 kΩ resistor calculate the output voltage off-set. (0.179 V or 0.197 V)

5.65 An op-amp has a slew rate of 1.5 V/μs and is used as an inverting amplifier with a gain of 6. Calculate the maximum signal input voltage if the frequency is 100 kHz. (0.28 V rms)

5.66 Calculate the output voltage V_o for each of the circuits shown in Fig. 5.221. (0.75, 7.37, 0.6, −0.83)

5.67 From first principles, calculate the input impedance to the circuit shown in Fig. 5.222. (−667 Ω)

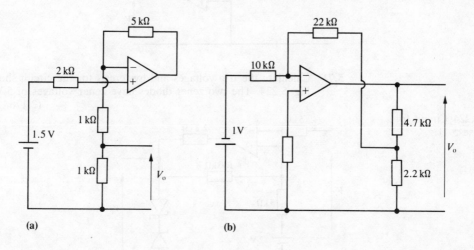

(a) (b)

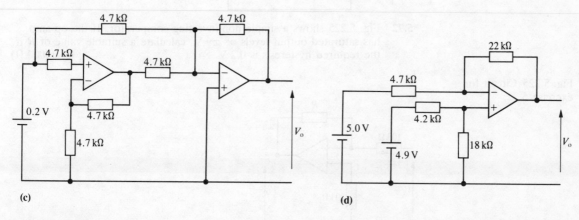

(c) (d)

Fig. 5.221 Circuits for exercise 5.66.

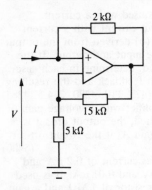

Fig. 5.222 Circuit for exercise 5.67.

Fig. 5.223 Circuit for exercise 5.70.

5.68 The circuit in Fig. 5.222 is used as a voltage to current converter by connecting a resistor R in series with the input and a 4 kΩ load between the non-inverting op-amp input and earth. Calculate the value of R and the range of V_i required to provide a current range of 0.05 mA to 6 mA. (667 Ω, 33 mV, 4 V)

5.69 With reference to the GIC circuit in Fig. 5.181, determine the equivalent input circuit if Z_1, Z_3 and Z_4 are 5 kΩ resistors and (a) Z_2 is a 1 mH inductor and Z_5 is a 0.1 μF capacitor or (b) Z_5 is a 2 kΩ resistor and Z_2 consists of a 10 kΩ resistor in parallel with a 0.02 μF capacitor. In each case, the signal frequency is 10 kHz. (−12.67 kΩ, (1 + j12.6) kΩ)

5.70 The circuit shown in Fig. 5.223 may be used as either an integrator or a differentiator. Determine the signal frequency ranges for which these operations apply and the maximum circuit gain. (above 3 kHz, below 300 Hz, 0.3)

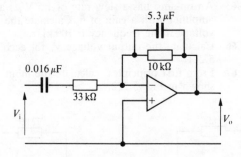

5.71 Calculate the trip voltages and hysteresis for the circuit shown in Fig. 5.224. The two zener diodes have zener voltages of 5 V. (2.176, 1.633, 0.54)

Fig. 5.224 Circuit for exercise 5.71.

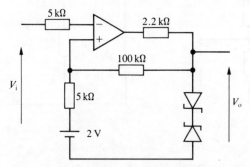

5.72 Fig. 5.225 shows a simple non-inverting comparator. If the op-amp has saturated output levels of ±9 V, calculate a suitable value of R if the required hysteresis is 0.2 V. (900 kΩ)

Fig. 5.225 Circuit for exercise 5.72.

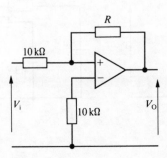

6 Waveform generation and conversion

The principal learning objectives of this chapter are to:

	Pages	Exercise
• Investigate various techniques used in the production of square or rectangular wave forms	523–33	6.1–6.4
• Analyse a range of circuits used to generate ramp and triangular wave forms	533–9	6.5
• Consider alternative methods for the production of staircase voltage wave forms	539–45	6.6
• Appreciate the principles of sinusoidal oscillator operation	545–6	6.7
• Analyse a range of RC and LC oscillator circuits used in the production of sinusoidal wave forms	546–58	

In the last chapter, we were concerned with the behaviour of amplifiers and signal processing using amplifiers, particularly op-amps. In the following sections, the emphasis is more on the various signal waveforms employed in electronic systems. The three basic forms are rectangular, sawtooth and sinusoidal, the first two of which become square and triangular if they are symmetrical. These different forms may also be converted from one to the other of they may be generated directly. They may also be combined in various ways to form more complex waveforms if required. Generation and conversion may employ a variety of devices and examples using transistors, logic gates and op-amps are examined in this chapter.

Generation of rectangular waves

Rectangular waves are usually obtained from circuits having two saturated or cut-off states. These may include transistors, logic gates, op-amps or special purpose integrated circuits. The device may be driven into the two states by an external signal (usually sinusoidal) or the complete system may have two states, like a flip-flop and switch between them with no external signal. Another possibility, the conversion of a train of short pulses into a rectangular wave, is considered in the section on timing in Chapter 4 (p.293). In this section, conversion from sine waves is considered first and then a range of rectangular waveform generators are examined.

Sine wave to square wave conversion

The circuits shown in Fig. 6.1 can all be driven, by a sine wave input, into saturated and cut-off states. In each case, output waveforms are shown and various features need further consideration.

Fig. 6.1 Square waves from sine waves: (a) Transistor circuit; (b) logic gate circuit; (c) and (d) op-amp circuits.

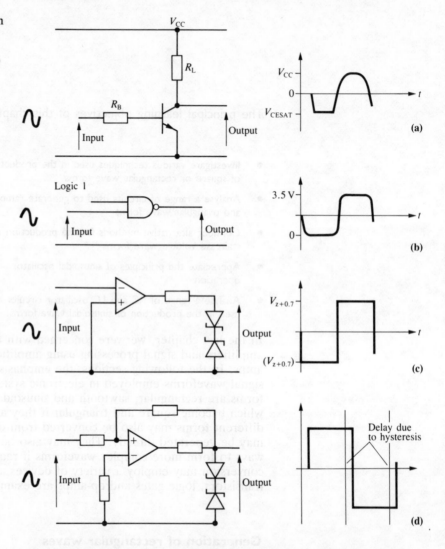

The first circuit, a saturated BJT is cut off for V_i less tha 0.5 V and is saturated when I_B is greater than $V_{CC}/h_{FE}R_L$. The input impedance at the base varies between infinity and a few tens of ohms in saturation. The output impedance is R_l in the cut-off state and a low value given by the slope of the output characteristic in the saturated state. The output square wave will only have steep sides if large input signals are used and two levels of output are fixed at V_{CC} and V_{CESAT}. The logic gate circuit shown in (b) has a smaller variation in input impedance and a higher gain between the two states. The minimum amplitude of the input signal must exceed the impedance range for the gate and the output levels are fixed. The output impedance can be low in both states. Faster edges will be obtained if a Schmitt input gate is used.

The op-amp in (c) is connected as an inverting zero-crossing detector. The very high open-loop gain ensures very steep sides to the square wave. The input impedance is very high and the output impedance is given by the dynamic resistance of the zener diodes. Output levels are normally positive and negative at levels selected by the zener diodes. Some special purpose comparators operate from a single power supply allowing one of the output levels to be zero. For low frequency signals, noise derived chatter may occur and if low amplitude input signals are used, the input voltage off-set of the op-amp can cause problems.

The last circuit (d) is a non-inverting comparator with hysteresis. Positive feedback provides even greater gain during switching from one state to the other but the maximum output slope is limited (as it is for circuit (c)) by the op-amp slew rate. The input impedance is determined by the external resistors and will be the same in both states. Hysteresis avoids the problem of chatter but results in a 'delay' between the input sine wave zero-crossing point and the square wave output. If precise zero crossing in one direction is required, a reference voltage in the inverting input lead can make one of the trip voltages exactly zero. The resulting output wave will however, be slightly asymmetrical.

Astable multivibrators

Astable multivibrators are free running rectangular wave generators. The timing of the two states is based on the principles discussed in the section on timing circuits in Chapter 4. The active devices may be transistors, logic gates or op-amps each having applications in different circumstances. Alternatively, integrated circuit timers like the 555 may be connected in an astable mode. In each case, the two times, and the resulting mark to space ratio can be chosen, although in some cases there is a limit to this ratio.

The BJT astable multivibrator

A circuit for the astable BJT multivibrator is shown in Fig. 6.2. The first state is T_{r1} on and saturated with T_{r2} off and the second state is T_{r2} on and T_{r1} off. Neither of these states is stable as after a time determined by the passive components, the circuit will always switch to the other state. For this reason, the states are referred to as quasi-stable states.

The starting point for the descriptive analysis is when one transistor, say T_{r1}, is turning from off to on. The collector voltage

Fig. 6.2 A BJT multivibrator.

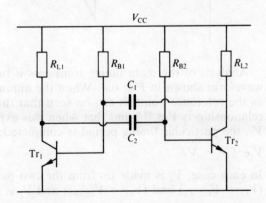

will fall from V_{CC} to V_{CESAT}, a negative change of nearly V_{CC}. This change is passed through C_2 (whose charge cannot change instantaneously) causing V_{BE2} to fall by the same amount. Prior to this action, T_{r2} will have been on with V_{BE2} at 0.7 V. V_{BE2} now falls to a negative value from +0.7 V by V_{CC} to nearly $-V_{CC}$ holding T_{r2} off. C_2 can now charge exponentially through R_{B2} towards $+V_{CC}$. When V_{BE2} reaches the threshold value of about 0.5 V, T_{r2} will start to conduct. The resulting fall in V_{CE2} is passed through C_1 reducing V_{BE1} and I_{C1}. This in turn causes a rise in in V_{CE1} and with it, in V_{BE2}. This is a cumulative effect switching T_{r2} on and T_{r1} off. In this case, V_{CE2} falls by nearly V_{CC} and takes V_{BE1} with it. The duration of the quasi-stable state is determined by C_1 charging through R_{B1}. The waveforms for both collectors and both bases are shown in Fig. 6.3. For the purpose of the description, it is assumed that $R_{B1} = R_{B2}$ and $R_{L1} = R_{L2} = R_B/10$; the two capacitors will have a ratio given by $C_1 = 3C_2$.

Fig. 6.3 Waveforms for the BJT multivibrator.

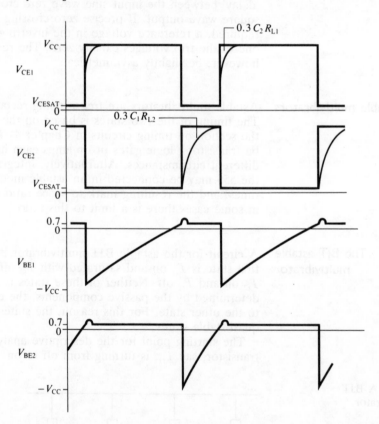

Analysis of the main timing transients is based on the detailed waveform shown in Fig. 6.4. When the aiming voltage (V_{CC}) is taken as the reference point, it can be seen that the exponential relationship is $V_1 e^{-t/CR}$ and that when this exponential has the value V_2, the particular timing period is completed; therefore

$$V_1 e^{-T/CR} = V_2 \qquad [6.1]$$

In each case, V_1 is made up from the two parts shown ($V_{CC} - V_{CESAT}$) and ($V_{CC} - V_{BESAT}$) and V_2 is simply ($V_{CC} - V_{BEth}$).

Fig. 6.4 Timing for the BJT
multivibrator.

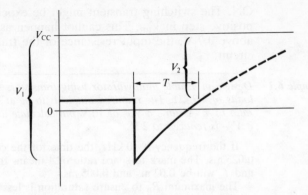

From equation [6.1] above, the time T_1 for which T_{r1} is off is
obtained from

$$\frac{V_1}{V_2} = e^{+T_1/C_1R_{B1}}$$

$$T_1 = C_1R_{B1}\log_e\frac{V_1}{V_2}$$

Substituting for V_1 and V_2

$$T_1 = C_1R_{B1}\log_e\frac{(2V_{CC} - V_{CESAT} - V_{BESAT})}{(V_{CC} - V_{BEth})}$$

Inserting typical values

$$T_1 = C_1R_{B1}\log_e\frac{(2V_{CC} - 0.8)}{(V_{CC} - 0.5)} \tag{6.2}$$

which approximates to

$$T_1 = C_1R_{B1}\log_e 2 \tag{6.3}$$

unless very low values of V_{CC} are used. Similarly, the time T_2 for
which T_{r2} is off is given by

$$T_2 = C_2R_{B2}\log_e 2 \tag{6.4}$$

Equations [6.3] and [6.4] give the time periods for the rectangular
output waveform but there are also some other transients of
interest.

For each collector waveform, when the transistor turns off, the
voltage might be expected to rise instantly to V_{CC}. The waveforms in
Fig. 6.3 show that the collector voltages rise exponentially towards
V_{CC}. This may be explained as follows: at the instant that T_{r2} turns
off (and T_{r1} turns on) capacitor C_1 is virtually uncharged with V_{BE1}
at the threshold of 0.5 V and V_{CE2} at V_{CESAT} of perhaps 0.1 V.
Cumulative switching takes place turning T_{r2} off, but before V_{CE2} can
rise, C_1 must charge through R_{L2}. Allowing three time constants
C_1R_{L2} or for the figures chosen $0.3\,C_1R_2$ (since $R_L = 0.1\,R_B$) we can
see that this is about half the duration of the *other* output pulse.
Similarly, the transient for V_{CE1} will be about half the duration of
the V_{CE2} output pulse. This relationship obtains to the particular R_L
to R_B ratio, but in general terms, the shorter output pulse will have
the longer transient.

The remaining transient is at the base when either transient turns

ON. The switching transient might be expected to cause a large positive step in V_{BE}. This cannot happen as if V_{BE} were to rise above 0.7 V, the input resistance of the transistor tends to a short circuit.

Example 6.1   *Design an astable multivibrator using transistors with h_{FE} of 120 and collector loads of 4.7 kΩ. The required output is to be at 40 kHz with a mark space ratio of 4. Assume a V_{CC} of 10 V but calculate also the error in the frequency if V_{CC} is reduced to 2 V.*

If the frequency is 40 kHz, the time for the complete cycle is 1/40 or 0.025 ms. The mark to space ratio of 4 means that the two time periods T_1 and T_2 will be 0.02 ms and 0.005 ms.

The maximum R_B to ensure saturation is less than $h_{FE}R_L$ or 120 × 4.7 kΩ. A high value of R_B will mean smaller timing capacitors and therefore, shorter output transients. A value of 560 kΩ would meet the h_{FE} requirement but this would not allow for any component or parameter tolerance. A choice of 330 kΩ would allow a sufficient margin.

Applying equations [6.3] and [6.4],

$$T_1 = 2 \times 10^{-5} = C_1 \times 330 \times 10^3 \times \log_e 2$$

therefore

$$C_1 = \frac{2 \times 10^{-5} \times 10^{12}}{330 \times 10^3 \times 0.69}\, \text{pF} = 88\ \text{pF}$$

and since $T_2 = 5 \times 10^{-6}$, $C_2 = 22$ pF.

The output transient time constants are respectively, $22 \times 10^{-12} \times 4.7 \times 10^3$ and $88 \times 10^{-12} \times 4.7 \times 10^3$ or 0.4 μs and 0.1 μs. Allowing for three time constants, the output waveform is shown in Fig. 6.5

Fig. 6.5 The output waveforms for example 6.1.

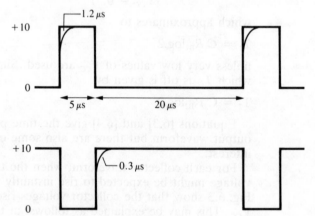

With the lower value V_{CC} of 2 V, the accurate equation [6.2] must be used.

$$T_1 = C_1 R_{B1} \log_e \frac{(2V_{CC} - 0.8)}{(V_{CC} - 0.5)}$$

$$= 88 \times 10^{-12} \times 330 \times 10^3 \frac{(4 - 0.8)}{(2 - 0.5)} = 22\ \mu\text{s}$$

Similarly

$$T_2 = 5.5\ \mu\text{s}$$

Thus the mark space ratio is unchanged but the frequency will be reduced to about 36 kHz, a 10% fall.

An op-amp astable multivibrator

The two-state requirement for a mulitvibrator can be provided by an op-amp in the comparator with hysteresis mode. A simple arrangement having equal mark space ratio is shown in Fig. 6.6. The output amplitude is limited by the d.c. power supply levels but these can of course by selected by zener diode circuits.

Fig. 6.6 An op-amp astable multivibrator.

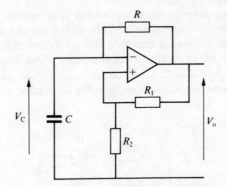

The positive feedback circuit can be seen to be the same as that in the inverting comparator with hysteresis. The effective trip voltage is that across the capacitor C which charges and discharges through R towards the positive and negative output levels of $+V_o$ and $-V_o$. The analysis is best understood by reference to the waveforms in Fig. 6.7.

Fig. 6.7 Waveforms for the op-amp multivibrator.

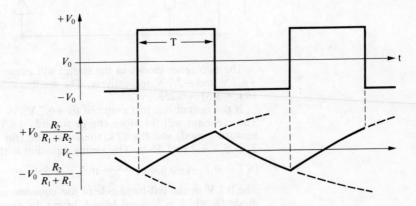

The two trip levels are $\pm\left(\dfrac{V_o R_2}{R_1 + R_2}\right)$ which may also be written $\pm\beta V_o$. At the instant that the output switches to $+V_o$, V_C will be at the negative trip level. C will now charge towards $+V_o$ and this voltage waveform, with reference to the $+V_o$ level is given by

$$V_C = V_o(1 + \beta)e^{-t/CR} \qquad [6.5]$$

When V_C passes the positive trip level, the output will switch to $-V_o$. Again with reference to the $+V_o$ level, this point is given by $V_o(1 - \beta)$. The duration of the output pulse is therefore given by

$$V_o(1 + \beta)e^{-T/CR} = V_o(1 - \beta) \qquad [6.6]$$

$$\frac{(1 + \beta)}{(1 - \beta)} = e^{T/CR}$$

and $T = CR \log_e \left(\frac{1 + \beta}{1 - \beta}\right) \qquad [6.7]$

This is, of course, the time for half a cycle and the frequency will be given by $1/2T$.

If an asymmetrical waveform is required, this can be achieved by the use of alternative resistor paths for the charge and discharge of C. The following example demonstrates this point together with the use of zener diodes to clamp the output levels.

Example 6.2   *Determine the output waveform for the astable multivibrator shown in Fig. 6.8.*

Fig. 6.8 Op-amp multivibrator circuit for example 6.2.

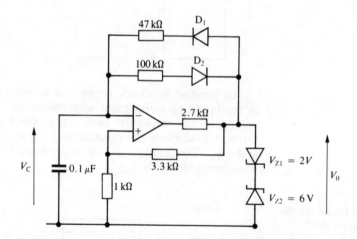

The two zener diodes in the circuit will cause the two output levels to be +6.7 V and −2.7 V respectively. The feedback factor β is given by $1/(1 + 3.3) = 0.233$.

If the output has just switched to +6.7 V, $V_C = \beta \times -2.7$ or −0.63 V. The capacitor will therefore charge towards +6.7 through D_1 (which is forward biased) and the 47 kΩ resistor. The trip level completing the period T_1 is $\beta \times 6.7$ or 1.56 V. The timing equation is thus given by

$$[6.7 - 0.7 -(-0.63)]e^{-T/CR} = (6.7 - 1.56)$$

The 0.7 V in the left-hand side of the equation is the volt drop across the diode D_1 which is forward biased during the capacitor charging period.

Rearranging $T_1 = CR \log_e \dfrac{6.63}{5.14} = 47 \times 10^3 \times 10^{-7} \times 0.255$

$$= 1.2 \text{ ms.}$$

For the second period T_2, the opposite voltage levels and the 100 kΩ resistor are used.

$$2.7 - 0.7 -(-5.6)e^{-T/CR} = (2.7 - 0.63)$$

Rearranging $T_2 = CR \log_e \dfrac{3.56}{2.07} = 5.4$ ms.

The waveforms are shown in Fig. 6.9 and the frequency and mark space ratio are 152 Hz and 4.5 respectively.

Fig. 6.9 Waveforms for example 6.2.

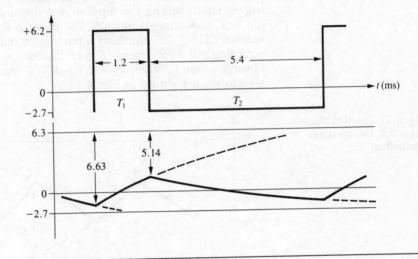

The 555 timer as an astable multivibrator

The use of a 555 timer as a monostable or one-shot was described in Chapter 4 with modified connections, the same monolithic integrated circuit may be used as an astable or free running multivibrator. The connection diagram and relative external components are shown in Fig. 6.10.

Fig. 6.10 The 555 timer connected as an astable multivibrator.

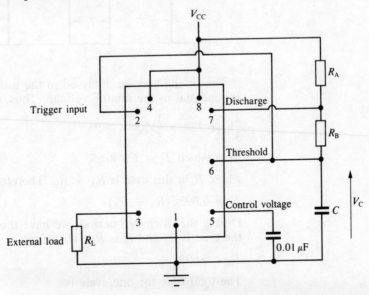

The operation depends on an internal bistable flip-flop. In one state, the discharge pin 7 is held at ground (through a saturated BJT) and the output, pin 3 is also low: in the second state, pin 7 is effectively open allowing C to charge and the circuit output is high at V_{CC} volts. Change of state of the flip-flop depends upon the voltages at the trigger input, pin 2 and the threshold input, pin 6, the two levels respectively being $V_{CC}/3$ and $2V_{CC}/3$. Thus at one point in the cycle, C has just discharged to $1/V_{CC}/3$ operating the trigger input, setting the flip-flop and opening the discharge input pin 7. C now charges towards V_{CC} through $R_A + R_B$. When V_C reaches $2V_{CC}/3$ the threshold input causes the flip-flop to reset returning pin 7 to ground. C now discharges through R_A alone towards ground. When V_C reaches $V_{CC}/3$, the cycle is repeated. The waveforms are shown in Fig. 6.11.

Fig. 6.11 The waveforms for a 555 timer astable multivibrator.

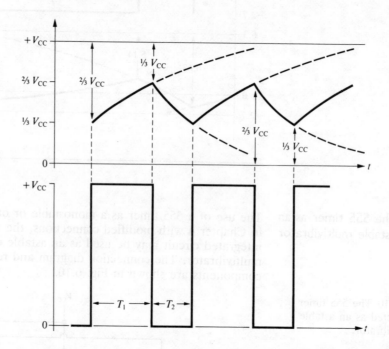

The timing may be analysed in the usual way by referring each exponential to the aiming voltage. Thus, during the charge period

$$\frac{2}{3}V_{CC}e^{-T/CR} = \frac{1}{3}V_{CC} \qquad [6.8]$$

from which $T_1 = CR \log_e 2$

where R in this case is $R_A + R_B$. Therefore

$T_1 = 0.693C(R_A + R_B)$

During the discharge period, we have the same equation as [6.8] but the time is T_2 and R is R_B. Therefore

$T_2 = 0.693CR_B$

The total time for one cycle is

$$T_p = 0.693C(R_A + 2R_B) \qquad [6.9]$$

and the frequency $f = \dfrac{1}{T_p} = \dfrac{1.44}{C(R_A + 2R_B)}$

Example 6.3　*An astable multivibrator is required to produce an asymmetrical output at 40 kHz with a mark space ratio of 3. If a 555 timer is to be used for this purpose, determine a suitable set of timing components.*

Since the frequency of 40 kHz, the total time T_p is $10^{-3}/40$ or 25 μs. As the mark space ratio is 3:1, $T_1 = 3T_2$ and $R_A + R_B = 3R_B$, or $R_A = 2R_B$. Substituting in equation [6.9] $25 \times 10^{-6} = 0.693C(2R_B = 2R_B)$ taking C as 0.01 μF

$$4R_B = \frac{25 \times 10^{-6}}{0.693 \times 10^{-8}} = 3608 \ \Omega.$$

therefore $R_B = 902 \ \Omega$, and $R_A = 2R_B = 1804 \ \Omega$.

This will be near the upper frequency limit of the 555 timer which, like other linear circuits will have a slew rate limitation on the output waveform.

Generation of triangular waveforms

In many electronic systems, a signal having a constant rate of change of voltage is required. Examples include time base generators for cathode ray tube display, electronic voltmeters and sweep frequency generators. Such ramp voltage signals may be free running with a true triangular waveform or they may be triggered by an external pulse. These alternatives are illustrated in Fig. 6.12.

Fig. 6.12 Ramp waveforms: (a) a triangular waveform; (b) a triggered ramp waveform.

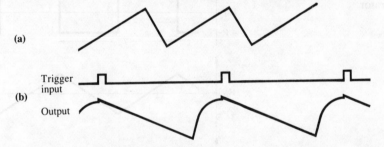

(a)

Trigger input

(b)

Output

In the triggered waveform, at the end of the linear ramp, the voltage falls back to a reference level where it remains until the next trigger pulse is applied.

If the requirement for linearity is not very precise, the output of an astable mulitvibrator will provide a free running triangular wave while a monostable can similarly provide a triggered ramp (see Figs 6.9 and 4.28 respectively). When a more linear slope is required this is usually obtained from a square wave and an integrator.

A free running triangle/ square wave generator

This circuit shown in Fig. 6.13, combines a symmetrical comparator with an integrator: the comparator output is a square wave and this also provides the input to the integrator. The integrator output is triangular and provides the input to the comparator. Another feature is the zener diode bridge used instead of the two zener diodes in the comparator circuits described in earlier sections.

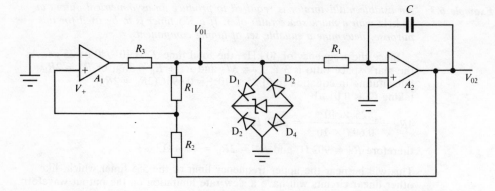

Fig. 6.13 A free-running
triangle/square wave
generator.

When amplifier A_1 output is positive, D_1 and D_4 conduct and V_{o1} is at the zener voltage V_z plus two diode volt drops. If A_1 output is negative, D_3 and D_2 conduct and V_{o1} has a negative value equal to the previous positive level.

For the operation of the complete circuit, consider the waveforms shown in Fig. 6.14. These have been drawn for the case where $R_1 = R_2$. At the instant when V_{o1} switches to $+V_i$, the non-inverting input to A_1 (V_+) must be equal to zero, going positive. Prior to this, V_{o1} has been at $-V_i$ and V_+ is a combination of this negative value and the value of V_{o2}. Since $R_1 = R_2$, the value of V_{o2} must have been $+V_i$.

Fig. 6.14 Waveforms for
the triangle/square wave
generator.

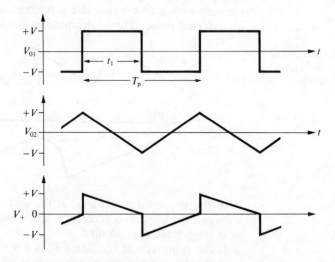

Starting from this point, the steady positive value of V_{o1} is integrated by A_2 to give V_{o2} as

$$V_{o2} = +V_i - \frac{V_i t}{CR_I}$$
[6.10]

For the next switch in V_{o1} to occur, V_+ must pass through zero going negative. Now, as $R_1 = R_2$,

$$V_+ = \frac{V_{o1} + V_{o2}}{2}$$

Thus, from equation [6.10] since $V_{o1} = V_i$

$$\frac{V_i + V_i - \dfrac{V_i T_1}{CR_I}}{2} = 0$$

where T_1 is the time of switching. Rearranging

$$2V_i = \frac{V_i T_1}{CR_I}$$

$$T_1 = 2CR_I$$

The second half cycle has the same duration for the symmetrical circuit. The period for the triangular V_{o2} and the square V_{o1} is therefore $T_p = 4CR$ and the frequency of operation is

$$f = 1/4CR \text{ Hz.}$$

Many variations of this basic generator are used; for example, variation of R_I can be used as a frequency control and the use of back to back unequal zeners as the reference will produce an asymmetrical output with unequal positive and negative slopes for V_{o2}. The amplitude of V_{o2} relative to V_{o1} can also be chosen by using non-equal ratios for R_1 and R_2. These features can be examined in the following example.

Example 6.4	*Determine the output waveform for the circuit shown in Fig. 6.15 and calculate the range of R_I required for a frequency range of 10 Hz to 1 kHz.*

The output waveforms for V_{o1} and V_{o2} will be similar to those shown in Fig. 6.14 but the timing will be asymmetrical resulting from the different positive and negative output levels due to the two zener diodes. Also, as R_1 and R_2 are unequal, the peak to peak value of V_{o2}, $(V_{o2+} - V_{o2-})$ will not be the same as the peak to peak value of V_{o1} $(V_{o1+} - V_{o1-})$ which is derived directly from the zener diodes.

The first step is to establish values for these output levels.

$$V_{o1+} = 6 + 0.7 = 6.7 \text{ V} \tag{6.11}$$

$$V_{o1-} = -3 - 0.7 = -3.7 \text{ V} \tag{6.11}$$

While V_{o1} is at V_{o1+},

$$V_{o2} = V_{o2+} = \frac{6.7t}{CR} \tag{6.12}$$

Since at the instant V_{o1} switches from -3.7 to $+6.7$, V_{o2} will have reached its maximum positive value and will start integrating negatively. Similarly, when V_{o1} is at V_{o1-},

$$V_{o2} = V_{o2-} + \frac{3.7t}{CR} \tag{6.13}$$

At the instant of switching, the non-inverting input will be at 0 V. A part circuit is drawn in Fig. 6.16 showing how V_{o2+} and V_{o2-} can be calculated from equations [6.11] and [6.12] above. From Fig. 6.16a, applying superposition,

$$-3.7 \times \frac{2.2}{2.2 + 4.7} + V_{o2+} \times \frac{4.7}{2.2 + 4.7} = 0$$

From which $V_{o2+} = 1.74$ V.

Similarly, from Fig. 6.16b

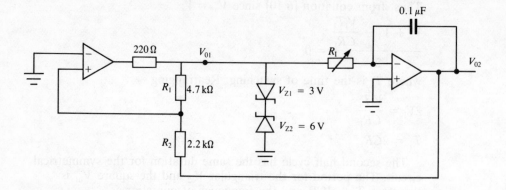

Fig. 6.15 Triangle/square wave generator circuit for example 6.3.

Fig. 6.16 Part circuits for the solution of example 6.3.

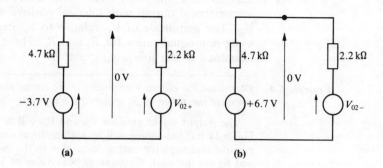

(a)   (b)

$$6.7 \times \frac{2.2}{2.2 + 4.7} + V_{o2} \times \frac{4.7}{2.2 + 4.7} = 0$$

$$V_{o2-} = -3.14 \text{ V}$$

The two time intervals T_1 and T_2 can now be found from equations [6.12] and [6.13]

$$1.73 - \frac{6.7T_1}{CR} = -3.14$$

$$T_1 = 0.727CR$$

and $-3.14 + \dfrac{3.7T_2}{CR} = 1.73$

giving

$$T_2 = 1.316CR$$

The total period $T_p = T_1 + T_2 = 2.043CR$

and the frequency of operation f is given by

$$f = \frac{1}{T_p} = \frac{0.489}{CR} \text{ Hz}$$

The capacitor C is given as $0.1 \ \mu F$, thus

$$f = \frac{4.89 \times 10^6}{R}$$

From which for 10 Hz, $R = 489 \text{ k}\Omega$, for 1 kHz, $R = 4.89 \text{ k}\Omega$.

The waveforms are shown in Fig. 6.17.

Fig. 6.17 Waveforms for example 6.3.

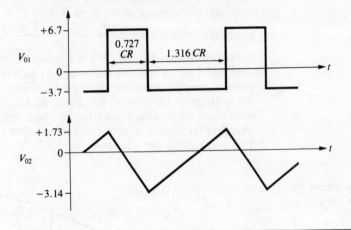

A bootstrap ramp generator

An alternative circuit for provision of ramp voltages with a rapid fly back is a bootstrap ramp generator. The basic form as shown in Fig. 6.18 requires an external timing pulse to initiate a ramp input, but a comparator and feedback arrangement can be added if a free running system is required. The active device here is a BJT connected as an emitter follower, but alternative follower amplifiers can be used.

Fig. 6.18 A bootstrap ramp generator.

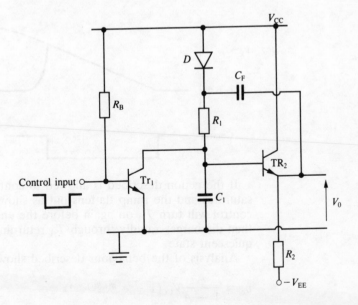

The operation is as follows; when the control input is positive, T_{r1} will be ON and saturated, C_1 will be discharged through T_{r1}. T_{r2} will be conducting in the linear region with base current supplied through R_1 and the diode D. V_o will be slightly negative (V_{CESAT} for $T_{r1} - V_{BE}$ for T_{r2}). C_F which has a large value of capacitance will be charged to approximately V_{CC} volts.

When the control input goes negative, T_{r1} turns off and C_1 starts to charge with a current $\dfrac{V_{CC} - V_d - V_{CESAT}}{R_1} - I_{B2}$, which approximates to V_{CC}/R_1. V_{C1} is the signal input to T_{r2} which is connected as an emitter follower (A_v nearly $+1$). As V_o rises, the voltage change is passed through C_F where it turns D off and raises the voltage at the top of R_1. Thus, as C_1 charges, the voltages at both ends of R_1 rise at nearly the same rate. This means that the charging current is nearly constant resulting in a linear ramp output. The waveforms are shown in Fig. 6.19.

Fig. 6.19 Waveforms for the bootstrap ramp generator.

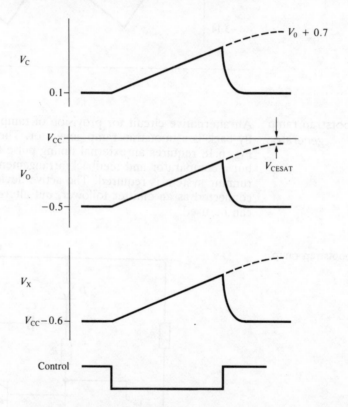

If the action described is allowed to continue, eventually T_{r2} saturates and the ramp flattens out as shown. In practice, the control will turn T_{r1} on again before the end of the linear range. C_1 then discharges rapidly through T_{r1} returning the circuit to its quiescent state.

Analysis of the behaviour described shows

$$v_o = \frac{A}{1 - A} V_{CC}(1 - e^{-t/(CR/1) - A}) \qquad [6.14]$$

For a voltage follower circuit, A approaches unity; this expression thus represents a very large gain and a very long time constant. The ramp output is, in effect, the first small section of a capacitor charging waveform.

Example 6.5 *The bootstrap ramp generator in Fig. 6.20 is controlled by a monostable multivibrator. Assuming the emitter follower to have a gain of 0.93, calculate a value for the timing capacitor C_T to ensure a 4 V peak to peak sawtooth waveform at V_o.*

Fig. 6.20 Ramp generator circuit for example 6.4.

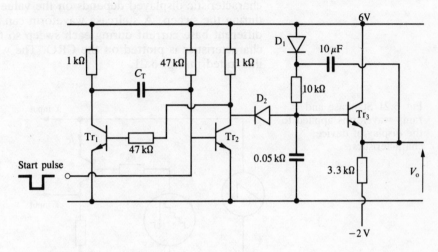

Applying equation [6.14]

$$v_o = \frac{0.93}{0.07} \times 6(1 - e^{-t/0.0005/0.07})$$

$$= 79.7(1 - e^{-t/7.14})$$

where t is in milliseconds.

This is required to rise to 4 V, therefore

$$\frac{4}{79.7} = 1 - e^{-T/7.14}$$

rearranging and solving for T,

$$T = 0.368 \text{ ms}$$

Thus the slope of the ramp is $4000/0.368 = 10.8 \times 10^3$ V/s. This duration is controlled by the time for which T_{t2} is off. This is given by $0.69 C_T R$ where R is 47 kΩ. Therefore

$$C_T = 0.0113 \ \mu\text{F}$$

If the original charging current of 6 V/10 kΩ had been maintained, a time of 0.368 ms would have resulted in a V_C of 4.416 V and V_o of 4.1 V. The non-linearity amounts to 0.1 in 4 or 2.5%. A practical amplifier would have a gain of 0.99 or even nearer 1. The corresponding linearity would be greatly improved.

Generation of staircase waveforms

Staircase waveforms have applications in D to A conversion (Chapter 7) and in various display systems such as device characteristic plotters. For this second application, taking a BJT as an example, the collector voltage is repeatedly swept from zero to some chosen maximum value. During each sweep, the collector voltage is applied to the X input to a CRO while a voltage derived from the collector current is applied to the Y input. The particular characteristic displayed depends on the value of the base current during the sweep. A staircase waveform can be used to provide a different base current during each sweep so that a family of characteristics is plotted on the CRO. The whole arrangement is illustrated in Fig. 6.21.

Fig. 6.21 Staircase and ramp waveforms applied to the display of device characteristics.

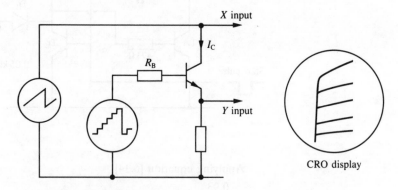

The collector sweep and the staircase must of course, be synchronised. The five steps indicated correspond to the base currents for the five characteristic curves shown. A small resistor in the transistor emitter lead provides a voltage proportional to I_C and this will be amplified before application to the *CRO* input.

Staircase waveforms can be generated in a variety of ways, the two to be discussed here being a diode pump and a system based on a digital counting circuit.

A diode pump staircase generator

The diode pump is an arrangement of capacitors and clamping diodes which, when supplied with a train of pulses, produces a staricase waveform by 'pumping up' the voltage on a capacitor. The basic arrangement is shown in Fig. 6.22 and the action may be described as follows: when the first pulse is applied, V_x tends to go positive, turning D_1 on and D_2 off. C_1 charges rapidly to V, the amplitude of the input pulse. At this stage, V_x will be approximately zero. At the end of the first pulse, the input returns to zero and V_x tends to $-V$ volts (the capacitor charge cannot change instantaneously). This turns D_1 off and D_2 on. The charge on C_1 is now shared between C_1 and C_2 but as C_2 will be much larger than C_1, V_o will rise to a voltage much smaller than the original V. $\left(V_{o1} = \dfrac{C_1}{C_1 + C_2} \times V\right)$. The next pulse results in a similar action, but in this case, at the end of the pulse, the charge shared between the two capacitors is the sum of the previous charge on C_2 and the new charge on C_1. As a result, V_o rises to a new value V_{o2} forming the second step of the staircase. This action will be more easily understood through a numerical example.

Fig. 6.22 A diode pump staircase generator.

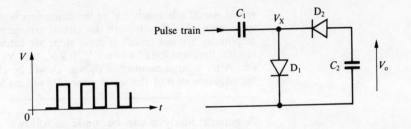

Pulse train →

Example 6.6

A diode pump employs capacitors of 0.1 μF and 0.9 μF respectively and a train of positive 10 V pulses are applied. Calculate the amplitude of the first five output steps and hence deduce a general expression for the output steps of a diode pump circuit.

On the first pulse, C_1 charges to

$$Q_1 = C_1 V \text{ or } 10^{-7} \times 10 \text{ C}$$

At the end of this pulse, the charge is shared between C_1 and C_2 to give

$$V_{o1} = \frac{Q}{C} = \frac{C_1 V}{C_1 + C_2} = \frac{10^{-7} \times 10}{(1 + 9)10^{-7}} = 1 \text{ V}$$

Now, $Q_2 = C_1 V + V_{o1} C_2$

$$= 10^{-6} + 0.9 \times 10^{-6} = 1.9 \times 10^{-6} C$$

This is once again shared to give

$$V_{o2} = \frac{Q}{C} = \frac{1.9 \times 10^{-6}}{1 \times 10^{-6}} = 1.9 \text{ V}$$

Then $Q_3 = C_1 V + V_{o2} C_2$

$$= 10^{-6} + 0.9 \times 1.9 \times 10^{-6} = 2.71 \times 10^{-6} C$$

making $V_{o3} = \dfrac{2.71 \times 10^{-6}}{10^{-6}} = 2.71 \text{ V}$

Further calculations show $V_{o4} = 3.439 \text{ V}$, $V_{o5} = 4.095 \text{ V}$ and $V_{o6} = 4.685 \text{ V}$.

The waveforms illustrating the process are shown in Fig. 6.23.

Fig. 6.23 Waveforms for the diode pump circuit in Ex. 6.5.

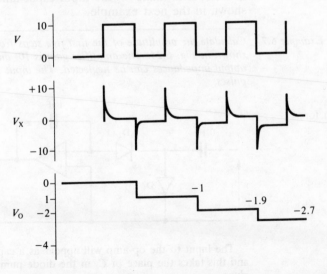

For each step, a short time constant is indicated; this is due to forward resistance of the conducting diode in each case. In a practical circuit, V_x

values would not reach $\pm V$ as the capacitors will charge too rapidly for this to be seen. A difficulty with this circuit arrangement is that the step amplitudes are not equal. If these steps are calculated from the above results, they are found to be 1 V, 0.9 V, 0.81 V, 0.729 V, 0.656 V, 0.591 V etc. After a large number of pulses, C_2 will be charged to $-V$ volts and on the negative step of the input pulse, D_2 will no longer be turned on.

A general analysis can be made as follows

$$Q_1 = C_1 V$$

therefore output $V_{o1} = \dfrac{Q}{C} = \dfrac{C_1 V}{C_1 + C_2}$

total charge $Q_2 = C_1 V + V_{o1} C_2$

$$V_{o2} = \frac{Q}{C} = \frac{C_1 V + V_{o1} C_2}{C_1 + C_2}$$

$$V_{o3} = \frac{C_1 V + V_{o2} C_2}{C_1 + C_2}$$

$$V_{o4} = \frac{C_1 V + V_{o3} C_2}{C_1 + C_2}$$

and the nth step, $V_{on} = \dfrac{C_1 V + V_{(n-1)} C_2}{C_1 + C_2}$

$$V_{on} = \frac{C_1}{C_1 + C_2}\left(V + \frac{C_2}{C_1} V_{(n-1)}\right) \qquad [6.15]$$

This result allows for a rapid calculation of the amplitude of a series of steps.

In the previous example, the steps were unequal; this effect can be reduced by using a C_2 value much greater than C_1 which, while making the steps more equal, will make them very small (from $C_1/(C_1 + C_2)$). Since they will then require amplification, the best solution is to make C_2 the input capacitance to an op-amp system as shown in the next example.

Example 6.7   *Calculate the amplitude of the first five steps from the output of the circuit shown in Fig. 6.24. The amplifier gain is 10^5 and the effects of input and output impedances can be neglected. The input to the circuit is a train of 5 V pulses.*

Fig. 6.24 Staircase generator circuit for example 6.6.

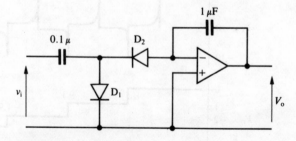

The input to the op-amp will appear as a capacitance of value $C_F(1 + A)$ and this takes the place of C_2 in the diode pump circuit. For the figures given

$$C_2 = 10^{-6}(1 + 10^5) = 0.1 \text{ F}$$

Substituting this, and the C_1 value in equation [6.15],

$$V_n = \frac{10^{-7}}{10^{-7} + 0.1}\left(V + \frac{0.1}{10^{-7}}V_{(n-1)}\right) \qquad [6.16]$$

$$= 10^{-6}(V + 10^6 V_{(n-1)}) \qquad [6.17]$$

This voltage is V', the amplifier input voltage; the final output voltage is found by multiplying equation [6.17] by 10^5.

$$V_{on} = 10^5 V_n = 0.1(V + 10^6 V_{(n-1)}) \qquad [6.18]$$

where V is the amplitude of the input pulses and for the first step $V_{(n-1)} = 0$.

$$V_{o1} = 10^5 V_1 = 0.1(5 + 0) = 0.5 \text{ V}$$

$$V_{o2} = 0.1\left(5 + 10^6 \times \frac{0.5}{10^5}\right) = 1.0 \text{ V}$$

$$V_{o3} = 0.1\left(5 + 10^6 \times \frac{1.0}{10^5}\right) = 1.5 \text{ V.}$$

Similarly, $V_{o4} = 2.0 \text{ V}$ and $V_{o5} = 2.5 \text{ V}$ etc. These steps now appear to be exactly equal. This results from the approximation in equation [6.16] when $10^{-7}/(10^{-7} + 0.1)$ is taken as 10^{-6}, an error of 0.0001%.

If repeated trains of steps are required, the capacitor C_F will need to be discharged at the end of each train. This can be accomplished by logic controlled FET circuits in parallel with C_F.

A logic controlled staircase generator

The system diagram shown in Fig. 6.25 includes the essential elements of this form of generator. The binary counter may be synchronous or asynchronous and it will have one of the circuits described on p.216. The number of flip-flops will determine the number of steps in the output waveform so that for n flip-flops there will be a 2^n steps. In the example shown, the 5 flip-flops will reproduce 32 steps. The duration of each step depends on the time between consecutive input pulses. These times would usually be equal but they could be varied if required by modification to the pulse train.

Fig. 6.25 A precision staircase waveform generator.

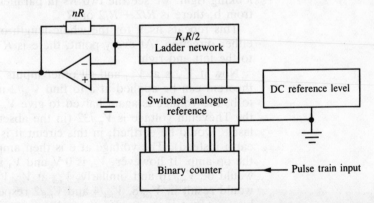

The counter output consists of both Q and $\bar{Q}$ for each flip-flop. The 0 and 1 output levels could be used directly but as there is a variation in flip-flop output level, precise levels can be provided by the switched analogue reference circuit shown in Fig. 6.26. This is shown for only two flip-flop outputs and additional circuits are provided for each flip-flop. The common reference levels are earth or 0 V and V_{ref} which can be provided by a zener diode. V_{ref} would be typically a few volts. Considering the circuit for V_O, either Q_O is 0 and $\bar{Q}_O$ is 1, or vice versa. If Q_O is 0, the 1 at $\bar{Q}_O$ turns T_{r2} on while Q_O holds T_{r1} off. Thus the output V_O is connected through the saturated T_{r2} to earth making $V_O = 0$ V (neglecting V_{CESAT}). When the condition is reversed, the 1 at Q_O turns T_{r1} on connecting V_O to V_{ref}. Thus the five outputs from the switched analogue reference will be at either 0 V or V_{ref} according to the state of count. Each Q at 0 produces 0 V output while each Q at 1 produces V_{ref} at the corresponding output.

Fig. 6.26 A switched analogue reference circuit.

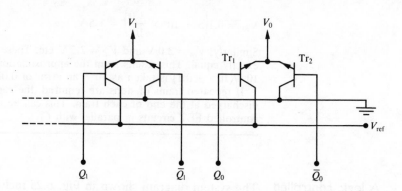

These analogue voltage signals are combined in the ladder network shown in Fig. 6.27. The accuracy of the final output depends on the precision of the ladder network. Only two values of resistance, R and $R/2$ are involved. The operation can be explained in terms of the superposition principle. The final output due to each single input can be calculated separately and then these results may all be added. Consider first the impedance levels. Starting at a and looking right, we see the two Rs in parallel giving $R/2$. Now looking from b, there is $R/2 + R/2$ or R.

This repeats itself for the whole length of the ladder looking in either direction. At every point, there is R down to the input and R to the left and right.

Now if V_O is at V_{ref} and all other inputs are at 0 V, Thévenin's theorem can be applied at a to find $V_{ref}/2$ in series with $R/2$. Moving to b, the voltage is again halved to give $V_{ref}/4$ and so on to e where the Thévenin voltage is $V_{ref}/32$ (in the absence of the op-amp, the last R would be earthed; in this circuit it is connected to the virtual earth instead). The voltage at e is then amplified by the gain n of the op-amp. If however, V_O is 0 V and V_1 is V_{ref}, the voltage at e would be $V_{ref}/16$ and similarly, V_{ref} at V_2, V_3 and V_4 individually would result in $V_{ref}/8$, $V_{ref}/4$ and $V_{ref}/2$ respectively. Each state in the binary count increases V_e, the voltage at e by $V_{ref}/32$ until after 31 input pulses, $V_e = 31/32V_{ref}$. The next pulse returns this to zero. Fig. 6.28 shows the final output if V_{ref} is 2 V and n, the op-amp gain is 5.

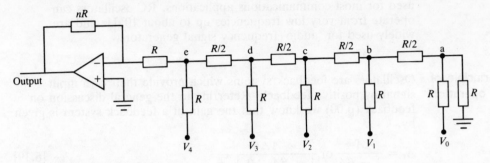

Fig. 6.27 An R, $R/2$ ladder network.

Fig. 6.28 Output waveforms for the precision staircase waveform generator.

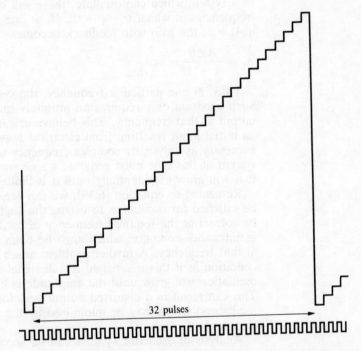

32 pulses

Generation of sine waves

Sinusoidal signals are required in many systems as test signals, carrier signals, tone generators, etc. The specific requirements range from a signal having fixed frequency with a wide tolerance in the actual frequency to circuits having variable frequency with a tuning range over a factor of 10^6. Oscillators at spot frequencies within the range 10^{-3} Hz to 10^{10} Hz are used but at the top end of this range, special devices and techniques must be used. For many applications, a tight specification on amplitude and distortion is also required.

Two main classes of sine wave generator or oscillator are used; these are LC oscillators and RC oscillators. LC oscillators are more convenient for higher-frequency ranges above 10^5 Hz and they are used for most communications applications. RC oscillators can operate from very low frequencies up to about 10^7 Hz and are widely used for 'audio' frequency signal generators.

The general principle of oscillator operation

Oscillators are feedback systems which provide their own input signal by positive feedback. Referring to the general discussion on feedback (p.00) we know that the gain of a feedback system is given by

$$A_f = \frac{A}{1 - \beta A} \text{ or } \frac{A\angle\theta}{1 - \beta A \angle\theta + \phi} \qquad [6.19]$$

In general, both $(\theta + \phi)$ and $|\beta A|$ change with a signal frequency. In a system which can oscillate, there will be one or more frequencies at which $\theta + \phi = 0°$. If, at one of these frequencies, $|\beta A| = 1$, the gain with feedback becomes

$$A_f = \frac{A\angle\theta}{1 - 1} = \infty$$

Thus, at one particular frequency, the system has infinite gain. Such a system only requires an infinitely small input to produce an output at that frequency. This behaviour can be justified in terms of an initial input resulting from electrical noise but this is not necessary as anlysis by complex frequency techniques shows that a system of this type must generate a sine wave with an amplitude that will grow exponentially until it is limited by system saturation.

Returning to equation [6.19], we can see that two conditions must be satisfied for oscillation to occur, the angle condition, which must be correct at the required frequency of oscillation, and the gain or maintenance condition which must be equal to or greater than one at that frequency. A further problem arises from the maintenance condition as if this is satisfied for all signal amplitudes, the oscillation will grow until the amplitude is limited by saturation. This can result in a distorted output waveform (particularly in RC oscillators) which may be minimised by the use of *negative* feedback circuitry.

Analysis of oscillator systems can be accomplished by a number of methods, the most appropriate depending on the type of circuit involved and on the information required from the analysis. For RC oscillators, the angle condition and the required gain are usually obtained by consideration of the feedback network in isolation: a suitable amplifier is then added to provide the necessary gain. With LC oscillators, it is usually best to write a set of general equations for the complete circuit and then, on the assumption that oscillation is occurring, the required frequency and maintenance conditions are extracted. Examples of both methods of analysis are given in the following sections.

A phase-shift RC oscillator

Referring to Fig. 6.29, the amplifier will in general have zero or 180° phase shift between input and output. If $\theta + \phi$ is to be 0°, then the feedback angle ϕ will also have to be 0° or 180° at the frequency of oscillation. In phase-shift oscillators, the network has an angle of 180° at the required frequency allowing the system to be completed with a simple inverting amplifier. Suitable phase-shift networks are shown in Fig. 6.30. In each case, three RC sections are used (since the maximum phase shift with two sections is only just 180° at a frequency where the network gain β is 0).

Fig. 6.29 The arrangement for a feedback sinusoidal oscillator.

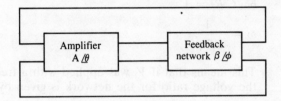

Fig. 6.30 Feedback networks for RC phase shift oscillators.

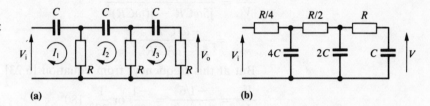

(a)   (b)

In the first circuit, equal C and R values are used but the final R may include the input impedance to the following amplifier. The input impedance to the network will also be high compared with the amplifier output impedance. The second network has staggered constants which may be used to minimise the required amplifier gain.

The first circuit can be analysed by mesh analysis as follows

$$V_i = I_1(R + 1/j\omega C) - I_2R$$
$$0 = -I_1R + I_2(2R + 1/j\omega C) - I_3R \qquad [6.20]$$
$$0 = -I_2R + I_3(2R + 1/j\omega C)$$

and $V = I_3R$ \qquad [6.21]

Solving equation [6.20] for I_3 by determinants

$$I_3 = \frac{\begin{vmatrix} (R + 1/j\omega C) & -R & V_i \\ -R & (2R + 1/j\omega C) & 0 \\ 0 & -R & 0 \end{vmatrix}}{\begin{vmatrix} (R + 1/j\omega C) & -R & 0 \\ -R & (2R + 1/j\omega C) & -R \\ 0 & -R & (2R + 1/j\omega C) \end{vmatrix}}$$

$$I_3 = \frac{(j\omega C)^3 R^2 V_i}{(j\omega CR)^3 + 6(j\omega CR)^2 + 5j\omega CR + 1}$$

substituting in equation [6.21] and writing $j^2 = -1$,

$$\frac{V_o}{V_i} = \frac{-j(\omega CR)^3}{1 - 6\omega^2 C^2 R^2 + 5j\omega CR - j(\omega CR)^3} \qquad [6.22]$$

Now remembering the required conditions for oscillation, the angle of the feedback expression must be either 0° or 180° (so that $\angle\beta A$ will be 0°). For equation [6.22], the angle of the numerator is −90°; the angle of the expression can only be 0° or 180° if the angle of the denominator is $\pm 90°$. For this to be satisfied, the real terms of equation [6.22] must be zero, i.e

$$1 - 6\omega^2 C^2 R^2 = 0$$

$$6\omega^2 C^2 R^2 = 1 \qquad [6.23]$$

$$\omega = \frac{1}{CR\sqrt{6}}$$

This means that if V_i was applied with a frequency of $\dfrac{1}{CR\sqrt{6}}$, then the voltage ratio for the network is given by

$$\frac{V_o}{V_i} = \frac{-j(\omega CR)^3}{j5\omega CR - j(\omega CR)^3}$$

$$= \frac{-\omega^2 C^2 R^2}{5 - \omega^2 C^2 R^2}$$

But at this frequency, from equation [6.23], $\omega^2 C^2 R^2 = 1/6$. Therefore

$$\frac{V_o}{V_i} = \frac{-1/6}{5 - 1/6} = \frac{-1}{29} \text{ or } \frac{1}{29} \angle 180°$$

Summarising the results of this analysis, for the network shown in Fig. 6.30, the phase shift will be 180° at a signal frequency of $\omega = 1/CR\sqrt{6}$, and at this frequency, the network gain $|\beta| = 1/29$.

A similar analysis, left for the reader, of the second circuit (Fig. 6.30b) shows that the same conditions occur at a frequency given by

$$\beta = \frac{2.06}{CR} \qquad [6.24]$$

and at this frequency, $|\beta| = 1/16$.

In both the circuits described above, the phase shift is 180° at the required frequency of oscillation. The amplifier required to complete the system must therefore be inverting to satisfy the phase condition. Fig. 6.31 shows an op-amp circuit using the second network to make a complete phase-shift oscillator.

Fig. 6.31 An op-amp *RC* phase shift oscillator.

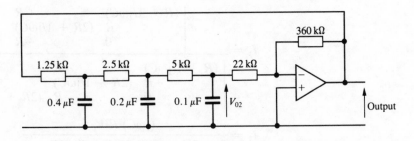

Equation [6.24] shows that the network provides the necessary phase condition at $\omega = 4120$ r/s or $f = 658$ Hz. The network will cause no appreciable loading of the amplifier as Z_{out} will be very low. The amplifier input impedance is 22 kΩ and this is in parallel with the final capacitor. This capacitive reactance at the oscillatory frequency is j2.4 kΩ and the effect of the 22 kΩ will be small causing only a slight shift in the oscillatory frequency. The amplifier gain is 360/22 or 16.4. This is slightly larger than the required 16 and this will ensure the start of oscillation when the circuit is switched on. The amplitude of oscillation will stabilise between the op-amp saturation levels with some distortion of waveform at V_{o1}. This being a low impedance point in the circuit, any external load changes will have little effect. V_{o2} however will be less distorted as the network will have greater attenuation to the higher frequency harmonics. Excessive loading at this point will either change the frequency or stop the oscillation.

Phase-shift oscillators provide useful fixed frequency circuits but if tuning of the frequency is required, the need to adjust three components simultaneously is inconvenient.

Wien bridge circuits   The Wien bridge oscillator is used in many forms, particularly in audio frequency signal generators. Tuning is more convenient than in the previous circuit and various techniques of gain stabilisation are employed to ensure an undistorted output waveform. As with the previous circuit, the starting point is to investigate the feedback network which may either be used by itself or as part of a bridge. These alternative arrangements are shown in Fig. 6.32.

Fig. 6.32 Wien bridge feedback networks: (a) Simple feedback network; (b) bridge network.

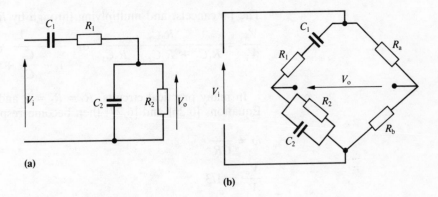

(a)

(b)

The circuit in Fig. 6.32 may be treated as a series parallel potential divider network as follows

$$V_o = V_i \times \cfrac{\cfrac{R_2/j\omega C_2}{R_2 + 1/j\omega C_2}}{R_1 + \cfrac{1}{j\omega C_1} + \cfrac{R_2/j\omega C_2}{R_2 + 1/j\omega C_2}}$$

Multiplying numerator and denominator by $(R_2 + 1/j\omega C_2)$

$$V_o = \cfrac{V_i \times R_2/j\omega C_2}{R_1 R_2 + \cfrac{R_1}{j\omega C_2} + \cfrac{R_2}{j\omega C_1} + \cfrac{1}{(j\omega)^2 C_1 C_2} + \cfrac{R_2}{j\omega C_2}}$$

Multiplying through by $(j\omega)^2$ and dividing by $R_1 R_2$

$$V_o = \cfrac{V_i \times j\omega/R_1 C_2}{(j\omega)^2 + j\omega\left(\cfrac{1}{R_2 C_2} + \cfrac{1}{R_1 C_1} + \cfrac{1}{R_1 C_2}\right) + \cfrac{1}{C_1 C_2 R_1 R_2}} \qquad [6.25]$$

As with the phase shift oscillator, the necessary angle condition for this expression is either 0° or 180°. Since the numerator is imaginary, the condition can only be satisfied if the denominator is also imaginary; this will occur at any frequency when the real parts of the denominator become zero. Therefore

$$(j\omega)^2 + \frac{1}{C_1 C_2 R_1 R_2} = 0$$

and since $j^2 = -1$,

$$\omega^2 = \frac{1}{C_1 C_2 R_1 R_2}$$

$$\omega = \frac{1}{\sqrt{C_1 C_2 R_1 R_2}} \qquad [6.26]$$

Also, from equation [6.25] the voltage ratio at this frequency is given by

$$\frac{V_o}{V_i} = \frac{j\omega/R_1 C_2}{j\omega \left(\dfrac{1}{R_1 C_1} + \dfrac{1}{R_2 C_2} + \dfrac{1}{R_1 C_2} \right)}$$

The $j\omega$ cancels, and multiplying through by $R_1 R_2 C_1 C_2$,

$$\frac{V_o}{V_i} = \frac{R_2 C_1}{R_1 C_1 + R_2 C_2 + R_2 C_1} \quad \text{or} \quad \frac{1}{1 + \dfrac{C_2}{C_1} + \dfrac{R_1}{R_2}} \qquad [6.27]$$

In many practical circuits, $R_1 = R_2 = R$ and $C_1 = C_2 = C$. Equations [6.26] and [6.27] then become respectively

$$\omega = \frac{1}{CR}$$

$$\frac{V_o}{V_i} = 1/3$$

The above results can all be applied to the bridge circuit shown in Fig. 6.32b. V_o in this case can only be in phase with V_i at the frequency given by equation [6.26]. The amplitude of V_o will now depend on the ratio R_a/R_b. In general

$$V_o = V_i \left(\frac{R_2 C_1}{R_1 C_1 + R_2 C_2 + R_2 C_1} - \frac{R_b}{R_a + R_b} \right) \qquad [6.28]$$

Practical Wien bridge oscillators

Equation [6.27] shows no phase change for V_o/V_i at the specified frequency. Thus the circuit will be completed by a non-inverting amplifier ($\theta + \phi = 0$) having a gain that is the reciprocal of the network attenuation. Taking the case of equal capacitors and resistors, the requisite gain is +3. In the op-amp circuit shown in Fig. 6.33 the necessary gain is provided by the negative feedback circuit $R_1 R_2$.

The gain V_o/V_i is $(1 + R_1/R_2)$ which will be 3 if $R_1 = 2R_2$. In practice, to allow for tolerance in the CR network, R_1 will be slightly larger than this making the gain greater than 3. This ensures oscillation but also results in a distorted output due to saturation.

Fig. 6.33 An op-amp Wien
bridge oscillator.

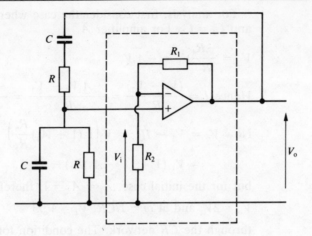

This may be overcome by the use of a temperature-sensitive resistor
(or a lamp) for R_2. Then, as the amplitude of the oscillation
increases, the dissipation and temperature of R_2 rise with consequent
reduction in the ratio R_1/R_2. This system will then stabilise with
exactly the required gain at an amplitude that is below saturation.

The tuning that is required for signal generators is usually
achieved by range switching the resistors and fine tuning over a 10:1
range with ganged capacitors. Small inaccuracies in the tracking can
result in a variation in the stabilised output amplitude (particularly
during tuning). An alternative approach is shown in the next circuit
in Fig. 6.34.

Fig. 6.34 A tunable Wien
bridge oscillator.

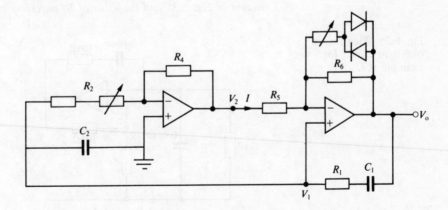

The objective in this circuit is to provide tuning by the adjustment
of a single component R_2. This resistor is not only part of the C_2R_2
parallel section of the postive feedback network but also controls
the forward gain V_2/V_1. The complete feedback network behaves
exactly like the original circuit in Fig. 6.32a as the non-inverting
input to A_2 is high impedance and the earth end of C_2 is connected
to R_2 through the virtual earth. R_6 provides the amplitude
stabilisation as for small signals, the diodes are both high
impedance. For large signals, the parallel R_7 reduces the gain A_2.

For analysis, first consider the case where $R_1 = R_2 = R_4$, $R_6 = R_7$ and $C_1 = C_2$. For amplifier A_1

$$V_2 = \frac{-R_4}{R_2}V_1 = -A_1V_1 \qquad [6.29]$$

Hence $I = \frac{(V_2 - V_1)}{R_5} = \frac{-A_1V_1 - V_1}{R_5} = \frac{-V_1(1 + A)}{R_5} \qquad [6.30]$

Now $V_o = V_1 - IR_6 = V_1\left(1 (1 + A_1)\frac{R_6}{R_5}\right)$

$$= V_1 (1 + (1 + A_1)A_2) \qquad [6.31]$$

but for the initial case, $A_1 = A_2 = 1$, therefore

$V_o = 3V_1$ and at $\omega = 1/CR$, $V_1 = V_o/3$

through the CR network. The condition for oscillation is thus satisfied. For the more general case, at $\omega = 1/\sqrt{C_1C_2R_1R_2}$ the feedback fraction β is $1/(1 + (C_2/C_1) + (R_1/R_2))$ or if $C_1 = C_2 = C$, at $\omega = 1/C\sqrt{R_1R_2}$, $\beta = 1/(2 + R_1/R_2)$. Thus, the necessary forward gain $= 2 + R_1/R_2$.

The actual forward gain from equations [6.29] to [6.31] is $(1 + A_2 + A_1A_2)$ where $A_2 = R_6/R_5$ and $A_1 = R_4/R_2$. so, if $R_5 = R_6$

Forward gain $= (2 + R_4/R_2)$

Now as R_2 is adjusted, the frequency of oscillation $\omega = 1/C\sqrt{R_1R_2}$ is changed and the maintenance condition is continually self correcting. A 100:1 range of R_2 will provide a decade or 10:1 tuning ratio.

Example 6.8   *Determine the frequency of oscillation for the Wien bridge oscillator circuit shown in Fig. 6.35 and the value of R_1 necessary to maintain oscillation.*

Fig. 6.35 Wien bridge oscillator circuit for example 6.7.

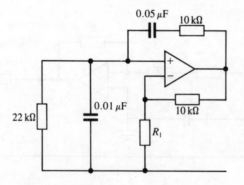

From equation [6.26],

$$\omega = \frac{1}{\sqrt{22 \times 10^3 \times 10^4 \times 10^{-8} \times 5 \times 10^{-8}}} = 3015 \text{ r/s}.$$

or $f = 480$ Hz.

The attenuation of the feedback network is given by equation [6.27]

$$\beta = \frac{1}{1 + \dfrac{0.01}{0.05} + \dfrac{10}{22}} = 0.604$$

The necessary gain is therefore $\dfrac{1}{0.604} = 1.65$.

The gain of the non-inverting amplifier is $\left(1 + \dfrac{10}{R_1}\right)$,

so $\dfrac{10}{R_1} = 0.65$

and $R_1 = 6.5 \text{ k}\Omega$.

LC oscillators

With RC oscillators, the necessary feedback phase relationship requires at least two RC network sections. The resulting loss in signal level is then made up by amplification. The phase relationship is only true at one frequency and unless several components are changed simultaneously, the change in signal loss must be compensated by a change in gain. If the gain is excessive, the oscillatory output will be distorted. With *LC* oscillators, only two components are required to provide the phase relationship at the chosen frequency. This will be the natural resonant frequency of the tuned *LC* circuit and in the ideal case, (no losses) the circuit could sustain oscillation without electronic circuitry. Tuning of the oscillatory frequency is accomplished by a change of one component only and excess gain is less of a problem as the network itself filters out the harmonics resulting from saturation distortion.

There are many different forms of *LC* oscillator and a variety of different approaches to their analysis are possible. In the following sections, the methods that are illustrated include: analysis from feedback theory, solution of the general circuit equations and the application of negative resistance. *LC* oscillators are usually used at higher frequencies than *RC* types and the amplifying element is often a single transistor which will have a better high-frequency performance than an op-amp.

The tuned collector oscillator

In the circuit shown in Fig. 6.36, R_1, R_2 and R_E provide the normal biasing and since these are decoupled by C_B and C_E they need not be considered in the analysis. The tuned collector load will have a maximum gain at the resonant frequency ($f_o = 1/2\pi\sqrt{1/LC - 4R^2/L^2}$) when the collector load is resistive with $Z_o = L/CR$.

Thus, at f_o $V_c = V_b \times \dfrac{-g_m L_c}{CR}$

and the current in the inductor L_c lags the voltage by nearly 90° since R is very much less than X_L. Therefore

$$I_C = \dfrac{-g_m L_c V_b}{CR \times j\omega L_c} = \dfrac{jg_m V_b}{\omega CR}$$

The voltage induced in inductor L_b is $\pm j\omega M I_c$; therefore

$$V_b = \pm j\omega M \times \dfrac{jg_m}{\omega CR} = \pm \dfrac{M g_m V_b}{CR}$$

Thus, provided the sign of M is correct and $M g_m / CR$ is greater than one, the circuit provides its own input signal and will oscillate

Fig. 6.36 A tuned collector oscillator.

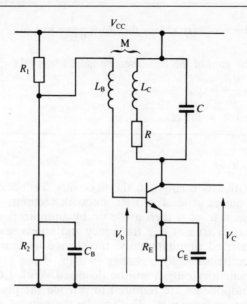

at the resonant frequency of the tuned collector load. This analysis ignores the fact that the base circuit impedance will be reflected into the collector circuit effectively increasing R. This in turn reduces the phase lag between V_c and I_c. These factors are small and will only result in a slight difference in the frequency of oscillation.

Direct feedback *LC* oscillators

A widely used family of oscillators employ direct feedback from a resonant load to the input. The principle is illustrated in Fig. 6.37.

The parallel *LC* circuit will have maximum impedance at the resonant frequency f_o. Since this impedance, L/CR is resistive, the gain of the amplifier will be a maximum (with load R_L in parallel with L/CR) and V_o will be 180° out of phase with V_{be}. A fraction of the voltage across the tuned circuit can be tapped off using a potentiometer as shown. This voltage, V_f is out of phase with V_{be} and therefore inphase with V_{eb}. Thus, if it is applied to the input terminals as shown, the correct phase relationship for oscillation is provided. In practical circuits the voltage tapping is provided, not from a resistive potentiometer but from a split coil (Hartley oscillator) or from a split capacitor (Colpitt's and Clapp oscillators)

Fig. 6.37 Circuit demonstrating the principle of direct feedback oscillators.

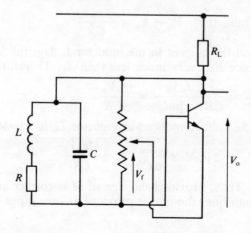

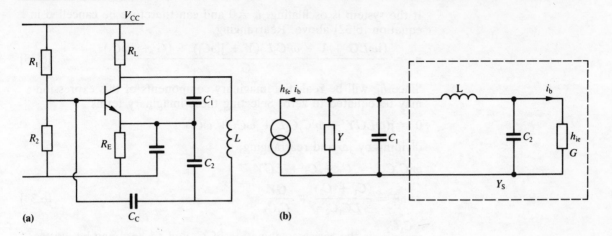

Fig. 6.38 A Colpitt's
oscillator; (a) Full circuit,
(b) small signal equivalent
circuit.

A Colpitt's oscillator

A common form of oscillator, the Colpitt's oscillator, uses an LC
network as the direct feedback network. A circuit is shown in Fig.
6.38a and the small signal equivalent is given in Fig. 6.38b

R_1, R_2 and R_E provide normal biasing and the collector load R_L
includes and external loading effects. C_C provides a.c coupling
between output and input and will have negligible reactance at the
oscillatory frequency. R_1 and R_2 are not shown in the equivalent
circuit but their effect can be included in an effective h_{ie} and an
effective h_{fe} (both reduced compared with the internal parameters of
the transistor). The load is shown as an admittance Y and h_{ie} as a
conductance G as this simplifies the form of the circuit equations.
The coil resistance has also been neglected; this will have little
effect upon the resonant frequency and will only modify the
maintenance condition if R_L is large. The best approach to analysis
is to find an expression for i_b in terms of $h_{fe}i_b$; then using the
condition that h_{fe} is real, the oscillatory frequency and maintenance
condition can be obtained. The small signal equivalent circuit can be
analysed by nodal analysis or simply by current splitting in series
and parallel admittances.

An expression for the series parallel combination shown as Y_S can
be written as

$$Y_S = \frac{(G + j\omega C_2)1/j\omega L}{G + j\omega C_2 + 1/j\omega L} = \frac{G + j\omega C_2}{j\omega LG - \omega^2 C_2 L + 1}$$

Now, by current division,

$$i_b = \frac{-h_{fe}i_b \times \dfrac{G + j\omega C_2}{j\omega LG - \omega^2 C_2 L + 1}}{(Y + j\omega C_1) + \dfrac{(G + j\omega C_2)}{j\omega LG - \omega^2 C_2 L + 1}} \times \frac{G}{G + j\omega C_2}$$

$$= -h_{fe}i_b \frac{G}{(j\omega LG + 1 - \omega^2 C_2 L)(Y + j\omega C_1) + (G + j\omega C_2)} \qquad [6.32]$$

If the system is oscillating, $i_b \neq 0$ and can therefore be cancelled in equation [6.32] above. Rearranging,

$$h_{fe} = \frac{(j\omega LG + 1 - \omega^2 C_2 L)(Y + j\omega C_1) + (G + j\omega C_2)}{G} \qquad [6.33]$$

Since h_{fe} will be real, the imaginary components of this expression may be equated to zero. Selecting these imaginary terms

$$0 = j(\omega LGY - \omega^3 C_1 C_2 L + \omega C_2 + \omega C_1)$$

dividing by $j\omega$ and rearranging,

$$\omega^2 C_1 C_2 L = C_1 + C_2 + LGY$$

$$\therefore \omega^2 = \frac{(C_1 + C_2)}{LC_1 C_2} + \frac{GY}{C_1 C_2} \qquad [6.34]$$

$\dfrac{C_1 C_2}{C_1 + C_2}$ is the series combination of C_1 and C_2 and may be written C_s. The term $GY/C_1 C_2$ indicates the way in which the frequency is 'pulled' by excessive loading. This term would be negligible for typical circuit values.

Refer to equation [6.33] to find the maintenance condition by equating the real terms

$$-h_{fe} = \frac{Y(1 - \omega^2 LC_2) - \omega^2 C_1 LG + G}{G}$$

$$h_{fe} = \omega^2 LC_1 - 1 + \frac{Y}{G}(\omega^2 LC - 1)$$

but from equation [6.34], $\omega^2 \simeq \dfrac{(C_1 + C_2)}{LC_1 C_2}$, and therefore,

$$h_{fe} = \left(\frac{C_1 + C_2}{C_2} - 1\right) + \frac{Y}{G}\left(\frac{C_1 + C_2}{C_1} - 1\right)$$

simplifying gives $h_{fe} = \dfrac{C_1}{C_2} + \left(\dfrac{Y}{G} \times \dfrac{C_2}{C_1}\right)$

and rewriting in impedance values,

$$h_{fe} = \frac{C_1}{C_2} + \frac{h_{ie}}{R_L}\left(\frac{C_2}{C_1}\right)$$

$$\frac{h_{fe}}{h_{ie}} = g_m = \frac{C_1}{h_{ie}C_2} + \frac{C_2}{R_L C_1}$$

from which the necessary d.c. conditions for the transistor may be established.

Where frequency stability is of the first importance, the coil may be replaced either by a series combination of L and C in the Clapp oscillator or by a piezo-electric crystal in the Pierce oscillator. Circuits for these are shown in Fig. 6.39. In each case, the resonant frequency will be determined by the series components around the resonant loop. If C_3 is small compared with C_1 and C_2, then $\omega^2 \simeq \dfrac{1}{LC_3}$ for the Clapp circuit and it will have the series resonant frequency of the crystal in the Pierce circuit.

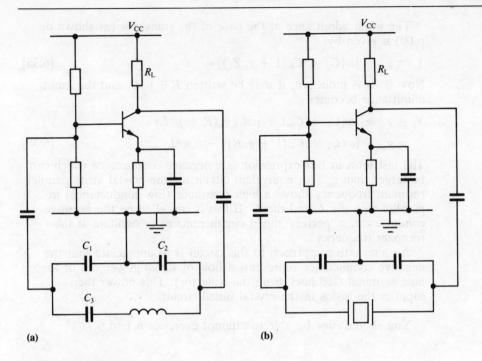

Fig. 6.39 (a) Clapp
oscillator circuit, (b) Pierce
oscillator circuit.

**Negative resistance
oscillators**

Another form of oscillator uses the capacitance (Miller) feedback
with a common emitter amplifier to produce a *negative* resistance
between base and earth. This is then used to cancel the losses in a
parallel tuned circuit (usually a crystal) which can then oscillate
freely at its *parallel* resonant frequency. A circuit showing the
arrangement is given in Fig. 6.40.

Fig. 6.40 A BJT negative
resistance oscillator.

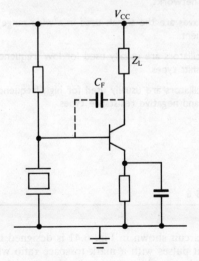

The input admittance at the base of the transistor (as shown on p.00) is given by

$$Y_i = g_{ie} + j\omega(C_{ie} + C_{bc}(1 + g_m Z_L)) \qquad [6.35]$$

Now if Z_L is inductive, it may be written $R + j\omega L$, and the input admittance becomes

$$Y_i = g_{ie} + j\omega(C_{ie} + C_{bc}) + j\omega C_{bc}g_m(R + j\omega L)$$
$$= g_{ie} + j\omega(C_{ie} + C_{bc}(1 + g_m R)) - g_m\omega^2 L C_{bc} \qquad [6.36]$$

The last term in this expression is a *negative conductance* which can be larger than g_{ie}. An equivalent circuit for the crystal at its parallel resonant frequency shows a high resistance (low conductance) in parallel with the L and the C. If this is cancelled by the negative conductance, a 'perfect' tuned circuit remains to oscillate at the resonant frequency.

An alternative approach to this circuit is to appreciate that the negative conductance represents a flow of signal power out of the base terminal (fed back from the collector). This power then supplies the losses in the crystal tuned circuit.

You should now be able to attempt exercises 6.1 to 6.7.

Points to remember

- Rectangular waves are used to provide 'clock' signals and coded control signals in digital circuits and may be generated from sine waves or directly using multivibrator types of circuit.

- Triangular waves are used in timing applications and may be generated directly using bootstrap methods or by integration of a rectangular wave.

- Staircase waves are used in instrumentation and analogue to digital conversion. They can be generated using a diode pump or by counter, analogue switch and ladder networks.

- Sine waves are the signals used in a wide range of communication and test equipment.

- RC oscillators are usually used for low frequencies and include Wien bridge and phase shift types.

- LC oscillators are usually used for high frequencies and inlcude Colpitt's, Hartley, Clapp and negative resistance types.

EXERCISES 6

6.1 The circuit shown in Fig. 6.41 is designed to produce positive going output pulses with a mark to space ratio which is adjustable with C_1.
(a) Calculate the output pulse duration, the frequency and the mark to space ratio if C_1 is 0.047 μF.

Fig. 6.41 Multivibrator circuit for exercise 6.1.

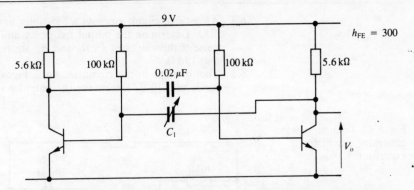

(b) Estimate the maximum value of C_1 if excessive distortion of the output pulse is to be avoided. (1.4 ms, 0.43, 213 Hz, 0.1 μF)

6.2 In the astable circuit shown in Fig. 6.42, R_2 is adjustable to modify the frequency of the output waveform.
(a) What is the frequency range if R_2 is adjusted between 4.7 Ω and 470 Ω.
(b) What value of R_2 is required for an output frequency of 1 kHz.
  (2.53 kHz, 97 Hz, 12.8 Ω)

Fig. 6.42 Op-amp astable circuit for exercise 6.2.

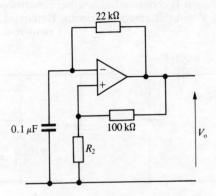

6.3 The circuit shown in Fig. 6.43 is required to produce a rectangular output waveform holding a negative level of -2 V for 2 ms and a positive level of $+10$ V for 8 ms. Determine the necessary values for the zener voltages and the resistors R_1 and R_2
  (1.3, 9.3, 15.2 kΩ, 126 kΩ)

Fig. 6.43 Rectangular wave generator circuit for exercise 6.3.

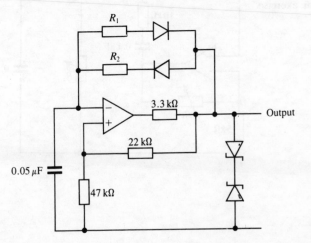

6.4   An astable circuit employs a 555 timer in the circuit shown in Fig. 6.10. Determine the output frequency and the duration of the positive pulse if the values of C, R_A and R_B are respectively 0.02 μF, 10 kΩ and 15 kΩ.   (1.8 kHz, 0.21 ms)

6.5   The op-amps in the circuit shown in Fig. 6.44 have saturated output levels of ± 10 V. Determine the output waveform V_o indicated.

(triangular, ± 2.5 V at 500 Hz)

Fig. 6.44 Triangular wave generator circuit for exercise 6.5.

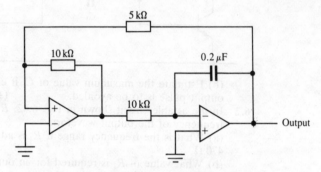

6.6   A train of 5 V pulses is applied to the circuit shown in Fig. 6.45. The output is connected to a voltage controlled switch set to operated at 1 V and this resets the circuit. Determine the output waveform.

(11 positive steps of approximately 0.095 V)

Fig. 6.45 Diode pump circuit for exercise 6.6.

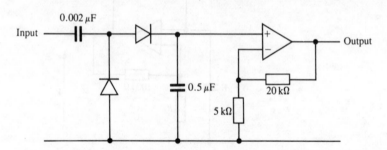

6.7   The circuit shown in Fig. 6.46 is required to oscillate in the range 100 Hz to 1 kHz. Determine the necessary range for the two variable resistors shown.   (25 kΩ to 250 Ω, 7 kΩ to 203 kΩ)

Fig. 6.46 Wien bridge oscillator circuit for exercise 6.7.

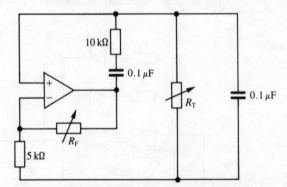

7 Analogue and Digital Conversion

The principal learning objectives of this chapter are to:

	Pages	Exercises
● Investigate techniques for conversion of digital signals into analogue form	562–77	7.1–7.8, 7.10–7.12
● Consider the use of transistor voltage, current switches and current sources	570–5	7.9, 7.13
● Consider the use of sample and hold circuits	577–80	7.14–7.16
● Investigate techniques for conversion of analogue signals into digitial form	580–90	7.17–7.27

There are situations where the world of analogue electronics meets the world of digital electronics and an interface between the two is necessary. For example, in a computer-based system the processing unit may need to use sampling signals from transducers which produce analogue signals. Transducers can monitor temperature, fluid flow, pressure, angular movement, etc., and produce voltages, or currents, that are a measure of the state of the element being monitored, i.e. value of temperature, rate of flow, etc. Such signals may be continuously varying as is the case for the analogue type signals and there is a need to convert these signals into digital form so that the processor can operate on the digital data. The process of converting an analogue signal into digital form involves a sequence which includes sampling, holding, quantising and encoding. The process of sample and hold is usually done in one circuit while quantisation and encoding is done together in another circuit. The latter process forms what is known as the **analogue to digital converter** (A/D Converter or simply ADC). If information is required to be returned to the analogue world after processing then the digital signal must be reconstituted to analogue form by a **digital to analogue converter** (D/A Converter or simply DAC). Circuits which achieve the sampling function and the conversion will be discussed in this chapter. Digital to analogue conversion will be dealt with first because its circuitry often features in the analogue to digital circuits.

Digital to analogue conversion

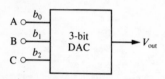

A o—— b_0

B o—— b_1 3-bit DAC   ——► V_{out}

C o—— b_2

Fig. 7.1 Block diagram of a three-bit DAC.

The DAC accepts data in digital form and converts it to a voltage or current which is proportional to the digital value. A basic block diagram of a three-bit DAC is shown in Fig. 7.1 where it is assumed that the output is a voltage level although current flow could just as easily be used.

With just three bits for the digital signal it follows that there can be only $8(2^3)$ states that can be represented by each combination of levels on the digital lines although the use of a fourth bit (the sign bit) can increase the level to 16 states. It is usual to classify the bits from the least significant bit (LSB) to the most signficant bit (MSB). In this case bit b_0 represents the LSB and bit b_2 the MSB. The transfer characteristic is shown in Fig. 7.2.

Fig. 7.2 Input/output characteristic for a three-bit DAC.

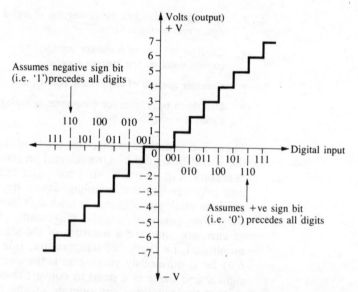

The choice of only three bits for the digital data means that the **resolution** of the DAC is poor. For example, if digital data 0101 produces a voltage of +5 V while the data 0110 produces +6 V the accuracy is only to $\pm\frac{1}{2}$ V. Thus the resolution of the DAC is limited to this value which is poor. This is often quoted as an **accuracy** of $\pm\frac{1}{2}$LSB, where LSB is in this case 1 V on the analogue output. Resolution can be improved by increasing the number of bits used to define a quantity and this is done in practice using up to 16 bits. A DAC with 16 bits can have $2^{16} = 65,536$ different levels of output voltage so that the change between levels is 1/65,536 of the full-scale output range. Using the range of Fig. 7.2 which gave a full-scale output voltage of ±7 V and converting it to use 16 bits gives a voltage difference between levels of 7/65,536 V or 0.107 mV instead of the 1 V as in the diagram. The accuracy of the DAC is a measure of the difference between the actual analogue output and the value it would be under ideal conditions. There are many factors contributing to accuracy including the linearity of the DAC (i.e.

equal increments in analogue output for equal changes in digital input value) which in turn depends on the precision of such components as resistors in the DAC circuit. Other factors affecting accuracy are tolerances on reference voltages, amplifier gains, etc., and the ability of the circuit to resist noise.

Example 7.1   *A three-bit DAC has a voltage output which, for an input of 0100, is +5 V. What will the output voltage level be for a digital input of 0011?*

Since 0100 has a decimal 'weighting' of 4 then V_{out} of +5 V is proportional to 4 and for each increment in the decimal value the voltage is given by 5/4 V or 1.25 V. Hence a digital input of 0011 has a decimal weighting of 3 and V_{out} for an input of 0011 is $3 \times 1.25 = 3.75$ V. This could be confirmed on the transfer characteristic of Fig. 7.2 by changing the voltage levels to increments of 1.25 V for each stage and not 1.0 V as drawn.

Binary weighted resistor DAC

Consider the three-bit circuit of Fig. 7.3 which consists essentially of a voltage reference source, binary weighted resistors and associated switches, together with a summing element which in this case is a resistor.

Fig. 7.3 DAC using binary weighted resistors.

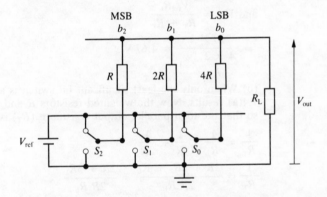

The position of the switches means that either a current flows to load via the binary weighted resistor because of V_{ref} or no current flows because the resistor is grounded. The resistors are weighted so that the current contribution of each switch in the logic 1 position (i.e. connected to V_{ref}) is different. The greatest contribution comes from the most significant bit (MSB) of value $\dfrac{V_{ref}}{R}$ while the least significant bit (LSB) contributes the smallest current $\dfrac{V_{ref}}{4R}$.

Example 7.2   *A three-bit binary weighted resistor DAC has $R = 2\ k\Omega$ with a reference voltage of 10 volts. What would the output voltage be (a) if all switches were set to logic 1? (b) if only the least significant bit switch is at logic 1? Assume $R_L = 1\ k\Omega$.*

(a) When *all* switches are at logic 1 the circuit of Fig. 7.4(b) is the result. The total resistance provided by the binary weighted resistors is the parallel combination of R, $2R$ and $4R$, i.e. R_T is given by

Fig. 7.4 Circuit for the calculation of V_{out} in a three-bit DAC: (a) when the LSB is 1, (b) when all bits are 1.

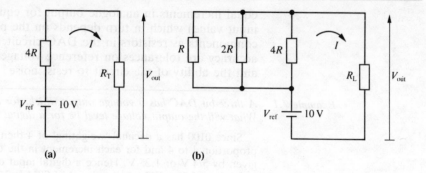

$$\frac{1}{R_T} = \frac{1}{R} + \frac{1}{2R} + \frac{1}{4R}$$

$$= \frac{4 + 2 + 1}{4R} = \frac{7}{4R}$$

hence $R_T = \frac{4R}{7} \ \Omega$

and $I = \dfrac{V_{ref}}{R_T + R_L}$

and $V_{out} = \dfrac{V_{ref}R_L}{R_T + R_L}$

$$= \frac{10 \times 1}{\dfrac{4 \times 2}{7} + 1} \approx 4.67 \text{ V}$$

(b) When only the least significant bit switch is at logic 1 the circuit of Fig. 7.4(a) results. Now the weighted resistors R and $2R$ are in parallel with R_L so that the total parallel output resistance (R_T) is given by

$$\frac{1}{R_T} = \frac{1}{R_2} + \frac{1}{R} + \frac{1}{2R}$$

$$\frac{1}{R_T} = \frac{2R + 2R_2 + R_L}{2R.R_L} = \frac{2R + 3R_L}{2R.R_L}$$

hence $R_T = \dfrac{2R.R_L}{2R + 3R_L}$

and $I = \dfrac{V_{ref}}{4R + R_T}$

and $V_{out} = \dfrac{V_{ref}.R_T}{4R + R_T}$

$$= \frac{V_{ref}}{4\dfrac{R}{R_T} + 1}$$

$$= \frac{10}{4 \times \dfrac{2}{4/7} + 1}$$

$$\left(\text{since } R_T = \frac{2R.R_L}{2R + 3R_L} = \frac{2 \times 2 \times 1}{(2 \times 2) + (3 \times 1)} = \frac{4}{7} \text{ k}\Omega \right)$$

hence $V_{out} = \dfrac{10}{14 + 1} = 0.67 \text{ V}$

The relative values for the resistors used in example 7.2 are unimportant. It should be clear from this example that the output voltage could be increased by simply increasing the value of R_L. The value of output switch positions and relative magnitude of V_{out} to V_{ref} is not a factor worth pursuing.

The resolution of the three-bit DAC of Fig. 7.3 is poor giving wide voltage output variations according to switch positions. The resolution could be improved by adding to the number of bits, with an extra resister and switch for each bit, and giving the correct weighting for each new resistance added. For an n-bit DAC using this technique the least significant bit (b_0) uses switch S_0 and has the highest value of resistance ($2^{n-1}.R$), the next bit (b_1) uses switch S_1 and has the next highest value of resistance ($2^{n-2}.R$) and so on until the most significant bit (b_{n-1}) is reached, using switch S_{n-1} with the lowest resistance of $2^{n-n}.R$ or simply R.

Imagine for a moment that each of the switch positions for S_0 through to S_{n-1}, when closed, allows current to flow via a short-circuited load, i.e. $R_L = 0$. This assumption makes it easier to calculate the total short-circuit current and once this is known it is a relatively simple matter to allow for a non-zero load. Let the short-circuit current be I_{sc} then

$$I_{sc} = V_{ref}\left(\frac{S_{n-1}}{R} + \frac{S_{n-2}}{2^{n-(n-1)}R} + \frac{S_{n-3}}{2^{n-(n-2)}R} + \cdots \frac{S_0}{2^{n-(n-(n-1))}R}\right)$$

$$= \frac{V_{ref}}{R}(S_{n-1}2^0 + S_{n-2}2^{-1} + S_{n-3}2^{-2} + \cdots S_0 2^{-(n-1)}) \qquad [7.1]$$

where S_k is the logic state taken by the switch and can be 1 or 0, and suffix k can have any value between $(n-1)$ and 0.

Since all resistors are effectively in parallel the total effective parallel resistance of the circuit is given by R_T where

$$R_T = \frac{2^{n-1}R}{2^n - 1} \qquad [7.2]$$

Thus the total short-cirucit current I_{sc} can be represented as a current generator of internal resistance R_T, i.e. a Norton equivalent circuit as shown in Fig. 7.5.

Fig. 7.5 Norton equivalent circuit. Used to give a value for I_{sc} and hence allow V_{out} to be found.

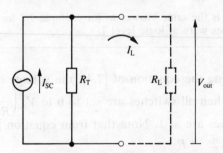

From Fig. 7.5 if the load R_L is short-circuited then the current through the short-circuited load is I_{sc} as established from equation [7.1]. If R_L is *not* zero then the amount of current through R_L(I_L) is a fraction of I_{sc} given by

$$I_L = I_{sc}\left(\frac{R_T}{R_L + R_T}\right) \qquad [7.3]$$

and since V_{out} is given by $I_L R_L$ then

$$V_{out} = I_{sc}\left(\frac{R_T}{R_T + R_L}\right)R_L$$

$$= I_{sc}\left(\frac{1}{1 + R_L/R_T}\right)R_L$$

$$= I_{SC}\left(\frac{R_L}{1 + \dfrac{R_L(2^n - 1)}{2^{n-1}R}}\right)$$

and substituting from equation [7.1] for I_{sc}

$$V_{out} = \frac{V_{ref}}{R}\left(\frac{R_L}{1 + \dfrac{R_L(2^n - 1)}{2^{n-1}R}}\right)(S_{n-1}2^0 + S_{n-2}2^{-1} + S_{n-3}2^{-2} + \ldots + S_0 2^{-(n-1)}) \qquad [7.4]$$

Equation [7.4] looks complicated but simplifies when actual values are used as the following example shows.

Example 7.3   *Using the formula given by equation [7.4], find V_{out} when $n = 3$, $V_{ref} = 10$ V, $R_L = 1$ kΩ and $R = 2$ kΩ if all switches are at logic 1.*

$$V_{out} = \frac{10}{2}\left(\frac{1}{1 + \left(\dfrac{1 \times 7}{4 \times 2}\right)}\right)(1 \times 2^0 + 1 \times 2^{-1} + 1 \times 2^{-2})$$

In this expression S_0, S_1 and S_2 are all 1. Hence

$$V_{out} = 5\left(\frac{1}{1 + \frac{7}{8}}\right)(1 + 0.5 + 0.25)$$

$$= 5\left(\frac{1}{15/8}\right)(1.75)$$

$$= 5\left(\frac{8}{15}\right)(1.75) = 4.67 \text{ V}$$

which is the same answer as for example 7.2 for that condition when all switches were at logic level 1.

Using the equation of [7.4], the voltage output can range from 0 V when all switches are set to 0 to $V_{ref}\left(\dfrac{R_L}{R_T + R_L}\right)$ when all switches are at 1. Note that from equation [7.2],

$$R_T = \frac{2^{n-1}.R}{2^n - 1}$$

In examples 7.2 and 7.3 the values chosen for the switch 0 and 1 lines give a positive voltage out for all conditions except $S_k = 0$. The output voltage levels could equally have been all negative, except for the conditions $S_k = 0$, by simply choosing V_{ref} to be a negative value, say -10 V. It follows then that by choosing $V_{ref(1)} = +5$ V, say, and $V_{ref(0)} = -5$ V the range of voltages can swing from negative through to positive giving a graph of V_{out} against the digital states similar to that of Fig. 7.2.

In practice the output current does not feed a load resistance directly but is buffered by an amplifier circuit. Typically an operational amplifier could be used with the circuit arrangement of Fig. 7.6.

Fig. 7.6 Use of an operational amplifier stage as an output buffer in a three-bit DAC circuit.

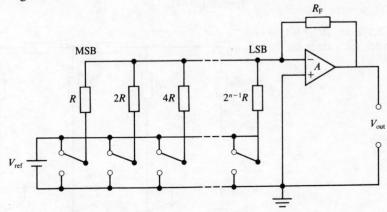

The use of an operational amplifier with a feedback resistor R_f gives a value of V_{out} of

$$V_{out} = \frac{-V_{ref}}{R}(S_{n-1}2^0 + S_{n-2}2^{-1} + S_{n-2}2^{-2} + \ldots S_0 2^{-(n-1)}) \qquad [7.5]$$

This equation assumes an infinite gain for the amplifier. The arrangement gives a negative value for V_{out} because of the inverting effect of the operational amplifier.

Example 7.4   *A three-bit DAC with resistance values $R = 2$ kΩ and reference voltage of +10 V is to be used with an operational amplifier of infinite gain and feedback resistance 6 kΩ. What is the output voltage if the binary input is 101?*

From equation [7.5]

$$V_{out} = \frac{-10 \times 6}{2}(1 + 0 + 0.25) \text{ V}$$

$$= \frac{-60}{2} \times 1.25$$

$$= -37.5 \text{ V}$$

A major disadvantage of the weighted resistor DAC is concerned with the wide range of resistor values necessary for a reasonable size converter. If, for example, ten bits are to be accommodated and the basic value for R is 2 kΩ then the resistance value for the least significant bit is 2 k$\Omega \times 2^9$ or 1.024 MΩ. It is difficult with such a range of resistance values to obtain components which have the necessary precision. Also with high values of resistance it is difficult to fabricate these components using integrated circuit techniques. There is a DAC which overcomes these problems and this will be discussed in the next section.

The R − 2R ladder DAC

In the arrangement of Fig. 7.7 there are only two values of resistance used although each bit position now requires two resistors instead of the single resistor per bit position of the binary weighted converter. A three-bit DAC is used in this circuit for convenience and for ready comparison with the previous type of converter. However, it can be expanded to *n*-bits as will be shown later. Once again a voltage reference is used and switch positions correspond to levels 1 or 0.

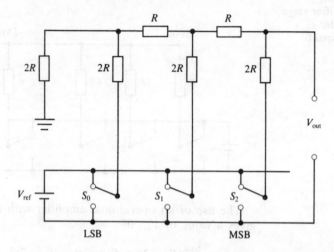

Fig. 7.7 Basic *R*/2*R* ladder network for a three-bit DAC.

The circuit may be analysed by the use of Thévenin's theorem. Assume that $S_0 = 1$, $S_1 = S_2 = 0$. The $R - 2R$ network to the right of the LSB position is then effectively the load for the circuit which can be redrawn as in Fig. 7.8(a).

The voltage across the open-circuit terminals of the circuit with the load removed is thus $V_{ref}/2$ V while the Thévenin series resistance is found by replacing the reference voltage generator by a short-circuit (this assumes zero internal impedance for the generator) and finding the total resistance when looking back from the open-circuit terminals with the load still removed. The value of the series resistance is thus R (i.e. $2R$ in parallel with $2R$) and the Thévenin equivalent circuit is shown in Fig. 7.8(b). Note that if this technique is applied at *any* bit position the Thévenin equivalent circuit is always the same as Fig. 7.8(b) since the total resistance to the left of any bit position is *always* $2R$ if the appropriate switch S is earthed. Suppose S_1 is being considered and S_0 is earthed; then to the left of bit position 1, $2R$ is in parallel with $2R$ giving a value of R; this is in series with R giving $2R$. This will always be the case regardless of the number of bit positions as long as the relevant switches are earthed. To find the contribution of each bit position to the output voltage it is necessary to examine the complete circuit i.e. replacing the 'load' and replacing the bit position at level 1 with the Thévenin equivalent circuit. This is done in Fig. 7.7(a), (b) and (c) for the LSB to MSB respectively for the three-bit DAC of Fig. 7.7.

It can be seen from Fig. 7.9 that each switch contributes its correct weighted value. Total output voltage for the circuit of Fig. 7.7 is thus

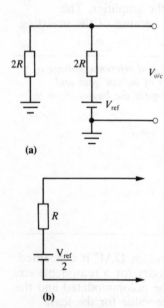

(a)

(b)

Fig. 7.8 Thévenin equivalent circuits for calculating effective reference voltage for the LSB position: (a) with only the LSB switch set to 1, (b) minimised circuit form.

$$V_{out} = V_{ref}(S_0 2^{-3} + S_1 2^{-2} + S_2 2^{-1}) \text{ V} \qquad [7.6]$$

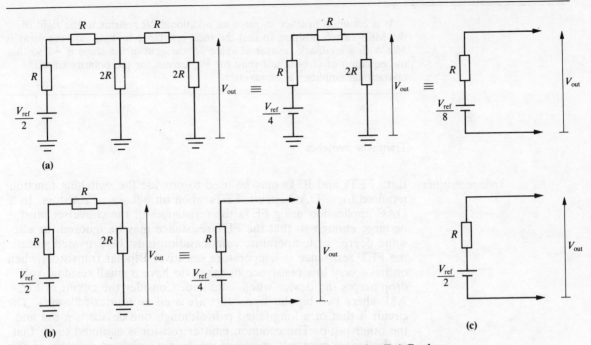

(a)

(b)

(c)

Fig. 7.9 Effective Thévenin circuits for each bit position of the three-bit $R/2R$ DAC: (a) MSB, (b) next most significant bit, (c) LSB.

In general if there are n bits in the DAC, then:

$$V_{out} = V_{ref}(S_0 2^{-n} + S_1 2^{-(n-1)} + \ldots S_{n-1} 2^{-1})$$

and once again if the load resistance R_L is to be taken into account then

$$V_{out} = V_{ref}\left(\frac{R_L}{R+R_L}\right)(S_0 2^{-n} + S_1 2^{-(n-1)} + \ldots S_{n-1} 2^{-1})$$

Example 7.5   *For the three-bit $R-2R$ ladder network of Fig. 7.7 calculate V_{out} for the switch value $S_0 = 1$, $S_2 = 1$ and $S_1 = 0$. Assume $V_{ref} = 10\,V$.*

From equation [7.6]

$$V_{out} = 10(0.125 + 0 + 0.5)\,V$$

$$= 10 \times 0.625\,V$$

$$= 6.25\,V$$

An operational amplifier could also be used with the $R-2R$ ladder network as the circuit of Fig. 7.10 shows.

Fig. 7.10 Three-bit $R/2R$ ladder network with output operational amplifier.

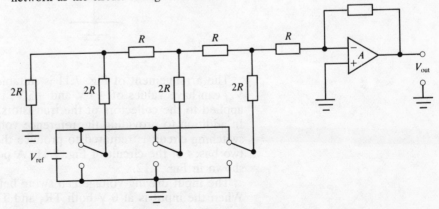

It is common practice to place an additional $2R$ resistor to the right of the MSB switch position so that the total resistance looking into the array is $3R$. With a feedback resistor of value $3R$ the gain of the stage is -1 so that the equation of [7.6] is still valid for V_{out} except for the polarity since the operational amplifier is an inverter.

Transistor switches

Voltage switches   Both FETs and BJTs may be used to provide the switching function required for a DAC circuit. FETs when on behave as resistors. In a DAC application using FETs the resistances of the converter must be large enough so that the FET resistance may be ignored, or else some degree of temperature compensation must be provided since the FET resistance is temperature sensitive. Bipolar transistors when on have very low resistance but they do have a small residual volt drop across the device when saturated. Consider the circuit of Fig. 7.11 where two bipolar transistors are used as emitter-followers. The circuit is that of a long-tailed pair although one device is n-p-n and the other p-n-p. The common emitter resistor is assumed to be that of the ladder network or one of the binary weighted resistors of the DAC circuit. The collector voltages are $+5$ V or -5 V, depending in the device, and the base voltage can be either $+5.7$ V or -5.7 V. Considering the base of TR_1 and TR_2 at $+5.7$ V then TR_1 is on with $V_E \approx +5$ V and TR_2 is off. When TR_2 conducts the voltage at the emitter is -5 V with both bases at -5.7 V. Thus TR_1 is held off.

Fig. 7.11 BJT voltage switch.

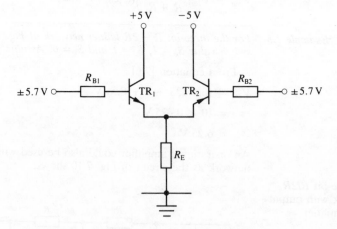

The arrangement of Fig. 7.11 is suitable for an application where V_{ref} can have values of $+5$ V and -5 V, i.e. the reference voltage is applied to the collectors of the transistors of the switching circuits. In addition to providing the reference voltage an additional switching circuit is required to produce the $+5.7$ V or -5.7 V at the two bases of the circuit of Fig. 7.11. A possible arrangement is shown in Fig. 7.12.

The input driving voltage can swing between 0 V and $+3$ V. When the input is at 0 V both TR_3 and TR_4 are on and saturated.

Fig. 7.12 An application of the voltage switch of Fig. 7.11 to a practical circuit.

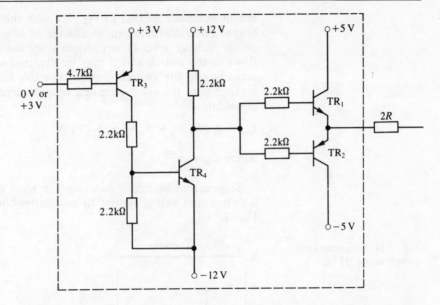

Since the collector of TR$_4$ is about -11.8 V the base current of TR$_2$, which is on, is enough to give approximately -5.7 V at TR$_2$ base and hence -5 V at the emitter. When the input is at $+3$ V, TR$_3$ and TR$_4$ are both off and the collector voltage of TR$_4$ is less than $+12$ V because of the base current from TR$_1$, which is on. The effect of this base current is to produce a voltage of about $+5.7$ V at TR$_1$ base and hence $+5$ V at the emitter. Thus the emitter voltage of TR$_1$/TR$_2$ will be ± 5 V depending on which device is conducting i.e either TR$_1$ (or TR$_2$) conducts while the other device is off.

The arrangement in Fig. 7.11 and 7.12 uses bipolar junction transistors. Similar circuits are possible using FETs and these circuits are often simpler because there is no offset voltage in the case of the unipolar device.

Current switches

A disadvantage of the voltage switch is that on changing from one voltage state to another, charging or discharging of stray capacitance is involved. Since most of the parasitic elements in the monolithic integrated circuits are capacitive, current switching is preferred. Figure 7.13 shows an $R - 2R$ network with current switching. The

Fig. 7.13 *R/2R* ladder network using current switching.

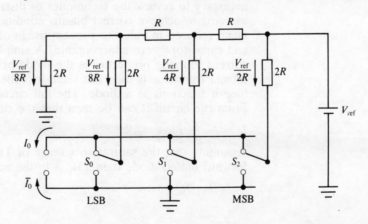

circuit is similar to that of Fig. 7.7 with the reference input and output lines interchanged. A change of switch state has little effect on the voltage level at any switch since the same current always flows to the switch and it may be diverted to the I_0 or $\overline{I}_0$ line according to the switch position. For this simple three-bit DAC arrangement the short-circuited output current is given by the equation

$$I_{0sc} = I_{ref}(S_0 2^{-3} + S_1 2^{-2} + S_2 2^{-1}) \text{ A} \qquad [7.7]$$

$$\text{where } I_{ref} = \frac{V_{ref}}{R}$$

The current switching function can most easily be achieved using a differential switch similar to an emitter-coupled gate as shown in Fig. 7.14.

Fig. 7.14 Basic current switch using BJTs.

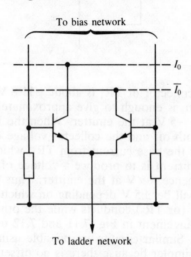

The effect of the bias network, controlled in turn by the state of the digital bit input, determines which transistor is connected to the ladder network. More details regarding the current switch can be seen later in this section when a practical DAC is examined.

Current sources   The advent of monolithic integrated circuit technology made it necessary to review the techniques of discrete circuits for current sourcing to achieve correct biasing conditions. Resistors had to be much reduced in value to prevent usage of large areas of the chip, and capacitors were unacceptable. A simple circuit which has current sourcing properties is the diode-biased transistor. In fact a second transistor is used but since its collector is connected to its base it functions as a diode. The full circuit is shown in Fig. 7.15. From the circuit it can be seen that the current I_1 is given by

$$I_1 = I_{C1} + I_{B1} + I_{B2} \qquad [7.8]$$

Assuming that the saturation voltage of TR_1 is less than that of a forward biased diode then TR_1 is in the active region and

$$I_{C1} = h_{FE1}I_{B1} \qquad [7.9]$$

Fig. 7.15 Diode-based
current source.

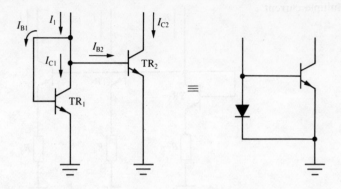

and

$$I_{C2} = h_{FE2}I_{B2} \qquad\qquad [7.10]$$

Rearranging equation [7.8] gives

$$I_{C1} = I_1 - I_{B1} - I_{B2}$$

and substituting for I_{C1} from equation [7.9] gives

$$h_{FE1}I_{B1} = I_1 - I_{B1} - I_{B2}$$

$$I_{B1} = \frac{I_1 - I_{B1} - I_{B2}}{h_{FE1}} \qquad\qquad [7.11]$$

If the transistors are assumed to be matched then $I_{B1} = I_{B2}$ and
equation [7.11] can be rewritten as

$$I_{B2} = \frac{I_1 - 2I_{B1}}{h_{FE1}}$$

and substituting for I_{B2} from equation [7.10] gives

$$\frac{I_{C2}}{h_{FE2}} = \frac{I_1 - 2I_{B1}}{h_{FE1}}$$

$$I_{C2} = \frac{h_{FE2}}{h_{FE1}}(I_1 - 2I_{B1}) \qquad\qquad [7.12]$$

If h_{FE} is large so that I_B is small and for well matched transistors
where $h_{FE1} = h_{FE2}$ then equation [7.12] reduces to

$$I_{C2} = I_1$$

Matching of h_{FE} for two such transistors can be very good but could
be further improved by adding a small resistance in series with the
emitter.

If several current sources are required it is feasible to use the
circuit of Fig. 7.16. If, in this circuit, $R_1 = R_2 = R_n$ then $I_1 = I_2 = I_n$.
However, the values of current may be deliberately made unequal
simply by altering the values of the emitter resistors. Since,
however, the transistor in a low-current arm of the arrangement
would be operating at a lower current density its gain would be less
than that of other transistors, producing an error. Nonetheless the
arrangement may be satisfactory for many applications.

Fig. 7.16 Multiple current sources.

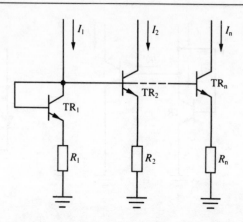

The use of constant current sources for a four-bit DAC is shown in the simple circuit of Fig. 7.17. In this case current sources related by the ratio 1, 2, 4 and 8 are connected via the switches into a summing resistor, or amplifier, to give an output voltage proportional to the sum of the currents.

Fig. 7.17 Four-bit DAC using current sources.

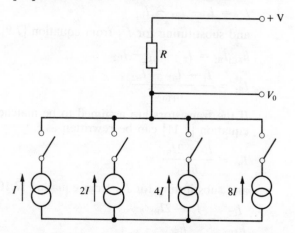

A major problem is the production of stable current sources that are accurately related to each other. A stable reference voltage, possibly derived from a zener diode circuit, may be applied to the bases of the transistors connected as current sources. If there is no compensation the base-emitter volt drop must be subtracted from the reference voltage so that

$$I = \frac{V_{ref} - V_{BE}}{R}$$ [7.13]

which assumes a current value of I amps and emitter resistor of R ohms.

The difference in voltage between V_{ref} and $V_{ref} - V_{BE}$ is not a problem since V_{ref} could be adjusted to compensate for the difference. What is a problem is the possible temperature drift that could occur in V_{BE} giving rise to error. This problem could be solved by the use of a diode, or transistor connected equivalent, in the voltage reference amplifier feedback path as shown in Fig. 7.18. This diode compensates for the base-emitter volt drop of all current source resistors over a wide temperature range.

Figure 7.18 also suggests a method for increasing the current

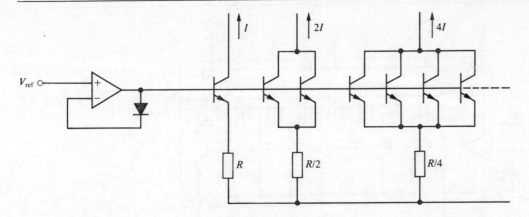

Fig. 7.18 A method of obtaining the required current source values using the BJT current source.

weighting of a particular arm of a current source arrangement while maintaining parameter matching. To double the current two single arms are simply connected together at the collectors; two resistors of value R in the parallel emitter leads could be combined as a single resistor of $R/2$ and so on for the other weighted current values. In practice it is not necessary to allow for more than one transistor in parallel; all that is needed is to scale the geometry of the device so that the same current density is present across each base-emitter junction *regardless* of the value of the current. This can be allowed for by presenting a larger emitter cross-sectional area for those arms designed to take larger current.

A practical DAC circuit

The LMDAC08 is an eight-bit high-speed current output DAC in monolithic form. It has complementary current outputs to allow differential output voltages of 20 V peak to peak with simple resistive loads. Reference to full-scale current matching is better than ±1 LSB. Power supply range is from ±4.5 V to ±18 V while the power consumption is only 33 mW at ±5 V. A basic block diagram is shown in Fig. 7.19(a) while the full equivalent circuit is shown in Fig. 7.19(b).

The relationship between the output currents and related circuit variables is given by

$$I_O + \bar{I}_O = \frac{+V_{ref}}{R_{ref}} \times \frac{255}{256} \, \text{mA}$$

$$= I_{ref} \times \frac{255}{256} \, \text{mA}$$

where I_O and $\bar{I}_O$ are complementary output currents, V_{ref} is reference voltage in volts and R_{ref} is reference resistance in kΩ. Typically for the LMDAC08 device the output current is 2 mA with a voltage reference of +10 V, $R_{ref} = 5$ kΩ.

The addition of an operational amplifier to the device output is a simple but effective means of providing a low impedance analogue output. This provides a buffer stage for the analogue output and additionally allows the output voltage swing to be adjusted using a load resistance R_L. The operational amplifier may require offset nulling and if so the null circuit should be adjusted so that the output is zero for all bits at the DAC input. Figure 7.20 shows a circuit arrangement to produce low impedance output.

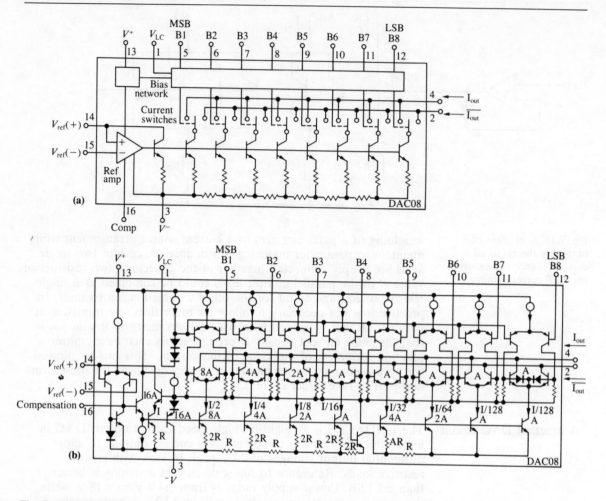

Fig. 7.19 LMDACO8 eight-bit current output DAC: (a) basic block diagram, (b) full equivalent circuit.

Fig. 7.20 Use of an operational amplifier to give a positive low impedance output.

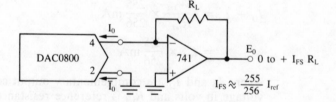

DAC specifications

Factors such as accuracy and resolution were discussed earlier. Manufacturers include the following in their DAC specifications:

Nonlinearity error

An ideal DAC would have equal increments of the analogue output for equal increments in the numerical value of the digital input. Nonlinearity refers to the deviation of the measured output from a straight line which extends over the complete range of the input.

Usually expressed as a fraction of the voltage equivalent of the LSB and should typically be less than $\pm\frac{1}{2}$ LSB.

Differential nonlinearity   This is due to the difference between the measured change and an ideal 1 LSB change between any two adjacent codes. A specified differential nonlinearity of ± 1 LSB maximum over the temperature range ensures monotonicity. A **monotonic** DAC is one whose analogue output remains the same or increases as the digital input increments. A DAC may be non-monotonic if the differential linearity error is greater than ± 1 LSB, i.e. the output could fall when the digital input increments.

Gain error   Gain is the ratio of the DAC's full-scale output value and the reference input voltage. An ideal 8-bit DAC for example would have a gain of 255/256. Gain error is a measure of the amount by which the device actual gain may deviate from the designed value; it is usually expressed as a percentage deviation from the normal value. Gain error can usually be minimised by using external preset resistors.

Settling time   This is the time taken for the output of a DAC to approach closely enough to its correct value after an input change has occurred. The settling time is caused by transients set up by voltage switching and the effect on the stray capacitance in the circuit. In terms of accuracy the settling time may define the time taken, after input changes, for the output to be within $\pm\frac{1}{2}$ LSB. Times vary according to device but the LMDACO8 mentioned earlier in the section has a settling time of 100 n seconds.

Temperature sensitivity   For a given fixed digital input value the output analogue value varies with temperature. This is mainly because of the sensitivity to temperature of the reference voltage, the resistors, the operational amplifier and even the amplifier offset voltage. Typical sensitivities can be as high as ± 50 ppm/° C but lower values exist for good quality converters. The LMDACO8 has a quoted figure of ± 10 ppm/° C.

Sample and hold circuits

Where it is required to sample an anlogue signal at a timing interval of T_s seconds *and* retain the amplitude of the sample for the time between sampling pulses then a sample and hold circuit must be used. The basic process is illustrated in the waveforms of Fig. 7.21.

The basic circuit that can perform the sample and hold task is shown in Fig. 7.22(a). A switch may be opened or closed by a control logic input and when closed the capacitor may charge to the value of the applied analogue signal. The switch should only be closed for a short time compared with the sampling interval T_s and for an even shorter time compared to the period of the analogue input so that during the sampling time the input is essentially

Fig. 7.21 Sample and hold
circuit waveforms.

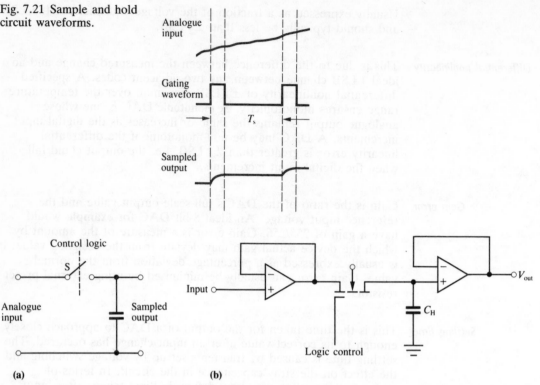

Fig. 7.22 Sample and hold
circuit: (a) basic
arrangement, (b) practical
arrangement using unity
gain voltage follower
stages.

constant in amplitude. Once the switch opens the value of the
sampled analogue input is held by the capacitor until the next
sampling occurs when the switch closes once more and the capacitor
charges to represent a new sampled value of the analogue input.

The circuit of Fig. 7.22(a) ignores the source resistance of the
circuit being sampled and the input resistance of the circuit
connected across the capacitor. One of the problems of the basic
circuit is that the source resistance of the analogue circuit is not
zero and the sampled waveform is not a precise copy of the
analogue signal; also with a 'loaded' capacitor the value of stored
charge may not be held constant during the sampling period T_s and
some charge could leak away through the input resistance of the
loading circuit. There is also the problem of causing the switch to
close at the right time and for the right amount of time.

The basic circuit could be implemented in the integrated circuit
form by the circuit of Fig. 7.22(b).

The input stage is a voltage follower with high input resistance
and low output resistance. The low output resistance means that C_H
can be charged quickly. The output stage is also a voltage follower
and acts as a buffer between C_H and the external circuit. If this
stage utilises FET devices then the device high input resistance
ensures that the capacitor does not readily discharge between
sampling pulses. In this circuit a MOSFET is used as a switch and is
normally an open-circuit, which isolates the input from the hold
capacitor, until a logic control pulse arrives to turn the MOSFET
on; while it is on the capacitor can charge to the value of the
analogue input.

A practical monolithic sample and hold circuit is the LF198/298/398 series, a functional block diagram of which is shown in Fig. 7.23. This device utilises mixed bipolar and FET (BIFET) technology to gain advantage of the best points of each type. It is claimed for this device that it has fast signal acquisition of less than 10 μs with a d.c. gain accuracy of 0.002% typical and with low output noise in the hold mode. The input stage is bipolar to provide low offset voltage and wide bandwidth and has provision for input offset adjust. Using a 1 μF hold capacitor a droop rate (amount by which output level falls with time) of 5 mV/min is specified and the device can operate from ±5 V to ±18 V making it TTL and CMOS compatible. Devices with acquisition times of less than 10 μs are currently available.

Fig. 7.23 The LF198 series sample and hold circuit. Functional diagram.

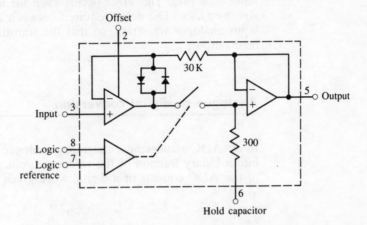

Sample error due to moving input signals may cause problems because of the finite phase delays through the circuit creating an input-output differential for fast moving signals. There is also an additional lag due to the 300 Ω series resistor on the chip so that the moment the hold command arrives, the hold capacitor voltage may be somewhat different from the actual analogue input signal. The effect of these delays may be opposite to the effect caused by delays in the logic which switches the circuit from sample to hold. Analogue delay is proportional to hold capacitor value while digital delay remains constant. A family of 'dynamic sampling error' curves are available for the device to assist the user to estimate errors.

A typical connection arrangement for the device is shown in Fig. 7.24.

Fig. 7.24 Typical connection diagram for the LF198 sample and hold circuit.

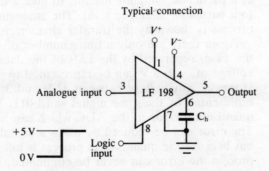

The following are some of the terms used in connection with sample and hold circuits. *Hold step*: The voltage step at the output of the sample and hold when switching from sample mode to hold mode with a steady (dc) input analogue input voltage. Logic swing is 5 V. *Acquisition time*: The time required to acquire a new analogue input voltage with an output step of 10 V. Acquisition time also includes the time for all internal nodes to settle so that the output assumes a proper value when switched to the hold mode. *Hold settling time*: The time required for the output to settle within 1 mV of final value after the hold logic command. *Dynamic sampling error*: The error introduced into the held output due to a changing analogue input at the time the hold command is given. Error is expressed in mV with a given hold capacitor value and input slew rate. This error occurs even for long sample times. *Aperture time*: The delay required between hold command and an input analogue transition, so that the transition does not affect the held output.

Analogue to digital conversion

In an ADC arrangement the input analogue voltage V_{in} is encoded into a binary fraction of the reference voltage V_{ref} so that the output of the ADC consists of a digital word given by

$$\frac{V_{in}}{V_{ref}} = b_0 2^{-1} + b_1 2^{-2} + \ldots b_{n-1} 2^{-n} \qquad [7.14]$$

where n is the number of bits in the digital word and b_0, b_1 etc. are binary bit coefficients which can have the logic value 0 or 1. A possible arrangement is shown in block diagram form for a three-bit ADC in Fig. 7.25.

The digital output from the ADC of Fig. 7.25 is in parallel since all bits representing the digital word equivalent of the analogue input voltage appear simultaneously. It is posssible to represent the digital word on single output line by generating the bits singly, starting with the MSB, and so on until the complete digital word has been taken out of the stage.

Because the analogue to digital conversion process is the opposite to that of the DAC process discussed earlier it is possible to use the 'staircase' voltage representation of Fig. 7.2 but in reverse. Now the analogue voltage is the input and the ADC 'quantises' that voltage, at a particular sampling instant, to give a digital word corresponding to a particular analogue level. The analogue to digital translation process is shown by the transfer characteristic of 7.26.

Again there are only a finite number of discrete levels (eight in this case) separated by the LSB of the data word. Thus an analogue voltage of, say, 3 V can be represented by the digital word 011. However, any voltage between $2\frac{1}{2}$ V and $3\frac{1}{2}$ V would also be represented by the same digital word 011. This gives rise to **quantisation error** in the ADC which can have a value of $\pm\frac{1}{2}$ LSB. The error can be reduced by using a digital word of greater length but because the quantisation process is inherent in the encoding process the error can never be eliminated, only minimised. In terms

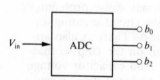

$V_{in} \longrightarrow$ ADC $\quad\circ\, b_0$
$\quad\circ\, b_1$
$\quad\circ\, b_2$

Fig. 7.25 Block diagram of a three-bit ADC.

Fig. 7.26 Input/output
characteristic for a three-bit
ADC.

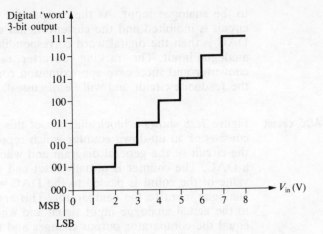

of an arbitrary analogue voltage value V_{in} where $V_{in} < V_{ref}$ the
quantisation error may be specified as V_{in} where

$$0 \leq |V_{in}| \leq \frac{V_{ref}}{2^n}$$

where n is the number of bits in the digital word.

Another problem with ADC circuits is time. Because the
conversion rate of the circuit is finite there must be a delay between
the time of sampling V_{in} and producing the output digital word. This
delay is known as the **conversion time**. In some circumstances where
the analogue input varies as a function of time the conversion time
could lead to an extra error and this must be taken into account
when designing the system.

There are many conversion techniques available for ADC
circuits and some of these will be discussed in some length in the following
sections.

Feedback ADC circuits

Feedback ADC circuits use a DAC circuit in a feedback loop
together with a comparator and a digital circuit as shown in Fig.
7.27. In this arrangement the digital circuit produces a digital
number which is fed to the DAC circuit which in turn produces an
analogue signal. The comparator has one input from the analogue
input signal and the other from the output of the DAC. The output
from the comparator changes when the DAC output becomes equal

Fig. 7.27 Block diagram of
a feedback ADC circuit.

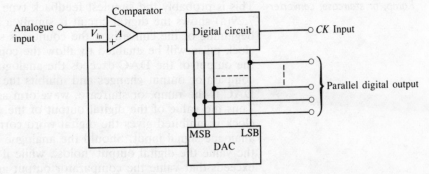

to the analogue input. As the comparator output changes the digital circuit is inhibited and the current digital word at the output of the DAC is then the digital word corresponding to the sampled analogue input. The tracking converter, ramp (or staircase) converter and successive-approximation converter are all examples of the feedback circuit and will be discussed.

Tracking ADC circuit   Figure 7.28 shows a block diagram of this type of converter. It consists of an up-down counter which represents the digital part of the circuit of the general diagram and whose output is connected to a DAC. The counter is initially reset and allowed to count; the value of the count is passed to the DAC which in turn produces an analogue input to the comparator. This analogue signal is compared to the actual analogue input signal and when these two values are equal the comparator output changes and temporarily stops the count. The value of the counter output at this time is thus the digital word corresponding to the analogue input V_{in}. The output state of the comparator determines the direction of the count (up or down) depending on whether the feedback voltage from the DAC is smaller or larger than the input voltage V_{in}. In this arrangement V_{ref} is an integral part of the DAC circuit. This type of converter is quite slow since $2^n - 1$ clock pulses need to be counted. However, if the input changes by a small amount or if the input variations are very slow then this type of circuit can track very quickly, i.e. within a few clock cycles.

Fig. 7.28 Block diagram of a feedback ADC circuit employing an up/down counter for tracking.

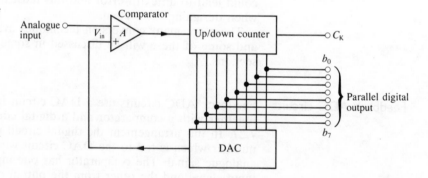

Ramp, or staircase, converter   This is probably the simplest feedback type of circuit. As Fig. 7.29(a) shows the digital circuit is simply a binary counter. At the beginning of the converison the counter is set to zero and a gated clock pulse will be enabled to allow the counter to increment until the output of the DAC exceeds the analogue input when the comparator output changes and inhibits the clock. The output of the DAC is the ramp, or staircase, waveform as shown in Fig. 7.29(b). Thus the value of the digital output of the counter at the instant the clock is inhibited gives the digital word corresponding to the analogue signal input. Should the analogue signal decrease below the value the digital output 'holds', while if the analogue input exceeds that value the comparator output again changes and enables

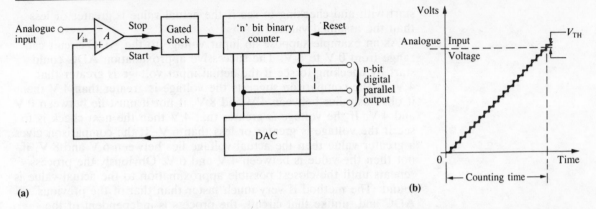

(a)   **(b)**

Fig. 7.29 Ramp type ADC circuit: (a) block diagram, (b) DAC output waveform.

the clock circuit so that the counter can once again increment until the DAC output again equals the analogue input. Thus the digital output in this circuit is equal to the maximum value of the analogue input between counter resets.

The conversion time in the ramp type ADC is not fixed but depends on the actual value of the analogue input voltage expressed as a fraction of the full-scale value. This can be expressed as

$$\text{conversion time} = \frac{V_{in}}{V_{ref}} 2^n T \qquad [7.15]$$

where n is the number of bits in the digital word and T is the period of the clock pulse.

Example 7.6

A ramp type ADC circuit has the following parameters; $n = 8$, $V_{ref} = +5.10$ V, clock frequency = 1 MHz. Find the digital word for an input voltage of 4.36 V and the conversion time taken to reach this value. Assume the threshold voltage for the comparator is 2 mV.

The DAC has an eight-bit input and can have a maximum output voltage given by V_{ref} or 5.10 V. Hence the maximum number of steps in the ramp function is $2^n - 1 = 255$ so that the step size is $5.10/255 = 20$ mV. The DAC output will therefore increment in 20 mV steps as the counter increments from zero. To change the output of the comparator when the analogue input is 4.36 V the output from the DAC must be *at least* 4.362 V. To reach a voltage of 4.362 V requires $4.362/20 \times 10^{-3}$ steps or 219 steps. Thus the digital word corresponding to an analogue input voltage of 4.36 V is the binary equivalent of 219, i.e. 11011011.

The conversion time required to reach this output state is given by equation [7.15] as

$$4.36/510 \times 256 \times 1 \ \mu s = 219 \ \mu s$$

It can be seen from this example that since the period of the clock pulse $T = \dfrac{1}{f} = 1 \ \mu s$ (when $f = 1$ MHz) then the conversion time is simply the number of steps, 219 in this case, multiplied by the clock period.

Successive approximation ADC

This type of ADC approximates to the analogue voltage value not, as in the case of the counter-controlled converter by increasing the count from zero until the value is found, but by assuming a value to start with and checking to see if the actual value is greater or less than the assumed value.

As an example suppose an input voltage to the ADC circuit can range from 0 V to 8 V. The successive approximation ADC could start by checking to see if the actual input voltage is greater than 4 V. If the comparison suggests the voltage is greater than 4 V then it obviously lies between 4 V and 8 V, if not it must lie between 0 V and 4 V. If the voltage *is* greater that 4 V then the next check is to see if the voltage is greater or less than 6 V. If the comparison gives a greater value then the actual voltage lies between 6 V and 8 V, if not then the value is between 4 V and 6 V. Obviously the process repeats until the closest possible approximation to the actual value is found. The method is very much faster than that of the previous ADC and, unlike that circuit, the process is independent of the value of the analogue input. In practice the successive approximation ADC would first place a logic 1 in the MSB of the converter, with all other bits set to 0, and compare the analogue output from a DAC with the actual analogue input voltage. If the analogue input voltage is greater than the DAC output signal then the MSB is left at 1 and the next most significant bit is set to 1 and the process repeated; if the analogue input voltage is less than the DAC output then the MSB is set to 0 before the next digit is set to 1. The process repeats until all the bits in the converter have been set, or not, according to the value of the input analogue voltage. A basic block diagram of a successive approximation ADC circuit is shown in Fig. 7.30. The circuit which decides whether to leave a bit at 1 or change it to 0 is the logic circuit block of this diagram. A storage register is also used and the bits in the register output can be set or reset according to the command of the output of the comparator at each step of the process. The process is under the control of an external clock and, as the following example shows, the process will only take *n* clock pulses if the ADC is an *n*-bit converter.

Example 7.7 *A 4-bit ADC of the type shown in Fig. 7.30 is to be used with a DAC of step size 1 V. If the input analogue voltage to the ADC is 5.2 V show the*

Fig. 7.30 Block diagram of a successive approximation ADC.

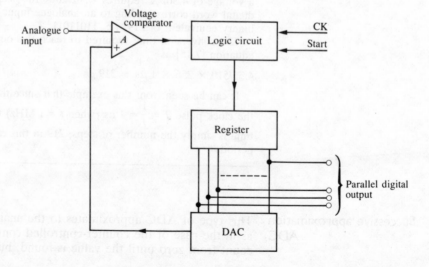

process by which the ADC achieves an approximate digital output corresponding to the input analogue voltage and find the value of the digital word.

Initially the register is reset and then the MSB of the register is set to 1 so that the digital word is 1000. This gives an output of 8 V since the DAC step size of 1 V. Since this value is greater than the analogue input voltage the MSB is reset to 0 and the next significant bit set to 1. The register now contains 0100 which gives an output from the DAC of 4 V.

Since the DAC output value is less than the analogue input voltage this bit remains at 1. The next significant bit is then set to 1. The register now contains 0110 which gives an output from the DAC of 6 V. This value is greater than the analogue input voltage so the bit is reset to 0 and the next bit (the LSB in this case) is set to 1. The register now contains 0101 which gives an output from the DAC of 5 V. This is the final value since although the value is less than the actual value for the analogue input voltage, it would be even more in error if the LSB were set to 0. Final register output digital word is thus 0101 which represents 5.2 V.

Example 7.8 *Using the data given in example 7.6 but assuming now the ADC circuit is of the successive approximation type, calculate the conversion time to give the output digital word.*

Example 7.6 used an 8-bit converter with a clock frequency of 1 MHz. Since for the successive approximation method the conversion time for the ADC is equal to the clock period times the number of bits, i.e. $n \times T$ seconds, then since $n = 8$ and $T = 1 \mu s$, the conversion time is 8 μs.

Compare the solution to example 7.8 with the value of 219 μs for the ADC of example 7.6. Also bear in mind the fact that 219 μs was not the slowest time for that circuit since the value of voltage used for the analogue input was 4.362 V and the circuit could have a maximum input voltage of 5.10 V using 255 steps. Thus the slowest time for the ADC circuit of example 7.6 would be 255 μs whereas it is always 8 μs for the successive approximation device regardless of the value of the analogue input voltage.

A possible arrangement of a three-bit successive approximation ADC is shown in Fig. 7.31. In this arrangement the time to complete an analogue to digital conversion is five time intervals since one time interval is used to actually read out the digital information while another is used to clear the register ready for the next conversion. The remaining three time intervals of course are for the actual conversion as discussed earlier. The flip-flops FFA to FFE form a modulo-5 ring counter. The flip-flops FF1 to FF3 are used to provide the register output bits with FF1 provididng the LSB and FF3 the MSB.

The ring counter will successively shift a logic 1 along the counter when clocked. In this case a logic 1 is placed in Q_A initially while Q_B to Q_E are all zero. Thus FF3 will be set while FF2 and FF1 will be reset. The output digital word presented to the DAC is thus 100 and the corresponding analogue output of the DAC is compared with the actual analogue input voltage at the comparator stage. The output of the comparator will be a 0 or 1 depending on whether the DAC output is greater than the analogue input or not. The 1 on the ring counter will be shifted to Q_B at the next clock interval and this will set FF2 to a 1. The same output will be fed to FF3 via a 2 input AND gate, the other input to the gate being from the

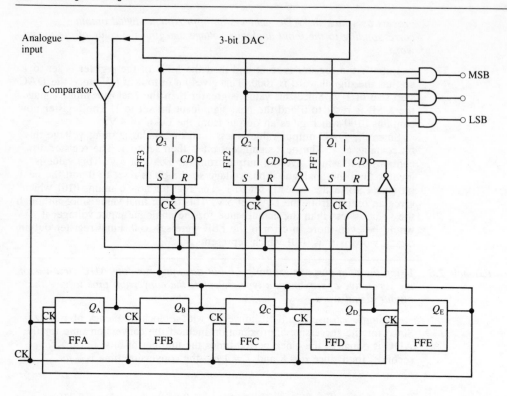

Fig. 7.31 A possible circuit arrangement for a three-bit successive approximation ADC.

comparator output. If the comparator output is a 1 then the output of FF3 is reset, otherwise the output of FF3 remains set.

The procedure is repeated for successive bit positions until the approximation is complete, the process taking three clock periods. When Q_E goes to logic 1 no comparison is being made and the digital word can be read.

During the comparison sequence the value of the analogue input should be held constant. Hence the analogue input to the comparator should be preceded by a sample and hold circuit. Since the timing for a sample and hold circuit requires a period of time for the sample and a longer period for the hold condition, the sampling could be done when the digital word is being read from the ADC circuit and the analogue signal could be held while the comparison process is being done in the ADC. Thus the Q_E output could be used as the logic input signal to the sample and hold circuit.

Dual-slope ADC

Dual-slope ADC requires no feedback. The principle of operation is best explained by the simplified circuit of Fig. 7.32(a). At the start of the conversion process the input of the analogue integrator is connected to a sample of the analogue input signal via a switch. The integrator thus falls at a rate given by

$$V_{O1} = -\frac{V_{in}}{CR} \text{ V/s} \qquad [7.16]$$

This assumes that the input signal is held constant at the switch input for the time the switch is in this position. Under these

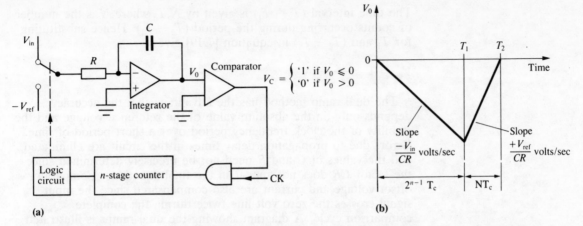

Fig. 7.32 Dual ramp ADC:
(a) block diagram, (b)
integrator output waveform.

conditions the output may fall for a set time T_1 so that using equation [7.16], the output voltage at the end of that time is

$$V_{O1} = -\frac{V_{in}T_1}{CR} \text{ V} \qquad [7.17]$$

The time T_1 is itself set by the time taken by a counter, initially reset, to cycle through all its possible values until it reaches a value where all bits are 0 except the MSB which is 1. This logic level on the MSB causes the switch position to change, disconnecting the input voltage and connecting a reference voltage. The reference voltage is opposite in polarity to the analogue input voltage so that the integrator output returns towards 0 V, i.e. a positive ramp voltage. When the output reaches 0 V the comparator senses this and changes state which inhibits the clock and stops the counter. The time at which this occurs is T_2 and since the voltage at the output at any time during the positive ramp is given by

$$V_{O2} = -V_{O1} + \frac{V_{ref}}{CR}(T - T_1) \text{ V} \qquad [7.18]$$

where T is any value between T_1 and T_2. At time T_2, when $V_{O2} = 0$ V, equation [7.18] becomes

$$0 = -V_{O1} + \frac{V_{ref}}{CR}(T_2 - T_1)$$

so that

$$V_{O1} = \frac{V_{ref}}{CR}(T_2 - T_1)$$

and substituting for V_{O1} from equation [7.17] gives

$$\frac{V_{in}T_1}{CR} = \frac{V_{ref}}{CR}(T_2 - T_1)$$

so that

$$V_{in} = V_{ref}\left(\frac{T_2 - T_1}{T_1}\right) \text{ V} \qquad [7.19]$$

The time T_1 is a constant since it is the time taken by the counter to reach the value 10 . . . 0 after initial resetting. Thus for an n-bit counter $T_1 = 2^{n-1} . T_c$ where T_c is the period of the clock pulses.

The time interval $(T_2 - T_1)$ is given by $N.T_c$ where N is the number of counts occurring during the period $(T_2 - T_1)$. Hence substituting for T_1 and $(T_2 - T_1)$ in equation [7.19] gives

$$V_{in} = V_{ref} \frac{N}{2^n - 1} \text{ V}$$

The dual ramp method has the advantage that the accuracy depends only on the absolute value of the reference voltage and the stability of the clock frequency period over a short period of time. Errors due to propagation delay times in the circuit are eliminated and the values of C and R need not be precisely determined since the factor CR does not appear in the final equations. Comparator offset voltage and current are also compensated since the input signal crosses the zero volt line twice during the complete comparison cycle. A diagram showing the dual ramps is illustrated in Fig. 7.32(b).

A practical ADC circuit

Some circuits are available purely as ADC circuits while others can have a dual function of providing ADC or DAC. Because of the versatility of the latter one of these devices will be discussed in more detail.

The ZN425E is an eight-bit dual mode ADC/DAC utilising an R-2R ladder network, an array of precision bipolar switches, an eight-bit binary counter and a 2.5 V precision voltage reference all on a single monolithic chip. A block diagram of the device is shown in Fig. 7.33.

The design of the ladder network gives full eight-bit accuracy using normal diffused resistors. The voltage reference used may be either the internal value or an external value which may be fixed or variable. A logic input select switch is incorporated which determines whether the precision switches accept the outputs from the binary counter or external inputs depending upon whether the control signal is respectively high or low. The converter is of the voltage switching type and uses an R-2R resistor ladder network with $R = 10 \text{ k}\Omega$. Each 2R element is connected to 0 V or V_{ref} by transitor switches specially designed for low offset voltage (typically 1 mV).

A possible DAC application is shown in Fig. 7.34. The ZN425E gives an analogue voltage output directly from pin 14 hence a current to voltage converting amplifier is not required. To remove the offset voltage and to calibrate the converter a buffer amplifier is necessary. Figure 7.34 shows a scheme using the internal reference voltage. The calibration procedure is:

(a) Set all bits low and adjust R_2 until V_{out} is zero.
(b) Set all bits high and adjust R_1 until V_{out} = nominal full-scale reading − 1 LSB.
(c) Repeat (a) and (b).

A possible ADC application is shown in Fig. 7.35. A counter type ADC can be constructed by adding a voltage comparator and a latch as in Fig. 7.35. On the negative edge of the CONVERT COMMAND pulse (15 μs minimum) the counter is set to zero and the STATUS latch to logic 1. On the positive edge the gate is opened, enabling clock pulses to be fed to the counter input of the device. The minimum negative clock pulse width to the ZN425E is

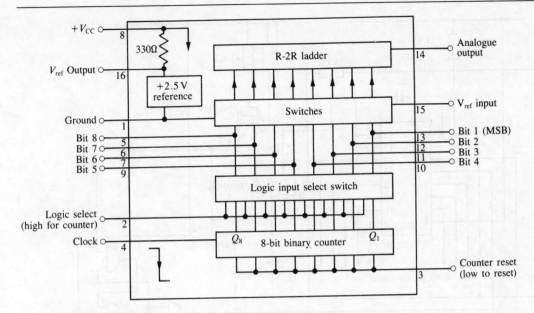

Fig. 7.33 Block diagram of
the ZN425E eight-bit dual
mode ADC/DAC.

Fig. 7.34 Eight-bit digital to
analogue converter.

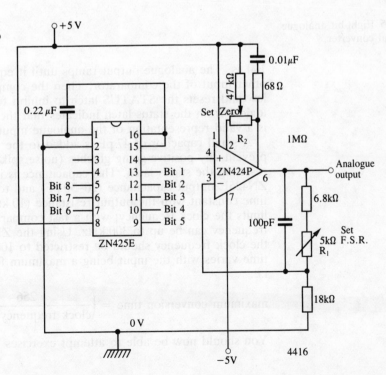

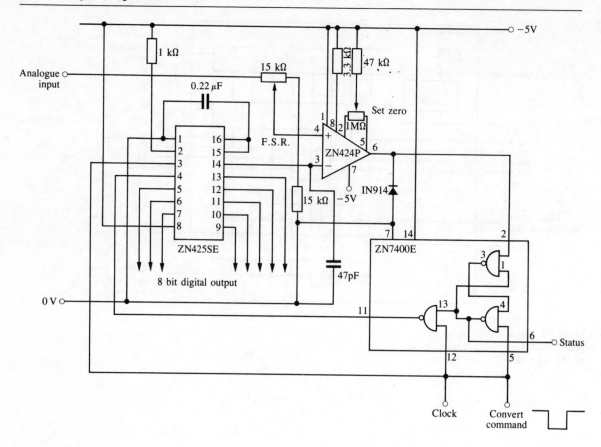

Fig. 7.35 Eight-bit analogue
to digital converter.

100 ns. The analogue output ramps until it equals the voltage on the
other input of the comparator. Then the comparator output goes
low and resets the STATUS latch to inhibit further clock pulses. A
logic 0 from the status latch indicates that the eight-bit digital output
is a valid representation of the analogue input voltage.

A small capacitor of 47 pF is added to the device output to
prevent any positive-going glitches (noise spikes) from prematurely
resetting the status latch. This capacitance is in parallel with the
ZN425E output capacitance (20–30 pF) and together they form a
time constant with the output resistance (10 kΩ). This time constant
limits the clock frequency; with a fast comparator the clock
frequency can be up to 300 kHz. Using the ZN424P as a comparator
the clock frequency should be restricted to 100 kHz. The conversion
time varies with the input being a maximum for a full-scale input,
i.e:

$$\text{maximum conversion time} = \left(\frac{256}{\text{clock frequency in Hz}}\right) \text{seconds}$$

You should now be able to attempt exercises 7.1 to 7.27

Points to remember

- Digital to analogue (DAC) conversion has a digital input and an analogue output.
- One form of DAC employs binary weighted resistors to give a 'weighted' output according to the state of switches at each bit position.
- Another form of DAC uses an R/2R ladder network to give a weighted output according to the state of switches at each bit position.
- The switches used in DAC can be produced using BJT or FET devices and may be voltage or current operated.
- Practical DAC circuits are available in integrated circuit form.
- Analogue to digital (ADC) conversion has an analogue input and a digital output.
- One form of ADC employs a feedback loop, a DAC and a digital circuit. The digital circuit produces an analogue signal, via the DAC, which is compared with the actual analogue input. When the signals are equal the output of the digital circuit gives the digital version of the analogue input.
- Practical ADC circuits are available in integrated circuit form.

EXERCISES 7

7.1 A three-bit DAC has a voltage output which, for an input of 0101, is 7.5 V. What will the output voltage level be for a digital input of 0100? (6 V)

7.2 A three-bit DAC uses binary weighted resistors with $R = 2$ kΩ. Reference voltage is 10 V. If the input is 101 what is the voltage at the output? Assume $R_L = 1$ kΩ. (3.85 V)

7.3 A four-bit DAC uses binary weighted resistors with a reference voltage of 10 V. If the MSB current increment is 2 mA what value of resistances are used? Find the output voltage if the input is 1011. Assume $R_L = R$ (5 kΩ, 4.78 V)

7.4 A four-bit DAC with resistance value of $R = 2$ kΩ and reference voltage of 10 V is to be used with an operational amplifier of infinite gain and feedback resistance 5 kΩ. What is the output voltage if the binary input is 0011? (−9.375 V)

7.5 Sketch a circuit to shown how an operational amplifier with a feedback resistance of value R_f could be used with a binary rated resistance network. If the reference voltage is 10 V show that the DAC obeys the relationship:

$$V_{out} = -20(S_{n-1}2^0 + S_{n-2}2^{-1} + \ldots S_0 2^{-(n-1)}) \text{ V}$$

7.6 For a three-bit R/2R ladder network of Fig. 7.7 calculate V_{out} if the switch positions are all set to 1. Assume $R = 20$ kΩ and $V_{ref} = 10$ V. (8.75 V)

7.7 Sketch a circuit to show how an operational amplifier could be used with an R/2R ladder network. If the reference voltage is 10 V show that the DAC obeys the relationship:

$$V_{out} = -10(S_0 2^{-n} + S_1 2^{-(n+1)} + \ldots S_{n-1} 2^{-1}) \text{ V}$$

7.8 An R/2R ladder network has a 10-volt reference and digitally operated switches to produce a four-bit DAC. If the short-circuit output current

is 3.75 mA when all switches are made, sketch a suitable circuit
together with resistance values.

7.9   A four-bit current output DAC of the type shown in Fig. 7.13 is to
be used with a reference current of 2 mA. If $V_{ref} = 10$ V find the
resistance values required for a ± 10 V full-scale analogue input range.
$\qquad$ (5 kΩ)

7.10   An LMDACO8 has a digital input of 01000100 and produces an
output current of $I_0 = 0.53125$ mA. What is the value of the DAC
reference current? What are the expected values for I_0 and $\overline{I}_0$ if the
digital input is 10011001? $\qquad$ (2 mA, 1.46 mA; 1.195 mA, 0.805 mA)

7.11   For a DAC briefly describe what you understand by:
(i) non-linearity error, (ii) differential non-linearity, (iii) gain error.

7.12   What do you understand by the term monotonic DAC? If an 8-bit
DAC has a differential linearity error of $\pm\frac{1}{2}$ LSB at 25 °C, while the
differential linearity error temperature coefficient is given as ± 50 ppm
of full-scale/°C, what is the temperature range over which the DAC
remains monotonic? $\qquad$ (-53 °C to $+103$ °C)

7.13   Discuss current and voltage switching and their effects on the
reference supply and conversion speed.

7.14   With the aid of simple sketches explain the function of a sample and
hold circuit. Draw a simple circuit that could achieve the sample and
hold function.

7.15   In the sample and hold circuit shown in Fig. 7.36 the values of R_1 and
R_2 are both 20 kΩ and the sample interval is 40 nsec. Find a suitable
value of C so that the output may track the input. $\qquad$ (0.2 pF)

Fig. 7.36 Sample and hold
circuit for exercise 7.15.

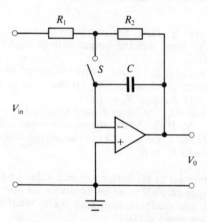

7.16   For a sample and hold circuit state briefly what you understand by:
(i) acquisition time, (ii) gain error, (iii) aperture time.

7.17   An eight-bit ADC has a full-scale input voltage range of 10 V. Find
the resolution of the circuit and quantisation error. An eight-bit ramp
type ADC has a resolution of 20 mV and a clock frequency of
2 MHz. Find the digital 'word' for an input voltage of 4.5 V and the
conversion time taken to reach this value.
$\qquad$ (39 mV, ± 20 mV; 11100001, 13 μs)

7.18   Draw a block diagram and explain the action of a tracking ADC.
State the advantages/disadvantages of this type of circuit.

7.19   Design an eight-bit tracking ADC circuit using a DAC generated
ramp and a bi-directional counter. State the equation for conversion
time for such a circuit in terms of defined circuit parameters.

7.20   Draw a block diagram and explain the action of a ramp ADC circuit.
State the advantages/disadvantages of this type of circuit.

7.21   Using an eight-bit counter, an eight-bit DAC and an operational
amplifier, design a ramp ADC circuit. Deduce the frequency of a

repetitive input waveform in terms of the counter clock frequency. Assume zero time is taken to reset after the required voltage is reached.

7.22 A ramp type ADC has the following parameters: $n = 6$, $V_{ref} = +10.2$ V, clock frequency = 1 MHz. Find the digital 'word' for an input voltage of 3.4 V and the conversion time taken to reach this value. Assume the threshold voltage for the comparator is 4 mV.

(010110, 21.34 μs)

7.23 Draw a block diagram and explain the action of a successive-approximation ADC. State the advantages/disadvantages of this type of circuit.

7.24 Show that for a successive-approximation ADC an *n*-bit conversion requires *n* comparison operations.

7.25 A four-bit ADC of the type shown in Fig. 7.30 is to be used with a DAC of step size 1 V. If the input analogue voltage to the ADC is 4.1 V show how the circuit achieves an approximate digital output and find the value of the digital 'word'.

7.26 Using the parameters of example 7.30 but applying them to a successive-approximation ADC, find the conversion time necessary to give the required output 'word'.

(6 μs)

7.27 Draw a block diagram and explain the action of a dual ramp ADC. What are the advantages/disadvantages of this type of circuit?

8 Power Supplies

The principal learning objectives of this chapter are to:

The electronic devices and circuits discussed in this text all require direct voltage power supplied. For a device or circuit where the required voltage and direct current drain from the power supply is low a battery could be used. The main disadvantages in the use of a battery are its limited life and the manner in which the battery internal resistance increases with age and deteriorating battery condition. The effect of the increased resistance is to lower the battery terminal voltage for a given load current; also with a battery the effect of varying load current will cause a change in the terminal voltage of the battery. An ideal power supply will not change its terminal voltage despite changes in input voltage or output current conditions. Thus the battery is far from ideal but is useful in limited applications. For larger amounts of power the source is nearly always the mains; for the bottom end of the larger power supply requirements a single-phase supply is usually sufficient whereas for the higher end of the scale multi-phase systems are used. Since the mains is alternating current (a.c.), **rectification** is necessary to obtain direct-current (d.c.). This may be preceded by a transformer stage if the amplitude of the d.c. voltage required is different from that of the mains voltage or if electrical isolation is required. Since the d.c. produced by rectification is pulsating, some form of smoothing would be required. Finally since for most circuits variations in d.c. voltage levels and changes in load current values must not be allowed to affect the d.c. voltage output level, some form of rectification is required. In block diagram form a complete mains to d.c. system with regulation would be as shown in Fig. 8.1.

Fig. 8.1 Block diagram of a mains to d.c. system with regulation.

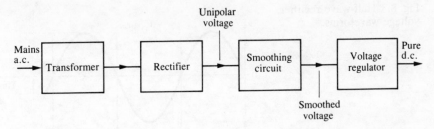

It is the purpose of this chapter to investigate circuits which may be used to form part or all the block diagram of Fig. 8.1 with reference where possible to available MSI circuits.

Rectification

The mains supply voltage is alternating sinusoidal voltage. For a single phase this would take the value 230 V at 50 Hz. This is a voltage which changes polarity during a cycle, i.e. it is positive in one half-cycle and negative in the second half-cycle and gives an average value of zero over the cycle. The requirement of the rectification stage is to provide a unipolar voltage, i.e. a voltage which is made up of sinusoidal half-cycles but each of the *same* polarity, either positive or negative. Rectification can be achieved using a p-n junction diode but the method of rectification can vary as the following sections will show.

Half-wave rectifier

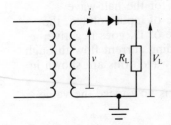

Fig. 8.2 Half-wave rectifier circuit.

The circuit of Fig. 8.2 assumes a sinusoidal voltage $v = V_{pk} \sin \omega t$ at the secondary of the transformer and that a rectifying diode is in series with a resistive load across which the output d.c. voltage is to be developed. In effect the secondary of the transformer is acting as a voltage generator and the voltage delivered to the circuit will be shared between the diode and the load.

In the circuit $v = v_D + iR_L$ where v_D is diode instantaneous volt drop and i is instantaneous current. When v is positive the diode is forward biased hence v_D is small, at peak value about 0.6 V, and current is delivered to the load. When v is negative the diode is reverse biased and since the reverse bias resistance is very much greater than R_L the instantaneous volt drop across the diode is equal to the applied voltage v. Only a small leakage current flows to the load and the effect can usually be ignored. Waveforms are shown in Fig. 8.3.

The average voltage is given by integrating the positive half-cycle over a complete cycle, i.e.

$$V_{dc} = \frac{1}{2\pi} \int_0^\pi (V_{pk} - V_D) \sin \omega t \, d\omega t$$

$$= \frac{V_{pk} - V_D}{\pi}$$

$$= 0.3185(V_{pk} - V_D) \text{ volts}$$

[8.1]

Fig. 8.3 Half-wave rectifier
voltage waveforms.

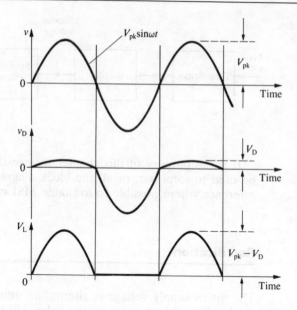

The amount of **ripple** (a.c. superimposed on the main d.c. voltage) is very large in this circuit.

Full-wave rectifier

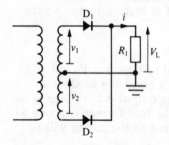

Fig. 8.4 Full-wave rectifier circuit.

By using a second diode, connected as shown in Fig. 8.4, it is possible to have current flowing in the load resistor for the whole of the input voltage cycle. The circuit basically consists of two half-wave rectifiers that conduct on alternate half-cycles.

The transformer secondary has a centre tap so that during the first half-cycle of input voltage the anode D_1 goes positive while the anode D_2 is negative. Hence D_1 conducts and the resulting current flow and voltage across R_L is the same as for the half-wave arrangement. During the second half-cycle of input voltage, the anode of D_2 goes positive while the anode of D_1 goes negative; hence D_2 conducts, while D_1 is off, providing current flow through and a volt drop across R_L as before. The waveforms are shown in Fig. 8.5.

The average value of the input voltage is now twice the value obtained using equation [8.1] so that

$$V_{dc} = \frac{2(V_{pk} - V_D)}{\pi}$$

$$= 0.637(V_{pk} - V_D) \text{ volts}$$

The amount of ripple is reduced compared with the previous circuit.

The bridge rectifier

This is a variation of the full-wave rectifier which does not require a centre tapped transformer. The circuit is shown in Fig. 8.6(a). During the half-cycle when point A is positive with respect to point B, diodes D_2 and D_4 are conducting while D_1 and D_3 are reverse biased. The arrangement is shown in Fig. 8.6(b) where D_1 and D_3 are omitted since they represent open circuits. During the half-cycle when point B is positive with respect to point A, diodes D_1 and D_3

Fig. 8.5 Full-wave rectifier
voltage waveforms.

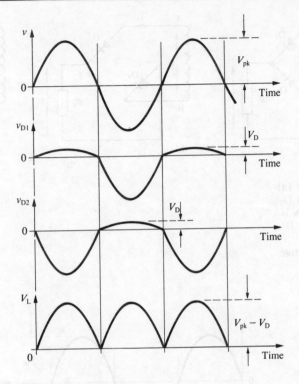

conduct while D_2 and D_4 are reverse biased. The arrangement is
shown in Fig. 8.6(c) where D_2 and D_4 are omitted since they
represent open circuits.

As can be seen the transformer winding is always connected to
the load but the sense of connection reverses with each reversal of
the alternating input voltage. There is a voltage output similar to
that of the full-wave rectifier but with a peak output voltage equal
to the peak input voltage from the secondary less $2V_D$ volts since
two diodes are conducting in series for each output half-cycle. The
resulting output waveform is shown in Fig. 8.7.

The half-wave rectifier is the simplest of the circuits discussed but
its average d.c. voltage output is low with considerable ripple. The
ripple component has a frequency the same value as the input
waveform, i.e. 50 Hz if mains supply is used.

The full-wave and bridge rectifiers both give improved average
d.c. voltage levels with reduced ripple. The ripple component in
these circuits is at *twice* the input frequency, i.e. 100 Hz if mains
supply is used. The full-wave rectifier needs a transformer with
twice the number of secondary turns as the half-wave circuit to give
the same output voltage while the bridge recitifier uses four diodes
so that the cost of both the full-wave and bridge rectifier circuits is
higher than the simple half-wave rectifier circuit. Full-wave bridge
rectifier circuits are normally available as a single package
containing the four diodes.

The diodes used in these power supply circuits must be operated
within their current ratings. Since the diodes in all circuits will at
some stage during a cycle of input be reverse-biased the reverse-
voltage breakdown value is important. The maximum reverse

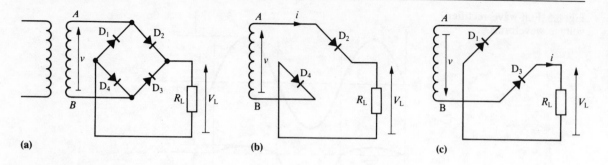

(a)   **(b)**   **(c)**

Fig. 8.6 Bridge rectifier (a) basic circuit, (b) diodes D_2 and D_4 conduct when A is positive with respect to B, (c) diodes D_1 and D_3 conduct when B is positive with respect to A.

Fig. 8.7 Bridge rectifier voltage waveforms.

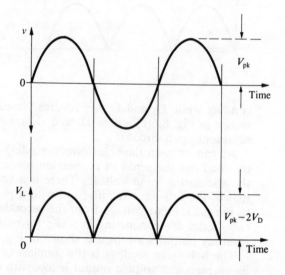

voltage across a diode is called the **peak inverse voltage** (PIV). In the half-wave rectifier the PIV is V_{pk} whereas in the full-wave rectifier the value of PIV is $2V_{pk}$ since, referring to Fig. 8.4, when D_2 is off, say, the voltage at its anode has a maximum value of $-V_{pk}$ while its cathode is at $+V_{pk}$. For a bridge rectifier the PIV is again only V_{pk} volts.

Smoothing circuits

Capacitor smoothing   The simplest way to 'smooth' the voltage output of a rectifier to minimise the ripples and give a better approximation to a d.c. voltage is to use a capacitor in parallel with the resistive load.

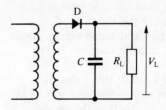

Fig. 8.8 Capacitive smoothing in a half-wave rectifier circuit.

Figure 8.8 shows a circuit for a half-wave rectifier. The effect of adding a capacitor is to prevent the output voltage from falling to zero as the diode cuts off. During the positive half-cycle of input voltage the diode conducts and its forward current will flow through the load as before but now some will flow to charge the capacitor. The output voltage can rise to the same peak voltage as before so the capacitor will charge to this value. As the input voltage begins to fall from its peak value the diode becomes reverse-biased since the voltage across the capacitor is now greater than the applied voltage. The capacitor now begins to lose its charge as it has a ready discharge path via the load resistor R_L. Thus the voltage across the capacitor, and hence the output voltage, falls exponentially with a time constant CR_L seconds. The instantaneous voltage across the capacitor and load at any time during the period the diode is cut off is given by

$$v_L = (V_{pk} - V_D)e^{-t/CR_L} \text{ volts}$$

If the time constant CR_L is chosen to be very much greater than the period of the supply waveform then the capacitor will not lose much of its charge before the diode conducts again and restores the output voltage to its peak value once more. Voltage and current waveforms are shown in Fig. 8.9. From the waveforms it can be seen that the output waveform consists of a mean value V_{dc} upon which is superimposed a ripple voltage, the peak value of which is

$$V_{ripple(pk)} = (V_{pk} - V_D) - V_{dc}$$

From Fig. 8.9, it would appear that the ripple voltage could be reduced by increasing the time constant so that C discharges more slowly and will therefore not have fallen by so much when the input voltage again causes C to charge. The time constant may be increased by using a larger value of C. However, there is a limit to the upper value that C can have. Since the diode only conducts when the input voltage v is greater than v_L then the lower the ripple voltage the smaller is the time available to recharge the capacitor. The current pulse delivered by the diode must therefore have a

Fig. 8.9 Voltage and current waveforms for a capacitor smoothed half-wave rectifier circuit.

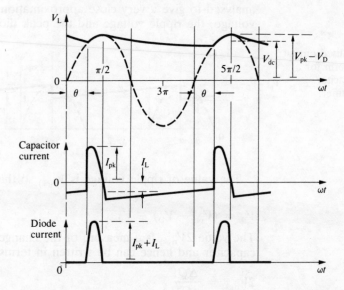

greater peak value to deliver the required load current (since the mean current passed by the diode must equal the mean load current). Rectifiers have peak current ratings and these values could be exceeded if the value of C is too large causing damage to the transformer or diode or both.

When the diode has charged C to a peak value this value is maintained at the cathode even though the anode voltage falls. The input voltage (anode voltage) can fall to $-V_{pk}$ so that the diode must be able to withstand a PIV of approximately $2V_{pk}$. The maximum reverse bias voltage which a diode can withstand is usually specified by the manufacturer so that suitable diodes able to withstand safely the PIV of $2V_{pk}$ can be selected.

Full wave and bridge rectifiers can be smoothed in much the same way as for a half-wave circuit with small, but important differences as listed below:

1. Ripple voltage for full and bridge rectifier circuits is at twice the frequency of the half-wave circuit. Thus the discharge time is reduced by half and the ripple voltage is reduced by the same amount.
2. In a full-wave circuit each diode will alternatively charge the capacitor and when the diode conducts it must pass enough current to charge C to its peak value. Since the capacitor will discharge less in the full-wave circuit then the peak current through each diode is less (approximately half the value of the half-wave case).
3. The peak inverse voltage for the full-wave case is $2V_{pk}$ and for the bridge rectifier it is V_{pk}.

To calculate the output d.c. voltage and the peak value of the ripple voltage is not a simple matter. Referring to the output voltage waveform of Fig. 8.9 for a half rectifier it can be seen that the diode conducts after an angle of θ radians and cuts off again at an angle of $\pi/2$ radians. The fall in output voltage after the diode cuts off is exponential and continues until $2\pi + \theta$ radians later when the diode conducts again; the process is repetitive. The output waveform may be simplified, without too much loss of accuracy, as Fig. 8.10 shows. This approximation to the output voltage can be analysed to give a very close approximation to the d.c. output voltage, the ripple voltage and the peak diode current.

Fig. 8.10 Idealised output voltage waveform for a half-wave capacitor smoothed rectifier.

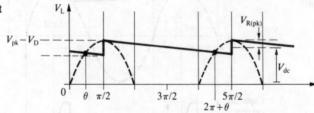

The value of ripple voltage is $V_{r(pk)}$ so that the d.c. output voltage is

$$V_{dc} = (V_{pk} - V_D) - V_{r(pk)} \qquad [8.2]$$

The value $2V_{r(pk)}$ is a measure of the change in voltage across the capacitor and hence can be written in terms of stored charge, i.e.

$$2V_{r(pk)} = \frac{\Delta Q}{C} \qquad [8.3]$$

Once the diode cuts off at $\pi/2$ (and at the next peak $5\pi/2$, etc.) the voltage across the capacitor is assumed to decrease linearly and the charge is lost over a period of $\pi/2$ to $5\pi/2$ or 2π radians. Since current is equal to the rate of change of charge the current delivered by the capacitor into the load during the period of 2π radians is given by

$$I_L = \frac{\Delta Q}{\Delta T} \text{ amps}$$

However, the change in time, ΔT, is the time of one cycle (2π radians) so that

$$\Delta T = \frac{1}{f}$$

where f is the frequency of the output waveform in Hz. Thus

$$I_L = \Delta Q f \text{ amps} \tag{8.4}$$

Substituting for ΔQ from equation [8.4] into equation [8.3] gives

$$2V_{r(pk)} = \frac{I_L}{fC} \tag{8.5}$$

and substituting for $2V_{r(pk)}$ from equation [8.5] into equation [8.2] gives

$$V_{dc} = (V_{pk} - V_D) - \frac{I_L}{2fC}$$

As stated earlier the current flowing in the load (I_L) during the period from $\pi/2$ to $5\pi/2$ ($5\pi/2$ to $9\pi/2$, etc.) when the diode is off, is given by the capacitor discharging through R_L. Hence

$$V_{dc} = I_L R_L \tag{8.6}$$

This equation assumes the value for I_L is *constant* during the discharge part of the capacitor cycle. This is clearly not so but little accuracy is lost if I_L is assumed to be the *average* value of load current during that period. Substituting for I_L from equation [8.6] gives

$$V_{dc} = (V_{pk} - V_D) - \frac{V_{dc}}{2fCR_L}$$

and rearranging gives

$$V_{dc} = \frac{(V_{pk} - V_D)2fCR_L}{1 + 2fCR_L} \tag{8.7}$$

The ripple component of the output voltage is triangular in the approximated version and has an effective a.c. value of

$$V_{r(rms)} = \frac{(V_{pk} - V_D) - V_{dc}}{\sqrt{3}}$$

The factor $\sqrt{3}$ allows for the triangular shape of the waveform. This can be shown with reference to the triangular waveform of Fig. 8.11. The value of the a.c. ripple voltage will fall linearly from $V_{r(pk)}$ to a value of $-V_{r(pk)}$, with reference to V_{dc}. Thus the instantaneous value of the a.c. ripple voltage can be given by the equation:

$$V_{r(pk)} - \frac{2V_{r(pk)}t}{T}$$

Fig. 8.11 Capacitor input
filter for a full-wave
rectifier circuit.

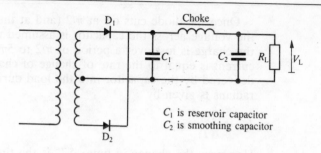

C_1 is reservoir capacitor
C_2 is smoothing capacitor

where T is the period of the fall in seconds (i.e. $T = \frac{1}{f}$ seconds
where f is the frequency of the waveform in Hertz) and t is the
instantaneous value of time which can have any value between 0
and T seconds.

Thus the rms value of the a.c. ripple can be found from

$$V_{r(rms)} = \sqrt{\frac{1}{T}\int_0^T \left(V_{r(pk)} - V_{r(pk)}\frac{2t}{T}\right)^2 dt}$$

$$= V_{r(pk)}\sqrt{\frac{1}{T}\int_0^T \left(1 - \frac{2t}{T}\right)^2 dt}$$

$$= V_{r(pk)}\sqrt{\frac{1}{T}\left[\frac{1}{3}\left(1 - \frac{2t}{T}\right)^3\left(-\frac{t^2}{T}\right)\right]_0^T}$$

(ignoring constant of integration)

$$= V_{r(pk)}\sqrt{\frac{1}{T}\left[\frac{1}{3}\left(1 - \frac{2T}{T}\right)^3\left(-\frac{T^2}{T}\right)\right]}$$

$$= V_{r(pk)}\sqrt{\frac{1}{T}\left[\frac{1}{3}(-1)(-T)\right]}$$

$$= \frac{V_{r(pk)}}{\sqrt{3}}$$

and since $V_{r(pk)} = (V_{pk} - V_D) - V_{dc}$

then

$$V_{r(rms)} = \frac{(V_{pk} - V_D) - V_{dc}}{\sqrt{3}} \quad \text{as shown earlier.}$$

Substituting for $(V_{pk} - V_D)$ and V_{dc} gives

$$V_{r(rms)} = \frac{V_{dc}}{2\sqrt{3}fCR_L} \text{ volts} \qquad [8.8]$$

Ripple factor has been mentioned earlier and is the ratio of rms
value of ripple voltage to the mean d.c. value. Using equations [8.7]
and [8.8] and rearranging gives

$$\text{ripple factor} = \frac{1}{2\sqrt{3}fCR_L} \qquad [8.9]$$

This could also be obtained directly from equation [8.8] by dividing
both sides by V_{dc}.

Finally to determine the peak diode current refer again to the
voltage output and capacitor current waveforms of Fig. 8.9. While
the diode is conducting the capacitor charging current has a value,
i_c, given by

$$i_c = C\frac{dv}{dt}$$

where v is the input voltage given by

$$v = (V_{pk} - V_D)\sin \omega t$$

Then

$$i_c = C(V_{pk} - V_D)\omega \cos \omega t \qquad \text{[8.10]}$$

for the period θ to $\pi/2$ radians.

The current delivered to the load during this same period is

$$i_L = \frac{(V_{pk} - V_D)\sin \omega t}{R_L}$$

and the diode current is

$$i_d = i_L + i_c$$

Since it has been assumed that $\omega C \gg R_L$ it follows that i_L is very small compared to i_c and hence i_d can be approximated to i_c only. The maximum value of i_c occurs at θ radians ($2\pi + \theta$ radians, etc.) when the capacitor begins to recharge so that to find $i_{d(max)}$ substitute θ for ωt in equation [8.10] i.e.

$$i_{d(max)} = C(V_{pk} - V_D)\omega \cos \theta$$

To determine θ, since the diode begins conducting at an output voltage $(V_{pk} - V_D) - 2V_{r(pk)}$ then

$$(V_{pk} - V_D)\sin \theta = (V_{pk} - V_D) - 2V_{r(pk)}$$

$$\sin \theta = 1 - \frac{2V_{r(pk)}}{(V_{pk} - V_D)}$$

Substituting for $V_{r(pk)}$ from equation [8.2] and for V_{dc} from equation [8.7] gives

$$\sin \theta = \frac{2fCR_L - 1}{2fCR_L + 1} \qquad \text{[8.12]}$$

from which θ may be calculated.

Example 8.1 *Find the peak value of ripple voltage in a half-wave rectifier when the peak output voltage is 20 V, load resistance is 250 Ω and capacitance is 1250 μF. Assume a 50 Hz supply.*

From equation [8.5]

$$V_{r(pk)} = \frac{I_L}{2fC} \text{ volts}$$

Since $V_{out} = 20$ V and $R_L = 250$ Ω then $I_L = \frac{20}{250} = 80$ mA

then

$$V_{r(pk)} = \frac{80 \times 10^{-3}}{2 \times 50 \times 1250 \times 10^{-6}}$$

$$= 0.64 \text{ V}$$

Example 8.2 *Using the parameters of example 8.1 but using a capacitor of 5000 μF, find the peak ripple voltage.*

I_L is still 80 mA. Hence

$$V_{r(pk)} = \frac{80 \times 10^{-3}}{2 \times 50 \times 5000 \times 10^{-6}}$$

$$= 0.16 \text{ V}$$

Example 8.3   *Using the parameters of example 8.1, find the d.c. output voltage and the ripple factor.*

From equation [8.7],

$$V_{dc} = \frac{(V_{pk} - V_D)2fCR_L}{1 + 2fCR_L}$$

and substituting values for C, R_L, f etc.,

$$V_{dc} = \frac{20 \times 2 \times 50 \times 1250 \times 10^{-6} \times 250}{1 + (2 \times 50 \times 1250 \times 10^{-6} \times 250)}$$

Hence

$$V_{dc} = \frac{20 \times 31.25}{1 + 31.25}$$

$$= \frac{625}{32.25} = 19.38 \text{ V}$$

Or $V_{dc} = V_{pk} - V_{r(pk)} = 20 - 0.64 \text{ V} = 19.36 \text{ V}$, using the value of $V_{r(pk)}$ from example 8.1. From equation [8.9],

$$\text{Ripple factor} = \frac{1}{2\sqrt{3} \times 50 \times 1250 \times 10^{-6} \times 250}$$

$$= 0.0185$$

Example 8.4   *A power supply using a half-wave rectifier is to have a d.c. output voltage of 25 V using a load resistance of 500 Ω. The ripple factor is not to exceed 0.015. Find a suitable value for C. What is the peak diode current for this circuit? Assume supply frequency is 50 Hz.*

Since from equation [8.9],

$$\text{ripple factor} \leqslant \frac{1}{2\sqrt{3}fCR_L}$$

then

$$C \geqslant \frac{1}{2\sqrt{3}fR_L \times \text{ripple factor}}$$

$$\geqslant \frac{1}{2\sqrt{3} \times 50 \times 500 \times 0.015}$$

$$\geqslant 770 \text{ } \mu\text{F}$$

Assume a value of 1000 μF can be used. To find peak diode current $(V_{pk} - V_D)$ and θ must be found. From equation [8.7]

$$(V_{pk} - V_D) = \frac{V_{dc}(1 + 2fCR_L)}{2fCR_L}$$

and using $C = 1000 \text{ } \mu\text{F}$ gives

$$(V_{pk} - V_D) = \frac{25 \times (1 + (2 \times 50 \times 1000 \times 10^{-6} \times 500))}{2 \times 50 \times 1000 \times 10^{-6} \times 500}$$

$$= \frac{25 \times 51}{50}$$

$$= 25.5 \text{ V}$$

From equation [8.12]

$$\sin \theta = \frac{2fCR_L - 1}{2fCR_L + 1}$$

$$\sin \theta = \frac{49}{51} = 0.9608$$

$$\theta = \sin^{-1} 0.9608 = 73.9°$$

Hence peak diode current from equation [8.11] is

$$i_{d(max)} = C(V_{pk} - V_D)\omega \cos \theta$$

$$= 1000 \times 10^{-6} \times 25.5 - 2\pi \times 50 \times \cos 73.9°$$

$$= 2.22 \text{ A}$$

This example highlights the problem regarding the peak current that must be delivered by the diode when conducting compared to its mean value. The mean load current in the example is V_{dc}/R_L or $25/500 = 50$ mA while the peak value is 2.22 A. The diode must be capable of handling this peak current.

The analysis above was carried out for the half-wave rectifier. The reasoning is the same for a full-wave rectifier except that the period of the ripple is halved. Thus the equation previously obtained for the half-wave rectifier can be used for the full-wave rectifier if $2f \cdot$ is used instead of f. For example

$$V_{dc} = \frac{(V_{pk} - V_D)4fCR_L}{1 + 4fCR_L} \qquad [8.13]$$

$$\text{ripple factor} = \frac{1}{4\sqrt{3}fCR_L} \qquad [8.14]$$

$$\sin \theta = \frac{4fCR - 1}{4fCR_L + 1} \qquad [8.15]$$

Example 8.5   *Using the data of the previous example, recalculate a suitable value for C and find the peak diode current assuming a full-wave rectifier is used. Supply frequency is still 50 Hz.*

$$\text{Ripple factor} = \frac{1}{4\sqrt{3}fCR_L} \text{ from equation [8.14]. Hence}$$

$$C = \frac{1}{4\sqrt{3}fR_L \times \text{ripple factor}}$$

$$= 385 \ \mu\text{F}$$

Thus value of capacitor is half the value needed for the half-wave circuit. Assume a value of 470 μF is used. Since, from equation [8.13]

$$(V_{pk} - V_D) = \frac{V_{dc}(1 + 4fCR_L)}{4fCR_L}$$

$$(V_{pk} - V_D) = \frac{25(1 + 100)}{100}$$

$$= 25.25 \text{ V}$$

and from equation [8.15]

$$\sin \theta = \frac{4fCR_L - 1}{4fCR_L + 1}$$

$$\sin \theta = \frac{99}{101} = 0.9802, \ \theta = 78.58°$$

Also, the peak diode current from equation [8.11] is

$$i_{d(max)} = C(V_{pk} - V_D)\omega \cos \theta$$

$$= 470 \times 10^{-6} \times 25.25 \times 2\pi \times 50 \times \cos 78.58°$$

$$= 0.738 \ A$$

The peak current for the full-wave case is significantly lower than the equivalent value for the half-wave case. This is due to reduced capacitance value and the reduced conduction period for the diode helping to reduce its peak current.

Ripple reduction

The amount of ripple voltage from a capacitor-smoothed rectifier may still be too great for some applications and, as has been mentioned, there are limitations on the method of reducing the ripple by increasing the capacitor value. An extra circuit is required which acts as a filter to suppress the a.c. but allows the d.c. to pass. A possible circuit is shown in Fig. 8.11 and consists of a series inductor-capacitor filter with the output taken across the capacitor. The inductor is often known as a **choke** and is selected because of low d.c. resistance but high reactance to the ripple frequency. This type of filter is effective but not widely used in transistor equipment because of the cost and large physical size of the choke. The circuit is known as a **capacitor input filter** and its effect in reducing ripple voltage at the output can best be explained by the following example.

Example 8.6

The circuit of example 8.5 is modified by the addition of a filter, consisting of a choke and a smoothing capacitor, inserted between the reservoir capacitor and the resistive load. If the choke has an inductance of 5 H with a series resistance of 20 Ω and the smoothing capacitance is 1000 μF, find V_{dc} at the output and the peak ripple voltage across the load.

Under d.c. conditions the reservoir and smoothing capacitors are effectively open circuits. The output voltage developed across C_1 is the same value as in example 8.5, i.e. 25 V. This is now applied across the d.c. resistance of the choke (r) and the load resistance (R_L) in series so that the new value of V_{dc} (given by V'_{dc}) is

$$V'_{dc} = V_{dc}\left(\frac{R_L}{r + R_L}\right)$$

$$= 25\left(\frac{500}{20 + 500}\right)$$

$$= 24.04 \ V$$

For the ripple voltage, frequency $2f = 100$ Hz, the series combination of X_L and X_{c2} allows a fraction only of the ripple voltage across C_1 to be developed across C_2. The reduction in ripple is given by

$$V'_{r(pk)} = V_{r(pk)}\left(\frac{X_{c2}}{X_L + X_{C2}}\right) \text{volts}$$

where $V'_{r(pk)}$ is the new ripple voltage after filtering, and $V_{r(pk)}$ is the value of the peak ripple voltage from the previous example. Since

$$V_{r(pk)} = \frac{V_{dc}}{4fCR_L} \text{ for the original circuit, then:}$$

$$V_{r(pk)} = \frac{25}{4 \times 50 \times 470 \times 10^{-6} \times 500}$$

$$= 0.53 \text{ V}$$

Since

$$X_{C2} = 1/2\pi f C_2 \text{ and } X_L = 2\pi f L \text{ then}$$

$$\frac{X_{C2}}{X_L + X_{C2}} = \frac{1/2\pi f C_2}{2\pi f L + 1/2\pi f C_2}$$

$$= \frac{1}{1 + 4\pi^2 f^2 L C_2}$$

Substituting the values for f, L and C_2 gives

$$\frac{X_{C2}}{X_L + X_{C2}} = \frac{1}{1 + 4\pi^2 \times 100^2 \times 5 \times 10^3 \times 10^{-6}}$$

so that

$$V'_{r(pk)} = \frac{V_{r(pk)}}{1975}$$

$$= \frac{0.53}{1975}$$

$$= 0.269 \text{ mV}$$

Because of the disadvantages of the choke a simple *RC* filter could be used with the resistor replacing the choke. The arrangement is shown in Fig. 8.12. The attenuation of the ripple is now given by the ratio

$$\frac{X_{C2}}{R + X_{C2}} = \frac{1}{1 + R/X_{C2}}$$

$$= \frac{1}{1 + 2\pi f C_2 R}$$

Fig. 8.12 *RC* filter circuit showing the potential divider effect of *R* and *C*.

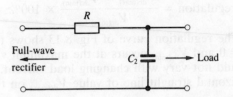

The value of *R* must be carefully chosen to give suitable ripple reduction but not allow too much of a d.c. volt drop across it.

Example 8.7 *The circuit of example 8.5 is to be modified by the inclusion of the RC filter between the rectifier and load. If the value of C_2 is 1000 μF and R is 50 Ω, find the new values of V_{dc} and $V_{r(pk)}$ across the output.*

From the previous example we know that without the *RC* filter the values for V_{dc} and $V_{r(pk)}$ are 25 V and 0.53 V respectively. With the filter the attenuation of the ripple signal is given by $\dfrac{1}{1 + 2\pi f C_2 R}$ and substituting for f, C_2 and R gives

$$\frac{1}{1 + 2\pi f C_2 R} = \frac{1}{1 + 2\pi \times 100 \times 1000 \times \times 10^{-6} \times 50}$$

$$= \frac{1}{1 + 10\pi}$$

$$= 0.031$$

Note that in the above calculation $f = 100$ Hz. The new ripple voltage after filtering

$$V'_{r(pk)} = 0.031 \times 0.53$$

$$= 0.0164 \text{ V}$$

$$= 16.4 \text{ mV}$$

Because the load resistance is 500 Ω, from the previous example, then the attenuation of V_{dc} is given by

$$V'_{dc} = V_{dc}\!\left(\frac{R_L}{R_L + R}\right)$$

$$= 25\!\left(\frac{500}{500 + 50}\right)$$

$$= 22.73 \text{ V}$$

As can be seen from example 8.7 the *RC* filter is not so effective as the *LC* filter but is quite suitable for many applications.

Regulation

The change in load voltage with changes in load current is called the **regulation**. It is usually defined as a percentage, i.e.

$$\% \text{ regulation} = \frac{V_{dc(max)} - V_{dc(min)}}{V_{dc(max)}} \times 100\% \qquad [8.16]$$

The regulation curve of Fig. 8.13 shows that $V_{dc(max)}$ occurs at $I_L = 0$ and $V_{dc(min)}$ occurs at the maximum value of I_L. Ideally V_{dc} should not vary with changing load current. This suggests a horizontal straight line of value $V_{dc(max)}$ on the graph of Fig. 8.13.

Fig. 8.13 Regulation curve for a filtered rectifier circuit.

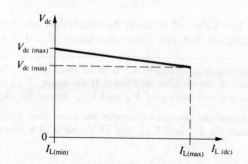

Since $\Delta V_{dc} = 0$ for finite changes in I_L then the internal resistance of the ideal power supply is $\Delta V_{dc}/\Delta I_L = 0$.

In practical circuit applications the finite output resistance of the power supply could cause problems since signal voltages could appear across the power supply and find their way to other signal circuits causing possible oscillations due to feedback. At high signal frequencies this is not a problem since the shunt capacitance of the rectifier circuit appears as a short-circuit at those frequencies.

For a full-wave rectifier the value of V_{dc} is given by:

$$V_{dc} = (V_{pk} - V_D) - \frac{I_L}{4fC}$$

This could be rewritten as:

$$V_{dc} = (V_{pk} - V_D) - I_L R_O \qquad [8.17]$$

where R_O is the output resistance of the power supply and if $R_O = 0$ then

$$V_{dc} = (V_{pk} - V_D)$$

always, regardless of the value of load current. Thus

$$R_O = \frac{1}{4fC}$$

where f is the frequency of the supply, assumed to be 50 Hz.

Example 8.8 | *For the full-wave rectifier of example 8.5 and using the value for C from that example, find the output resistance of the circuit.*

Since $C = 470\ \mu F$ for the circuit

Then $R_O = \dfrac{1}{4 \times 50 \times 470 \times 10^{-6}}\ \Omega$

Thus $R_O = 10.64\ \Omega$.

The value for R_O in the previous example could be reduced by the use of a larger capacitor but as stated earlier there is an upper limit to capacitor values because of the peak currents involved when the diodes conduct.

The regulation curve of Fig. 8.13 is shown as a straight line over all of its range. In practice, output resistance could vary depending on the value of load current and the regulation curve may not be a straight line. In any event the value of output resistance found in example 8.8 is far too high and extra circuits are required to reduce the value of R_O to more satisfactory values.

Voltage regulators

It is desirable for a power supply output voltage to remain constant regardless of load current variations. Factors such as variations in the input voltage to the power supply should also not affect the circuit output. A block diagram of a voltage regulator is shown in Fig. 8.14.

Fig. 8.14 Block diagram of
a voltage regulator.

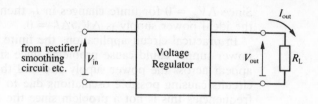

Stability factor (S) for a voltage regulator is a measure of the
effectiveness of the regulation. The stability factor is given by

$$S = \frac{\Delta V_{out}}{\Delta V_{in}}$$

for a constant output current I_L. Ideally S should be 0 and in
practice, for good regulators, the value can be from 0.005 to 0.0002.

The output resistance of the regulator should also be zero ohms
(assuming the regulator is a voltage generator).

Typically

$$R_O = \frac{\Delta V_{out}}{\Delta I_L}$$

for a constant input voltage. In a practical regulator R_O should be
less than 1 Ω, sometimes considerably less.

Zener diode regulator

The zener diode is a silicon diode specially doped to give certain
breakdown conditions under reverse bias. That part of the diode
characteristic is shown in Fig. 8.15. Operated in the region between
$I_{z(max)}$ and $I_{z(min)}$, large current swings can be achieved for small
changes in zener voltage. Thus the device has a low output
resistance; values of 4 Ω to 10 Ω are common. A basic zener
regulator is shown in Fig. 8.16.

Provided V_{in} is such as to produce zener breakdown then a

Fig. 8.15 Reverse
characteristic of a zener
diode.

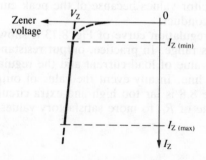

Fig. 8.16 Basic zener
regulator circuit.

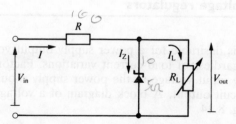

stabilised V_{out} is produced even if V_{in} and/or I_L should vary. The effect of variations in these parameters can be explained as follows:

1. If V_{in} should, say, increase then the voltage across the zener diode tends to increase and a larger current flows through the zener and since,

$$I = I_L + I_z \qquad\qquad [8.18]$$

then an increase in I_z causes a similar increase in I, assuming I_L is constant. Thus there is a larger volt drop across the ballast resistor R and this allows V_{out} to remain substantially constant.

2. If I_L should, say, increase due to a reduction in R_L then since V_{out} $(= V_z)$ tends to decrease, the zener will pass a reduced current. From equation [8.18] if a reduction in I_z balances the increase in I_L then I remains constant and the volt drop across R remains constant. Thus assuming V_{in} has not varied, V_{out} will be substantially constant.

For both the above considerations of varying load current and/or input voltage neither must be allowed to cause the zener voltage to fall below its minimum value or exceed its maximum value. If I_z falls below $I_{z(min)}$ then the zener behaves like an ordinary reverse-biased diode of large resistance. If I_z goes above $I_{z(max)}$ then the zener would exceed a maximum power rating since $I_{z(max)} = P_{diss}/V_z$. Values for power dissipation and V_z are quoted for all zener diodes by the manufacturer.

Example 8.9   *A 10 V, 500 mW zener diode is to be used in a stabiliser circuit. Find a suitable value of ballast resistor if V_{in} is 18 V. Load resistance is 1 kΩ. Minimum diode current is 3 mA. Find the stabilisation factor for the circuit over the correct diode operating conditions if V_{in} should rise above 18 V. Assume dynamic resistance of the diode is 10 Ω.*

$$V_{out} = 10\text{ V so that } I_L = \frac{10}{1 \times 10^3} = 10\text{ mA}$$

Minimum zener current = 3 mA

Volt drop across ballast resistor $= V_{in} - V_{out}$

$$= 18 - 10$$

$$= 8\text{ V}$$

So that $R = \dfrac{V_{in} - V_{out}}{I} = \dfrac{8}{13}\text{ kΩ} = 615\ \Omega.$

As V_{in} increases, I increases and this must flow through the diode.

Maximum possible value of $I = I_{z(max)} + I_L$

$$= \frac{P_{diss}}{V_z} + I_L$$

$$= \left(\frac{500}{10} + 10\right)\text{ mA}$$

$$= 60\text{ mA}$$

If I_z changes from 3 mA to 50 mA then $I_z = 47$ mA. Since resistance of diode (r_d) is 10 Ω then

$$\Delta V_z = \Delta I_z r_d$$

$$= 47 \times 10^{-3} \times 10$$

$$= 0.47\text{ V}$$

So that

new value of V_{out} = 10.47 V

new value of v_{in} = 10.47 + (60 × 10⁻³ × 615)

$\qquad\qquad$ = 10.47 + 36.9

$\qquad\qquad$ = 47.37 V

Hence

ΔV_{in} = 47.37 − 18

$\qquad$ = 29.37V

and stability factor

$$S = \frac{\Delta V_{out}}{\Delta V_{in}}$$

$$= \frac{0.47}{29.37}$$

$$= 0.016$$

The zener stabiliser will also give good ripple voltage reduction. Because the ballast resistor R is in series with the zener diode resistance r_d the output ripple voltage will be reduced in proportion to the values of the resistors

$$V_{r(out)} = V_{r(in)}\left(\frac{r_d}{r_d + R}\right) \text{ volts}$$

Example 8.10 *If the zener stabiliser of example 8.9 has a ripple input voltage of 0.6 V (peak), what will be the ripple voltage output from the circuit?*

Since for the stabiliser of example 8.9 r_d = 10 Ω and R = 615 Ω, then

$$V_{r(out)} = V_{r(in)}\left(\frac{10}{10 + 615}\right) \text{ volts}$$

$$= 0.6\left(\frac{10}{625}\right) \text{ volts}$$

$$= 9.6 \text{ mV}$$

Zener diodes are available with breakdown voltages of hundreds of volts and power dissipations in tens of watts. However, such diodes are not cheap and in general zener diode circuits of Fig. 8.16 are used as regulators for low power levels. For larger regulators a zener diode could still be used but only to provide a reference source.

Series voltage regulator The stabilisation efficiency of a zener diode is improved if the current through the diode is a fraction of the load current. This can be achieved using the circuit of Fig. 8.17.

The circuit uses a bipolar transistor TR_1 in series with a load, hence the circuit description of 'series' regulator. Since the volt drop between base and emitter is typically 0.6 V for the emitter currents in the range of operation of the device it follows that the emitter voltage is 0.6 V less than the base voltage which is now fixed by the

Fig. 8.17 Simple series
voltage regulator circuit.

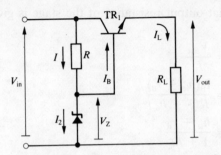

zener diode. The transistor is in fact operating as an emitter
follower which promises to give a low value of regulator output
resistance.

Figure 8.18 shows the *h*-parameter equivalent circuit of the series
regulator. The collector of TR_1 is shown connected to a.c. earth
since it is assumed that V_{in} is a constant d.c. voltage. The effects of
h_{re} and h_{oe} have been ignored in the equivalent circuit. To find the
output resistance of the circuit, R_L is removed and replaced by a
voltage generator of value V_{out} while the voltage generator V_z is
replaced by a short-circuit. The arrangement is shown in Fig. 8.19.

Fig. 8.18 Equivalent circuit
of the simple series voltage
regulator.

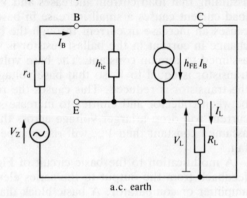

Fig. 8.19 Equivalent circuit
of the series voltage
regulator modified to find
the output resistance of the
circuit.

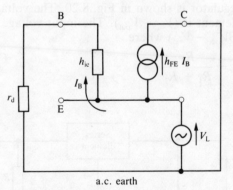

From Fig. 8.19 the value of the voltage generator V_{out} is given by

$$V_{out} = I_B(h_{ie} + r_d)$$

while the output current I_L is given by

$$I_L = I_B + h_{fe}I_B$$
$$= I_B(1 + h_{fe})$$

so that output resistance of the stage is given by the ratio of V_{out} to I_L

$$R_O = \frac{V_{out}}{I_L}$$
$$= \frac{I_B(h_{ie} + r_d)}{I_B(1 + h_{fe})}$$
$$= \frac{h_{ie} + r_d}{1 + h_{fe}} \qquad [8.19]$$

The actual value of R_O will obviously depend on the device used and the manufacturer's specified values for h_{ie} and h_{fe} over the output current range of the regulator. As an example consider a device where at the specified load current, $h_{ie} \approx 300\ \Omega$ and $h_{fe} \approx 200$. Then assuming $r_d = 10\ \Omega$, from equation [8.19]

$$R_O = \frac{300 + 10}{201} \approx 1.55\ \Omega$$

The way in which the series element provides compensation for, say, changes in load current can be explained qualitatively by assuming that load current increases and V_{out} falls. The increase in load current causes a small increase in base current which in turn causes an increase in current through the ballast resistor. Since the change in current in the ballast resistor is small, V_{in} and V_z are assumed to remain constant. The base voltage of the series transistor is equal to V_z so that base voltage is constant and V_{BE} for the transistor is reduced. This causes the resistance of the transistor between collector and emitter to increase so that the increased load current will drop a larger voltage across the transistor. Since V_{in} is assumed constant then V_{out} will rise compensating for the original fall.

A modification to the basic circuit of Fig. 8.17 is to provide feedback from the output to the series element via an error amplifier or comparator. A basic block diagram of the feedback regulator is shown in Fig. 8.20. The voltage across the series element is $(V_{in} - V_{out})$. The input voltage to the error amplifier is $(\beta V_{out} - V_{ref})$ where

$$\beta = \frac{R_2}{R_1 + R_2}$$

Fig. 8.20 Block diagram of a series voltage regulator with feedback.

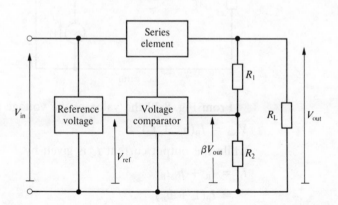

The output of the series element is an amplified version of the input

$$(V_{in} - V_{out}) = A(\beta V_{out} - V_{ref}) \qquad [8.20]$$

where A is the gain between the input to the error amplifier and the output of the regulator.

For *changes* in voltage, equation [8.20] can be rewritten as

$$\Delta(V_{in} - V_{out}) = \Delta(\beta V_{out} - V_{ref})A \qquad [8.21]$$

but if V_{ref} is constant, $\Delta V_{ref} = 0$, so that equation [8.21] becomes

$$\Delta V_{in} - \Delta V_{out} = A\beta\Delta V_{out}$$

or

$$\Delta V_{in} = (1 + A\beta)\Delta V_{out}$$

$$\therefore \Delta V_{out} = \frac{\Delta V_{in}}{(1 + A\beta)} \qquad [8.22]$$

Since stability factor is the ratio of change in V_{out} to the change in V_{in}, it follows from equation [8.22] that stability factor is improved by a factor of $(1 + A\beta)$.

To find output resistance of the series regulator divide both sides of equation [8.22] by ΔI_L to give

$$\frac{\Delta V_{out}}{\Delta I_L} = \frac{\Delta V_{in}/\Delta I_L}{(1 + A\beta)}$$

But $\Delta V_{out}/\Delta I_L$ is the output resistance of the regulator (R_O), and $\Delta V_{in}/\Delta I_L$ is the output resistance of the filter (r_O). So that

$$R_O = \frac{r_O}{(1 + A\beta)}$$

Thus the output resistance of the regulator is reduced by a factor of $(1 + A\beta)$ also.

It could also be shown that the ripple voltage component of the filter output is also reduced by a factor of $(1 + A\beta)$.

A practical arrangement of the feedback series voltage regulator is shown in Fig. 8.21. In this circuit TR_2 acts as a differential amplifier. Holding V_E of TR_2 constant via the zener diode means that changes in base voltage can cause changes in I_E and hence I_C. The resistor R_3 connected to the supply line can produce a change in the volt drop depending on TR_2 collector current which affects the base

Fig. 8.21 Practical example of a series voltage regulator with feedback.

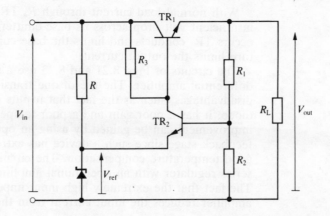

voltage of TR$_1$ since $V_{C2} = V_{B1}$ in this circuit. Since TR$_1$ is connected as an emitter-follower then the voltage change at TR$_1$ base is passed to the output. Thus if the output voltage rises, because of changes in load conditions, the rise is passed to the base of TR$_2$. Emitter and collector currents of TR$_2$ will rise (since V_{BE} of that device has increased). The increase in I_C of TR$_2$ produces an increased volt drop across R_3 hence V_{C2} and V_{B1} fall. Thus V_{out} falls compensating for the original rise.

The circuit of Fig. 8.21 may be modified to provide overload protection. Although it is likely that there would be a fuse in the transformer primary circuit of the rectifier, the time taken for the fuse to 'blow' under overload would be inadequate for protection of electronic circuits. An electronic overload protection system can be produced simply with a bipolar transistor used as a switch. A basic arrangement is shown in Fig. 8.22.

The necessary base-emitter voltage is produced by the load current flowing through the resistance R_B. The value for R_B can be chosen so that for normal load current the device is off because the volt drop across the resistor is less than the minimum value of 0.5 V needed to turn the device on. Only if the load current becomes excessive will the transistor conduct and this can be utilised to limit the output current of the regulator. Figure 8.23 shows the original circuit of Fig. 8.21 with the overload transistor incorporated.

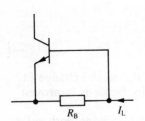
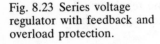

Fig. 8.22 An overload protection transistor.

Fig. 8.23 Series voltage regulator with feedback and overload protection.

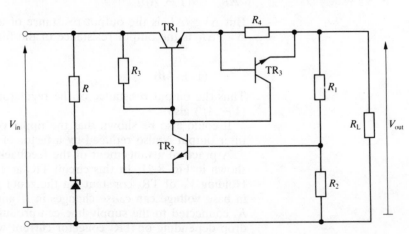

With normal load current through R_4 TR$_3$ is off because of an insufficient volt drop across its base-emitter junction. If an overload occurs TR$_3$ conducts and limits the base current to TR$_1$ which in turn limits the output current.

The circuits of Figs. 8.21 and 8.23 use a single transistor as the differential amplifier. The use of one transistor for this purpose has disadvantages, such as the fact that it puts current into the zener diode, it has a poor gain and a poor temperature coefficient. A big improvement can be gained by using an operational amplifier as the feedback stage since such a device has extremely high gain and very good temperature compensation. The circuit of Fig. 8.24 shows a series regulator with an operational amplifier for feedback control. The fact that the extremely high input impedance of the operational amplifier reduces the input current from the bleeder resistance chain means that a potentiometer can be used to make the output voltage variable.

If V is the actual input to the feedback amplifier, then

$$V = V_{ref} - \beta V_{out} \tag{8.23}$$

The amplified output is applied to TR_1 base. If the gain of the feedback amplifier is high then V can be small and, taken to the limit, equal to zero. If $V = 0$ then

$$V_{ref} = \beta V_{out}$$

$$V_{out} = \frac{V_{ref}}{\beta} \tag{8.24}$$

The value of β is variable because of the potentiometer R_v. When the potentiometer slider is at the bottom of its travel

$$\beta = \frac{R_2}{R_1 + R_2 + R_v} \tag{8.25}$$

while at the top of its travel

$$\beta = \frac{R_2 + R_v}{R_1 + R_2 + R_v} \tag{8.26}$$

Example 8.11   *A voltage regulator is to use the circuit of Fig. 8.24. Values for the bleeder chain resistors are: $R_1 = 1\ \Omega$, $R_2 = 2\ k\Omega$ and $R_v = 2\ k\Omega$. Find the range of output voltages available. Assume $V_{ref} = 5.6\ V$. What would be the effect on the output voltage if R_1 were short-circuited?*

Fig. 8.24 Series voltage regulator using an operational amplifier in the feedback circuit.

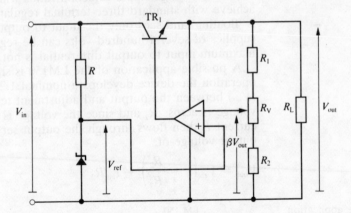

From equation [8.25] when the slider of the potentiometer is at the bottom of its travel

$$\beta = \frac{2}{1 + 2 + 2} = 0.4$$

so that, from equation [8.24]

$$V_{out} = \frac{5.6}{0.4} = 14.0\ V$$

From equation [8.26] when the slider of the potentiometer is at the top of its travel

$$\beta = \frac{2 + 2}{1 + 2 + 2} = 0.8$$

so that from equation [8.24]

$$V_{out} = \frac{5.6}{0.8} = 7.0 \text{ V}$$

Hence the range of output voltage is from 7.0 V to 14.0 V.

If R_1 were short-circuited the new values of β would be:
(From equation [8.25])

$$\beta = \frac{2}{4} = 0.5 \text{ giving } V_{out} = \frac{5.6}{0.5} = 11.2 \text{ V}$$

(and from equation [8.26])

$$\beta = \frac{4}{4} \approx 1.0 \text{ giving } V_{out} = V_{ref} = 5.6 \text{ V}.$$

Hence the range of output voltages would be from 5.6 V to 11.2 V.

A practical voltage regulator

The LM150/250/350 series are adjustable three-terminal positive voltage regulators capable of supplying in excess of 3 A over a 1.2 V to 33 V output range. Only two external resistors are necessary to set the output voltage. The circuits offer full overload protection.

Normally no capacitors are needed unless the device is situated far from the input filter capacitors in which case an input bypass is needed. An optional output capacitor may be added to improve transient response. The adjustment terminal can be bypassed to achieve very high ripple rejection ratios which are difficult to achieve with standard three-terminal regulators. Because the LM150 is 'floating' and sees only the input to output differential voltage, supplies of several hundred volts can be regulated as long as the maximum input to output differential is not exceeded.

A possible application of the LM150 is shown in Fig. 8.25. In operation the device develops a nominal 1.25 V reference voltage (V_{ref}) between the output and adjustment terminal. The reference voltage is across R_1 and since the voltage is constant a constant current I_1 then flows through the output set resistor R_2 giving an output voltage of:

$$V_{out} = V_{ref}\left(1 + \frac{R_2}{R_1}\right) + I_{adj}R_2 \tag{8.27}$$

Fig. 8.25 One application of the LM150 voltage regulator.

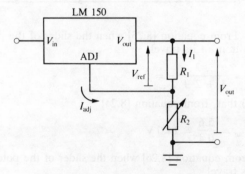

Since the 50 μA current from the adjustment terminal represents an error term, the LM150 has been designed to minimise I_{adj} and make it constant with line and load changes. To achieve this all quiescent operating current is returned to the output establishing a minimum load current requirement. If there is insufficient load on the output, the output will rise.

The adjustment terminal can be bypassed to ground to improve ripple regulation. The capacitor prevents the ripple from being amplified as the output voltage is increased. With a 10 μF bypass capacitor 86 dB ripple rejection is obtainable at any output level.

The load regulation of the device is good but certain precautions need to be taken for best performance. The current set resistor R_1 (usually 240 Ω) should be tied directly to the device output rather than near the load to eliminate line drop from appearing effectively in series with the reference and degrading the regulation. Figure 8.26 shows the effect of the series resistance between the device and R_1. As an example consider a 15 V regulator with a 0.05 Ω resistance between the regulator and load. This will give a load regulation due to line resistance of 0.05 $\Omega \times I_L$ whereas if the set resistor R_1 is connected near the load the effective line resistance will be 0.05 $\Omega(1 + R_2/R_1)$ which with the figures quoted would be 11.5 times worse.

Fig. 8.26 LM150 regulator with line resistance in the output lead.

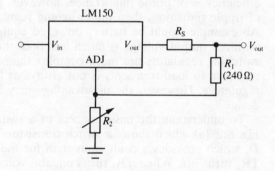

When external capacitors are used with a regulator it may be necessary to add protection diodes to prevent the capacitors from discharging through low current points into the regulator. A 10 μF capacitor has a low enough internal series resistance to produce 20 A spikes when shorted. Even though the current surge is of short duration there may be sufficient energy to damage the i.c. When an output capacitor is connected to a regulator and the input is short-circuited, the output capacitor will discharge into the regulator output. The value of the discharge current depends on the regulator voltage, capacitor value and the rate of decrease of V_{in}. In the LM150 a surge of 25 A can be sustained without damage. For output capacitors of 25 μF or less there is no need for diodes to be used. The bypass capacitor on the adjustment terminal can discharge through a low current junction when either the input or output is short-circuited. The LM150 has an internal 50 Ω resistor to limit the peak discharge current. Protection is unnecessary if the output voltage is less than 25 V or if the capacitance value is less than 10 μF. Figure 8.27 shows the regulator with protection diodes should they be necessary.

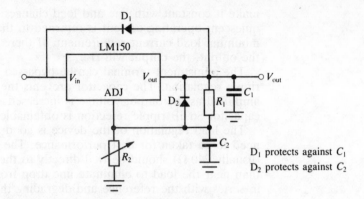

Fig. 8.27 LM150 regulator
with protection diodes.

D_1 protects against C_1
D_2 protects against C_2

Switching regulators

The regulators discussed so far have the disadvantage of being comparatively inefficient. The series resistor must be operated in the active region and takes the full load current so that its power dissipation is very high. In those circumstances where efficiency is of minor importance but good ripple reduction is required then the previously discussed series regulators are perfectly adequate. Where efficiency is of prime importance, however, perhaps to the detriment of ripple reduction, then the switching regulator is a good choice. An example would be battery powered equipment where the required output voltage is much less than the battery voltage. The switching regulators are more complex than linear regulators and the response to load transients is not always as fast as with linear regulators. However, the disadvantage may be overcome by good design.

To understand the basic concept of a switching regulator refer to Fig. 8.28(a) which shows a switch transistor TR_1 and a 'catch' diode D_1 which provides a continuous path for the inductor current when TR_1 turns off. When TR_1 turns on, the voltage at its collector will be nearly equal to the unregulated input voltage. When TR_1 turns off, the magnetic field, produced in the coil L, begins to collapse driving the collector voltage of TR_1 to ground where it is clamped by D_1. The voltage waveform on TR_1 collector is shown in Fig. 8.28(b). L and C act as a filter and the output waveform will be the average value of the switch waveform V_1. Neglecting volt drop across the transistor and diode

Fig. 8.28 Switching
regulator: (a) basic circuit,
(b) voltage waveform.

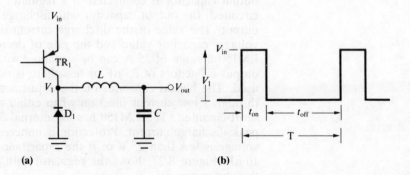

$$V_{\text{out}} = V_{\text{in}}\left(\frac{t_{\text{on}}}{T}\right) \text{ volts}$$

and is independent of load current. Thus, for example, if the input voltage is 10 V and the switch transistor has a 50% duty cycle, then the average voltage at TR_1 collector is 5 V. The waveform is filtered by L and C to appear as 5 V d.c. at the output.

If V_{out} is not of the correct mean value it can be compensated for by altering the switching duty cycle of TR_1. If the value for t_{on} compared to t_{off} increases in a duty cycle T then the mean value of V_{out} increases and vice versa.

Figure 8.29 shows a self-oscillating switching regulator. A_1 is an operational amplifier with inputs of V_{out} and V_{ref}. The reference voltage V_{ref} is equal to the desired value of output voltage. The resistors R_1 and R_2 (with $R_1 \gg R_2$) give a small amount of positive feedback at high frequencies to make the circuit self oscillating. At lower frequencies where the attenuation of the LC filter is less than the attenuation of the R_1/R_2 divider there is net negative feedback to the operational amplifier inverting input.

Fig. 8.29 Self-oscillating switching regulator circuit.

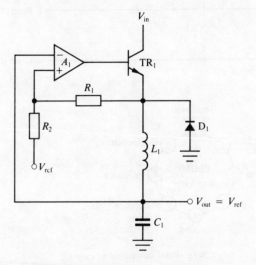

When the circuit is first switched on, $V_{\text{out}} < V_{\text{ref}}$ so that TR_1 is turned on. Then through R_1 raises the voltage on the non-inverting input slightly above the reference voltage. The circuit remains switched on until the ouput rises to this voltage. The amplifier now goes into the active region, switching TR_1 off. At this point, the reference voltage seen by the amplifier is lowered by feedback through R_1 and the circuit will stay off until the output voltage drops to this lower value. Thus the output voltage oscillates about the reference voltage. The amplitude of the oscillation is nearly equal to the voltage fed back through R_1 to R_2 and can be made reasonably small. Thus the ripple can be reasonably small.

Switching regulator circuits will be described that use the LM100 integrated voltage regulator as the control element. This device contains the voltage reference, the operational amplifier and the circuitry for driving a p-n-p switch transistor all on a single chip.

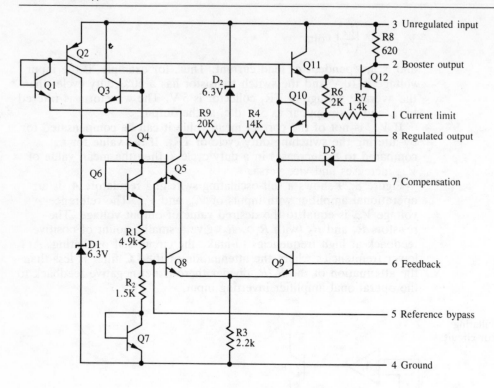

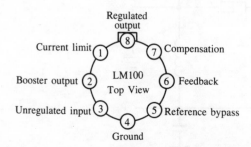

Note: Pin 4 connected to case

Fig. 8.30 Schematic diagram of the LM100 integrated voltage regulator.

Figure 8.30 shows a schematic diagram of the LM100 together with its connection diagram. The voltage reference portion starts with a breakdown diode D_1 which is supplied by a current source from the unregulated input (one of the collectors of Q_2). The output reference diode is buffered by an emitter–follower Q_4 followed by a diode-connecting transistor Q_6. This arrangement gives a positive temperature coefficient of 7 mV/°C. A resistor divider reduces the voltage and the temperature coefficient to exactly compensate for the negative temperature coefficient of Q_7. The output is 1.8 V temperature compensated.

The transistor pair Q_8 and Q_9 form the input stage of the operational amplifier, the gain of which is high because of a current source, one of Q_2's collectors. The output of this stage drives a compound emitter-follower Q_{11}/Q_{12}. The output of Q_{12} is used to drive the p-n-p switch transistor. Q_{10} is also used to limit the output current of Q_{12} to the correct value for driving a p-n-p transistor connected on the booster output. This current is determined by a resistor placed between the current limit and regulated output terminals.

Q_5, Q_3 and Q_1 are bias stabilisation components for Q_2 to give the required collector current values. R_4, R_9 and D_2 start the regulator and D_3 is a clamp diode which stops Q_9 from saturating when it is switching. Figure 8.31 shows a switching regulator circuit utilising the LM100. Feedback to the inverting input of the operational amplifier is via a resistive divider which can be used to set the output voltage in the range 2 to 30 V. R_3 determines the base drive for the switch transistor Q_1 providing it with sufficient drive to saturate it with maximum load current. R_4 works into the 1 kΩ impedance at the reference terminal producing the positive feedback. C_2 helps to minimise output ripple by causing the full ripple to appear on the feedback terminal. C_3 removes fast risetime transients which would otherwise be coupled to pin 5 via the shunt capacitance of R_4. C_3 must be small so as not to integrate the waveform at this point. The circuit of Fig. 8.31 is suitable for output currents up to 500 mA, the limit being set by the output current available from the LM100 to saturate the switch transistor Q_1.

Fig. 8.31 Switching regulator using the LM100.

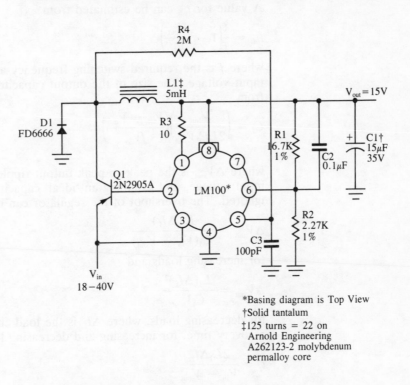

*Basing diagram is Top View
†Solid tantalum
‡125 turns = 22 on
Arnold Engineering
A262123-2 molybdenum
permalloy core

The optimum switching frequency for this circuit has been determined to be between 20 kHz and 100 kHz. At high frequencies switching losses in Q_1 and D_1 become excessive and both devices should be fast switching to minimise these losses.

The output ripple of the regulator at the switching frequency is mainly determined by R_4. The peak-peak ripple output is nearly equal to the peak-peak voltage fed back to pin 5 of the LM100.

Since the input resistance at pin 5 is approximately 1000 Ω, the voltage will be

$$V_{ref} \approx \frac{1000V_{in}}{R_4} \qquad [8.28]$$

In practice the ripple will be somewhat larger than this. When the switch transistor shuts off, the inductor current will be greater than the load current so that the output voltage will continue to rise above the value required to shut off the regulator. The value of the inductor must be chosen so that the current through it does not drastically change during the switching cycle. If it does then Q_1 and D_1 must be able to handle peak currents significantly larger than the load current. The change in inductor current can be written as

$$\Delta I_L = \frac{V_{out}t_{off}}{L}$$

For peak current to be about 1.2 times maximum load current then

$$L_I = \frac{2.5V_{out}t_{off}}{I_{out(max)}} \qquad [8.29]$$

A value for t_{off} can be estimated from

$$t_{off} = \frac{1}{f}\left(1 - \frac{V_{out}}{V_{in}}\right) \qquad [8.30]$$

where f is the required switching frequency and V_{in} the nominal input voltage. The size of the output capacitor can be determined from

$$C_I = \left(\frac{V_{out}}{2L_1\Delta V_{out}}\right)\left(\frac{V_{in} - V_{out}}{fV_{in}}\right)^2 \qquad [8.31]$$

where ΔV_{out} is the peak to peak output ripple and V_{in} the nominal input voltage, and assuming an 'ideal' capacitor, i.e. capacitor loss is ignored. The overshoot of the regulator can be determined from

$$\Delta V_{out} = \frac{L_1(\Delta I_L)^2}{C_1(V_{in} - V_{out})}$$

for increasing loads and

$$\Delta V_{out} = \frac{L_1(\Delta I_L)^2}{C_1 V_{out}}$$

for decreasing loads, where ΔI_L is the load current transient. The recovery time, for increasing and decreasing loads respectively, is

$$t_r = \frac{2L_1\Delta I_L}{V_{in} - V_{out}}$$

$$t_r = \frac{2L_1\Delta I_L}{V_{out}}$$

To improve the load transient response a larger peak to average current ratio in the switch transistor and catch diode is required. Reducing the value of inductance given in equation [8.29] by a factor of two will reduce the overshoot by four times and halve the response time. The output capacitance should be doubled to maintain the switching frequency constant.

Example 8.12 — *Find suitable values of L, C_1 and R_4 for a switching regulator circuit of Fig. 8.30 which is required to deliver 15 V at a maximum current of 300 mA from a 28 V supply. Assume a 40 kHz switching frequency and an output ripple of 14 mV peak to peak.*

From equation [8.28]

$$R_4 \approx \frac{1000 \cdot V_{in}}{V_{ref}}$$

$$= \frac{1000 \times 28}{14 \times 10^{-3}} = 2\ \text{M}\Omega$$

From equation [8.30]

$$t_{off} = \frac{1}{40 \times 10^3}\left(1 - \frac{15}{28}\right)$$

$$= 11.6\ \mu s$$

From equation [8.29]

$$L_1 = \frac{2.5 \times 15 \times 11.6 \times 10^{-6}}{300 \times 10^{-3}}$$

$$= 1.45\ \text{mH}$$

From equation [8.31]

$$C_1 = \left(\frac{15}{2 \times 1.45 \times 10^{-3} \times 14 \times 10^{-3}}\right)\left(\frac{28 - 15}{40 \times 10^3 \times 28}\right)^2$$

$$= 49.775\ \mu F$$

Voltage multipliers

These are simple and inexpensive circuits for producing higher voltages without the need for step-up transformers. However, they would need to be restricted to very light current loads.

Voltage doubler — The circuit of a half-wave voltage doubler is shown in Fig. 8.32. In this circuit when the a.c. input is negative going at the left-hand plate of C_1, diode D_1 is forward biased and C_1 is allowed to charge

Fig. 8.32 Half-wave voltage doubler circuit.

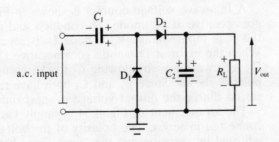

up to peak voltage with polarity shown. D_2 during this time is reverse biased. When the input to the left-hand plate of C_1 goes positive, on the positive-going part of the input a.c. cycle, D_1 is reverse biased but D_2 is now forward biased allowing capacitor C_2 to charge up with polarity shown. The magnitude of the voltage developed across C_2 depends on the magnitude of the input peak voltage *and* the stored charge on C_1 so that the d.c. output votage across C_2 is nearly twice the peak input voltage. The circuit has a high ripple factor and R_L is constantly discharging C_2.

Let us examine the voltage across diode D_1 during one complete cycle of input a.c. variations. It is assumed that capacitor C_1 has already been charged to a value such that the cathode of D_1 is at V_{pk} with respect to the anode. Refer to Fig. 8.33 which shows the voltage across D_1 for a complete input cycle. At point 1: Voltage across input of the circuit is zero. However voltage across D_1 is V_{pk} since C_1 has been charged to the peak value from a previous cycle of input. At point 2: Voltage across input of the circuit is such that the left-hand plate of C_1 is $-V_{pk}$ with respect to the common input line. Thus anode D_1 is effectively $+V_{pk}$ with respect to the left-hand plate of C_1 so that D_1 is an effective short-circuit and the cathode of D_1 is at 0 V.

At point 3: As for point 1 above.

At point 4: Voltage across input of the circuit now has the left-hand plate of C_1 at $+V_{pk}$ with respect to the common input line. But voltage across C_1 is V_{pk} so that the cathode of D_1 is at $2V_{pk}$ volts with respect to the cathode.

At point 5: As for point 1 above. (And so on.)

Fig. 8.33 Voltage across diode D_1 during one cycle of input to the voltage doubler circuit.

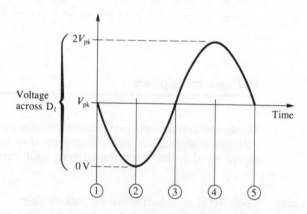

Hence the voltage across D_1 has a peak value of $2V_{pk}$ and it is to this value that C_2 will charge.

A full-wave voltage doubler is shown in Fig. 8.34. When the input goes positive at D_1 anode, D_1 conducts and allows C_1 to charge to the peak input voltage V_{pk}. Diode D_2 is reverse biased at this time. When the input at D_1 anode goes negative D_1 is reverse biased but D_2 is forward biased allowing C_2 to charge up to V_{pk} with the polarity shown. Since C_1 and C_2 both have the same polarity of stored charge the output voltage is approximately $2V_{pk}$ as before.

The arrangement of Fig. 8.34 is simply two half-wave rectifiers connected in series. The polarity of the voltage output could be adjusted by repositioning the 'earth' line. If the earth is connected to the top of R_L (i.e. the cathode of D_1 instead of the anode of D_2)

Fig. 8.34 Full-wave voltage doubler circuit.

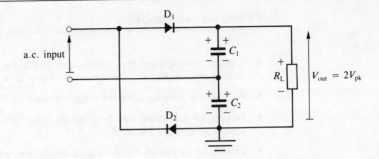

then the output voltage is $-2V_{pk}$. If the earth connection is made at the junction of C_1 and C_2 then separate 'loads' could be used for each capacitor, giving output voltages of $+V_{pk}$ across C_1 and $-V_{pk}$ across C_2.

Voltage trebler

This is basically a half-wave voltage doubler and an extra half-wave rectifier connected in series. The circuit is shown in Fig. 8.35. In this arrangement C_2 charges up to $2V_{pk}$ while C_3 charges to V_{pk}. Since the capacitors are in series then the total output voltage is $3V_{pk}$.

Fig. 8.35 Voltage trebler circuit.

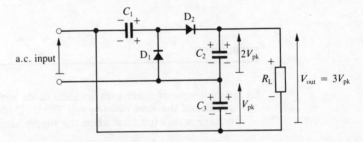

Voltage quadrupler

This is two half-wave doublers connected in series. The circuit is shown in Fig. 8.36. The circuit action for each of the half-wave doublers is the same as described earlier in this section giving a voltage across each output capacitor (C_2 and C_4) of $2V_{pk}$ with the polarity shown. Since C_2 and C_4 are in series the total output votage is thus $4V_{pk}$. In this circuit, as for the others, if R_L is high so that load current drawn is small, the circuit operates efficiently.

Fig. 8.36 Voltage quadrupler circuit.

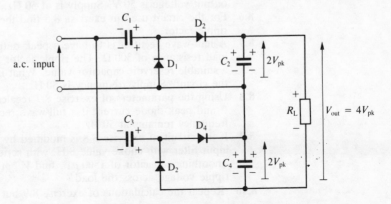

You should now be able to attempt exercises 8.1 to 8.30.

Points to remember

- Circuits utilising electronic components and active devices require direct current (d.c.) power suplies from a battery or mains via a rectifier circuit.

- Rectification may be achieved using half-wave or full-wave circuits.

- Full-wave rectification may be achieved using two or four diodes; the latter is known as a bridge rectifier.

- Smoothing to reduce the a.c. ripple on the d.c. output voltage may be achieved by a smoothing capacitor; ripple may be reduced still further by a filter circuit.

- Regulation of a power supply is a measure of the change in load voltage with changing load current.

- Voltage regulators help maintain power supply output voltage constant despite changes in input voltage and/or load current.

- Practical voltage regulators are available in integrated circuit form.

- Switching regulators may be used instead of 'ordinary' regulators and are more efficient.

- D.c. voltage levels in a circuit may be increased by the use of voltage multipliers although they are only useful for light current loads.

EXERCISES 8

8.1 A half-wave rectifier uses an ideal diode with a forward resistance of 100 Ω and the load resistance is 750 Ω. Deduce the d.c. and rms voltage across the load when the supply voltage is 250 V peak.
(70.2 V, 110.3 V)

8.2 If the circuit in exercise 8.1 uses a full-wave rectifier what would the values of the d.c. and rms voltages become? (140.4 V, 156 V)

8.3 If the circuit of exercise 8.1 uses a bridge rectifier what would the values of the d.c. and rms voltages become? (125.6 V, 139.6 V)

8.4 If an input voltage is given by $v = V_{max}.\sin \omega t$, the forward resistance of the diodes is r and the load resistance is R_L, deduce an expression for the load current in a full-wave bridge rectifier.

8.5 Find the peak value of ripple voltage in a half-wave rectifier using a reservoir capacitor of 1000 μF. Load resistance is 200 Ω and the peak output voltage is 50 V. Supply is at 50 Hz. (2.5 V)

8.6 For the circuit used in exercise 8.5 find the d.c. output voltage and the ripple factor. (47.6 V, 0.029)

8.7 A half-wave rectifier is to have a peak output voltage of 20 V with a load resistance of 300 Ω. The ripple factor must not exceed 0.015. Find a suitable reservoir capacitor value. What is the peak diode current for the circuit? Supply frequency is 50 Hz. (1283 μF, 2.6 A)

8.8 Using the parameters of exercise 8.7 recalculate the required values for C and peak diode current if a full-wave rectifier is to be used. Supply frequency remains at 50 Hz. (642 μF, 0.92 A)

8.9 If the circuit of exercise 8.8 is modified by the addition of a choke input filter with choke value 8 H, with series resistance of 30 Ω and a smoothing capacitor of 1500 μF, find V_{dc} at the output and the peak ripple voltage across the load. (18.18 V, 0.055 mV)

8.10 Repeat the calculations of exercise 8.9 but use a choke input filter of

inductance value 3 H and series resistance 10 Ω and a smoothing
capacitor of 1000 μF. (19.35 V, 0.22 mV)

8.11 Repeat the calculations of exercise 8.9 but use an *RC* filter (of the
type shown in Fig. 8.12) of resistance value 20 Ω and capacitance
1500 μF. (18.75 V, 13 mV)

8.12 The zener diode regulator of Fig. 8.16 is to be implemented using a
supply voltage of 18 V to 24 V with a 12 V zener diode. The zener
minimum current is 1 mA. The load current must not exceed
0.50 mA. Find a suitable value for the ballast resistor. (4 kΩ)

8.13 Draw the output characteristic of a zener diode and discuss how the
device works. How may the diode be incorporated into a simple
regulator circuit?

8.14 For the circuit of exercise 8.12 what would be the power rating of the
zener diode and the percentage regulation for the circuit given that
the internal resistance of the diode is 1 Ω. (36 mW, 0.0167%)

8.15 In the zener regulator circuit of Fig. 8.16 the diode voltage is 10 V
with a resistance of 5 Ω. The ballast resistor is 100 Ω. Calculate the
output resistance of the regulator and deduce the change in output
voltage when the load current changes by 25 mA. (5 Ω, 125 mV)

8.16 If the circuit of exercise 8.15 has a load resistance of 250 Ω what is
the stability factor? If the supply voltage changes by 10 V what is the
change in voltage at the output? (0.0189, 189 mV)

8.17 A zener regulator has a zener diode with a voltage rating of 8 V. The
input voltage is 25 V and the ballast resistance 250 Ω. If the diode
current is not to fall below 1 mA what is the minimum value of the
diode power rating? Between what values may the load resistance
vary while still maintaining correct voltage regulation?
(544 mW, 120 Ω → ∞)

8.18 Using the circuit if Fig. 8.17 with a 12 V zener diode of minimum
current 1 mA, a transistor of $h_{fe} = 99$ and a supply that varies
between 15 V and 20 V, calculate a suitable value for *R* if the load
current is not to exceed 2 A. (143 Ω)

8.19 If the transistor of exercise 8.18 has $h = 800$ Ω and diode resistance
of 10 Ω what is the output resistance of the stage? (8.1 Ω)

8.20 Repeat exercises 8.18 and 8.19 if the transistor is replaced by one
with $h_{fe} = 49$ and $h_{ie} = 300$ Ω.

8.21 Explain, with the aid of a block diagram, how a comparator circuit
added to a series voltage regulator improves the output resistance and
ripple component value at the output.

8.22 With reference to Fig. 8.23 explain how the circuit is protected
against current overload.

8.23 Using the circuit of Fig. 8.24 explain how the circuit compensates for
possible changes in input voltage and load current to keep the output
voltage sensibly constant.

8.24 A voltage regulator using the circuit of Fig. 8.24 has $R_1 = 2$ kΩ,
$R_2 = 4$ kΩ and $R_v = 5$ kΩ. What is the range of output voltages
available? Assume $V_{ref} = 8$ V. (9.78–22 V)

8.25 With reference to the i.c. voltage regulator connections of Fig. 8.25
and Fig. 8.26 explain how the connection of the current set resistor R_1
could affect the load regulation. If $R_1 = 200$ Ω and R_2 can have a
maximum value of 2 kΩ what would be the difference in load
regulation between the two connections? Assume $R_s = 0.05$ Ω.
(Increased by a factor of 11)

8.26 Explain, with reference to a basic switching regulator circuit of
Fig. 8.28, how the value of V_{out} is obtained. If V_{in} is 20 V, t_{on} is
2 μsec and *T* is 20 μsec what is the value of V_{out}? (2 V)

8.27 Explain the operation of the switching regulator circuit of Fig. 8.29.
What is the purpose of the inductor L_1 and capacitor C_1?

8.28 Explain the operation of the switching regulator of Fig. 8.31. What is the function of C_2 and C_3? What would be a suitable value for the upper limit of output current for such a circuit?   (500 mA)

8.29 A switching regulator of the type shown in Fig. 8.31 is designed to deliver 15 V with maximum current 400 mA from a 25 V supply. The switching frequency is 50 kHz. $L_1 = 1.5$ mH, $C_1 = 47\ \mu$F. What would be the expected peak-peak output ripple voltage?   (6.8 mV)

8.30 Explain the operation of the voltage doubler circuit of Fig. 8.32.

9 Controlled Rectifier Systems

The principal learning objectives of this chapter are to

There are applications where it is required to vary the amount of power taken by a load; variation in a.c. and d.c. control and lamp dimmer circuits are possible examples. The basic concept is to place a switch in the circuit and allow the state of the switch, in a given input cycle, to control the amount of power taken by the load.

Current is allowed to flow to the load but only during the time the switch is closed and that may be for a fraction of the input cycle. There would be no power loss in the ideal switch giving good efficiency. There are many devices that could perform the switch function; among these are the silicon controlled rectifier (SCR) and the triode a.c. semiconductor (triac). These devices and others that may play a part in the triggering of the switches are considered in the following sections.

Controlled rectification

Assume a half-wave rectifier output is to be used to supply current to a load. The output of the rectifier gives the expected half-cycle of sinusoidal output once every cycle except that conduction of the rectifier diode is not allowed to begin at the start of the cycle but after an angular measure of θ radians has occurred. The resulting current waveform is shown in Fig. 9.1.

If the angle θ can be varied from 0 to $\pi/2$ radians (or even from 0 to π radians) then the mean value of current taken by the load can be varied as can the rms value. The following analysis allows the mean and rms currents to be derived.

Fig. 9.1 Control of current
to a load by variation of a
firing angle θ.

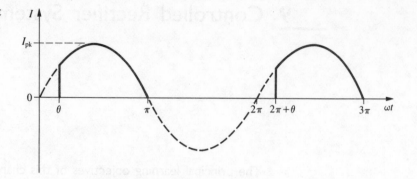

The load current is given by
$$i = I_{pk} \sin \omega t \qquad \theta \le \omega t \le \pi$$
$$\text{or } i = \qquad 0 \qquad 0 \le \omega t \le \theta \text{ and } \pi \le \omega t \le 2\pi \qquad [9.1]$$

the average value of i is given by

$$I_{dc} = \frac{1}{2\pi} \int_0^\pi I_{pk} \sin \omega t \, d\omega t$$

$$= \frac{1}{2\pi} \left[-I_{pk} \cos \omega t \right]_\theta^\pi$$

$$= \frac{I_{pk}}{2\pi} (1 + \cos \theta) \qquad [9.2]$$

The rms load current can be expressed in terms of θ by

$$I_{rms} = \frac{I_{pk}}{2} \sqrt{\frac{2(\pi - \theta) + \sin 2\theta}{2}} \qquad [9.3]$$

If θ could be varied from 0 to π radians then, from equations
[9.2] and [9.3], both I_{dc} and I_{rms} would vary from a maximum value
to zero. Thus any system which could allow the angle θ to be varied
in the rectifier system would achieve the required control.

Silicon controlled rectifier

The silicon controlled rectifier (SCR), often called a **thyristor**, is the
semiconductor equivalent of the thyratron tube. In fact the name
thyristor derives from the term *thyra*tron trans*istor*. Another name
for the device is the reverse blocking triode thyristor. Here it is
referred to as the SCR for simplicity. The device is a p-n-p-n
structure with three terminals as shown in Fig. 9.2(a). The symbol is
shown in Fig. 9.2(b).

Theory of operation   For analysis purposes the four-layer device of Fig. 9.2(a) can be split
into two interconnected transistors as shown in Fig. 9.2(b). In fact
the SCR could be replaced by a two-transistor model using an n-p-n
and a p-n-p transistor and this has been done in Fig. 9.3 using a
suitable source voltage generator and resistance and a resistive load.
Referring to this model, if $V_G = 0$ then TR$_1$ and TR$_2$ are off and

Fig. 9.2 Silicon controlled rectifier: (a) basic construction, (b) device symbol, (c) two-transistor equivalent arrangement.

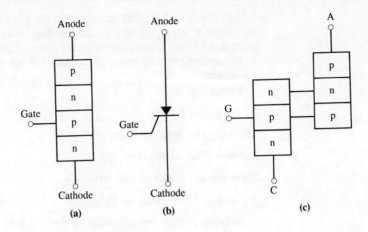

Fig. 9.3 Analysis of the two-transistor equivalent of the SCR.

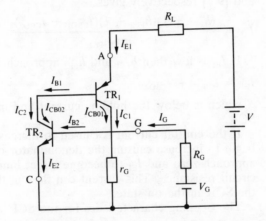

ideally there is no current in the load resistor. Some leakage current will, however, flow through the two devices in practice. Under the circumstances the SCR could be considered as an open switch. The small gate resistance r_a shown in Fig. 9.3 is often provided to lower the gate input impedance and hence give some protection against noise spike inputs to the gate which might otherwise cause false switching. The effect of r_G is ignored in the following analysis.

If V_G is applied, a small current I_{B2} flows into TR_2 base and via transistor action I_{C2} flows where $I_{C2} = h_{FE}I_{B2}$. Since now the emitter-base junction of TR_1 is forward biased then TR_1 turns on. The collector current from TR_1 flows into the base of TR_2 so that the effective base current, and hence the collector current, of TR_2 increases so that the base, and hence the collector, current increases in TR_1. The process repeats until TR_1 and TR_2 saturate. If V_G is again made equal to zero volts then, although $I_G = 0$, the base current of TR_2 is produced by TR_1 and the base current of TR_1 is produced by TR_2. Thus the gate can initiate conduction, when the saturated transistors represent a closed switch, but returning the gate to zero cannot halt the conduction process. The only way the transistors can be turned off is to reduce the collector current to a value which will not allow the regenerative action described above to continue. To achieve the reduction in collector current to a value similar to the leakage current the anode and cathode terminals would have to be short-circuited. The load current must be reduced to below a value known as the **holding current** which is that current necessary to maintain the regenerative action. A current below the holding current will cause TR_1 and TR_2 to go into the **blocking condition**.

From Fig. 9.3

$$I_{E1} = I_{C1} + I_{C2}$$

allowing for leakage current I_{CBO1}

$$I_{C1} = h_{FE1}I_{C2} + (h_{FE1} + 1)I_{CBO1} \qquad [9.4]$$

allowing for leakage current I_{CBO2}

$$I_{C2} = h_{FE2}I_{C1} + (h_{FE2} + 1)I_{CBO2} \qquad [9.5]$$

Solving for I_{E1} by substituting for I_{C1} and I_{C2} from equations [9.4] and [9.5] respectively gives

$$I_{E1} = \frac{(h_{FE1} + 1)(h_{FE2} + 1)(I_{CBO1} + I_{CBO2})}{1 - h_{FE1}h_{FE2}} \qquad [9.6]$$

If I_{E1} is low then h_{FE1} and h_{FE2} approach zero and

$$I_{E1} \approx I_{CBO1} + I_{CBO2}$$

which is below the holding current and insufficient to turn on the SCR.

If the emitter current is caused to increase then h_{FE1} and h_{FE2} begin to increase causing the denominator of equation [9.6] to approach zero and I_{E1} to become large, limited only by the external circuit resistance. This current can flow in the external circuit with the SCR in the on state.

The opening characteristics for the SCR are illustrated in Fig. 9.4.

Fig. 9.4 Voltage-current characteristic of the SCR.

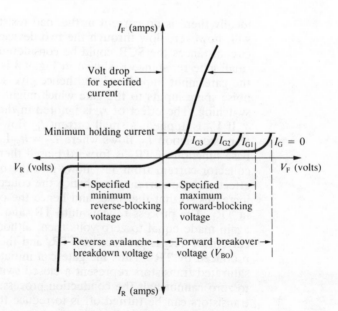

In the forward blocking voltage condition with $I_G = 0$ (i.e. no gate current) and a small anode-cathode voltage, only a small leakage current flows in the device. If, however, the anode-cathode voltage is increased above the maximum forward blocking voltage, the device will break down into forward conduction. For most applications the SCR is turned on with a gate trigger current (I_G). The value of I_G necessary to cause forward breakdown to occur depends on the anode-cathode voltage; the lower the value of anode-cathode voltage the greater the gate current required while smaller gate currents are required if the anode-cathode voltage is at a higher value. Thus $I_{G3} > I_{G2} > I_{G1}$ in the diagram of Fig. 9.4. Once fired the SCR remains on until the current is allowed to fall below the minimum holding current value.

Any anode-cathode voltage variations in the reverse blocking voltage region will not cause conduction. However, the device applied reverse voltage should never be allowed to exceed the minimum reverse blocking voltage as the device would then avalanche into a reverse conduction situation with no control.

Operation of the device in the forward voltage condition can be explained with reference to the diagram of Fig. 9.2(a). When connected in series with a load resistance and a d.c. source, say, and with the anode positive with respect to the cathode, no current flows if the voltage is below the maximum forward blocking voltage (assuming there is no gate current). Under these conditions the two outer p-n junctions are forward biased allowing the leakage current to flow readily. The centre n-p junction is reverse biased and prevents conduction other than the leakage current. Increasing the forward voltage accelerates the electrons forming the leakage current to higher velocities. As the velocity increases a critical value is reached where avalanching occurs which breaks down the centre p-n junction. The smaller resistance offered by this centre p-n junction will provide a smaller volt drop so that the forward voltage across the device falls and then increases with current as shown in Fig. 9.4.

With a lower forward voltage the SCR can be 'fired' by injecting charge carriers into the gate region. Because the centre n-p junction is reverse biased there is a depletion region associated with the junction. It is the function of the injected charge carriers to neutralise the local space charge associated with the depletion region so that avalanching can occur earlier, in terms of forward voltage, than it would otherwise do. The action of the gate current can be broadly summarised as follows. Holes are injected into the gate p-type region and these holes neutralise the negative space charge on the p-side of the junction. This tends to encourage electrons from the cathode to migrate to the p-side of the junction to compensate for the injected holes, the migration of the electrons causing avalanching as described earlier. Only a relatively small gate current is necessary to allow a very much greater device current to flow but once switched on there is no gate mechanism that can reverse the process. The external voltage must be removed or reduced so that the device current falls below the holding current.

In the reverse voltage condition the two outer p-n junctions are reverse biased and the centre n-p junction is forward biased. Only if the reverse bias is high enough can the leakage current cause avalanching to occur.

Manufacturers of SCRs provide data specifications for their devices that include such parameters as maximum voltage, current and power ratings; additionally there are specifications relating to the gate circuit such as gate trigger voltage and current and gate turn-on and turn-off time. The SCR is liable to damage if the gate parameters are exceeded and great care is needed to ensure the gate triggering is executed safely. Figure 9.5 shows the operating limits for a given member of a family of SCRs at a given operating temperature. Shown on this characteristic is the maximum d.c. gate voltage that will not turn on any SCR within that family; this voltage is usually specified at the rated operating temperature of the device and may be as low as 0.1 V. Noise spikes could therefore be a problem in the gate circuit and practical circuits must take this into account.

Fig. 9.5 Curve of gate current against gate voltage for a family of SCR devices.

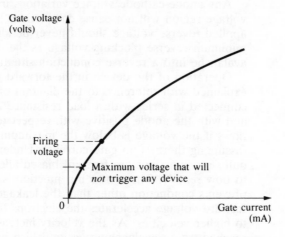

SCR circuits   The basic SCR operation can control power to the load in d.c. and a.c. circuits. In the simple circuit of Fig. 9.6(a) d.c. control is possible.

With the switch S_1 open no gate current can flow and the SCR is off. R_G will serve to limit the gate current. When the switch S_1 is closed gate current flows and the SCR fires, allowing load current to flow through the device. Opening the switch now will have no effect on the conduction which can only be stopped by opening S_2. As current falls to zero then the circuit can be reactivated by once again closing S_1 and S_2. The switches may be either manually or electrically operated.

The SCR is being used as a chopper and the average output voltage depends on either:
(1) the time for which the SCR conducts t_{on} in a given fixed period T, or
(2) the time between firings, t_{off}, for a fixed conduction time t_{on}.
(1) above is known as variable mark to space ratio control and its effect is shown in Fig. 9.6(b). (2) above is known as variable frequency control and its effect is shown in Fig. 9.6(c).

Example 9.1   *An SCR is used in the circuit of Fig. 9.6(a) to switch a d.c. supply of 100 V. If S_2 is opened momentarily and then closed once every 20 msec, find the mean output voltage if, via S_1, the t_{on} time of the SCR is 6 msec. To what value is the mean output voltage changed if S_1 is closed for 15 msec?*

The average voltage in a given period T of the switched supply is

$$V_{dc(ave)} = V_{dc}\left(\frac{t_{on}}{T}\right)$$

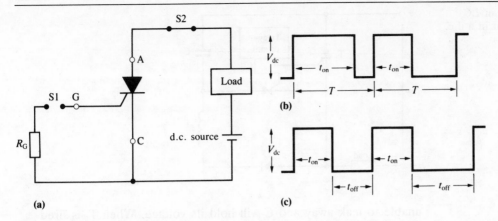

(a)

(b)

(c)

Fig. 9.6 SCR circuit controlling direct current flow to a load: (a) basic arrangement, (b) variable mark to space ratio control, (c) variable frequency control.

Hence if t_{on} is 6 msec

$$V_{dc(ave)} = 100 \times \frac{6}{20}$$
$$= 30 \text{ V}$$

and if t_{on} is 15 msec

$$V_{dc(ave)} = 100 \times \frac{15}{2}$$
$$= 75 \text{ V}$$

Practical circuits for d.c. control are shown in Figs 9.7 and 9.8. For the circuit of Fig. 9.7 when SCR T_1 is on the **commutating capacitor** C charges to V_{dc}, with the polarity shown, via resistor R. When SCR T_2 is fired then since the volt drop across T_2 is small the left-hand plate of C is effectively earthed and the full reverse voltage of the capacitor is placed across T_1 turning it off. The average load current is thus determined by the trigger circuit which controls the time for which T_1 is on (and T_2 off) and T_2 is on (and T_1 off).

Fig. 9.7 Use of a commutating capacitor in a d.c. circuit.

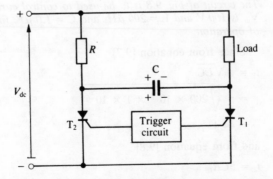

For the circuit of Fig. 9.8 if it is assumed that initially C is uncharged then when first switching T_2 on the capacitor can charge to the full supply voltage via T_2 and the load resistor. As the capacitor charges its charging current falls and turns off T_2 when the current is near zero. With both T_1 and T_2 off the charge on C is

Fig. 9.8 Use of an *LC* oscillatory circuit in a d.c. circuit.

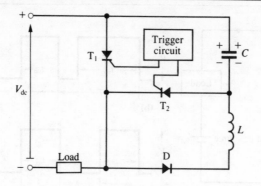

unable to leak away and *C* will hold its voltage. When T_1 is fired then load current flows via T_1. However, with T_1 conducting *C* now has a discharge path via T_1, *D* and *L*. Because of the presence of *L* an oscillation is initiated and *C* discharges and charges again negatively; because of the diode, only one half-cycle of oscillation can occur and this new charge is trapped until T_2 is again fired. The time required to charge *C* to $-V_{dc}$ is approximately given by:

$$t_1 = \frac{T}{2} = \frac{1}{2f} = \pi\sqrt{LC} \text{ seconds} \qquad [9.7]$$

When T_2 is fired the reverse voltage across *C* is applied to T_1 switching if off. While T_2 is on the capacitor is able to recharge to the supply voltage and switches T_2 off again with trapped charge across *C* waiting for the next firing of T_1. The approximate time required to discharge *C* via the load resistance is given by:

$$t_2 = 5CR_L \text{ seconds} \qquad [9.8]$$

where R_L is the load resistance.

Care must be taken to ensure that the timing of firing for T_1 and T_2 allows the capacitor to fully charge or discharge otherwise circuit operation is impaired. The following example should make this clear.

Example 9.2   *The circuit of Fig. 9.8 is to be used to control current to a load of 50 Ω. If V_{dc} is 100 V and L = 200 μH, and C = 1.0 μF, find the maximum frequency of operation.*

Since from equation [9.7]

$$t_1 = \pi\sqrt{LC}$$

$$= \pi\sqrt{200 \times 10^{-6} \times 1 \times 10^{-6}}$$

$$= 44.43 \text{ μsec.}$$

and from equation [9.8]

$$t_2 = 5CR_L$$

$$= 5 \times 1 \times 10^{-6} \times 50$$

$$= 250 \text{ μsecs}$$

SCR T_1 must be on for greater than t_1 to allow the capacitor to acquire its negative charge and SCR T_1 must be off for greater than t_2 to allow the capacitor to recharge to $+V_{dc}$. Thus minimum period of switching frequency

$T_{\min}$ is $t_1 + t_2$ and

$T_{\min} = 44.43 + 250$

$\qquad = 294.43 \ \mu\text{secs}$

and the maximum switching frequency $f_{\max} = \dfrac{1}{T_{\min}}$

$$= \frac{1}{294.43 \times 10^{-6}}$$

$$= 3.4 \text{ kHz}.$$

An a.c. control circuit is shown in Fig. 9.9. Resistors R_1 and R_2 act as a voltage divider network so that as the a.c. voltage increases positively from zero a voltage can be set to fire the device. R_2 could be made variable to adjust the firing voltage. Diode D prevents a negative voltage from being applied to the gate terminal. Once fired current can flow to load while the source is in its positive half-cycle. At the end of the positive half-cycle of source voltage the voltage across the SCR has fallen to zero causing the device to stop conducting. When the source voltage goes negative the SCR is reverse biased and will not conduct as long as the peak value of the a.c. source will not cause avalanching. Waveforms are shown in Fig. 9.10. This form of control is known as **phase-control** since the values for load current and voltage are controlled by varying the phase of the gate pulse relative to the source voltage.

Fig. 9.9 Basic SCR circuit controlling alternating current to a load.

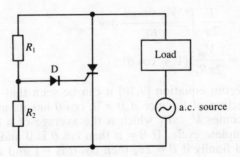

The advantage of a.c. control is that the fluctuation in the source voltage will itself cause the device to switch off as the positive half-cycle ceases. The value of θ, the phase angle at which conduction commences, can be controlled by correct setting of R_1 and R_2 to be a value in the range of nearly 0 to π radians. The use of the circuit of Fig. 9.7 is of limited practical application but is useful for mathematical analysis.

Referring to the load current waveform of Fig. 9.10 it can be seen that if the SCR is triggered by a gate current pulse at time $t = \theta/\omega$ or if the voltage at the gate, given by $v = V_{\text{pk}} \sin \theta$, is sufficient to cause firing then the load current is given by

$$i = \frac{V_{\text{pk}} \sin \theta}{R_{\text{L}}} \text{ for } \theta \leqslant \omega t \leqslant \pi \qquad [9.9]$$

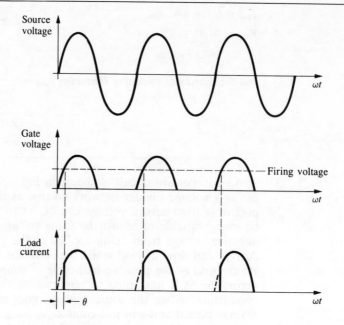

Fig. 9.10 Voltage and current waveforms illustrating *phase control*.

The mean d.c. value of current is the average value of i over a cycle and is given by

$$I_{dc} = \frac{1}{2\pi} \int_\theta^\pi i \, d\omega t$$

$$= \frac{1}{2\pi} \int_\theta^\pi \frac{V_{pk} \sin \omega t}{R_L} \, d\omega t$$

$$= \frac{V_{pk}}{2\pi R_L} (1 + \cos \theta) \qquad [9.10]$$

From equation [9.10] it can be seen that the value of I_{dc} is a function of θ, since if $\theta = 0$, $\cos \theta$ has its maximum value and I_{dc} becomes $V_{pk}/\pi R_L$ which is the average of a half-sinusoid over a complete cycle. If $\theta = \pi$ then $\cos \theta$ is 0 and I_{dc} becomes $V_{pk}/2\pi R_L$ and finally if $\theta = 2\pi$, then $\cos \theta$ is -1 and I_{dc} is zero.

The rms value of the load current is given by

$$I_{rms} = \sqrt{I_{dc}^2} = \sqrt{\left[\frac{1}{2\pi} \int_0^\pi \frac{V_{pk}^2 \sin^2 \omega t}{R_L^2} \, d\omega t \right]}$$

$$= \frac{V_{pk}}{R_L} \sqrt{\left[\frac{1}{2\pi} \int_0^\pi \frac{(1 - \cos 2\omega t)}{2} \, d\omega t \right]}$$

$$= \frac{V_{pk}}{R_L} \sqrt{\left[\frac{1}{2\pi} \left(\frac{\omega t}{2} - \frac{\sin 2\omega t}{4} \right) \right]_\theta^\pi}$$

$$= \frac{V_{pk}}{R_L} \sqrt{\left[\frac{1}{2\pi} \left(\frac{\pi}{2} - \frac{\theta}{2} + \frac{\sin 2\theta}{4} \right) \right]}$$

$$= \frac{V_{pk}}{R_L} \sqrt{\frac{1}{4\pi} \left(\pi - \theta + \frac{\sin 2\theta}{2} \right)} \qquad [9.11]$$

With this arrangement since a mean d.c. flows in the load there is d.c. power dissipated in the load given by

$$P_{dc} = I_{dc}^2 R_L$$

$$= \frac{V_{pk}^2 (1 + \cos \theta)^2}{4\pi^2 R_L} \qquad [9.12]$$

while the a.c. power in the load is given by

$$P_{ac} = I_{rms}^2 R_L$$

$$= \frac{V_{pk}^2}{R_L}\left(\frac{1}{4\pi}\left(\pi - \theta + \frac{\sin 2\theta}{2}\right)\right) \qquad [9.13]$$

Example 9.3   *In the circuit of Fig. 9.9 the peak value of supply voltage is 250 V while the load resistance is 50 Ω. Find the mean value of load current for a firing angle of θ = 75°. What is the d.c. load power dissipation caused by this current? What is the a.c. load power?*

From equation [9.10]

$$I_{dc} = \frac{V_{pk}}{2\pi R_L}(1 + \cos \theta)$$

$$= \frac{250}{2\pi \times 50}(1 + \cos 75°)$$

$$\approx 1.0 \text{ A}$$

From equation [9.12]

$$P_{dc} = \frac{V_{pk}^2 (1 + \cos \theta)^2}{4\pi^2 R_L}$$

$$\approx 50 \text{ W}$$

(or $P_{dc} = I_{dc}^2 R_L = 1^2 \times 50 = 50$ W). From equation [9.13]

$$P_{ac} = \frac{V_{pk}^2}{R_L}\frac{1}{4\pi}\left(\pi - \theta + \frac{\sin 2\theta}{2}\right)$$

$$= \frac{250^2}{50}\left(\frac{1}{4\pi}\left(\pi - \frac{5\pi}{12} + \frac{\sin 150°}{2}\right)\right) \quad \text{(since } 75° = 5\pi/12 \text{ radians)}$$

$$= 1250\left(\frac{1}{4\pi}\left(\frac{7\pi}{12} + \frac{0.5}{2}\right)\right)$$

$$= 1250\left(\frac{7}{48} + \frac{1}{16\pi}\right)$$

$$= 1250(0.146 + 0.020)$$

$$= 1250(0.166) = 207.5 \text{ W.}$$

The circuit of Fig. 9.9 is a half-wave phase controlled rectifier. This circuit suffers from the disadvantage that there is a mean, or direct, current from the source hence a d.c. power dissipation. A full-wave circuit can be produced using a bridge rectifier and a single SCR as shown in Fig. 9.11.

Switch S acts as an off/on control and R_1/R_2 divider is set to give the required firing level. The SCR acts on the 'd.c.' side of the rectifier with a firing angle of θ but because the load is on the 'a.c.' side of the rectifier the waveforms are as shown in Fig. 9.12. The load current and voltage are symmetrical about the horizontal axis and so have a mean value of zero. This eliminates the disadvantage of the half-wave circuit.

Fig. 9.11 Basic full-wave
rectifier circuit for
controlling alternating
current to a load.

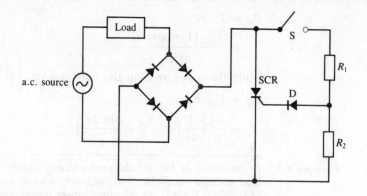

Fig. 9.12 Current and
voltage waveforms for the
full-wave rectifier control
circuit.

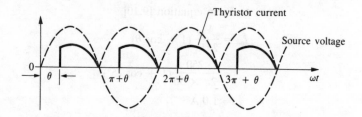

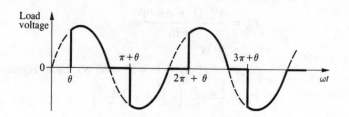

Example 9.4 *Using the circuit of Fig. 9.11 with a peak supply voltage of 250 V and a load resistance of 100 Ω, find the a.c. power taken by the load if the firing angle is 75°.*

Using equation [9.11] but applying it for the the *two* half-cycles of input waveform gives

$$I_{rms} = \sqrt{\frac{1}{2\pi} \int_{\theta}^{\pi} \frac{V_{pk}^2 \sin^2 \omega t}{R_L^2} \, d\omega t + \frac{1}{2\pi} \int_{\pi+\theta}^{2\pi} \frac{V_{pk}^2 \sin^2 \omega t}{R_L^2} \, d\omega t}$$

However, each half-cycle is identical so the above equation can be rewritten as:

$$I_{rms} = \sqrt{\frac{1}{\pi} \int_{\theta}^{\pi} \frac{V_{pk}^2 \sin^2 \omega t}{R_L^2} \, d\omega t}$$

Or, using the final form of equation [9.11]

$$
\begin{aligned}
I_{rms} &= \frac{V_{pk}}{R_L} \sqrt{\frac{1}{2\pi} \left(\pi - \theta + \frac{\sin 2\theta}{2} \right)} \\
&= \frac{250}{100} \sqrt{\frac{1}{2\pi} \left(\pi - \frac{5\pi}{12} + \frac{0.5}{2} \right)} \\
&= 2.5 \sqrt{\left(0.5 - \frac{5}{24} + \frac{1}{8\pi} \right)}
\end{aligned}
$$

$$= 2.5 \sqrt{(0.5 - 0.21 + 0.04)}$$

$$= 2.5 \sqrt{0.33} \text{ A}$$

And $P_{ac} = I_{rms}^2 R_L$

$$= (2.5 \sqrt{0.33})^2 \times 100$$

$$= 6.25 \times 0.33 \times 100$$

$$= 206.25 \text{ W}$$

The circuits described so far use static switching techniques. Static switching is where a constant, or possibly varying, d.c. signal is used to fire the SCR. The major advantage of static switching is that a small gate current can be used to control a large load current.

It has already been mentioned that the simple circuit of Fig. 9.9 has disadvantages. A further limitation of that circuit is that it is not possible to obtain SCR conduction angles of less than 90°. The circuit of Fig. 9.13(a) overcomes this problem by providing a delayed gate trigger signal. The firing angle can be adjusted by the phase shift network RC which permits an SCR firing angle from approximately 0° to 180° giving greater flexibility in the power levels applied to the load.

The phase-shift action is produced by the CR network since an alternating voltage applied to a series CR network produces a lagging capacitive voltage as the phasor diagram of Fig. 9.13(c) shows. The phase angle θ_1 by which the capacitive voltage lags the source voltage depends on the value of resistance, the smaller the resistance the greater the value of phase delay.

Fig. 9.13 Delayed gate trigger circuit: (a) basic arrangement, (b) voltage waveforms, (c) phase difference between applied voltage and capacitor voltage, (d) voltage waveforms showing the delay between source and capacitor voltage.

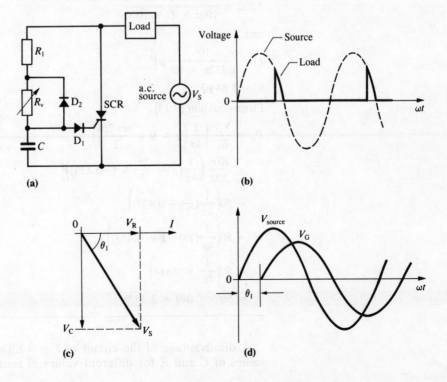

As the waveforms of Fig. 9.13(d) show the voltage applied to the trigger circuit (across the capacitor) can be delayed up to 90° on the source voltage and thus the true value of the firing angle can be extended by this amount.

Circuit action of Fig. 9.13(a) can be explained as follows. When the source voltage is in its positive half-cycle, D_1 is forward biased allowing the capacitor voltage to be applied to the SCR gate. When the gate firing voltage is reached the SCR conducts and allows current to the load. Since the SCR is on, capacitor C will discharge with a small time constant CR_1 via D_2 and R_1 as the source voltage falls back to zero rapidly removing the charge on C before the next a.c. cycle.

Example 9.5

For the circuit of Fig. 9.13(a) the total resistance $R_1 + R_v$ is 1 kΩ. Find the value of capacitor C necessary to give a firing angle of 120°. What will be the a.c. power to the load for this condition if $V_{source} = 100$ V and $R_L = 200$ Ω? Assume that without the phase delay circuit the firing angle is 60° and circuit operating frequency is 50 Hz.

For a firing angle of 120° the phase delay circuit must have a delay of 60°. From Fig. 9.13(c) it can be seen that:

$$\tan \theta = \frac{V_c}{V_R} = \frac{I X_C}{I R} = \frac{X_C}{R}$$

$$\tan \theta = \frac{1}{2\pi f C R}$$

$$\tan \theta = \frac{1}{2\pi \times 50 \times C \times 1000}$$

hence

$$1.732 = \frac{1}{100\pi \times C \times 1000}$$

and

$$C = \frac{10^6}{1.732\pi \times 10^5} \mu F$$

$$C = 1.84 \ \mu F$$

From equation [9.13]

$$P_{ac} = \frac{V_{pk}^2}{R_L} \left(\frac{1}{4\pi} \left(\pi - \theta + \frac{\sin 2\theta}{2} \right) \right)$$

$$= \frac{100^2}{200} \left(\frac{1}{4\pi} \left(\pi - \frac{2\pi}{3} + (-0.433) \right) \right)$$

$$= 50 \left(\frac{1}{4\pi} \left(\frac{\pi}{3} - 0.433 \right) \right)$$

$$= 50 \left(\frac{1}{4\pi} (1.0472 - 0.433) \right)$$

$$= 50 \left(\frac{1}{4\pi} \times 0.614 \right)$$

$$= 50 \times 0.049 = 2.44 \text{ W}$$

A disadvantage of the circuit of Fig. 9.13(a) is matching the values of C and R for different values of source voltage i.e. a circuit designed for, say, 100 V would not necessarily be suitable at, say, 400 V. This problem could be overcome by the use of the circuit of Fig. 9.14(a) which uses a step-down centre-tapped transformer in the trigger circuit. The centre-tapped transformer, R_v and C_1 form a bridge circuit; the voltage between points B and D can be varied in phase angle with respect to the SCR anode-cathode voltage, so as to phase shift the firing angle and vary the load current. If A is in phase with the SCR voltage, AC may be drawn as a phasor with B at its centre as shown in Fig. 9.14(b). The voltage drops across C_1 and R_v must, together, equal the voltage AC and since V_c must be in quadrature with V_R then the locus of D must be a semi-circle with AC as the diameter. The phasor BD is thus a variable-phase constant amplitude triggering voltage, lagging the SCR voltage.

Fig. 9.14 Variable phase-shift SCR a.c. control: (a) circuit arrangement, (b) phasor diagram showing voltages in the phase delay circuit.

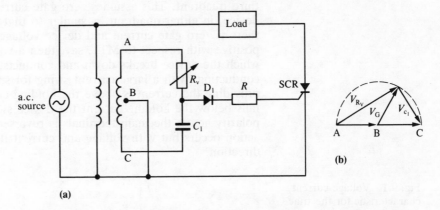

This arrangement allows a standard phase-delay gate trigger for different supply voltages.

The triac

The triac is a three terminal device for controlling a.c. in either direction. Its name derives from its original title of *triode ac semiconductor device*. The basic construction of the device is shown in Fig. 9.15(a) from which it can be seen that there are two main

Fig. 9.15 Triac: (a) basic construction, (b) device symbol.

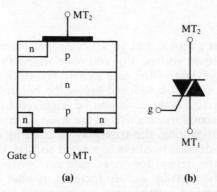

terminals (MT1 and MT2) and a gate terminal. The structure of the layers gives an n-p-n-p-n arrangement which is effectively two parallel SCRs connected in opposite directions to allow current flow to be bilateral. Unlike the SCR which requires positive gate signals, the triac can be triggered by positive and negative gate signals. The device symbol is shown in Fig. 9.15(b).

Theory of device operation

Figure 9.16 shows the voltage-current characteristic of a triac. It can be seen that device current is a function of the voltage applied between main terminals MT1 and MT2. With main terminal MT2 positive with respect to MT1 the current is in the first quadrant while if MT1 is positive with respect to MT2 the current is in the third quadrant. This assumes zero gate current. The action of the device, in either quadrant, is similar to that of the SCR in that if there is zero gate current and device voltage is increased with MT2 positive with respect to MT1, say, then a voltage V_{BO} is reached at which the device breaks down and conducts. The device remains conducting, with a large current swing for small voltage swings, providing the current is above the holding current. If the current falls below the holding current the device switches off. If the voltage polarity across the main terminals is reversed the same switching action occurs but with voltage and current flow in the opposite direction.

Fig. 9.16 Voltage-current characteristic for the triac.

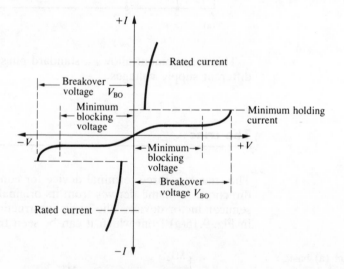

Just as for the SCR, applying gate current to the triac reduces the breakover voltage V_{BO} and once triggered the device current is independent of the gate signal. The device remains on until the main current is reduced below the holding current value. However, the triac has the ability to be triggered by a positive or negative gate signal irrespective of the main terminal polarity. In addition to gate triggering, the triac may be fired by exceeding V_{BO} although most devices would be operated so that V_{DRM} is never exceeded so that the device does need gate current to conduct. The triac may also be fired by a sharp increase in what is called the dV/dt rating. An applied voltage does not have to exceed the value of V_{BO} but if its rate of rise at the device terminals exceeds a critical dV/dt value then a charging current can exist in device internal capacitances. If this charging current exceeds the required gate current the triac will fire. Ratings for dV/dt are specified by triac manufacturers and vary from about 10 V/μs to 100 V/μs depending on the device. A triac may be protected against the sharp dV/dt transients by a series R and C network placed in parallel with the device as shown in Fig. 9.17. Such a network, known as a **snubber circuit**, delays the transient rise across the device as C charges towards the maximum value of the input voltage.

Fig. 9.17 Simple triac circuit with 'snubber' arrangement.

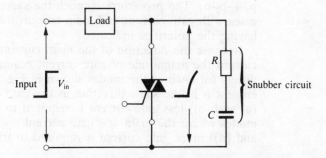

The design of snubber circuits is complicated by the need to take into account such factors as the line voltage and line current and the type of load (i.e. loads need not be resistive, in fact many loads are inductive or capacitive) as well as the rated value of dV/dt of the device.

If the triac is caused to conduct by either a value of dV/dt above the critical value or an increase in applied voltage above V_{BO}, no damage should result to the device. However, such methods are not normally used to switch triacs and their occurrence would suggest incorrect functioning with possible detrimental results to the circuit being controlled.

With reference to the action of the gate terminal for both positive and negative gate currents there must be four conduction modes taking into account the device can have positive and negative applied voltages across the main terminals. The action of each mode can best be explained with reference to Fig. 9.18.

Fig. 9.18 Triac triggering modes: (a) gate and MT$_2$ positive with respect to MT$_1$, (b) gate and MT$_1$ negative with respect to MT$_2$, (c) gate and MT$_2$ negative with respect to MT$_1$, (d) gate and MT$_1$ positive with respect to MT$_2$.

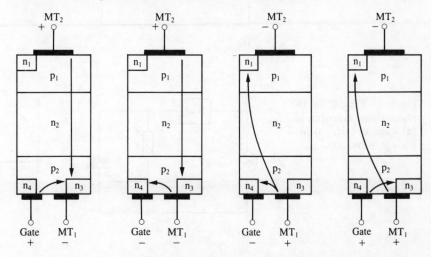

For each of the illustrations in Fig. 9.18, the polarity of the main terminals is as marked and the gate trigger polarity is always referenced to MT1. Because in (a) and (b) MT2 is positive with respect to MT1 the operation will be in the first quadrant whereas in (c) and (d) MT2 is negative with respect to MT1 so that operation will be in the third quadrant. Considering Fig. 9.18(a), MT2 and the gate are both positive with respect to MT1. Initial gate current flows via gate terminal to the p-type layer p_2 across into n_3 and out via MT1. The gate current causes current multiplication, as described for the SCR, and the triac switches on. When conducting the direction taken by the main current is as shown taking the path p_1-n_2-p_2-n_3. The procedure is much the same for the other three cases with gate current and main current taking the paths and having the polarities indicated.

Because the direction of the main current influences the gate current the magnitude of gate current required to fire the triac differs for each of the modes shown in Fig. 9.18. Where main current is in the same direction as the gate current (as in diagrams (a) and (c)) less gate current is required to trigger the device. For modes where the main and gate current oppose (as in diagrams (b) and (d)) more gate current is required to trigger.

Triac circuits

The SCR of Fig. 9.11 used a bridge rectifier and a single SCR to control the a.c. flow to a load. The same result could be achieved using the circuit of Fig. 9.19. The resistance R is used to vary the phase angle θ up to about 90° maximum. Should it be required to increase the firing angle up to about 180° then CR phase delay circuits can again be used as Fig. 9.20 shows.

Fig. 9.19 Triac 0° to 90° phase delay gate arrangement (a) basic circuit, (b) load voltage waveform.

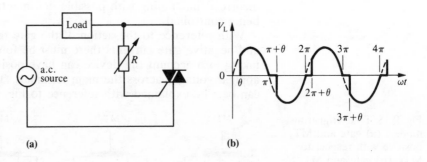

(a)   (b)

Fig. 9.20 Triac 0° to 180° phase delay arrangement; (a) basic circuit, (b) load voltage waveform.

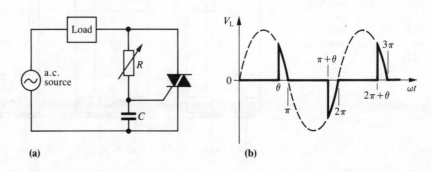

(a)   (b)

The simple circuits of Figs 9.19 and 9.20 are not really practical since it is impossible to calibrate R to give accurate firing angles due to the gate signal levels for triggering varying between triacs in a given family.

The diac

Theory of operation

This device is a two terminal bidirectional diode capable of conducting, or blocking current, in either direction. The device can be n-p-n or p-n-p but unlike the bipolar transistor the construction is uniform and the n-type and p-type doping levels are the same at both junctions. Figure 9.21 shows device construction, symbol and voltage-current characteristic.

Fig. 9.21 Diac: (a) basic construction, (b) symbol, (c) voltage-current characteristic.

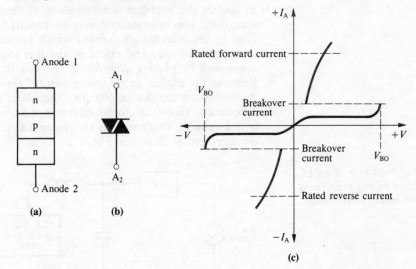

Applying a voltage of either polarity across the device produces one p-n junction forward biased and the other p-n junction reverse biased. At small voltage levels only, a small leakage current flows; as the voltage is increased to the breakover point (V_{BO}) the reverse biased junction avalanches and the resulting conduction pattern is similar to that of the SCR. The diac is suitable for trigger applications for thyristor control circuits since it can present a fast-rising trigger signal to the SCR giving a good turn-on of load current. The arrangement is shown in Fig. 9.22 for a phase delay circuit using a triac.

Fig. 9.22 Diac used as a trigger element for a triac phase delay circuit.

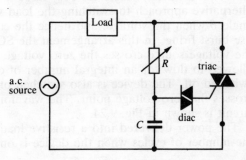

In this circuit the diac will have a high impedance with only a small leakage current flowing until C has reached a certain threshold level, determined by the phase-control resistor R, when the diac fires. This in turn switches on the triac. The circuit is similar to that of Fig. 9.20(a) but the diac improves the triggering of the triac.

Radio frequency interference

Because of the fast switching action of SCRs and triacs when they begin to conduct into resistive loads, the circuit current rises to its final value very rapidly. The sudden transient generates harmonics which could be a source of interference to other equipments. Because the amplitude of the harmonics generated decreases with frequency the problem is most acute at the lower end of the r.f. spectrum. The interference may be radiated or carried via power lines to the affected equipment which has the same power source for its supply voltage. The effect of the line induced interference can be minimised by placing a choke in series with the SCR or triac load. Because the choke, in opposing current flow, slows down the rise of current when the device switches the effect of harmonic interference is reduced. However, to be effective the choke must be very large and a suitable alternative is to use an LC filter. A possible arrangement is shown in Fig. 9.23.

Fig. 9.23 Use of an LC filter to minimise radio frequency interference.

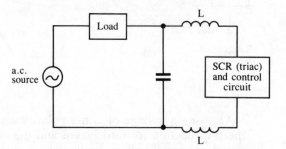

Burst firing

The use of LC filters to suppress radio frequency interference (RFI) is not economically viable for very large load currents. An alternative approach to switching the load voltage at some firing angle θ when the voltage, and hence the current, may be large is to use burst firing. In this arrangement the SCR or triac is switched as the voltage source crosses the zero voltage point and current is allowed to flow for an integral number of cycles before the device is switched off. The device is also switched off as the voltage source crosses the zero voltage point. The waveform for a switched triac circuit is shown in Fig. 9.24.

The power delivered into a resistive load depends on the ratio of the number of cycles when the device is on to the total number of

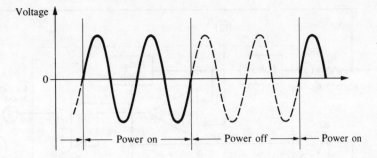

Fig. 9.24 Voltage waveform illustrating the concept of burst firing.

cycles in a switching cycle. From Fig. 9.24 the switching cycle is four cycles of which only two supply power to the load. Hence the average applied power to the load in this case is 2/4 × 100% or 50%.

A circuit for utilising the zero-crossing technique is shown in Fig. 9.25. A suitable trigger circuit for this circuit is available in monolithic i.c. form. The RCA-CA3058/CA3059/CA3079 are zero-voltage switches operating from an a.c. source voltage of 24, 120 or 230 V at 50, 60 or 400 Hz. The CA3059/CA3079 are 14 in DIL (dual in line) packs with plastic encapsulation while the CA3058 is a 14 pin ceramic DIL package.

Fig. 9.25 Basic circuit for burst firing control utilising a zero-crossing switch for the trigger circuit.

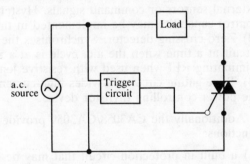

The devices are multistage circuits employing a diode limiter, a zero-crossing (threshold) detector, a differential comparator and an output driver stage. The d.c. operating voltages for the stages are provided by an internal power supply of sufficient current capability to drive external circuits such as other i.c.'s. The output trigger pulses can be applied directly to the gate of an SCR or triac. The CA3058/CA3059 also feature an interlock circuit that prevents the gating pulse to the SCR should an external sensor be inadvertently shorted or opened. An external inhibit connection is available so that an external signal can inhibit the output drive. Apart from this feature which is not present on the CA3079 the three devices are electrically identical. The block diagram of the zero-voltage switch as the trigger element in a triac controlled switching circuit is shown in Fig. 9.26.

From Fig. 9.26 it can be seen that the zero-voltage switch incorporates four functional blocks. Details of the blocks are as follows:

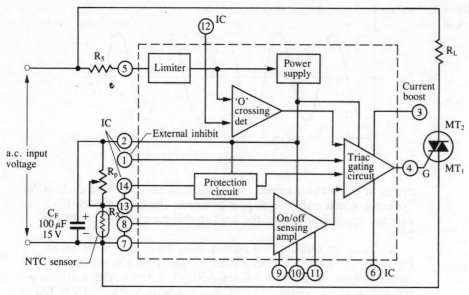

Fig. 9.26 CA3058 series zero voltage switch functional block diagram.

Negative temperature coefficient

(1) Limiter-power supply. Allows operation directly from an a.c. supply.
(2) Differential on/off sensing amplifier. Tests the condition of external sensors or command signals. Hysteresis or proportional control capability may be implemented in this section.
(3) Zero-crossing detector. Synchronises the output pulses of the circuit at a time when the a.c. cycle is at a zero-voltage point, thus eliminating RFI when used with resistive loads.
(4) Triac gating circuit. Provides high-current pulses to the gate of the power-controlling thyristor device.

Additionally the CA3058/CA3059 provide the following auxiliary functions:

(1) a built in protection circuit that may be activated to remove the drive from the triac if the sensor opens or shorts.
(2) thyristor firing may be inhibited through the action of an internal diode gate connected to terminal 1.
(3) high power d.c. comparator action is provided by overriding the action of the zero-crossing detector. This may be achieved by connecting terminals 7 and 12 together. Gate current to the thyristor is continuous when terminal 13 is positive with respect to terminal 9.

Miscellaneous thyristor devices

The major components of the thyristor family, namely the SCR and the triac, together with the diac have been covered in earlier sections. However, there are a variety of switching devices, mainly used in the trigger circuits for the SCR or triac, that are related to the thyristor family. Some of these devices will be discussed briefly in this section.

The Shockley diode

Sometimes called a reverse-blocking diode thyristor, this is a four layer device, as shown in Fig. 9.27(a), with an anode and cathode connection but no gate connection. Just as for the SCR, analysis of the Shockley diode could be made in terms of two interconnected transistors, i.e. a p_1-n_1-p_2 and n_1-p_2-n_2 device. Device symbol and voltage-current characteristic is shown in Fig. 9.27(b) and (c) respectively. When a forward voltage is applied (i.e. anode positive with respect to cathode) only the n_1-p_2 junction is reverse biased and at a breakover voltage the junction avalanches to give device conduction. In the reverse bias condition two junctions are reverse biased and could avalanche if the breakover voltage is exceeded in a similar manner to the SCR. Application of the device is for SCR triggering and in timing circuits but to a great extent its application has been superseded by alternative components. It may still be found, however, in older existing equipments.

Fig. 9.27 Shockley diode:
(a) basic construction, (b) symbol, (c) voltage current characteristic.

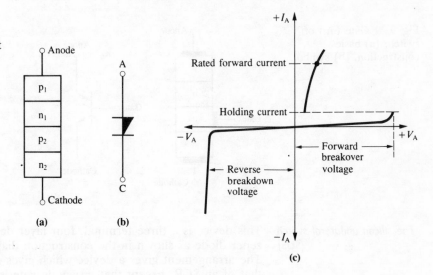

The silicon controlled switch

Possibly known as a reverse blocking tetrode thyristor, this is a four terminal, four layer device similar in construction to the SCR except that triggering can occur using either of two gate terminals. The device construction and symbol are shown in Fig. 9.28. The voltage-

Fig. 9.28 Silicon controlled switch: (a) basic construction, (b) symbol.

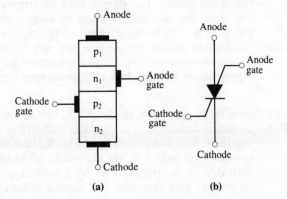

current characteristic is similar to that of the SCR. Using a positive signal to the cathode gate or a negative signal to the anode gate will result in device conduction. To stop conduction the device may be momentarily shorted between anode and cathode *or* a negative signal could be applied to the cathode gate *or* a positive signal to the anode gate. Applications for the device include driver circuits for displays or relays, counters, registers and pulse generator circuits.

The gate turn-off switch (GTO), or gate-controlled switch (GCS)

This is a three terminal, four layer device similar in construction and operation to the SCR. Conduction can occur, like the SCR, with a positive signal to the gate terminal but, unlike the SCR, a negative signal to the gate will turn this device off. Device construction and symbol is shown in Fig. 9.29. Applications are mainly in power switching and inverters.

Fig. 9.29 Gate turn off switch: (a) basic construction, (b) symbol.

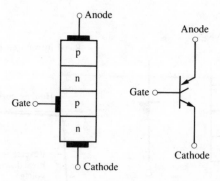

The silicon unilateral switch (SUS)

This device is a three terminal, four layer device with an added zener diode as shown in the construction diagram of Fig. 9.30(a). The arrangement gives a device which gives similar performance to that of an SCR, except that an anode gate is used instead of the cathode gate of the SCR. The SUS is intended to be used with current flow from anode to cathode since a reverse current could damage the device. When a positive voltage is applied between the anode and cathode no conduction will occur as long as the applied voltage is less than the switching voltage. The rating of the switching voltage is that of the p-n junction closest to the anode *plus* that of the zener diode. The device may be triggered with the main voltage less than the switching voltage by connecting a d.c. path between the gate and cathode terminals. Again, like the SCR, once conduction occurs the device can only be switched off by reducing anode-cathode voltage to give a current below the holding current. Applications include low-voltage trigger circuits and timing circuits.

The silicon bilateral switch (SBS)

Similar in operation to the SUS, this device will permit conduction in either direction; it consists essentially of two SUS circuits on a single chip as the diagram of Fig. 9.31(a) shows. The p-n-p structure can be used as the diode according to which SUS circuit is activated. It is triggered on by a switching voltage of either polarity or via the gate terminal in a similar manner to the SUS.

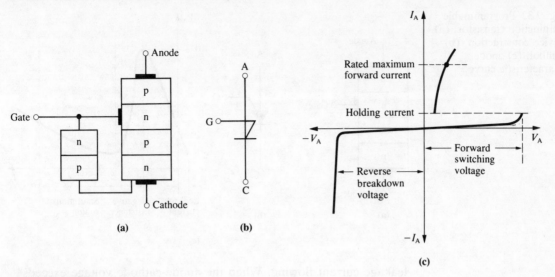

Fig. 9.30 Silicon unilateral switch: (a) basic construction, (b) symbol, (c) voltage-current characteristic.

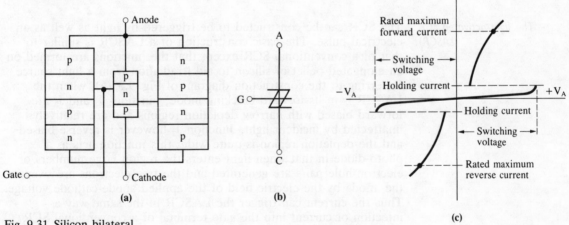

Fig. 9.31 Silicon bilateral switch: (a) basic construction, (b) symbol, (c) voltage-current characteristic.

Applications include low-voltage triggering circuits, voltage-level sensing circuits and pulse generators.

The programmable unijunction transistor (PUT)

This has operational characteristics similar to those of the unijunction transistor discussed in chapter 2. The construction shown in Fig. 9.32(a) reveals a four layer device with a gate terminal. The device has a voltage applied to its gate terminal via a potential divider network and providing the anode-cathode voltage is less than the gate voltage the PUT is non-conducting with only a small

Fig. 9.32 Programmable unijunction transistor: (a) device construction, (b) symbol, (c) anode characteristic curve.

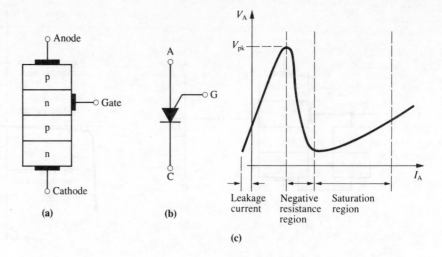

leakage current flowing. When the anode-cathode voltage exceeds the gate voltage by about 0.7 V the device conducts and anode current increases as shown in Fig. 9.32(c). The voltage at which conduction occurs depends on the gate voltage and since this can be varied the device is programmable. Applications include relaxation oscillators, phase-control trigger circuits and timers.

The light-activated SCR
(LASCR)

The SCR can be constructed to be triggered by light as well as an electrical pulse. The basic construction of a LASCR is similar to that of a conventional SCR except that the junctions are formed on an elongated pellet of silicon to aid irradiation from a light source. Referring to the construction diagram of Fig. 9.33(a), when the LASCR is in its forward blocking mode, junctions J_1 and J_3 are forward biased with narrow depletion regions and are relatively unaffected by incident light. Junction J_2 however is reverse biased and the depletion region is quite wide; this junction acts as a photo-diode in that when light enters the region large numbers of electron-hole pairs are generated and the free electrons are swept to the anode by the electric field of the applied anode-cathode voltage. Thus the current can trigger the LASCR in the same way as injection of current into the gate terminal of a conventional SCR. Light activated SCRs are available mainly in low current ranges and

Fig. 9.33 Light activated SCR: (a) basic construction, (b) symbol.

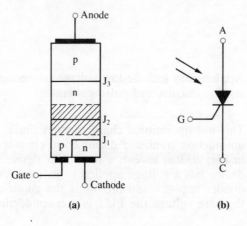

may have a window so that an external light source may be used or the light source may be encapsulated within the device package.

You should now be able to attempt Exercises 9.1 to 9.21.

Points to remember

- Controlled rectification allows variation in the amount of power taken by a device in a given time.
- The thyristor or SCR may be switched into conduction by a suitable forward voltage, or by gate-injected current, and remains in conduction until the voltage across the device falls sufficiently to reduce device current below a 'holding' value.
- The SCR allows control of power to a load for d.c. and a.c. sources.
- The SCR may be operated in phase control mode or burst-firing mode.
- The triac is similar in operation to the SCR but allows conduction in a forward *and* a reverse direction and can be triggered by positive *or* negative gate signals.
- Triacs may be used instead of SCRs for full-wave rectifier circuits and offer simpler circuitry.
- The diac is a two-terminal device which is basically a bi-directional diode allowing conduction, or not, in either direction.
- Radio frequency interference (RFI) could be produced by the fast switching of SCRs and triacs. The effect of RFI could be reduced by an LC filter or the use of burst firing techniques.

EXERCISES 9

9.1 Draw a two-transistor equivalent circuit of a thyristor and use it to explain device operation as gate voltage increases positively from 0 V. Derive an expression for device current in the forward blocking voltage mode.

9.2 Sketch a typical thyristor characteristic and use it to explain the terms: (i) maximum forward-blocking voltage, (ii) minimum holding current, (iii) minimum reverse-blocking voltage.
Explain what effect there is on the device characteristic if gate current is injected into the device.

9.3 Draw a basic circuit that uses an SCR to control power to the load of a d.c. circuit. Explain how your circuit works.

9.4 The basic circuit of Fig. 9.6(a) is used to switch a d.c. supply of 250 V. What would be the required t_{on} time of the SCR if switch S_2 is opened and closed at a frequency of 50 Hz and the mean output voltage is to be 100 V. (8 ms)

9.5 Using the circuit of Fig. 9.8, the value of C is 2 μF. What would be a suitable value for L if the maximum frequency of operation must not exceed 5 kHz? The load voltage may be taken as 100 V and load resistance of 10 Ω is used. (500 μH)

9.6 Using the circuit of Fig. 9.9 and assuming a sinusoidal source voltage, deduce an equation for the mean current in the load over a cycle and show that the rms current in the load is given by:

$$I_{rms} = \frac{V_{pk}}{R_L} \frac{1}{4\pi} \left(\pi - \theta + \frac{\sin 2\theta}{2} \right) \text{ amps}$$

where θ is the firing angle for the device.

9.7 Using the circuit of Fig. 9.9 with a peak supply voltage of 100 V and load resistance of 100 Ω find:
(i) mean load current for a firing angle of 60°, (ii) d.c. load power dissipation, (iii) a.c. load power.   (0.24 A, 5.7 W, 24.28 W)

9.8 Repeat exercise 9.7 using same parameters for supply voltage, load resistance and firing angle but using a full-wave circuit similar to that of Fig. 9.11.   (0, 0, 48.56 W)

9.9 Draw a basic circuit using an SCR to supply control power to a load with a delayed gate trigger. Explain the action of your circuit and suggest reasons why a delayed gate trigger may be a useful idea in practice.

9.10 For the circuit of Fig. 9.13(a) it is required to produce a firing angle of 150° with $C = 2 \mu F$ and $R = 100 \Omega$. If V_{source} is 100 V at 50 Hz and $R_L = 100 \Omega$ and without the phase delay circuit the firing angle is 75°, what would be the value of the variable resistance R_v to achieve the required firing condition?   (326.45 Ω)

9.11 Using the construction of a triac as shown in Fig. 9.15 explain how the device works and sketch a typical output characteristic. Assume for simplicity that no gate current flows.

9.12 Use the basic triac circuit of Fig. 9.11 and explain how the circuit operates. What do you understand by the term 'snubber' circuit and why would you expect it to be used in a triac circuit?

9.13 Draw the constructional details of a diac and describe the theory of operation of the device. Draw and label a typical VI characteristic for such a device.

9.14 Show, by means of a simple sketch, how a diac may be used to trigger a triac in a controlled rectifier circuit. Explain how the circuit operates.

9.15 What do you understand by the term 'radio frequency interference' associated with controlled rectifier circuits? Suggest a method by which the effect might be minimised.

9.16 Describe how an SCR may be controlled by:
(i) phase control, (ii) burst firing.
State under what conditions burst firing might be preferred to phase control.

9.17 Draw the basic construction and explain the action of a Shockley diode.

9.18 Draw the basic construction and explain the action of a silicon controlled switch.

9.19 Draw the basic construction and explain the action of a silicon bilateral switch.

9.20 Draw the basic construction and explain the action of a programmable unijunction transistor.

9.21 Draw the basic construction and explain the action of a light-activated SCR.

Appendix 1 Analysis of Electrical Networks

Electrical networks consist of interconnected branches, each branch being an electrical element. The elements may be passive components such as resistors, impedances or admittances or they may be voltage or current generators. The generators may be independent, as for example a battery or an external signal connected to the network, or they can be dependent on the voltage or current in another part of the circuit. An example of an electrical network is shown in Fig. A1.1.

Fig. A1.1 An electrical network of impedances and generators.

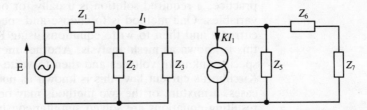

Notation In the network shown, there are six passive component branches, an independent voltage generator E, and a dependent current generator KI_1 whose current therefore depends upon the current I_1 flowing in impedance Z_3.

In such a network, there is a different current flowing in each branch and a different voltage or potential difference between each pair of points or nodes in the network. Where a branch consists of a passive impedance Z or admittance Y, the branch current and the voltage across the branch are related by Ohm's Law:

$$V = \pm IZ \text{ or } I = \pm VY$$

There results are $\pm$ unless the **relative sense of measurement** of the voltage and current are defined. This definition is commonly made by the use of arrows as shown in Fig. A1.2.

Fig. A1.2 Notation for voltages and currents in electrical networks.

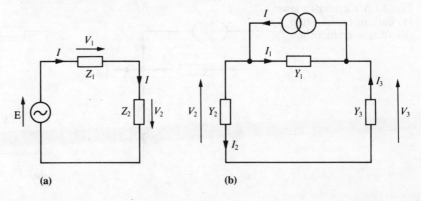

(a)   (b)

In the first circuit, the generator E is applied to the series circuit of Z_1 and Z_2. The same current I flows in both branches Z_1 and Z_2 (and in E since all three are in series). A sense of direction for I has been chosen and indicated by the arrows in a clockwise direction. The voltages V_1 and V_2 will be respectively $\pm IZ_1$ and $\pm IZ_2$. The sign + or − is determined by the relative sense of measurement in each case as follows; if the voltage arrow is in the opposite direction to the current arrow $V = +IZ$; if the voltage and current arrows have the same direction, $V = -IZ$. Thus in the first example, $V_1 = -IZ_1$ and $V_2 = +IZ_2$.

In the second circuit (A1.2b), a current generator I is applied to an admittance Y_1 in parallel with Y_2 and Y_3 in series. Once again chosen current and voltage senses are indicated by arrows and following the same rule;

$$I_2 = +V_2Y_2 \quad \text{and} \quad I_3 = -V_3Y_3.$$

Analysis

The complete analysis of a network consists of finding the current in each branch and the voltage between each pair of nodes. In practice, a required solution is usually for only one or two of the variables. One method is to choose and specify the unknown branch currents and then to write equations using Kirchhoff's voltage law; this is known as **mesh analysis**. Another method is to choose and specify unknown voltages and then to write equations using Kirchhoff's current law; this is known as **nodal analysis**. In some cases, a mixture of the two methods may be used. In each case, the resulting equations are solved simultaneously to give values for the chosen unknowns in terms of the independent generators (applied signals) and the circuit components.

Kirchhoff's laws

The current law states: the algebraic sum of the currents entering a node or junction is zero. (This is because such a node cannot *store* current.) This law must be used in terms of the current directions and is illustrated in the part circuit of Fig. A1.3a. Since 'entering' means 'flowing towards', the relationship for this circuit is:

$$I + I_1 - I_2 - I_3 = 0$$

or $I + I_1 = I_2 + I_3$

This may also be stated: the sum of the currents entering a junction is equal to the sum of the currents leaving the junction.

Fig. A1.3 Kirchhoff's laws:
(a) current relationships,
(b) voltage relationships.

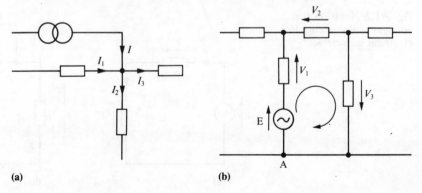

(a)   (b)

The voltage law states: the algebraic sum of the voltages taken around any closed loop of a network is zero. (This is because a particular point in a network cannot be at two different voltages at the same time.) This law must be used in terms of chosen voltage directions and is illustrated in Fig. A1.3b. Starting at point A and going around the loop in the direction indicated:

$E + V_1 - V_2 + V_3 = 0$ (going with the voltage arrow is positive, against it is negative).

Mesh analysis

For mesh analysis, the unknown branch currents are chosen and equations are written using the voltage law. To avoid having too many unknowns (and too many simultaneous equations to solve) the chosen currents can be related to each other by inspection using the current law. This is illustrated in the numerical example shown in Fig. A1.4. There are three branches, each having a current which could be individually labelled leading to three unknowns requiring three equations for solution. Instead, choose and label the two battery currents as shown. The remaining branch current in the 8 Ω resistor must be $(I_1 - I_2)$. The IZ voltage arrows (V_A, V_B, V_C) can also be inserted as shown so that the voltage equations can be written for each loop in the network.

Left hand loop, $6 - V_A - V_B = 0$
Right hand loop, $3 + V_B - V_C = 0$
Outside loop, $6 - V_A - V_C + 3 = 0$
V_A, V_B and V_C can be expressed in terms of the unknown currents I_1 and I_2 and the component values.

$V_A = 4I_1$, $V_B = 8(I_1 - I_2)$, $V_C = 12I_2$.

Fig. A1.4 A circuit labelled for solution by Kirchhoff's law methods.

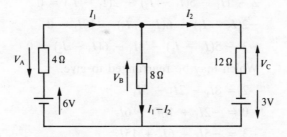

Substituting these in the above equations will give three new equations with only two unknowns (I_1 and I_2), thus only two of these need be used. Taking the first two,

$6 - 4I_1 - 8(I_1 - I_2) = 0$

$3 + 8(I_1 - I_2) - 12I_2 = 0$

rearranging, $6 = 12I_1 - 8I_2$

$3 = -8I_1 + 20I_2$

Solving by determinants or by elimination, $I_1 = 0.818$ A and $I_2 = 0.477$ A.

These results can now provide the complete solution as any other branch current or voltage between two points can be found using the Ohm's Law relationships.

Although the method illustrated in the above example is often useful when the required solution is a single branch current, the selection of the original unknown currents can sometimes lead to too many unknowns. A convenient alternative is to use Maxwell's cyclic currents as the unknowns. This assumes a current that is associated with each loop or mesh as shown in Fig. A1.5. (Note that all the mesh currents flow in the same direction, clockwise.)

Fig. A1.5 Circuit with loop or mesh currents for solution by mesh analysis.

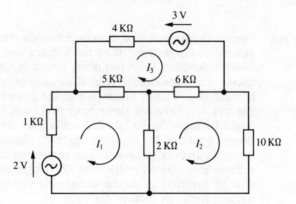

Now, when a branch is common to two meshes, the current in that branch is the difference between the two mesh currents. For example, in the 5 kΩ resistor, flowing from left to right, the current is $I_1 - I_3$; similarly in the 6 kΩ resistor, the current is $I_3 - I_2$ and in the 2 kΩ resistor, it is $I_1 - I_2$.

The voltage equations may be written in the same way as before:

$$2 - 1I_1 - 5(I_1 - I_3) - 2(I_1 - I_2) = 0$$

$$-2(I_2 - I_1) - 6(I_2 - I_3) - 10I_2 = 0$$

$$-3 - 5(I_3 - I_1) - 4I_3 - 6(I_3 - I_2) = 0$$

Which may be rearranged to give:

$$2 = 8I_1 - 2I_2 - 5I_3$$

$$0 = -2I_1 + 18I_2 - 6I_3$$

$$-3 = -5I_1 - 6I_2 + 15I_3$$

Comparison between these equations and the original circuit shows that a simple rule can be used to write the set of equations directly by inspection of the circuit as follows. Having inserted the clockwise mesh currents, for each loop in turn, equate any voltage generators (in the direction of the loop current) to that loop current times the sum of the impedances around the loop minus each adjacent loop current times the shared impedance.

Application of this rule to the three meshes provides directly the necessary three equations in the most convenient form for solution by determinants (or by substitution and elimination).

Nodal analysis   For nodal analysis, the unknown node voltages are chosen and specified with reference to a chosen node at zero potential. Equations are then written using Kirchhoff's current law. The most convenient form for the passive components is admittance but these can be expressed as reciprocals of the branch impedances if desired. Also, since current equations are written, current generators are the more convenient form. An alternative form using voltage generators is possible however and is illustrated in a later example.

Consider the admittance network shown in Fig. A1.6.

Fig. A1.6 Admittance circuit for solution by nodal analysis.

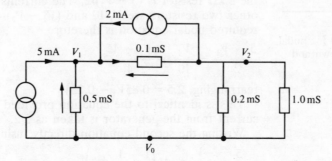

The branch components given by their admittances have equivalent impedance values between 1 kΩ and 10 kΩ. The circuit has three nodes (shown by black dots), a node being a point in the circuit where two or more branches are connected. One of these, the bottom one, is selected as the reference V_0 at 0 V and the other two are labelled V_1 and V_2 respectively (the unknown node voltages). Thus, the voltage across the 0.5 mS component is $(V_1 - 0) = V_1$ and that across the 0.1 mS is $(V_1 - V_2)$, both with the direction indicated by the arrows. The current in each of these branches will be simply $+(V \times Y)$ with a direction opposite to that of the voltage arrow.

Now, using the current law, we can write the basic nodal equations. At node V_1,

$$5 \text{ mA} + 2 \text{ mA} - V_1 \times 0.5 \text{ mS} - (V_1 - V_2) \times 0.1 \text{ mS} = 0$$

Since mA, mS and volts are used throughout, we can simply write:

$$5 + 2 - 0.5V_1 - 0.1(V_1 - V_2) = 0$$

and for node V_2, $-2 + 0.1(V_1 - V_2) - 0.2V_2 - 1V_2 = 0$

Collecting terms and rearranging,

$$7 = 0.6V_1 - 0.1V_2$$
$$-2 = -0.1V_1 + 1.3V_2$$

which are very similar in form to the final mesh equations. These may be solved to give $V_1 = 11.6$ V and $V_2 = -0.65$ V.

A rule for writing these equations directly by inspection of the circuit is as follows: Having labelled the unknown node voltages (and the reference); for each node in turn, equate the sum of the currents entering the node from generators to that node voltage times the sum of all the admittances connected to that node, minus each adjacent node voltage times the interconnecting admittance.

Application of this rule to each of the nodes in the circuit shown results in the two equations given above.

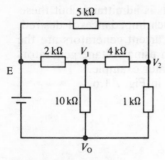

Fig. A1.7 Circuit for nodal analysis with known and unknown voltages.

Many networks requiring analysis are shown with voltage generators which may be converted by Norton's theorem (see below) to current generators in preparation for nodal analysis. With some arrangements however, such conversion is not possible and a method involving known and unknown voltages is simple to apply. Consider the circuit shown in Fig. A1.7.

This circuit has a reference V_0 and three other nodes at $5V$, V_1 and V_2. To write an equation for node V_1, the current entering from the $2 k\Omega$ resistor is $(5 - V_1)/2$. The currents leaving through the other two resistors are $V_1/10$ and $(V_1 - V_2)/4$ respectively. The required nodal equation is therefore,

$$\frac{5 - V_1}{2} = \frac{V_1}{10} + \frac{V_1 - V_2}{4}$$

rearranging, $2.5 = 0.85V_1 - 0.25V_2$

This is identical to the equation provided by the rule above if the current from the generator is taken as $5/2$.

Writing the second equation directly, using the rule,

$$\frac{5}{5} = \frac{-V_1}{4} + V_2\left(\frac{1}{5} + \frac{1}{4} + 1\right)$$

or $1 = -0.25V_1 + 1.45V_2$

Network simplification

The methods of mesh and nodal analysis can always be applied and even very complicated networks will yield a set of simultaneous equations. If the resulting number of unknowns is more than three, the algebraic manipulation required for a final solution increases by an order of magnitude for each additional unknown. In most cases, network simplification can be used either to lead directly to the required solution or to reduce the number of unknowns prior to application of mesh or nodal analysis. The two most useful techniques are known as **Thévenin's theorem** and **Norton's theorem** and each may be used to simplify a two terminal network including passive components and generators to a single generator and a single impedance (admittance). The principle is illustrated in Fig. A1.8.

In each case, the passive component Z_T or Y_N has the value that would be measured or calculated 'looking into' the original network with all the internal generators replaced by their internal impedances (suppressed). The Thévenin voltage E_T is the voltage that can be measured or calculated as appearing across the network when the terminals are open circuit (no external load). The Norton current I_N is the current that would flow in a short circuit placed across the terminals of the network.

Consider the circuit shown in Fig. A1.9.

First apply Thévenin at the points XX looking left. With the generator suppressed,

$$Z_T' = \frac{40 \times 50}{90} = 22.2 \,\Omega.$$

For E_T', we must consider the current that would flow around the loop and the resulting volt drop across the $40\,\Omega$ resistor.

$$E_T' = 5 + \frac{(10 - 5) \times 40}{90} = 7.22 \text{ V.}$$

Fig. A1.8 Network
simplification by Thévenin
and Norton's theorems.

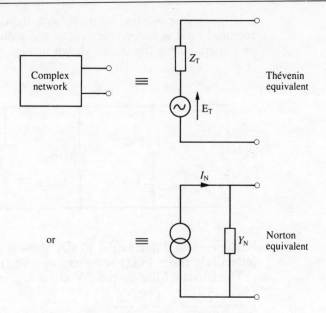

Fig. A1.9 Circuit for
solution by Thévenin's
theorem.

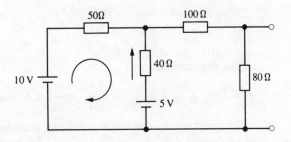

The left hand part of the circuit can now be replaced by a 7.22 V
generator in series with 22.2 Ω and Thévenin can then be reapplied
at the output terminals.

Now, $Z_T = \dfrac{80 \times 122.2}{80 + 122.2} = 48.3 \ \Omega.$

$E_T = \dfrac{7.22 \times 80}{80 + 122.2} = 2.86 \ \text{V.}$

The Norton equivalent can be obtained directly from these
results,

$Y_N = \dfrac{1}{48.3}$ or 20.7 mS

$I_N = \dfrac{2.86}{48.3} = 0.0592 \ \text{A.}$

The equivalence can be confirmed by connecting an equal load
(10 Ω for example) across the two forms and finding the same
output voltage by calculation.

One further simplification which is sometimes useful is to apply
the **superposition principle**. By this principle, if a network has more
than one generator and the required information is the current in a
particular branch, then the required current can be calculated due to

each generator in turn, by itself, with the others suppressed and the required total is simply the sum of the individual currents. This can be illustrated for the circuit shown in Fig. A1.10.

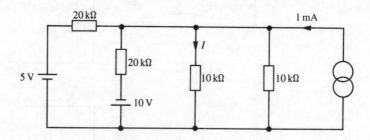

Fig. A1.10 Circuit for solution by the superposition principle.

The current I due to the 10 mA alone is given by: $I_1 = 1/3$ mA (effectively three 10 kΩ resistors in parallel).

The current I due to the 5 V alone is given by:

$$I_2 = \frac{2.5}{10 + 5} \times \frac{1}{2} = \frac{2.5}{30} \text{ mA.}$$

The current I due to the 10 V alone is given by:

$$I_3 = \frac{-5}{10 + 5} \times \frac{1}{2} = \frac{-5}{30} \text{ mA.}$$

By superposition,

$$I = I_1 + I_2 + I_3 = \left(\frac{1}{3} + \frac{2.5}{30} - \frac{5}{30}\right) \text{ mA} = 0.25 \text{ mA.}$$

In conclusion, electrical networks including generators and passive components can always be analysed by application of mesh and nodal analysis. The resulting simultaneous equations will be simple (non-differential) if the signals are d.c. or steady state a.c. If the networks are complex, the large number of simultaneous equations make solution difficult and it can be advantageous to use network simplifying techniques such as Thévenin and Norton's theorems either instead of before applying mesh or nodal analysis.

Appendix 2 Number Systems and Arithmetic Processes

The decimal numbering system uses the **base** of 10 and any number in this system can be represented as a series of multiples of powers of ten, for example the number 345_{10} consists of:

$3 \times 100 + 4 \times 10 + 5 \times 1$ or

$3 \times 10^2 + 4 \times 10^1 + 5 \times 10^0$

Any number can be expressed in any base system using a convention similar to that above with the most significant digit to the left.

From the above expression for the decimal number 345, it can be seen that:

$345_{10} = 3 \times n^2 + 4 \times n^1 + 5 \times n^0$

where n is the base of the system, 10 in this case.

Consider the **octal** system where the base is 8; this means that only digits 0 to 7 inclusive can exist. Similarly there is **binary**, with a base of 2, where digits 0 and 1 exist. Finally consider the **hexadecimal** system where the base is 16; this raises a problem insofar as the digits 0 to 15 inclusive would appear to be necessary giving two digits for a normal numeric system. The problem is resolved by using letters to represent the two digit numbers i.e. $A \equiv 10, B \equiv 11, \ldots F \equiv 15$.

Any of the systems mentioned obey the rule:

$(XYZ)_n = (X.n^2 + Y.n + Z)_{10}$

so that it is simple to convert from any base to decimal base. For example:

$101_2 = (1.2^2 + 0.2 + 1)_{10} = 4 + 1 = 5_{10}$

$577_8 = (5.8^2 + 7.8 + 7)_{10} = 320 + 56 + 7 = 383_{10}$

$5C9_{16} = (5.16^2 + 12.16 + 9) = 1280 + 192 + 9 = 1481_{10}$

Similarly it is possible to convert a number to the base of 10 to a number of any other base by successive division of the decimal number by the base of the required new number.

For example:

$2\overline{)5}_{10}$	=	2	remainder	1
$2\overline{)2}$	=	1	remainder	0
$2\overline{)1}$	=	0	remainder	1
		hence binary number is 101		
$8\overline{)383}_{10}$	=	47	remainder	7
$8\overline{)47}$	=	5	remainder	7
$8\overline{)5}$	=	0	remainder	5
		hence octal number is 577		

$$16\overline{)1481}_{10} \quad = \quad 92 \qquad \text{remainder} \qquad 9$$
$$16\overline{)92} \quad = \quad 5 \qquad \text{remainder} \qquad 12(\text{C})$$
$$16\overline{)5} \quad = \quad 0 \qquad \text{remainder} \qquad 5$$

hence hexadecimal number is 5C9

From these examples it is clear that division continues until the original number has a zero resultant, with or without a remainder, and the new number required is obtained by reading the column of digits in the remainder column from *bottom to top*.

Octal and hexadecimal numbers are of value in computer use since they have rapid conversion to binary form. Since octal is to the base of 8 and $2^3 = 8$, then 3 binary digits (or bits) can represent a single octal digit. Similarly since $2^4 = 16$, then 4 bits can represent a single hexadecimal digit.

Thus:

$$4352_8 = 100\ 011\ 101\ 010_2 \qquad \text{and}$$

$$75AF_{16} = 0111\ 0101\ 1010\ 1111_2$$

The use of hexadecimal notation is useful for applications involving microprocessors since bits are commonly collected in groups of 8, called a **byte**, which can be represented by a two digit hexadecimal number.

A code already used in Chapter 4 is the BCD (binary coded decimal) where a decimal number may be represented in binary form with each digit expressed by 4 bits, but only using 10 of the possible 16 combinations.

i.e. $75_{10} = 0111\ 0101_{BCD}$

The *true* binary equivalent of 75_{10} is:

$$2\overline{)75}_{10} \quad = \quad 37 \qquad r \qquad 1$$
$$2\overline{)37} \quad = \quad 18 \qquad r \qquad 1$$
$$2\overline{)18} \quad = \quad 9 \qquad r \qquad 0$$
$$2\overline{)9} \quad = \quad 4 \qquad r \qquad 1$$
$$2\overline{)4} \quad = \quad 2 \qquad r \qquad 0$$
$$2\overline{)2} \quad = \quad 1 \qquad r \qquad 0$$
$$2\overline{)1} \quad = \quad 0 \qquad r \qquad 1$$

hence binary number is 1001011

Fig. A2.1 Gray or cyclic code. Only 8 segments are shown and consequently 3 bits in the code. 16 segments would require 4 bits, etc.

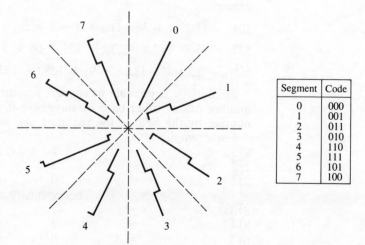

Segment	Code
0	000
1	001
2	011
3	010
4	110
5	111
6	101
7	100

The disadvantage of BCD is that while with hexadecimal code (using all combinations of the 4 bits) the 2 digit number can have a maximum value FF($=225_{10}$), with BCD, the 2 digit number can have a maximum value 99($=153_{10}$).

Another code that has useful properties is the Gray, or cyclic code which is popular for mechanical systems where a shaft position is required to generate a binary number. In this code, the binary number is arranged so that if a number is increased, or decreased, by one, only one bit of the code changes. Thus possible ambiguities that could occur if the shaft is lying across two segments are eliminated. Figure A2.1 shows the representation of a 3-bit Gray code.

Signed numbers

Considering a binary number of say 8 bits, then the decimal range using these 8 bits is from 0 to 225 (00000000 to 11111111). However, it is possible to give a sign to the number so that for the 8 bits it is possible to count from -128 to $+127$. Three basic systems are used:

Sign bit The most significant bit (MSB) of the binary number can be used to indicate the sign, i.e. 0 for a positive number and 1 for a negative number. The other 7 bits are used to represent a normal unsigned integer number.

Two's complement The true complement of a number may be found by subtracting the number from the largest number that can be represented, i.e. 472 can be subtracted from 999 to give a nine's complement of 527. For a binary system the one's complement is obtained by inverting all binary digits, i.e. replacing 1's with 0's and vice-versa. The *base* complement is found by adding 1 to the true complement, i.e. ten's complement is nine's complement + 1 and two's complement is one's complement + 1.

Two's complement (or ten's complement) can be used to represent negative numbers. For the binary system consider the 8-bit numbers which are to be split into positive and negative groups. The positive number range is from 0 to 127 (00000000 to 01111111) which is identical to the signed bit notation. However, the negative numbers are obtained by using the two's complement of the positive numbers i.e.

-1 is the two's complement of $+1$ (00000001)　　= 11111110 + 1
　　　　　　= 11111111
and the largest negative number is　　= 10000000 (-128).

The concept of two's complement might be more easily understood by considering the 'block' of numbers represented by an 8-bit binary number as being on the curved surface of a cylinder so that the number increases from 0 to 255 as the cylinder rotates, say, clockwise and then starts from 0 again. If the cylinder is stationary on 0 and rotates clockwise, the number increases and is said to be positive whereas if the cylinder rotates anti-clockwise from 0 it goes to 255, 254, 253, etc., and these numbers can be considered negative. Thus numbers 1 to 127 can be achieved with clockwise rotation from 0 (positive numbers) and 255 to 128 (equivalent to −1 to −128) achieved by anti-clockwise rotation from 0.

Two's complement is important in digital computers since it allows subtraction to occur in the same way as for addition i.e. subtraction +3 is the same as adding −3.

Excess codes These are created by introducing an offset so that the largest negative number is represented by all zeros. In fact the excess code is equivalent to the values given by the two's complement with only the MSB complemented. Consider Excess 4 code shown in Table A2.1.

Table A2.1 Table of excess 4 code for denary numbers 3 to −4

Denary	Excess 4 code
3	111
2	110
1	101
0	100
−1	011
−2	010
−3	001
−4	000

From this it can be seen that the number N has the binary number represented by the unsigned number $N + 2^{n-1}$, where n is the number of bits the value of the excess will increase.

Binary fractions

A binary fraction (the part of a number to the right of the base point) can be produced in the same way as for integer numbers, i.e.

2^3	2^2	2^1	2^0	2^{-1}	2^{-2}	2^{-3}	2^{-4}	
8	4	2	1	$\frac{1}{2}$	$\frac{1}{4}$	$\frac{1}{8}$		denary value
1	0	1	1	0	1	0	0	$= 1011.01_2 = 11.25_{10}$

hence:

Negative fractions can be treated as before but the base point must be placed so that the sign bit is always to its left. The magnitude of a number depends on the number of bits allocated to the integer part since restricting the number of bits in the fraction will limit the accuracy.

Floating point representation

For a particular number of bits used to represent a number, the extent of the values of numbers is limited, i.e. for 8 bits, the number represented can extend from 0 to 255 denary. This limitation can be resolved by representing the number by a fraction with a separate counter to locate the base point, i.e.

$$154.25_{10} = 0.15425 \times 10^3$$

$$-0.0067_{10} = 0.67 \times 10^{-2}$$

and $1101.0101_2 = 0.11010101 \times 2^4$

In general the number is given by:

$$\pm m.b^e$$

where m is the mantissa, e the exponent and b the base.

If the mantissa is chosen so that the digit immediately after the base point is the most significant non-zero digit, as shown above, the number is said to be **normalised**.

It is obvious from the above that the range of numbers that it is possible to represent with, say, 8 bits, is now very much increased and with greater accuracy.

Binary arithmetic

For unsigned numbers the rules of arithmetic apply as follows:

Addition

A	B	A + B	
0	0	0	
0	1	1	
1	0	1	
1	1	1	Carry 1.

Example:

binary		denary
01101001	Augend	105
+ 11011	Addend	+27
10000100	Sum	132

Subtraction

A	B	A − B	
0	0	0	
0	1	1	Borrow 1
1	0	1	
1	1	0	

Example:

binary		denary
01101001	Minuend	105
− 11011	Subtrahend	− 27
01001110	Difference	78

Multiplication

A	B	A × B
0	0	0
0	1	0
1	0	0
1	1	1

Example:

binary		denary
01101001	Multiplicand	105
× 11011	Multiplier	×27

```
   01101001
   01101001
   00000000
   01101001
   01101001
```

| 101100010011 | Product | 2835 |

The 'intermediate' numbers in the above example are called 'partial products'. In this example the product exceeds 8 bits so that either *two* bytes would be necessary to specify that number or the choice of values for the multiplicand and multiplier must be such that the product does not exceed 255.

Division

This may be accomplished using the long division.
Example:

$$
\begin{array}{r}
11.111000 \\
11011)\overline{01101001} \\
\underline{11011} \\
0110011 \\
\underline{11011} \\
0110000 \\
\underline{11011} \\
0101010 \\
\underline{11011} \\
011110 \\
\underline{11011} \\
000110 \\
\text{etc.}
\end{array}
\qquad
\begin{array}{r}
3.888. \\
27)\overline{105} \\
\underline{81} \\
240 \\
\underline{216} \\
240 \\
\underline{216} \\
24 \\
\text{etc.}
\end{array}
$$

The binary number, with 105 as the Dividend, 27 the Divisor and the answer the Quotient, given as the quotient is 11.111000 which equals 3.875 which is only approximately correct using 6 binary places for the fraction. More places must be allowed if greater accuracy is required.

Signed numbers

Sign and magnitude In this case since the MSB is used to represent a sign then a test of the sign bit must be undertaken to determine the process involved. For example, for a subtraction process involving A and B:
(a) If A and B positive and magnitude of A greater than B, then subtract B from A.
(b) If A and B positive but magnitude of A less than B, subtract A from B and set sign bit of answer to indicate negative.
(c) If A positive, B negative, *add* A and B.
(d) If A negative, B positive, *add* A and B and set sign bit of answer to indicate negative.
(e) If A negative, B negative, subtract A from B.
  Also for multiplication and division, the moduli can be multiplied, or divided, and the sign bit set according to the sign values of A and B.

Two's complement

In this case no test is required to determine sign and magnitude and for addition and subtraction the routine is the same as for unsigned numbers.

For multiplication and division several techniques may be used:

(1) By treating the numbers as though they were unsigned and adding a correction factor if either or both are negative.

(2) By converting negative numbers to sign and magnitude; multiply the magnitudes and evaluate sign of answer. Complement and add 1 if sign is negative to put answer back into two's complement form.

(3) By using special algorithms which are variations of the 'shift and add' technique.

Index